Variational Principles of Topology

# Mathematics and Its Applications (*Soviet Series*)



Volume 42

# Variational Principles of Topology

## Multidimensional Minimal Surface Theory

*by*

A. T. Fomenko

*Department of Mathematics and Mechanics,*
*Moscow State University, Moscow, U.S.S.R.*

KLUWER ACADEMIC PUBLISHERS

DORDRECHT / BOSTON / LONDON

**Library of Congress Cataloging in Publication Data**

```
Fomenko, A. T.
   [Topologicheskie variafsionnye zadachi.  English]
   Variational principles in topology : multidimensional minimal
 surface theory / by A.T. Fomenko.
      p.   cm.
   Revised translation of: Topologicheskie variafsionnye zadachi.
   Includes bibliographical references.
   ISBN 0792302303 (alk. paper)
   1. Topology.  2. Variational inequalities (Mathematics)
 I. Title.
 QA611.F6313  1990b
 514--dc20                                             90-4169
```

ISBN 0-7923-0230-3

---

Published by Kluwer Academic Publishers,
P.O. Box 17, 3300 AA Dordrecht, The Netherlands.

Kluwer Academic Publishers incorporates
the publishing programmes of
D. Reidel, Martinus Nijhoff, Dr W. Junk and MTP Press.

Sold and distributed in the U.S.A. and Canada
by Kluwer Academic Publishers,
101 Philip Drive, Norwell, MA 02061, U.S.A.

In all other countries, sold and distributed
by Kluwer Academic Publishers Group,
P.O. Box 322, 3300 AH Dordrecht, The Netherlands.

*Printed on acid-free paper*



This is the revised translation of the original book
ВАРИАЦИОННЫЕ МЕТОДЫ В ТОПОЛОГИИ
Published by Nauka Publishers © 1982

Translated from the Russian by O. Efimov
Typeset by Rosenlaui Publishing Services, Inc.



Printed in the Netherlands

# SERIES EDITOR'S PREFACE

'Et moi, ..., si j'avait su comment en revenir, je n'y serais point allé.'

Jules Verne

The series is divergent; therefore we may be able to do something with it.

O. Heaviside

One service mathematics has rendered the human race. It has put common sense back where it belongs, on the topmost shelf next to the dusty canister labelled 'discarded non-sense'.

Eric T. Bell

Mathematics is a tool for thought. A highly necessary tool in a world where both feedback and nonlinearities abound. Similarly, all kinds of parts of mathematics serve as tools for other parts and for other sciences.

Applying a simple rewriting rule to the quote on the right above one finds such statements as: 'One service topology has rendered mathematical physics ...'; 'One service logic has rendered computer science ...'; 'One service category theory has rendered mathematics ...'. All arguably true. And all statements obtainable this way form part of the raison d'être of this series.

This series, *Mathematics and Its Applications*, started in 1977. Now that over one hundred volumes have appeared it seems opportune to reexamine its scope. At the time I wrote

> "Growing specialization and diversification have brought a host of monographs and textbooks on increasingly specialized topics. However, the 'tree' of knowledge of mathematics and related fields does not grow only by putting forth new branches. It also happens, quite often in fact, that branches which were thought to be completely disparate are suddenly seen to be related. Further, the kind and level of sophistication of mathematics applied in various sciences has changed drastically in recent years: measure theory is used (non-trivially) in regional and theoretical economics; algebraic geometry interacts with physics; the Minkowsky lemma, coding theory and the structure of water meet one another in packing and covering theory; quantum fields, crystal defects and mathematical programming profit from homotopy theory; Lie algebras are relevant to filtering; and prediction and electrical engineering can use Stein spaces. And in addition to this there are such new emerging subdisciplines as 'experimental mathematics', 'CFD', 'completely integrable systems', 'chaos, synergetics and large-scale order', which are almost impossible to fit into the existing classification schemes. They draw upon widely different sections of mathematics."

By and large, all this still applies today. It is still true that at first sight mathematics seems rather fragmented and that to find, see, and exploit the deeper underlying interrelations more effort is needed and so are books that can help mathematicians and scientists do so. Accordingly MIA will continue to try to make such books available.

If anything, the description I gave in 1977 is now an understatement. To the examples of interaction areas one should add string theory where Riemann surfaces, algebraic geometry, modular functions, knots, quantum field theory, Kac-Moody algebras, monstrous moonshine (and more) all come together. And to the examples of things which can be usefully applied let me add the topic 'finite geometry'; a combination of words which sounds like it might not even exist, let alone be applicable. And yet it is being applied: to statistics via designs, to radar/sonar detection arrays (via finite projective planes), and to bus connections of VLSI chips (via difference sets). There seems to be no part of (so-called pure) mathematics that is not in immediate danger of being applied. And, accordingly, the applied mathematician needs to be aware of much more. Besides analysis and numerics, the traditional workhorses, he may need all kinds of combinatorics, algebra, probability, and so on.

In addition, the applied scientist needs to cope increasingly with the nonlinear world and the extra mathematical sophistication that this requires. For that is where the rewards are. Linear models are honest and a bit sad and depressing: proportional efforts and results. It is in the non-linear world that infinitesimal inputs may result in macroscopic outputs (or vice versa). To appreciate what I am hinting at: if electronics were linear we would have no fun with transistors and computers; we would have no TV; in fact you would not be reading these lines.

There is also no safety in ignoring such outlandish things as nonstandard analysis, superspace and anticommuting integration, $p$-adic and ultrametric space. All three have applications in both electrical engineering and physics. Once, complex numbers were equally outlandish, but they frequently proved the shortest path between 'real' results. Similarly, the first two topics named have already provided a number of 'wormhole' paths. There is no telling where all this is leading - fortunately.

Thus the original scope of the series, which for various (sound) reasons now comprises five subseries: white (Japan), yellow (China), red (USSR), blue (Eastern Europe), and green (everything else), still applies. It has been enlarged a bit to include books treating of the tools from one subdiscipline which are used in others. Thus the series still aims at books dealing with:

- a central concept which plays an important role in several different mathematical and/or scientific specialization areas;
- new applications of the results and ideas from one area of scientific endeavour into another;
- influences which the results, problems and concepts of one field of enquiry have, and have had, on the development of another.

Perhaps the best known variational problem in topology is the (two- and higher-dimensional) Plateau problem, which, in the two-dimensional case, asks for a minimal surface, a soap film, spanning a given boundary wire. It is a very famous problem and it is also a problem to which the present author has made fundamental contributions (and he received the prize of the Moscow Mathematical Society for some of his earlier contributions to the subject).

The present volume is a comprehensive and systematic treatment of, especially, higher-dimensional variational problems with special emphasis on the author's new concepts of stratified surfaces and volumes, new methods for proving minimality, and a number of interesting interrelations between Bott periodicity, vector fields on spheres and totally geodesic surfaces in Lie groups. It is by an absolute master in the subject and I welcome it with truly great pleasure in this series.

The shortest path between two truths in the real domain passes through the complex domain.

J. Hadamard

La physique ne nous donne pas seulement l'occasion de résoudre des problèmes ... elle nous fait pressentir la solution.

H. Poincaré

Never lend books, for no one ever returns them; the only books I have in my library are books that other folk have lent me.

Anatole France

The function of an expert is not to be more right than other people, but to be wrong for more sophisticated reasons.

David Butler

Bussum, January 1990

Michiel Hazewinkel

# CONTENTS

# PREFACE

The present work is devoted to the study of multidimensional variational problems, related to multidimensional Riemannian volume functionals and also the Dirichlet functional, defined on "submanifolds with singularities" in the ambient Riemannian manifold. The serious difficulties arising in the attempt to develop quantitative methods for variational problems in the cases where the domain of definition is multidimensional are well-known. There are no analogues of Morse theory for stationary points of multidimensional functionals even in comparatively simple cases. The study of absolutely minimal (or globally minimal) extremals is difficult, because we should consider not small, but large variations in order to prove global minimality, which requires topological and analytical methods. This circumstance revealed itself especially clearly in solving the multidimensional Plateau problem which not only had to be "posed correctly" but required the introduction of new analytic ideas. In order to solve variational problems of Plateau type (minimization of surface volume), all this led to the creation of synthetical, topological and analytic machinery bordering on several branches of mathematics, viz., algebraic topology, functional analysis, the theory of differential equations, groups and Lie algebras.

The physical soap film spanning a fixed boundary wire in three-dimensional Euclidian space is a model of a so called minimal surface , i.e., any sufficiently small perturbation increases its area. The mathematical proof of the existence of such a surface for any rectifiable contour in three-dimensional space was just the subject matter of the known Plateau problem in dimension two, solved by J. Douglas, T. Radó, and profoundly investigated by R. Courant (see the survey in [85]). The parametrized minimal surface may possess self-intersections and other singular points; however, if the contour has no singularities, then the film admits continuous parametrization by a two-dimensional manifold with boundary, i.e., it is its image under a certain continuous mapping. For surfaces of dimensions greater than two, the *multidimensional Plateau problem* was posed, viz., let a closed, smooth $(k-1)$-dimensional submanifold -"contour" $A$ be fixed in a Riemannian manifold $M$, and $\{X\}$ be the class of all such films $X$ embedded into $M$ and with boundary $A$ that each admits continuous parametrization, i.e., is representable as the image of a certain manifold with boundary; in other words, $X = f(W)$, where $W$ is a certain $k$-dimensional manifold with boundary

$\partial W$ homeomorphic to $A$, and $f : W \to M$ a continuous mapping coinciding on $\partial W$ with the fixed homeomorphism onto $A$. The question arises: Is it possible to find a film $X_0$ which would be a reasonably minimal film in $\{X\}$? In the sequel, for brevity, we will call such a formulation of the problem *classical*, and use the term in order to stress the analogy with the two-dimensional formulation by J. Douglas and T. Radó (see above). If we weaken the classical concept of manifold film with boundary and greatly extend the notion of film and its boundary (in particular, neglecting parametrization), then the Plateau problem can be formulated in the language of the usual chains and homologies, viz., it will consist of finding a minimal film annihilating the fundamental homology cycle of the manifold - "contour" $A$. In this statement, the solution of the problem was obtained in [12]–[14], [51], [131]–[132], [138], [145], [351]–[353]. The natural homological formulation was given in the pioneering works by H. Federer, W. Fleming and E. Reifenberg. The methods for the study of homological minimal surfaces in this direction turned out to be extremely powerful. The developed homological variational theory of measure was crowned with a series of remarkable results. In particular, H. Federer, W. Fleming, F. Almgren, E. De Giorgi, E. Ciusti, R. Harvey, H. Lawson, M. Miranda, E. Reifenberg, C. Morrey, S.-T. Yau and W. Meeks proved fundamental existence theorems for minimal homological surfaces such as currents, chains, varifolds, etc., brilliantly solving the Plateau problem in each class of homological surfaces. See the bibliography at the end of the book, where we also have included the works investigating various properties of minimal surfaces, so that the reader can orientate himself in the vast literature on the subject. Needless to say, our list certainly does not claim completeness. However, the described translation of Plateau's problem into the language of homological chains, i.e., in terms of the usual homology, does not permit us to solve the problem of the existence of a minimal "manifold with boundary," immediately, since the existence of a minimal one in the class $\{X\}$ of films which are continuous images of manifolds with boundary does not generally follow from that of a solution in the class of the usual homologies. The principal difficulty arising in attempt to minimize the $k$-dimensional volume functional, when the deforming film tends to the position of least volume, for $k > 2$ consists of processes of gluing and collapsing which lead to pieces of small dimensions, not influencing the $k$-dimensional volume of the film, but playing an important part in its topological structure. (In the two-dimensional case, there is no such difficulty; see its discussion in §3). This makes it necessary to consider at once all, and not only one, dimensions similarly to formulating the problem in terms of chains (see above), and also to minimize not only the greatest volume, but also all of lesser dimensions. Thus, the very nature of Plateau's problem made the author introduce a new concept of stratified surface and its stratified volume. It turns out that the Plateau problem can be equivalently reformulated in the language of spectral bordism theory (see Chapter 2). The complete solution of the problem of finding a minimal surface in the class of images of the spectra of "manifolds with boundary" is given in the author's papers [148], [151], [152], [155], [156], [158], [161], and [162]. It was found that the variational problem of finding the absolutely minimal surface is solvable not only for spectral bordism classes but also for quite a large class of boundary conditions determined by an arbitrary, so-called extraordinary (co)homology theory (Chapter 2). Besides, the problem of realizing non-trivial (co)cycles by means of minimal surfaces of least volume can be also solved.

The following three principal topics are discussed in the book:

(1) New concepts of stratified surfaces and their stratified volumes are introduced. The theory of these objects is constructed; in particular, a method for minimizing stratified volume in any topologically fixed class of surfaces is worked out. As a corollary, the spectral Plateau problem in the class of mappings on the spectra of manifolds with fixed boundary is solved.

(2) A new method for the proof of the global minimality of concrete surfaces in Riemannian manifolds is offered, and new examples of absolutely minimal surfaces are obtained.

(3) An interesting relation between the known Bott periodicity, topologically non-trivial, totally geodesic surfaces in Lie groups, and the number of independent vector fields on spheres.

The modern machinery of multidimensional variational problems is quite extensive; therefore, we have confined ourselves only to considering the method for minimizing stratified volume and that for determining the topological properties of minimal surfaces and their explicit construction. They were elaborated by the author in [148], [150]–[158], [161], and [162], where the multidimensional Plateau problem is solved in terms of spectral bordism theory (the relative results being treated in Chapters 2, 3 and 6). The theories of integral currents, chains and varifolds treated, e.g., in [13], [14], [148], [135], [150]–[158], [161], [162] and [391], were left outside the scope of the book.

The composition of the material and the whole architecture of the book are based by the author on a special course read by him to students and postgraduates at the Faculty of Mathematics and Mechanics of Moscow University.

We endeavoured to define and comment on all basic concepts restored to in the text, however, not striving for complete self-reliance. In particular, preliminary familiarity with basic geometric ideas of [115] will make it easier for the reader to assimilate the discussed methods and results. The constructive existence theorems for globally minimal surfaces have been formulated in the first chapter, because of the relation to certain further constructions; however, their proofs are placed in the last chapter in order not to interrupt the natural course of events when important topological and geometrical corollaries are proved by the very existence of minimal solutions.

In the first chapter, we discuss the Euler–Lagrange equations, properties of extremals for the Riemannian volume and Dirichlet functionals. The relation of local to global minimality is established.

The second chapter is devoted to stating and solving the multidimensional Plateau problem in terms of extraordinary spectral (co)homology theory; in particular, spectral bordism theory. Here, we formulate basic existence theorems for globally minimal surfaces and briefly sketch their proof, giving it completely in Chapter 6.

In the third chapter, we supply a geometric method for effectively discovering globally minimal surfaces in concrete manifolds. It turns out that an exact universal lower estimate on the volume of any topologically non-trivial minimal surface can be found for any Riemannian manifold, often permitting us to establish the global minimality of concrete submanifolds, e.g., in symmetric spaces.

Chapter 4 contains the complete classification of locally minimal, totally geodesic submanifolds realizing non-trivial (co)cycles and elements of homotopy groups in symmetric spaces; in particular, this enables us to classify stationary points of the Dirichlet functional on spaces of mappings of discs to symmetric spaces.

The machinery worked out permits us to obtain known topological Bott periodicity, proceeding from the extremals of two-dimensional and eight-dimensional Dirichlet functionals not in two or eight steps as is obtained if we apply Morse theory to geodesics, but in one (see Chapter 5). We also consider the three known geometrical problems of variational calculus, viz., that of finding minimal solutions which would be invariant under the action of a compact Lie group (including the cone problem), that of existence of non-linear functions whose graphs in Euclidean space are minimal surfaces, and that by of realization via a harmonic mapping of the homotopy classes of mappings of a manifold into a manifold. In particular, we list here the results of S. N. Bernstein, H. Lawson, W. Hsiang, F. Almgren, J. Simons, E. Bombieri, E. De Georgi, E. Giusti, R. Smith, J. Eels, A. I. Pluzhnikov, A. V. Tyrin et al.

Chapter 6 contains the detailed account of methods for minimizing the stratified volume functional, and a constructive method for finding globally minimal surfaces. Here, we solve Plateau's problem, i.e., completely move the basic existence theory for globally minimal surfaces of non-trivial topological type in arbitrary spectral bordism class.

The book is designed for the students and postgraduates of Faculties of Mathematics, and also for specialists interested in multidimensional variational calculus and minimal surface theory.

The author thanks Kluwer Academic Publishers and personally Prof. Dr. Michiel Hazewinkel and Dr. David J. Larner for their support.

A.T. Fomenko

# CHAPTER 1

# Simplest Classical Variational Problems

## §1 Equations of Extremals for Functionals

Variational problems form one of the most important classes of mathematical problems that originated from mechanics and physics in studying motion and stability. Thus, e.g., geodesic lines, minimal (soap) films spanned by a frame, equations of motion of mechanical systems, etc., are solutions to the corresponding variational problems.

We begin with the general concept of a functional. Consider a bounded domain $D$ with smooth or piecewise smooth boundary $\partial D$ in Euclidean space $\mathbb{R}^k$. Let $x^1, \ldots, x^k$ be Cartesian coordinates in $\mathbb{R}^k$, and $f(x^1, \ldots, x^k) = f(x^\alpha) = f^1(x^\alpha), \ldots, f^n(x^\alpha) = \{f^i(x^\alpha)\}$ be all possible smooth vector functions on $D$. We call $D$ the domain of parameters $x^1, \ldots, x^k$. We assume that $f$ determines a mapping of $D$ into a certain Riemannian manifold $M$. Then the space $F = F(D)$ of all such mappings is "infinite-dimensional." We will consider various functionals $I$ on $F$. Meanwhile, it is useful to bear in mind a certain analogy with the properties of the usual functions. We describe it below.

Let a smooth function $L(x^\beta, p^i, q^i_\alpha)$ of three groups of variables $1 \leqslant \alpha, \beta \leqslant k$, $1 \leqslant i \leqslant n$ be given. We call it a Lagrangian. Construct a functional $I[f], f \in F(D)$ by the rule $I[f] = \int_D L(x^\beta, f^i(x^\beta),\ f^i_{x^\alpha}(x^\beta))d\sigma_k$, where $\int$ denotes a $k$-integral with respect to a $k$-dimensional domain $D$, and $d\sigma_k = dx^1 \wedge \cdots \wedge dx^k$ the standard $k$-dimensional Euclidean volume form in $\mathbb{R}^k$. In writing $I(f)$, we will abbreviate it to $\int_D L(x^\beta, f^i, f^i_{x^\alpha})d\sigma_k$, where $f^i_{x^\alpha} = \frac{\partial f^i}{\partial x^\alpha}$ are partial derivatives.

We illustrate by example of the arc-length functional $\int_0^1 \sqrt{g_{ij}(\gamma(t)) \frac{dy^i}{dt}\frac{dy^j}{dt}}\,dt$, where $D = [0, 1] \subset R^1$, $k = 1$, and $\gamma(t) = f(t) = (y^1(t), \ldots, y^n(t))$ is a curve in $R^n$ supplied with the Riemannian metric $g_{ij}(y)$. As it is customary with the tensor form of writing, we assume that summation is over the repeating subscripts and superscripts. The Lagrangian is of the form $L = \sqrt{g_{ij}(y)\dot{y}^i\dot{y}^j}$. If the curve $\gamma(t)$ is given on the plane $R^2$ explicitly as $y = f(t)$, then $L = \sqrt{1 + (f_t)^2}$.

Let us consider another example, namely, the area functional $I[f] = \iint_D \sqrt{EG - F^2}du\,dv$, where the parameters $u$ and $v$ vary in a domain $D$ on $R^2$, and the mapping $f : D \to R^3$ specifies a two-dimensional surface $M^2 \subset R^3$ supplied with the induced Riemannian metric $ds^2 = Edu^2 + 2Fdudv + Gdv^2$, $E = \langle f_u, f_u \rangle$, $F = \langle f_u, f_v \rangle$, $G = \langle f_v, f_v \rangle$, and $L = L(f_u, f_v)$. If $M^2$ is given as a graph $z = z(u, v)$, then $I[f] = \iint_D \sqrt{1 + z_u^2 + z_v^2}du\,dv$.

While studying functionals, we shall give first priority to their "critical points." By analogy with the usual functions, consider those of one or two variables $\alpha(t)$ or $\alpha(u, v)$. To a great extent, their behaviour is determined by the number and position of those points $t_0(\text{or}(u_0, v_0))$ at which $\alpha_t'(t_0) = 0(\text{or}\,\alpha_u(u_0, v_0) = \alpha_v(u_0, v_0) = 0)$, i.e., $\text{grad}(\alpha) = 0$. Points at which $\text{grad}(\alpha) = 0$ are usually said to be *critical*, or *stationary*, of $\alpha$. For a function of two variables, there are, e.g., points of maximun, minimun, and nondegenerate saddle points among critical ones (Fig. 1). In mechanics, the knowledge of points at which potential energy attains its minimun is important for finding the equilibrium of a mechanical system.

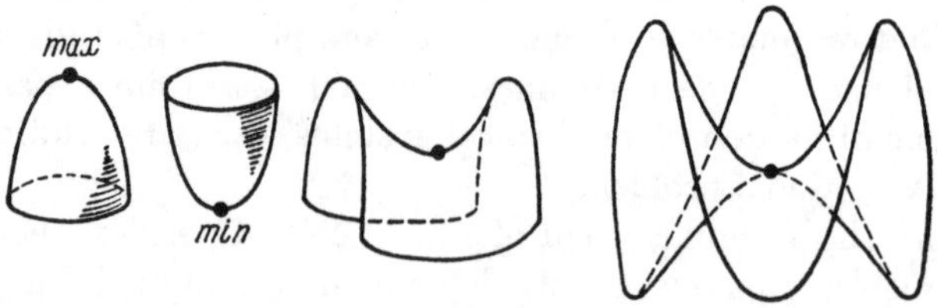


Nondegenerate critical points

Quadratic saddle (of order 2)

**Fig. 1**

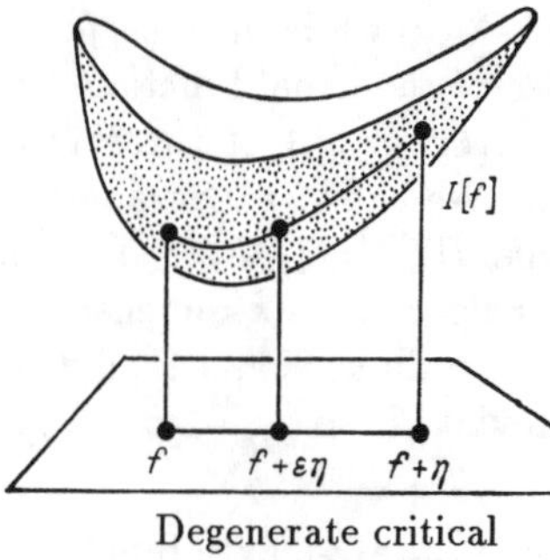


Degenerate critical point of order 3

**Fig. 2**

Similarly, in investigating a functional of form $I[f]$, much attention is paid to finding those stationary "points," i.e., functions $f_0$, at which the functional attains its minimum, or maximum, or possesses a saddle point. For a precise definition, it must be determined first of all what is a "directional derivative" of the functional $I$ at a certain "point" $f \in F$, i.e., an analogue of the equation $\text{grad}(\alpha) = 0$ obtained. If, at a certain point $(u_0, v_0) \in R^2$, a direction $a = (a^1, a^2)$ is given, then the derivative of the function $\alpha(u, v)$ in direction $a$ is of form $\frac{d\alpha}{da} = \langle a, \text{grad}(\alpha)\rangle = a^1\alpha_u + a^2\alpha_v$. Thus, in the case of the usual function, a point $(u_0, v_0)$ is critical if and only if $\frac{d\alpha(u_0,v_0)}{da} = 0$ for any direction $a$. Meanwhile, the derivative $\frac{d\alpha}{da}$ can be calculated as $\frac{d\alpha(x)}{da} = \lim_{\varepsilon\to 0} \frac{1}{\varepsilon}[\alpha(x + \varepsilon a) - \alpha(x)]$, where $x = (u, v) \in R^2$. It is in this form that we generalize the concept of the directional derivative to the case of functionals. Consider a "point" $f \in F(D)$ and a sufficiently small vector function $\eta \in F(D)$ such that $\eta|_{\partial D} \equiv 0$ (it being sometimes required that $\eta \equiv 0$ in a neighbourhood of $\partial D$). We call such functions $\eta$ perturbations of the function $f$. Suppose we shift from the "point" $f$ to a "point" $f + \varepsilon\eta$ (Fig. 2), i.e., we consider a small perturbation (variation) of the original function $f$; $\eta$ specifies the "displacement direction" from the "point" $f$ just like the vector $a$ has specified the direction of displacement from the point $(u, v)$. We then construct the expression $\frac{1}{2}(f + \varepsilon\eta - I[f])$. Passing to the limit, we obtain a function which we denote by $\frac{dI[f]}{d\eta} = \lim_{\varepsilon\to 0} \frac{1}{\varepsilon}(I[f + \varepsilon\eta] - I[f])$, and call the *derivative of the functional $I$ at the point $f$ in direction $\eta$*. The function $\frac{dI[f]}{d\eta}$ is sometimes represented in the form $\frac{dI[f]}{d\eta} = \int_D \langle \frac{\delta I}{\delta f}, \delta f\rangle d\sigma_n$, where $\delta f = \eta = (\eta^1, \ldots, \eta^k)$ and the vector $\frac{\delta I}{\delta f} = (\frac{\delta I}{\delta f^i})$ defined by the previous equality is called the variational derivative of the functional $I[f]$. Further proceeding on the analogy of the usual functions, we give the following:

**Definition 1.1.** A function $f_0 \in F$ is said to be stationary (or extremal, or critical) for a functional $I$ if $\frac{\delta I[f_0]}{\delta f} \equiv 0$ for any perturbation $\delta f = \eta \in F$, such that $\eta \equiv 0$ in a neighbourhood of the boundary $\delta D$.

We now pass to the analytic investigation of the derivative $\frac{\delta I[f_0]}{\delta f}$. We have

$$\delta I[f] = I[f + \varepsilon\eta] - I[f] = \int_D [L(x^\beta; f^j + \varepsilon\eta^j; f^i_{x^\alpha} + \varepsilon\eta^i_{x^\alpha}) - L(x^\beta; f^j; f^i_{x^\alpha})] d\sigma_k.$$

Expanding the integrand into the Taylor series, we obtain

$$\delta I[f] = \int_D \left[\sum_{j=1}^n \frac{\partial L}{\partial f^j} \varepsilon\eta^j + \sum_{i=1}^n \sum_{\alpha=1}^k \frac{\partial L}{\partial f^i_{x^\alpha}} \varepsilon\eta^i_{x^\alpha} + o(\varepsilon)\right] d\sigma_k$$
$$= \varepsilon \int_D \sum_{i=1}^n \left[\frac{\partial L}{\partial f^i}\eta^i + \sum_{\alpha=1}^k \frac{\partial L}{\partial f^i_{x^\alpha}}\eta^i_{x^\alpha}\right] d\sigma_k + \int_D o(\varepsilon)\, d\sigma_k.$$

By integrating by parts, we get

$$\delta I[f] = \varepsilon \int_D \sum_{\alpha=1}^{k} \frac{\partial}{\partial x^\alpha}\left(\frac{\partial L}{\partial f^i_{x^\alpha}} \eta^i\right) d\sigma_k + \varepsilon \int_D \sum_{i=1}^{n} \left(\frac{\partial L}{\partial f^i} - \sum_{\alpha=1}^{k} \frac{\partial}{\partial x^\alpha}\left(\frac{\partial L}{\partial f^i_{x^\alpha}}\right)\right) \eta^i \, d\sigma_k + \int_D o(\varepsilon)\, d\sigma_k.$$

Since we assume that all functions involved in computations are smooth (or piecewise smooth), integration with respect to the variable $x^\alpha$ in the first integral can be separated from integrating with respect to the other variables $x^1, \ldots, \hat{x}^\alpha, \ldots, x^k$ (by the theorem on changing the order of integration). The sign ^ indicates the omitted coordinate. We derive

$$\int_D \frac{\partial}{\partial x^\alpha}\left(\frac{\partial L}{\partial f^i_{x^\alpha}} \eta^i\right) d\sigma_k = \int_{x^1, \ldots, \hat{x}^\alpha, \ldots, x^k} \left[\int_P^Q \frac{\partial}{\partial x^\alpha}\left(\frac{\partial L}{\partial f^i_{x^\alpha}} \eta^i\right) dx^\alpha\right] d\sigma_{k-1}$$

(the points $P$ and $Q$ depend on $x^1, \ldots, \hat{x}^\alpha, \ldots, x^k$; $d\sigma_{k-1} = dx^1 \wedge \cdots \wedge \hat{dx}^\alpha \wedge \ldots dx^k$ ). Since the variables $x^1, \ldots, \hat{x}^\alpha, \ldots, x^k$ can be regarded as parameters in the interior integral $\int_P^Q$,

$$\int_D \frac{\partial}{\partial x^\alpha}\left(\frac{\partial L}{\partial f^i_{x^\alpha}} \eta^i\right) d\sigma_k = \int_{x^1, \ldots, \hat{x}^\alpha, \ldots, x^k} \left[\frac{\partial L}{\partial f^i_{x^\alpha}} \eta^i \Big|_Q - \frac{\partial L}{\partial f^i_{x^\alpha}} \eta^i \Big|_P\right] d\sigma_{k-1} = 0,$$

because $\eta^i(Q) = \eta^i(P) = 0$ (see above). Thus,

$$\delta I[f] = \varepsilon \int_D \sum_{i=1}^{n} \left[\frac{\partial L}{\partial f^i} - \sum_{\alpha=1}^{k} \frac{\partial}{\partial x^\alpha}\left(\frac{\partial L}{\partial f^i_{x^\alpha}}\right)\right] \eta^i \, d\sigma_k + \int_D o(\varepsilon)\, d\sigma_k.$$

Hence,

$$\frac{\delta I[f]}{\delta f^i} = \frac{\partial L}{\partial f^i} - \sum_{\alpha=1}^{k} \frac{\partial}{\partial x^\alpha}\left(\frac{\partial L}{\partial f^i_{x^\alpha}}\right),$$

since

$$\lim_{\varepsilon \to 0} \frac{1}{\varepsilon} \int_D o(\varepsilon)\, d\sigma_k = 0.$$

If $f_0 \in F$ is a stationary (extremal) function for a functional $I[f]$, then, for any function $\eta$ where $\eta = 0$ near $\partial D$, the identity holds

$$\int_D \sum_{i=1}^{n} \eta^i \left[\frac{\partial L}{\partial f_0^i} - \sum_{\alpha=1}^{k} \frac{\partial}{\partial x^\alpha}\left(\frac{\partial L}{\partial f^i_{0, x^\alpha}}\right)\right] d\sigma_k = 0,$$

whence

$$\frac{\delta I[f]}{\delta f^i} = \frac{\partial L}{\partial f_0^i} - \sum_{\alpha=1}^{n} \frac{\partial}{\partial x^\alpha}\left(\frac{\partial L}{\partial f^i_{0,x^\alpha}}\right) = 0 \quad (1 \leqslant i \leqslant n).$$

This system of differential equations is said to be Euler–Lagrange for the functional $I$. We combine the previous results into a theorem.

**Theorem 1.1.** A function $f_0 \in F$ is an extremal function for a functional $I[f]$ if and only if it satisfies the Euler–Lagrange equation system.

## §2 Geometry of Extremals

### 2.1. The Zero-Dimensional and One-Dimensional Cases

In the case of the usual functions $f(x)$ considered on a smooth manifold $M^n$, the extremal (critical) point equation for a function $f$ is of the form grad $f = 0$. If all the critical points are nondegenerate (i.e., the Hessian of the function $f$ is nondegenerate; e.g., see [115], [276]), then, as it turns out, the topology of the manifold $M$ is determined, to a considerable extent, by the number of critical points and their indices; at any rate, we can make out of which cells this manifold is glued together. The corresponding procedure is given by Morse theory, recounted, e.g., in J. Milnor's book [276].

The next step with respect to the dimension of the variational problem leads us to functions defined on the space of smooth (or piecewise smooth) curves. The length and action functionals

$$l_a^b(\gamma(t)) = \int_a^b \sqrt{g_{ij}\dot{\gamma}^i\dot{\gamma}^j}\, dt, \qquad S_a^b(\gamma(t)) = \int_a^b g_{ij}\dot{\gamma}^i\dot{\gamma}^j\, dt$$

belong to the most important one-dimensional functionals defined on the space of piecewise smooth curves in a Riemannian manifold. Their extremals are geodesics related to the natural parameter in the case of the action functional, and geodesics related to an arbitrary (piecewise smooth) parameter in the case of the length functional (see, e.g., [115]). Therefore, the knowledge of extremals for these functionals also permits us to restore substantially the topological structure of the manifold on which the curves and functionals are considered. "One-dimensional theory" is more complicated than "zero-dimensional theory," since now we have to deal with conjugate points on geodesics whose index is much more difficult to compute than that of critical points of smooth functions.

In the present book, we will consider multidimensional functionals (mostly, area and volume functionals) for which the dimension of $D$ (see §1) is greater than two. The latter circumstance makes the theory somewhat more complicated, but the obtained results turn out to be more flexible from the standpoint of applications. In the following, we shall strictly distinguish between the concepts of local and global minimality. In the one-dimensional case, there are, in general, many geodesics of different length, joining two fixed points $P$ and $Q$ on a Riemannian manifold $M$. As is generally known, all these trajectories (see, e.g., [115]) are locally minimal, i.e., are the shortest trajectories between any two of their sufficiently close points; on the other hand, globally minimal trajectories joining the points $P$ and $Q$ are only geodesics of least length. For example, the standard equator on the standard 2-sphere is locally minimal as a one-dimensional extremal, but not globally minimal. The concept of global minimality for volume functionals will be defined below.

## 2.2. Some Examples of the Simplest Multidimensional Functional. The Volume Functional

First, we consider the so-called soap films spanning closed wire frames. The corresponding mathematical description of such surfaces is given by minimal surfaces. Recall the definition of the volume functional on a Riemannian manifold $M^n$. In the simplest case of an open bounded domain $D$ with piecewise smooth boundary in $\mathbb{R}^n$, the integral $\int_D dx^1 \wedge \cdots \wedge dx^n$ is taken as the volume $\mathrm{vol}_n(D)$ (see, e.g., [115]). If $D$ is embedded into a smooth Riemannian manifold $M^n$ with Riemannian metric $g_{ij}(x)$ and is contained in one chart $U$, then $\int_U \sqrt{\det A} dx^1 \wedge \cdots \wedge dx^n$ is taken as $\mathrm{vol}_n D$, where $A = (g_{ij}(x))$ is the metric tensor matrix defining the first fundamental form. If, however, $D$ is contained in several charts, then its volume is defined as the integral of the indicated exterior differential volume form; meanwhile, a certain smooth partition of unity has to be fixed; it is proved in the course of analysis that the volume does not depend on the choice of this partition.

In special cases, the above definition of volume turns into the classical one. Thus, if $M^n = \mathbb{R}^n$, $x^1, \ldots, x^n$ are Cartesian coordinates and $g_{ij} = \delta_{ij}$, then, obviously, we obtain the definition of the usual Euclidean volume of a domain.

Further, if $M^n$ is a smooth submanifold in $R^N$ and $g_{ij}$ is the induced Riemannian metric, then the above definition coincides with another intuitive idea of the volume of a surface. Consider the local coordinates $x^1, \ldots, x^n$ within the limits of one chart on $M^n$, and represent $D$ as the union of arbitrarily small parallelepipeds $\Pi_i$, each of which is formed by the surfaces given by $x^i = \mathrm{const}$ (Fig. 3). Then, we can assume that $\mathrm{vol}_n(D) = \sum_i \mathrm{vol}(\Pi_i)$. We shall prove that $\mathrm{vol}_n(\Pi_i) = \sqrt{\det A} dx^1 \ldots dx^n$, where $dx^1, \ldots, dx^n$ denote the lengths of the edges of $\Pi_i$. Consider the vectors $a_1, a_2, \ldots, a_n$ which are tangent to the manifold $M^n$ at a point $x \in D$, and which are velocity vectors of coordinate lines passing through $x$. Denote a parallelepiped placed in the tangent plane and spanned by the vectors $a_1, \ldots, a_n$, by $\tilde{\Pi}(a_1, \ldots, a_n)$. Then the arbitrarily small parallelepiped $\Pi$ can be regarded as spanned by the vectors $(dx^1)a_1, \ldots, (dx^n)a_n$ (Fig. 4). We shall prove that $\mathrm{vol}_n(\tilde{\Pi}) = \sqrt{\det A}$. It will then immediately follow that

$$\mathrm{vol}_n(\Pi) = \mathrm{vol}\,\Pi((dx^1)a_1, \ldots, (dx^n)a_n) = \\ = (\mathrm{vol}\,\tilde{\Pi}(a_1, \ldots, a_n))\,dx^1 \ldots dx^n = \sqrt{\det A}\,dx^1 \ldots dx^n.$$

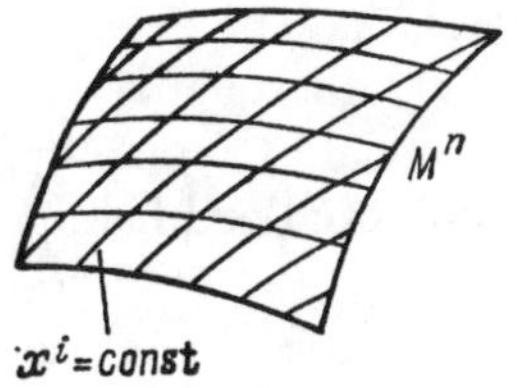


Fig. 3

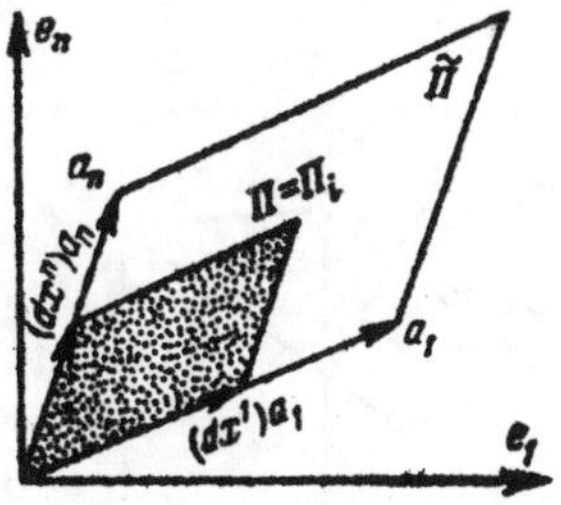


Fig. 4

Indeed, consider an orthobasis $e_1, \ldots, e_n$ in the tangent plane to $M^n$. Let $B : (e_1, \ldots, e_n) \to (a_1, \ldots, a_n)$ be a linear transformation in this plane, which is generated by the above curvilinear system of coordinates $x^1, \ldots, x^n$, and which transforms the unit cube in the tangent plane into the parallelepiped $\tilde{\Pi}(a_1, \ldots, a_n)$. If $A = (g_{ij})$, it follows from the transformation law for the fundamental form that $A = BB^T$ (where $T$ denotes transposition), i.e., $\det B = \sqrt{\det A}$. On the other hand, it is known from linear algebra that $\mathrm{vol}\,\tilde{\Pi}(a_1, \ldots, a_n) = \det B$, which completes the proof. Thus, the above general definition of the volume of a domain on a submanifold in $\mathbb{R}^N$ coincides with the intuitive idea of the volume obtained by summing up the volumes of arbitrarily small Euclidean parallelepipeds.

Consider a smooth hypersurface $V^{n-1} \subset \mathbb{R}^n$ given, e.g., as the graph of $x^n = f(x^1, \ldots, x^{n-1})$. Let the domain of the function $f$ be a bounded region in $\mathbb{R}^{n-1}$. We take the volume functionals $\mathrm{vol}(f) = \int_D \sqrt{|\det A|}\,d\sigma_{n-1}$ on the space of all these functions, where the matrix $A = (g_{ij}(x))$, $x \in D$ is the induced Riemannian metric on the surface $V^{n-1}$, and $d\sigma_{n-1} = dx^1 \wedge dx^2 \wedge \cdots \wedge dx^{n-1}$, $x^1, \ldots, x^{n-1}$ being Euclidean coordinates. The Lagrangian $\sqrt{|\det A|}$ can be explicitly written in terms of $f$. Let $d\tau^{n-1}$ be the $(n-1)$-dimensional volume form on $V$; then $\mathrm{vol}(f) = \int_D d\tau^{n-1}$. Suppose $P \in V^{n-1}$, $n(P)$ is the unit vector perpendicular to $V$ at $P$, $\alpha(P)$ is the angle between $n(P)$ and $e_n = (0, \ldots, 0, 1)$ (Fig. 5). Then

$$\mathrm{vol}\,(f) = \int_D d\tau^{n-1} = \int_D \frac{d\sigma_{n-1}}{\cos\alpha\,(P)}.$$

Further,

$$\cos\alpha\,(P) = \langle e_n,\ n\,(P)\rangle = \left(1 + \sum_{i=1}^{n-1} (f_{x^i})^2\right)^{-1/2}.$$

Thus,

$$\mathrm{vol}\,(f) = \int_D \sqrt{1 + \sum_{i=1}^{n-1} (f_{x^i})^2}\, dx^1 \wedge \ldots \wedge dx^{n-1}.$$

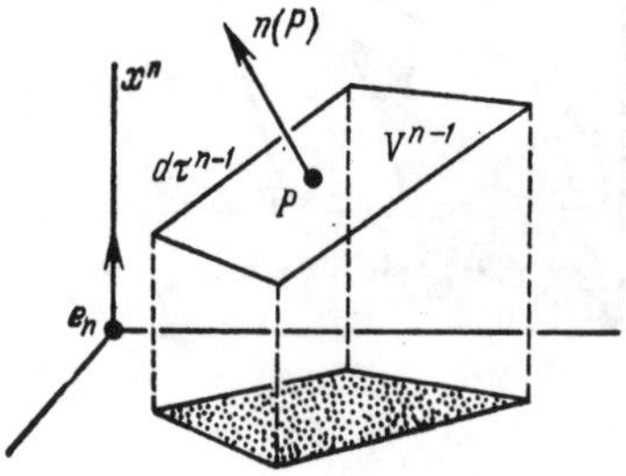


**Fig. 5**

Consider the extremal surfaces $V^{n-1}$ for the volume functional vol($f$), i.e., graphs of the extremal functions $x^n = f(x)$, $x \in D$. The Euler–Lagrange equation is then of the form

$$\sum_{i=1}^{n-1} \frac{\partial}{\partial x^i}\left(f_{x^i}\left[1 + \sum_{j=1}^{n-1} (f_{x^j})^2\right]^{-1/2}\right) = 0.$$

**Definition 2.2.1.** Surfaces which are extremal for the volume functional vol($f$) are said to be locally minimal.

For example, minimal surfaces can be modelled in $\mathbb{R}^3$ by means of soap films spanning a closed wire frame $f$ (in the absence of gravity), in which case vol($f$) coincides with the area of the surface.

**Lemma 2.2.1.** *For a two-dimensional minimal surface given as a graph of* $z = f(x, y)$ *in* $\mathbb{R}^3(x, y, z)$, *the Euler–Lagrange equation is*

$$(1+f_x^2)f_{yy}-2f_{xy}f_xf_y+(1+f_y^2)f_{xx}=0.$$

**Proof.** We have

$$\frac{\partial}{\partial x}(f_x\cdot[1+(f_x)^2+(f_y)^2]^{-1/2})+\frac{\partial}{\partial y}(f_y\cdot[1+(f_x)^2+(f_y)^2]^{-1/2})=0.$$

By differentiating and collecting similar terms, we obtain the required statement.

The equation of the minimal surface $V^{n-1}$ can be written in the language of the local embedding invariants of the surface in $\mathbf{R}^n$.

**Theorem 2.2.1.** *Let $V^{n-1}\subset\mathbf{R}^n$ be a smooth hypersurface. The mean curvature $H$ is identically zero if and only if $V^{n-1}$ is representable in the neighbourhood of each of its points as the graph of the volume functional extremal function, i.e., as a solution to the minimal surface equation.*

Thus, the condition $H\equiv 0$ is just that for minimum of $V^{n-1}\subset\mathbf{R}^n$. The proof is reduced to the direct computation of the mean curvature $H=\mathrm{Sp}(A^{-1}Q)$ of the graph of $x^n=f(x)$ (where $A,Q$ are the matrices of the first and second fundamental form, respectively) and verification of the coincidents of $H=0$ with the Euler–Lagrange equation.

We are not going to perform this computation for the general case and only consider the special situation of a two-dimensional surface in $\mathbf{R}^3$. We specify locally a surface $V^2\subset\mathbf{R}^3$ by the radius vector $r=r(u,v)$. Then $\mathrm{vol}(r)=\int_{D(u,v)}\sqrt{EG-F^2}du\,dv$, where $A=\begin{pmatrix}E&F\\F&G\end{pmatrix}$ is the matrix of the first fundamental form and $E=\langle r_u,r_u\rangle$, $F=\langle r_u,r_v\rangle$, $G=\langle r_v,r_v\rangle$. The mean curvature is $H=(EG-F^2)^{-1}(GL+2FM+EN)$, where $Q=\begin{pmatrix}L&M\\M&N\end{pmatrix}$ is the matrix of the second fundamental form $L=\langle r_{uu},n\rangle$, $M=\langle r_{uv},n\rangle$, $N=\langle r_{vv},n\rangle$, and $n$ is the unit vector perpendicular to the surface. We select (locally) so-called conformal (isothermal) coordinates on $V^2$. Let the two-dimensional surface in $\mathbf{R}^3(x,y,z)$ be given parametrically as $x=x(p,q)$, $y=y(p,q)$, $z=z(p,q)$, where $p,q$ vary in a certain domain of $\mathbf{R}^2$. Then the induced metric

$$ds^2=E\,(dp)^2+2F\,dp\,dq+G\,(dq)^2,\quad g=EG-F^2>0.$$

arises.

**Theorem 2.2.2.** *Let $E,F,G$ be three real analytic functions of variables $(p,q)$. Then we can introduce new local coordinates $u,v$ such that the metric $dl^2$ takes the form $dl^2=f(u,v)(du^2+dv^2)$ with respect to these. We call such coordinates isothermal, or conformal.*

The proof can be seen in [115]. Isothermal coordinates are not determined uniquely.

Thus, we will assume that $(u, v)$ are conformal coordinates on the surface $V^2 \subset \mathbf{R}^3$. With respect to the conformal coordinates $f = 0$, $E = G$; therefore,

$$\operatorname{vol}(r) = \iint_D \sqrt{\langle r_u, r_u \rangle \cdot \langle r_v, r_v \rangle}\, du\, dv$$
$$= \iint_D \sqrt{(x_u^2 + y_u^2 + z_u^2)(x_v^2 + y_v^2 + z_v^2)}\, du\, dv.$$

Furthermore, $H = \frac{1}{E}(L + N) = \frac{1}{E}\langle r_{uu} + r_{vv}, n\rangle = \frac{1}{E}\langle \Delta r, n\rangle$, where $\Delta$ is the Laplace operator. Consider the Euler–Lagrange equations for $\operatorname{vol}(r)$ with respect to $u, v$. Meanwhile, it should be remembered that the possibility of forming the Euler–Lagrange equation with respect to conformal coordinates follows from the fact that we can regard all the functions $r(u, v) + \varepsilon\eta(u, v)$ related to conformal coordinates by means of perturbation $\eta$ in varying the functional $\operatorname{vol}(r)$, for which it suffices to introduce them on each $r(u, v) + \varepsilon\eta(u, v)$ (as a matter of fact, the coordinates $(u, v)$ are, in general, not conformal any more on the perturbed surface $r + \varepsilon\eta$). The coordinates $(u_\varepsilon, v_\varepsilon)$ can be regarded as smoothly dependent on $\varepsilon$. The Euler–Lagrange equations then take the form

$$\frac{\partial}{\partial u}(x_u) + \frac{\partial}{\partial v}(x_v) = 0, \quad \frac{\partial}{\partial u}(y_u) + \frac{\partial}{\partial v}(y_v) = 0, \quad \frac{\partial}{\partial u}(z_u) + \frac{\partial}{\partial v}(z_v) = 0,$$

i.e., $\Delta r = 0$. The radii vectors $r$ satisfying the latter equations are said to be ***harmonic***. Thus, with respect to conformal coordinates, the "minimality" of the radius vector, i.e., its extremality, for the areal functional means its harmonicity. We can speak of the harmonicity of $r(u, v)$ only with respect to some other coordinate system. Varying coordinates may, in general, lead to disturbing the harmonicity property, and since $\Delta r = 0$, $H = \frac{1}{E}\langle \Delta r, n\rangle = 0$, thus proving Theorem 2.2.1 in one direction. Conversely, let $H \equiv 0$. We have to prove that $\Delta r = 0$ with respect to conformal coordinates. Since $\langle \Delta r, n\rangle = 0$, it suffices to verify two other equalities $\langle \Delta r, r_u\rangle = 0$, $\langle \Delta r, r_v\rangle = 0$. Hence, it will follow that $\Delta r = 0$. In fact, the vectors $n, r_u, r_v$ form a frame at any regular point of the surface $r(u, v)$ by definition of the surface. Due to the choice of coordinates, $E = G$ and $F = 0$, i.e., $\langle r_u, r_u\rangle = \langle r_v, r_v\rangle = 0$. By differentiating with respect to $u$ and $v$, we obtain $\langle r_{uu}, r_u\rangle = \langle r_{uv}, r_v\rangle$, $\langle r_{uv}, r_u\rangle = \langle r_{vv}, r_v\rangle$, $\langle r_{uu}, r_v\rangle + \langle r_u, r_{uv}\rangle = 0$, $\langle r_{uv}, r_v\rangle + \langle r_u, r_{vv}\rangle = 0$.

We should verify the identities $\langle r_{uu}, r_u\rangle + \langle r_{vv}, r_u\rangle = 0$, $\langle r_{uu}, r_v\rangle + \langle r_{vv}, r_v\rangle = 0$.

Obviously, they follow from the previous system. Thereby we have proved not only Theorem 2.2.1 but also the following.

**Proposition 2.2.1.** *A two-dimensional surface in three-dimensional Euclidean space is described by the minimal radius vector if and only if the mean cur-*

***vature is identically zero. With respect to conformal coordinates, the minimal radius vector becomes harmonic.***

The topological structure of minimal surfaces $V^2 \subset \mathbb{R}^3$ is rather complicated; in particular, if the boundary frame is fixed, then, in general, many "soap films" may span it, and there is no uniqueness theorem for a solution to the differential equation $H \equiv 0$, or $\Delta r = 0$ (see the examples in Figs. 6 and 7). The solutions to the equations $H = 0$, $\Delta r = 0$ may possess singularities, an example being the triple Möbius strip (see Fig. 8). This minimal film contains a singular circle filled with triple singular points.

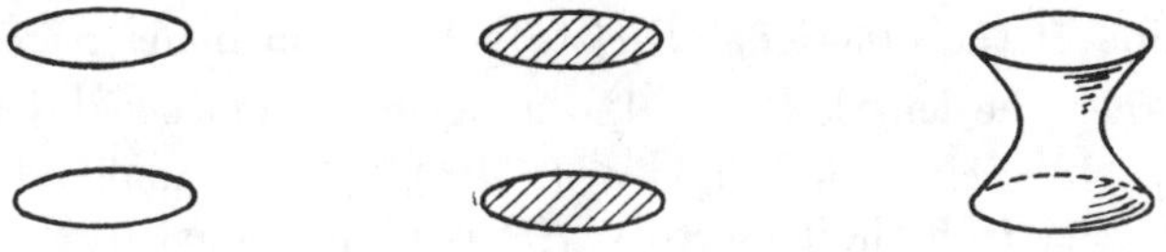

Outline A   Two discs   Catenoid

**Fig. 6**

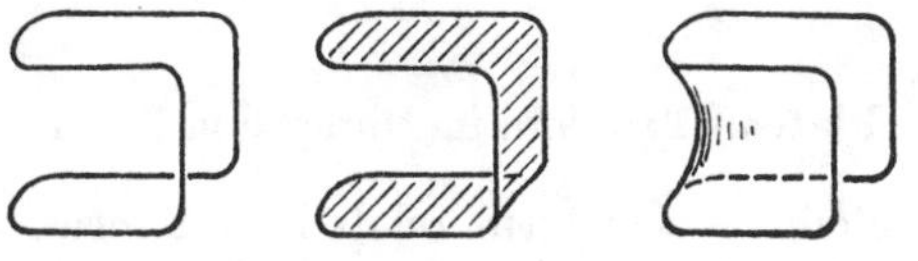

**Fig. 7**

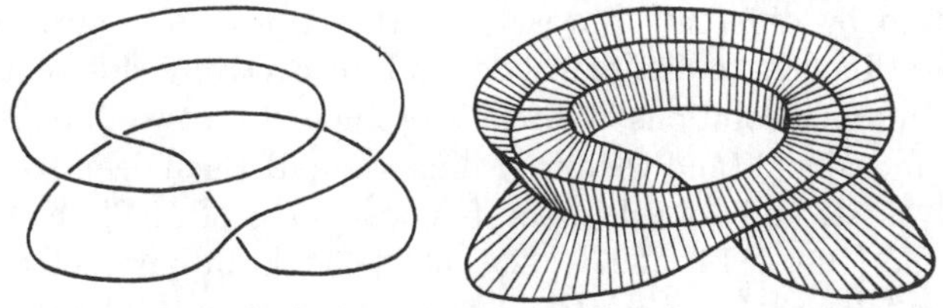

**Fig. 8**

The harmonic radii vectors are solutions of the Euler–Lagrange equation for another two-dimensional functional, the Dirichlet functional. Consider the three-dimensional radius vector $r(u, v)$, with the coordinates $u, v$ being arbitrary. We call $D[r] = \int_{D(u,v)} \frac{E+G}{2} du\, dv$ the *Dirichlet functional*, where $E, G$ are the coefficients of the first fundamental form for the surface $r(u, v)$, and $L(r_u, r_v) = \frac{1}{2}(x_u^2 + y_u^2 + z_u^2 + x_v^2 + y_v^2 + z_v^2)$; therefore, the Euler–Lagrange equation in vector form is $\Delta r = 0$, and its solutions are harmonic vectors. Since $\frac{E+G}{2} \geqslant \sqrt{EG - F^2}$, $D[r] \geqslant \mathrm{vol}(r)$ for any piecewise smooth radius vector $r(u,v)$, and the equality holds if and only if $E = G$, $F = 0$, i.e., with respect to conformal coordinate system. Thus, any extremal of the functional $D(r)$ for which the coordinates $(u,v)$ turned out to be conformal is an extremal for the functional $\mathrm{vol}(r)$, the converse not holding. In order to obtain all the extremals of $\mathrm{vol}(r)$, we should consider all those of $D[r]$, i.e., harmonic radii vectors, select them so that $(u,v)$ turn out to be conformal, and subject $(u,v)$ to an arbitrary regular change. The harmonic radius vector $r(u)$ for which $(u,v)$ are not conformal will not, in general, describe a minimal surface, an example being $r(u,v) = (u, v, \mathrm{Re}\, f(u+iv))$, the graph of the real (or imaginary) part of the nonlinear complex analytic function $f(u+iv)$.

The relation of the functional $D[r]$ to $\mathrm{vol}(r)$ is in many respects similar to that between the length $l_a^b$ and the action functionals $S_a^b[\gamma] = \int_a^b |\dot{\gamma}|^2 dt$ on a path $\gamma$. It is clear that $(l_a^b)^2 \leqslant (b-a)S_a^b$, and equality holds if and only if parameter $t$ on the trajectory $\gamma(t)$ is proportional to the arc length (such extremals $l_a^b[\gamma]$ being geodesic if the parameter is natural). This is related to the length and volume functionals being invariant relative to the regular change of variables, whereas the action and Dirichlet functionals are not invariant.

### 2.3. The Classical Plateau Problem in Dimension 2

The "Plateau problem" is the term related to a series of problems of the study of the extremals and absolute minima of the $k$-dimensional volume functional $\mathrm{vol}_k$ defined on $k$-dimensional surfaces embedded in an $n$-dimensional Riemannian manifold $M^n$, and satisfying various topological boundary conditions. In the history of variational problems of this type, several periods characterized by different approaches to the concepts of surface, boundary, or minimization can be distinguished, and, respectively, different methods for obtaining minimal solutions. Plateau's problem for a two-dimensional surface in three-dimensional (and then in $n$-dimensional) Eucilidean space was posed historically first, and solved in the famous works of J. Plateau, J. Douglas, and T. Radó [109]–[114], [327]–[329], [346]–[347]. In parametric form, it can be posed as follows. Let $f = f(u,v)$ be the radius vector of a two-dimensional surface $X^2$ in $M^n = \mathbb{R}^n$, i.e., $f : U \to \mathbb{R}^n$ (locally) determining a regular mapping of the domain $U \subset \mathbb{R}^2$ into $\mathbb{R}^n$; then $\mathrm{vol}_2 f(U) = \int_U \sqrt{EG - F^2}\, du dv$, and the extremals of this functional are solutions to simultaneous Euler–Lagrange equations

$$\frac{\partial}{\partial u}\left(\frac{\partial g}{\partial\left(\frac{\partial x^k}{\partial u}\right)}\right) + \frac{\partial}{\partial v}\left(\frac{\partial g}{\partial\left(\frac{\partial x^k}{\partial v}\right)}\right) = 0, \quad g = \sqrt{EG - F^2}, \quad 1 \leqslant k \leqslant n; \; x^1, \ldots, x^n$$

being Cartesian coordinates on $\mathbb{R}^n$. The "least-area" Plateau problem poses the question whether it is possible to find a surface $X_0^2 = f_0(U)$ (and a mapping $f_0$) such that it has a prescribed contour $\Gamma$, the boudary curve (i.e., a system of disjoint circles embedded in $\mathbb{R}^n$) as its boundary, its area being the least among all other surfaces $X^2 = f(U)$ bounded by $\Gamma$ (e.g., see [347]). In addition, the problem of finding a minimum in a given homotopy class of surfaces with prescribed boundary, which are homotopic to each other, was also considered. Comparatively long ago, the existence of a continuous mapping $f_0$ of a two-dimensional surface $M^2$ into $\mathbb{R}^3$ was proved, which transforms the boundary $\partial M^2$ of $M^2$ in to a prescribed contour $\Gamma \subset \mathbb{R}^3$ and minimizes the two-dimensional Dirichlet functional (see its definition above); in particular, it also minimizes the two-dimensional area functional $\mathrm{vol}_2$. The result was obtained by J. Douglas and T. Radó. The film $X_0^2 = f_0(M^2)$, certainly, may possess self-intersections and other singular points, depending on the configuration of $\Gamma$). Though the film does admit continuous parametrization by $M^2$ and is its continuous image in $\mathbb{R}^3$ under $f_0$, considered as a subset in $\mathbb{R}^3$, it may be nonhomeomorphic to $M^2$. The mapping $f_0$ defines a generalized minimal surface.

The literature on the two-dimensional and related problem is enormous; here, we are not able to even touch upon many of the profound and interesting investigations; therefore, we refer the interested reader to fundamental works and surveys, e.g., by J. Douglas [113], T. Radó [347], H. Federer [135], C. Morrey [287], [289], J. Nitsche [305], R. Osserman [310], J. Simons [391], M. Beeson and A. Tromba [39], S. Hildebrandt and H. Kaul [201], and others (see Bibliography). The above result regarding the existence of a two-dimensional minimal solution for a prescribed countour was extended by C. Morrey [289] to the case of those embedded in an arbitrary Riemannian manifold. Namely, let $\Gamma = \bigcup_{i=1}^{k} S_i^l$ be a set of $k$-oriented, rectifiable, closed and disjoint curves in a Riemannian manifold $M^n$, and $U$ a two-dimensional domain of type $k$, its boundary being the union of $k$ circles (where $U$ can be regarded as a disc with $k-1$ holes). Then, under certain simple conditions for $\Gamma$, there exists a mapping $f_0 : U \to M$ such that the boundary $\partial U$ of $U$ can be homeomorphically mapped onto the set of $\Gamma$, with $f_0$ being harmonic, conformal, and minimizing the Dirichlet and two-dimensional area functionals. In particular, the area $\mathrm{vol}_2(X_0^2)$ of the surface $X_0^2 = f_0(U)$ is the least among all $\mathrm{vol}_2(f(U))$, where $f$ ranges over all possible piecewise smooth mappings of $U$ of type $k$ into $M^n$ with fixed boundary $\partial U \to \Gamma$.

The catenoid, Scherk, and Enneper surfaces are classical examples of two-dimensional minimal surfaces. Their relation (both generalized and regular) to harmonic mappings and complex analytic functions (Weierstrass' representation) is well known (see R. Courant's, McShane's, J. Nitsche's, and G. Darboux's works). Critical-point theory is applied to two-dimensional minimal surfaces in M. Morse's, C. Tompkins's [290]–[294], R. Courant's, and M. Schiffman's works [81]–[86]. The problem of the number of different minimal surfaces with the same boundary was studied by P. Lévy, R. Courant and F. Morgan in [284], F. Tomi in [409], L. Simon and R. Hardt in [387]–[388], M. Beeson and A. Tromba in [38]–[39], and R. Böhme and A. Tromba in [48]–[50]. Many existence theorems for minimal two-dimensional surfaces were proved in the class of *generalized surfaces*. Hence the question: When is a generalized minimal surface regular, i.e., a surface in the usual sense? The works by R. Osserman [314], H. Alt [22]–[23], R. Gulliver [185], R. Gulliver and F. Lesley [186], R. Gulliver, R. Osserman and H. Royden [187], and Meeks and S.-P. Yau [270] are devoted to its solution. The papers of M. Grüter, S. Hildebrandt and J. Nitsche [184] study, in particular, the properties of minimal surfaces with variable boundary conditions; and Gauss maps of minimal surfaces are treated by R. Osserman, Xavier and E. Bombieri.

## 2.4. The Second Fundamental Form on the Riemannian Submanifold[1]

Let $f : M^k \to W^n$ be a smooth embedding of a smooth manifold $M^k$ into a smooth Riemannian manifold $W^n$. We assume that $W^n$ is orientable, connected, and without boundary, and denote the tangent bundle to a manifold $M$ by $TM$, the tangent plane to $M$ at a point $m \in M$ by $T_m(M)$, the scalar product of two vectors $x, y \in T_m(M)$, induced by the Riemannian metric $g_{ij}$, by $\langle x, y\rangle$, and the Riemannian connection on $TM$, i.e., the symmetric connection corresponding to the induced Riemannian metric by $\nabla$ (e.g., see [115]). Recall that the translation associated with this connection preserves scalar products. As usual, for an arbitrary tensor field $P$, we will denote the covariant derivative of this tensor field $P$ along a vector field $X$ given on $M$, by $\nabla_X(P)$. If $x$ denotes the value of $X$ at a point $m$, i.e., the vector $x = X(m) \in T_m(M)$, then $\nabla_x P$ is the covariant derivative of $P$ along $x$ at $m$. Since $\nabla$ is the tensor operation, $\nabla_x P$ is a tensor and depends on $m$, having the same type as the original tensor field $P$. Because the connection $\nabla$ is Riemannian and preserves scalar products, for any two vector fields $P, Z$ on $M$, the identity $\nabla_x\langle P, Z\rangle = \langle \nabla_x P, Z\rangle + \langle P, \nabla_x Z\rangle$ holds.

We now turn to a submanifold $f(M^k) \subset W^n$, again denoting it for simpler notation by $M^k$. Then the normal bundle $NM$ is determined uniquely as well as the tangent bundle $TM$ (i.e., the plane $N_m^{n-k}(M)$ orthogonal to the plane $T_m^k(M)$ is determined at each point $m \in M$). The Riemannian metric on $W^n$ induces the scalar product not only on $TM$, but also on $NM$; besides, two natural Riemannian connections induced on $TM$ and $NM$ by the embedding $M \to W$ arise. To describe them, suppose that $Y$ is a smooth vector field on $M$, $x \in T_m(M)$ is an arbitrary vector, and let, by definition, $\nabla_x(Y) = (\overline{\nabla}_x Y)^T$, where $\overline{\nabla}$ denotes the Riemannian symmetric connection defined on the ambient manifold $W$, i.e., on $TW$, whereas $(\ )^T$ denotes the orthogonal projection onto the plane $T_m(M)$.

**Lemma 2.4.1.** *The above operation $\nabla$ is a torsion-free connection on $TM$ and is uniquely determined by the Riemannian metric on $M$ which is induced*

---

[1] See the remarkable survey of the second-fundamental-form properties in J. Simons [391].

*by the embedding $M \to W$.*

The Riemannian connection on the normal bundle $NM$ is defined similarly. Consider an arbitrary $C^\infty$–cross section $V$ of the fibration $NM$, i.e., specify the normal vector $V(m) \in N_m(M)$. Due to the smoothness of $V$, we thereby obtain a smooth vector field defined on $M$. Let $x \in T_m(M)$. By definition, we put $\nabla_x V = (\overline{\nabla}_x V)^N$, where $(\ )^N$ is the orthogonal projection onto the plane $N_m(M)$ orthogonal to $T_m(M)$.

**Lemma 2.4.2.** *The above operation $\nabla$ is the torsion-free Riemannian connection on $NM$, and uniquely determined by the Riemannian metric induced on $M$.*

We now consider the construction of the second fundamental form.

Let $x \in T_m(M)$.

**Definition 2.4.1.** Let $v \in N_m(M)$. We include the vector $v$ in an arbitrary smooth vector field $V$ on a manifold $W$ so that $V$ may be orthogonal to the submanifold $M$ in a certain neighbourhood of the point $m \in M$. We also define a linear mapping $A^v : T_m(M) \to T_m(M)$ according to the formula $A^v(x) = -(\overline{\nabla}_x V)^T$. It turns out to be symmetric (see the proof below), and therefore, determines a certain bilinear form $A^v$ which we call the second fundamental form on the submanifold $M \subset W$.

We have actually defined a whole family $A$ of forms $A^v$, in which the vector $v \in N_m(M)$ plays the role of parameter, viz., $A = \{A^v, v \in N_m(M)\}$.

**Lemma 2.4.3.** *The form $A^v$ is defined correctly, i.e., is independent of how we include the vector $v$ in the vector field $V$ on $W$, and smoothly depends on all its arguments.*

Thus, $\langle A^v(x), y\rangle = \langle A^v(y), x\rangle$.

Q.E.D.

Equivalently, $A$ may be interpreted as a bilinear symmetric form on the tangent space $T_m(M)$ with values in the normal space $N_m(M)$. Indeed, if $x$, $y \in T_m(M)$, then we can define a form $B(x,y) \in N_m(M)$ by $\langle B(x,y), v\rangle = \langle A^v(x), y\rangle$.

**Lemma 2.4.4.** *Let $x$, $y \in T_m(M)$. We include the vector $y$ in a smooth vector field $Y$ on $W$, which is tangent to a submanifold $M$. Then the equality $B(x,y) = (\overline{\nabla}_x Y)^N$ holds, i.e., we have to covariantly differentiate $Y$ in the direction of the vector $x$, and project the obtained result, the vector at the point $m$, onto the normal plane.*

**Proof.** Since

$$\langle B(x,y), v\rangle = \langle A^v(x), y\rangle = -\langle(\overline{\nabla}_x V)^T, y\rangle = -\langle\overline{\nabla}_x V, y\rangle$$
$$= -\langle\overline{\nabla}_x V, Y\rangle = -\overline{\nabla}_x\langle V, Y\rangle + \langle V, \overline{\nabla}_x Y\rangle$$
$$= \langle V, \overline{\nabla}_x Y\rangle = \langle(\overline{\nabla}_x Y)^N, v\rangle,$$

for $\langle V,Y\rangle \equiv 0$. Hence, $\langle B(x,y) - (\overline{\nabla}_x Y)^N, v\rangle = 0$ for any $v \in N_m(M)$, and since $B(x,y) \in N_m(M)$, $B(x,y) = (\overline{\nabla}_x Y)^N$.

Q.E.D.

By means of the form $A$, we can now introduce the concept of mean curvature $H$ of submanifold $M$, which generalizes that of the scalar mean curvature for a hypersurface.

**Definition 2.4.2.** Consider the second fundamental form as a form $B$ on a tangent space $T_m(M)$, $m \in M$, with values in a normal space $N_m(M)$. Since the scalar product (see above) is defined on $T_m(M)$, we can consider the trace of $B$, which is a certain vector belonging to $N_m(M)$ at each point $m$. Thus, the trace of $B$ is represented as a certain section $H$ of the normal bundle $NM$. It is called the mean curvature of the embedded submanifold $M \subset W$. If $e_1, \dots, e_k$ is a certain orthobasis for the plane $T_m(M)$, then $H = \sum_{i=1}^k B(e_i, e_i) \in N_m(M)$.

**Remark.** If $M$ is a hypersurface in $N$, Definition 2.4.2 coincides with that of the scalar mean curvature $H = Sp(\mathfrak{G}^{-1}Q)$, where $\mathfrak{G}$ is the matrix of the first fundamental form, and $Q$ of the second (verify!).

## 2.5. Local Minimality

Let $M$ be a submanifold in $W$. We now define the concept of isotopic variation.

**Definition 2.5.1.** Let a smooth homotopy $f_t : M \to W$, $0 \leqslant t \leqslant 1$ be given, so that each mapping $f_t$ is an embedding, $f_0 = f$, and $f$ is the original embedding of $M$ into $W$. Then $f_t$ is called an isotopic variation, $F : M \times [0,1] \to W$, $F = \{f_t\}$. The variation $f_t$ of the embedding $f$ induces a smooth vector field $E$ which is defined on $F(M \times [0,1])$, and is the image of the standard vector field $\frac{\partial}{\partial t}$ on the cylinder $M \times [0,1]$. We shall be interested in the restriction of this field to the submanifold $M$, i.e., $E(m) = dF\left(\frac{\partial(m,0)}{\partial t}\right)$, with $dF$ being the differential of the mapping $F$. $E(m)$ determines two cross sections, viz., $E^T(m)$ of the bundle $TM$ and $E^N(m)$ of $NM$, $m \in M$, for which it suffices to project $E(m)$ orthogonally onto $T_m(M)$ and $N_m(M)$, respectively. It is clear that they are smooth. Regard $E^T$ as a vector field on $M$. Since the natural $k$-dimensional volume form induced by the underlying Riemannian metric is defined on $M$, the field $E^T$ uniquely determines an exterior differential form $\theta(E^T)$ of degree $k-1$ (by means of the operator $*$; see [115]). Consider the $k$-dimensional volume $\mathrm{vol}_k(f_t M)$ of the submanifold $f_t M$, and assume for simplicity that $M$ is compact. We have $v_k(t) = \mathrm{vol}_k(f_t M) < \infty$. Thus, each isotopic variation $F(M)$ is associated with a smooth function $v_k(t)$ defined on $0 \leqslant t \leqslant 1$.

**Proposition 2.5.1 (First variation formula).** *Let $M$ be a compact submanifold in $W$, and $v_k(t) = \mathrm{vol}_k(f_t M)$ be the $k$-dimensional volume function on the submanifold $f_t M$. Then*

$$v_k'(0) = -\int_M \langle E^N, H\rangle + \int_{\partial M} \theta(E^T)$$

*holds, where $\partial M$ is the boundary of the manifold $M$, the first integral of the function $\langle E^N, H\rangle$ is with respect to $M$ and the $k$-dimensional Riemannian volume form of the form $\theta(E^T)$, the second integral is with respect to the $(k-1)$-dimensional submanifold $\partial M$ and relative to the $(k-1)$-dimensional volume induced by the underlying metric.*

We leave the proof to the reader, its simplest method, probably, consisting of the introduction of local coordinates and subsequent straightforward computation (e.g., see Eisenhart's textbook [127]).

We now give the important definition of a locally minimal submanifold.

**Definition 2.5.2.** A sumbmanifold $M^k \subset W^n$ is said to be locally minimal if its mean curvature $H$ vanishes at all points $m \in M$.

If $k = 1$, then one-dimensional locally minimal submanifolds are geodesic in $W$ (e.g., see [115]). However, if $k = n - 1$, then the locally minimal hypersurface $M^{n-1} \subset W^n$ is a minimal submanifold in the sense of §2.2. And if $k < n - 1$, then the local minimality condition means that the mean curvature vector $H$ vanishes. Just as in §2.2., there exists an immediate relation between the vanishing of the mean curvature vector and that of the derivative $v_k'(0)$ of the volume function.

**Theorem 2.5.1.** *A compact submanifold $M^k \subset W^n$ is locally minimal (i.e., $H \equiv 0$) if and only if $v_k'(0) \equiv 0$ for any isotopic variation of $M$ that vanishes on the boundary $\partial M$.*

**Proof.** In view of Proposition 2.5.1., we have

$$v_k'(0) = -\int_M \langle E^N, H\rangle + \int_{\partial M} \theta(E^T).$$

Since $f_t(\partial M) \equiv f(\partial M)$ for all $t$, $0 \leqslant t \leqslant 1$, we have $\Theta(E^T) \equiv 0$ on $\partial M$, i.e., $v_k'(0) = -\int_M \langle E^N, H\rangle$. If $H = 0$, then $v_k'(0) = 0$. Conversely, if $H \neq 0$ at a certain point $m_0 \in M$, then we can always choose a variation whose support is concentrated near $m_0$ such that $v_k'(0)$ is other than zero, which is contrary to $v_k'(0) = 0$, and the theorem is thus proved.

Thus, the submanifolds on which the mean curvature vector is identically zero are the extremals of the volume functional. We have already proved this statement earlier in §2.2 for hypersurfaces. In other words, we have shown that the Euler–Lagrange equation system for the $k$-dimensional volume functional defined on $k$-dimensional submanifolds in $W^n$ is equivalent to $H \equiv 0$. The term "local minimality" means that the volume of a submanifold "remains unaltered to a first approximation," i.e., the first derivative of the volume vanishes for variations that are infinitesimal in amplitude and support; at the same time, if a variation is finite, then volume may decrease as in the case of the standard equator of a two-dimensional sphere, which is contractable to a point on the sphere. In the following, we introduce the concept of the globally minimal submanifold whose volume does not decrease any more under any variation "as large as we please," which, certainly, will require the definition of the notion "large variation."

As an example of locally minimal submanifolds, we specify the important class of the so-called totally geodesic submanifolds.

**Definition 2.5.3.** A submanifold $M \subset W$ is said to be totally geodesic if each geodesic of $M$ with respect to the Riemannian structure and connection induced by the underlying Riemannian metric is a geodesic in the manifold $W$.

**Theorem 2.5.2.** *A submanifold $M \subset W$ is totally geodesic if and only if its second fundamental form is identically zero.*

**Proof.** It is obvious that the vanishing of the second fundamental form is equivalent (see the above definition) to the coincidence of parallel displacements with respect to the Riemannian structures of $M$ and $W$. Since the geodesics may be defined as trajectories whose velocity field is preserved under parallel displacement, it follows that each geodesic on $M$ is also a geodesic from the standpoint of the ambient manifold $W$. Conversely, if $M$ is a totally geodesic submanifold, then we obtain from $B(x, y) = (\overline{\nabla}_x Y)^N$ (see Lemma 2.4.4) that $B \equiv 0$. The theorem is thus proved.

In the following, we will study in detail the class of totally geodesic submanifolds realizing topologically nontrivial homology or homotopy classes in symmetric spaces. The reader can acquaint himself with other properties of minimal surfaces (such as conjugate boundaries, Jacobi fields, index) in J. Simons's paper [391].

### 2.6. First Examples of Globally Minimal Surfaces

In the simplest case, we consider the free homotopy classes of closed loops on a Riemannian manifold $W$. Take the class of piecewise smooth mappings $f : S^1 \to W$ of the circle $S^1$ into $W$, which are homotopic to each other. It is generally known that there exists a mapping $f_0$, the length of whose trajectory $f_0(S^1)$ is the least among the lengths of the curves $f(S^1)$, where $f$ is homotopic to $f_0$. In other words, if $f_t : S^1 \to W$ is an arbitrary smooth homotopy of the mapping $f_0$, i.e., an arbitrary variation, then the length of the trajectory $f_t(S^1)$ is not less than that of $f_0(S^1)$, which means that the length functional attains absolute minimum in the prescribed homotopy class on $f_0$.

As a second and more nontrivial example, consider complex submanifolds in a Kähler manifold, and take a considerably larger class than the above homotopy class as variations, or deformations, of a submanifold $M$.

**Definition 2.6.1.** Let $M^k \subset W^n$ be a smooth, compact, orientable, and closed submanifold. We will say that its bordism deformation is given if there is a $(k+1)$-dimensional smooth, compact, and orientable submanifold $Z^{k+1}$ with boundary $\partial Z = M \cup (-P)$, where $P$ is a certain smooth, compact, orientable, and closed submanifold (film) in $W$, and $-P$ denotes $P$ with opposite orientation induced by that of $Z$ (see Fig. 9). We call $P$ a bordism variation of $M$. In the case of a noncompact submanifold $M \subset W$, we will say that its bordism deformation is given if there is a submanifold (film) $P$ coinciding with $M$ outside a certain compact domain, and also a $(k+1)$-dimensional submanifold $Z$ with piecewise smooth boundary $\partial Z \subset M \cup (-P)$.

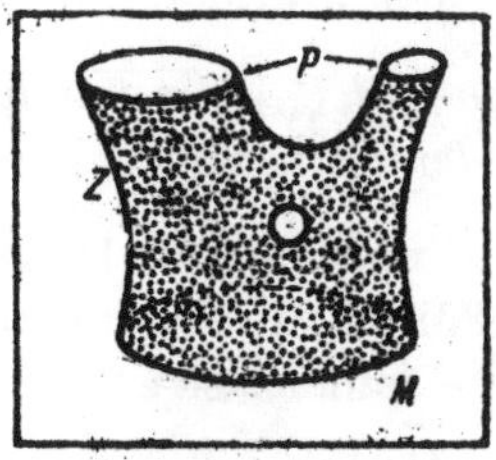


**Fig. 9**

Recall that a complex manifold $W$ is said to be *Kähler* if its Riemannian metric $g_{ij} dz^i \, d\bar{z}^j$ (with $z^1, \ldots, z^n$ being local coordinates) defines a closed exterior differential form $\omega = g_{ij} dz^i \wedge d\bar{z}^j$ of degree 2.

**Theorem 2.6.1 (H. Federer).** *Let $W$ be a Kähler manifold of complex dimension $n$, and $M \subset W$ its complex $k$-dimensional submanifold. Consider all possible real bordism deformations in $W$, so that the film $Z^{2k+1}$ is a real $(2k+1)$-dimensional submanifold in $W$. Let $P^{2k}$ be a bordism variation of the submanifold $M$. Then the volume $\mathrm{vol}_{2k}(M)$ is not less than $\mathrm{vol}_{2k}(P)$ if $M$ and $P^{2k}$ are compact. Otherwise, we mean the volumes of those domains on $M$ and $P$ where $M$ and $P$ are noncoincident. Moreover, if the volume of $P$ coincides with that of $M$, then $P$ is also complex in $W$.*

Thus, in Kähler manifolds, compact, complex submanifolds are globally minimal, i.e., are extremals for the volume minimum multidimensional variational problem. For example, in $\mathbb{C}P^n$ (complex projective space), globally minimal submanifolds are those complex in $\mathbb{C}P^k$. Complex Hermitian space $\mathbb{C}^n$ is also a Kähler manifold; therefore, any complex submanifold $X$ in it is minimal with respect to any perturbations fixed outside a certain bounded domain in $X$. Recall that all complex submanifolds in $\mathbf{C}^n$ are noncompact. We now turn to the proof of the theorem.

**Lemma 2.6.1.** *For an arbitrary exterior 2-form $\omega$ on $\mathbf{R}^{2n}$, we can select such an orthonormal basis $e_1, \ldots, e_{2n}$, with respect to which $\omega$ is of the form $\lambda_1\omega_1 \wedge \omega_2 + \cdots + \lambda_n\omega_{2n-1} \wedge \omega_{2n}$, where $\lambda_1, \ldots, \lambda_n$ are nonnegative numbers and $\omega_1, \ldots, \omega_{2n}$ is the basis dual of $e_1, \ldots, e_{2n}$ for $\mathbf{R}^{2n}$.*

**Proof.** Consider an arbitrary orthonormal basis $e'_1, \ldots, e'_{2n}$ for $\mathbf{R}^{2n}$. For an arbitrary given 2-form $\omega$, we construct the matrix $A = (a_{ij})$, where $a_{ij} = \omega(e'_1, e'_j)$. We call it associated with $\omega$. Since $\omega$ is completely determined by its values on the basis vectors, it is also completely determined by the matrix $A$ (with respect to the corresponding orthonormal basis $e'_1, \ldots, e'_{2n}$). It is clear that $A$ is skew-symmetric. Therefore, there exists an orthonormal basis $e_1, \ldots, e_{2n}$ for it such that $A$ is of the form

$$A = \begin{pmatrix} \begin{array}{cc|} 0 & \lambda_1 \\ -\lambda_1 & 0 \end{array} & & 0 \\ \hline & \ddots & \\ \hline 0 & & \begin{array}{|cc} 0 & \lambda_n \\ -\lambda_n & 0 \end{array} \end{pmatrix},$$

where $\lambda_1, \ldots, \lambda_n$ are nonnegative numbers. Let $\omega_1, \ldots, \omega_{2n}$ be the basis dual of $e_1, \ldots, e_{2n}$. It is obvious that $\omega = \sum_{\alpha=1,3,\ldots,2n-1} \lambda_\alpha\omega_\alpha \wedge \omega_{\alpha+1}$. The lemma is thus proved.

**Lemma 2.6.2 (Wirtinger Inequality).** *Let the Hermitian metric $g_{ij}$ be given on the space $\mathbf{R}^{2n} = C^n$, and $\omega$ the exterior 2-form corresponding to the metric $g_{ij}$, i.e., determined by $\omega(v_1, v_2) = \langle iv_1, v_2\rangle$. Put $\sigma_k = \frac{1}{k!}\omega^k = \frac{1}{k!}\underbrace{\omega \wedge \omega \wedge \cdots \wedge \omega}_{k}$. Then the inequality $|\sigma_k(v_1, \ldots, v_{2k})| \leqslant 1$ holds, where $v_1, \ldots, v_{2k}$ is an arbitrary orthonormal vector system in $\mathbf{R}^{2n}$. Besides, the equality $|a_k(v_1, \ldots, v_{2k})| = 1$ is valid if and only if $v_1, \ldots v_{2k}$ span a complex subspace in $\mathbf{R}^{2n}$, i.e., if the real linear closure of the vectors $v_1, \ldots v_{2k}$ is invariant under multiplication by $i$.*

**Proof.** Consider the case $k = 1$. Let $v_1, v_2 \in \mathbf{R}^{2n}$ be orthogonal vectors of length 1. It is clear that $|\omega(v_1, v_2)| = |\langle iv_1, v_2\rangle| \leqslant |iv_1| \cdot |v_2| = 1$. The equality holds if and only if $\pm v_2 = iv_1$, i.e., if the vectors $v_1, v_2$ span a complex one-dimensional subspace of real dimension 2. Let us now consider the general case. Let $V \subset \mathbf{R}^{2n}$ be a subspace spanned by $v_1, \ldots v_{2k}$. We denote the restriction $\omega|_V$ by $\tilde{\omega}$. According to Lemma 2.6.1, we can choose an orthonormal basis $e_1, \ldots e_{2k}$ in the subspace $V$ and $\omega_1, \ldots, \omega_{2k}$ dual of it so that $\tilde{\omega} = \lambda_1\omega_1 \wedge \omega_2 + \cdots + \lambda_k\omega_{2k-1} \wedge \omega_{2k}$, where $\lambda_1, \ldots, \lambda_k$ are nonnegative real numbers. It is clear that $\tilde{\omega}(e_{2p-1}, e_{2p}) = \lambda_p(1 \leqslant p \leqslant k)$. Hence, according to the case $k = 1$, we obtain that $\lambda_p \leqslant 1$, with $\lambda_p = 1$ if and only if $ie_{2p-1} = \pm e_{2p}$. Denote the restriction of the form $\sigma_k = \frac{1}{k!}\omega^k$ to $V$ by $\tilde{\sigma}_k$. Then

$$|\tilde{\sigma}_k(e_1, \ldots, e_{2k})| = \left|\frac{1}{k!}\tilde{\omega}^k(e_1, \ldots, e_{2k})\right| = \lambda_1 \ldots \lambda_k \leqslant 1,$$

with the equality holding if and only if $\lambda_p = 1 (1 \leqslant p \leqslant k)$, if $ie_{2p-1} = \pm e_{2p} (1 \leqslant p \leqslant k)$. The latter, i.e., means precisely that $V$ is a complex subspace of $\mathbf{R}^{2n}$, and the lemma is thus proved.

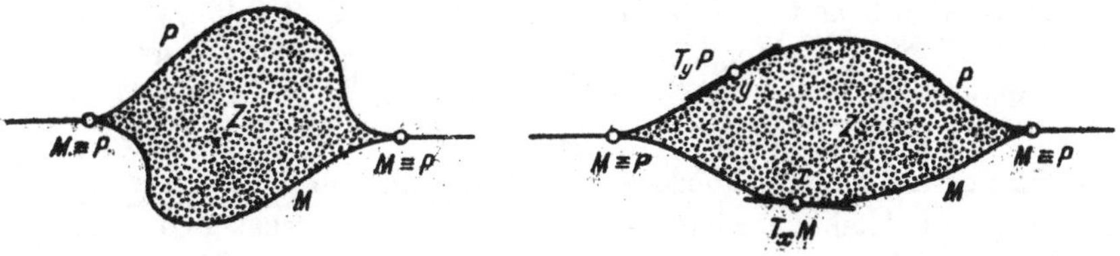


**Fig. 10**

**Proof of Theorem 2.6.1.** Let $\varphi$ be an exterior form of degree $l$ on $\mathbf{R}^{2n}$, $V$ a linear $l$-dimensional subspace of $\mathbf{R}^{2n}$, $v_1, \ldots, v_l$ and $v_1', \ldots, v_l'$ arbitrary orthonormal bases for $V$ and of one orientation class. It immediately follows from the transformation law for an $l$-dimensional exterior form on an $l$-dimensional space, viz., multiplication by the determinant of a linear transformation, that $\varphi(v_1, \ldots, v_l) = \varphi(v_1', \ldots, v_l')$. Therefore, $\varphi$ can be defined correctly as a function on the set of classes of oriented orthonormal bases for the same subspace; meanwhile, varying the subspace will lead to varying $\varphi$. In other words, $\varphi$ on the Euclidean space $\mathbf{R}^{2n}$ determines a function denoted by the same letter $\varphi$ on the real Grassmannian manifold $\widehat{G}_{2n,l}$ of oriented $l$-dimensional subspaces in $\mathbf{R}^{2n}$. We denote by $\widehat{V}$ the class of oriented orthonormal bases for the subspace $V$, i.e., a point from $\widehat{G}_{2n,l}$. Now, let $M$ be a complex submanifold in $W$, and $P$ a bordism variation. As before, we denote by $T_x(M)$ (resp. $T_y(P)$) the space tangent to $M$ (resp. $P$) at a point $x$ (resp. $y$). Let $Z$ be a submanifold in $W$, so that $\partial Z = M \cup (-P)$ (Fig. 10), $\omega = \frac{i}{2} g_{ij}\, dz^i \wedge d\bar{z}^j$ be the above closed 2-form on $W$, and $\sigma_k = \frac{1}{k!}\omega^k$. It is then obvious that also $d\sigma_k = 0$. By the Stokes theorem, we have $0 = \int_Z d\sigma_k = \int_{\partial Z} \sigma^k = \int_{M\cup(-P)} \sigma_k = \int_M \sigma_k - \int_P \sigma_k$, i.e., $\int_M \sigma_k = \int_P \sigma_k$. Denote the $2k$-dimensional exterior volume forms on $M$ and $P$ by $dx$ and $dy$, respectively. Then $\int_M \sigma_k = \int_M \sigma_k(\widehat{T}_x M)dx$, $\int_P \sigma_k = \int_P \sigma_k(\widehat{T}_x P)dy$. Since $M$ is a complex submanifold in $W$, i.e., $T_x(M)$ is a complex subspace in $T_x(W) = \mathbf{C}^n = \mathbf{R}^{2n}$, we have according to Lemma 2.6.2 that $\sigma_k(\widehat{T}_x M) \equiv 1$, $\sigma_k(\widehat{T}_y P) \leqslant 1$ (recalling that $P$ is not necessarily a complex submanifold). Hence,

$$\mathrm{vol}\,(M) = \int_M dx = \int_M \sigma_k(\hat{T}_x M)\,dx = \int_P \sigma_k(\hat{T}_y P)\,dy \leqslant \int_P dy = \mathrm{vol}\,(P).$$

Thus, the first part of the statement of the theorem is proved. Further, it is obvious that the equality $\mathrm{vol}(M) = \mathrm{vol}(P)$ holds if and only if $\sigma_k(\widehat{T_y}(P)) \equiv 1$. According to Lemma 2.6.2, the latter identity is equivalent to $T_y(P)$ being a complex subspace for (almost) all points $y \in P$, which means that $P$ is a complex manifold in $W$. The proof is completed.

It can be seen from the above that the assumption regarding the submanifolds $M \subset W$ being nonsingular is inessential. The proof in terms of subsets of complete measure also holds for algebraic complex surfaces $M \subset W$, i.e., given by a system of polynomial equations on $W$, in spite of the fact that they can possess singular points such as conic ones, in which case the condition for $M$ and $P$ being bordant should be replaced by a weaker condition, viz., that $M$ should be homologous to $P$ in the group $H_{2k}(W, \partial M)$, or $M$ and $P$ should determine the same element (see the details below). The submanifold $CP^k (1 \leqslant k \leqslant n)$ realizes in $CP^n$ a generator for the group $H_{2k}(CP^n, Z) = Z$, and is globally minimal in this homology class.

The ideas realized by the Wirtinger lemma and the Federer theorem were then developed by the author (for the criterion of globally minimal surfaces in symmetric spaces see [153],[165], and [170]), Ɖao Chông Thi (for the criterion of multivarifold global minimality, see [95], [97]–[98]). The remarkable "calibrated fibration theory" was worked out by R. Harvey and H. Lawson in [195]–[196]. Important results along these lines were then obtained by Le Hong Van [244], [245], [451], and [452].

# CHAPTER 2

# MULTIDIMENSIONAL VARIATIONAL PROBLEMS AND EXTRAORDINARY (CO)HOMOLOGY THEORY

## §3 The Multidimensional Plateau Problem and Its Solution in the Class of Mapping on Spectra of Manifolds with Fixed Boundary

### 3.1. The Classical Formulations (Finding the Absolute Minimum)

After the two-dimensional Plateau problem had been solved (see J. Douglas, T. Radó, C. Morrey, R. Courant, and others), the problem of solving the similar variational problem in higher dimensions arose. Consider a Riemannian manifold $M^n$, and fix a $(k-1)$-dimensional smooth, compact, and closed submanifold, or "contour" $A^{k-1}$. Assume that there exists a compact, smooth submanifold $X^k \subset M^n$ such that $A$ is its boundary. Then the Riemannian volume $\text{vol}_k(X) < \infty$ is defined. Consider the class $\{X\}$ of all such submanifolds $X$, i.e., $\partial X = A, X \subset M$. The question is whether there exists a submanifold $X_0$ whose volume is the least, or $\text{vol}_k X_0 = \inf_{X \in \{X\}} \text{vol}_k X$. It is clear that the answer is negative in this simplest formulation, since it is easy to give examples where the film $X$, while striving to occupy a position with least volume, starts to contract, and the minimum is attained on the "surface with singularities," but not on the submanifold. Therefore, the formulation should be more precise to take the natural effect of gluing into account.

Consider all possible pairs $(W, f)$ of smooth, compact manifolds $W$ of dimension $k$ with boundaries $\partial W$ homeomorphic to $A$, and of continuous (or piecewise smooth ) mappings $f : W \to M$ which are the identity on $\partial W$, i.e., coincident with the fixed embedding homeomorphism $i_0 : A \to i_0(A) \subset M$; in the following, we will omit the notation $i_0$.

**Problem A.** Can we find a pair $(W_0, f_0)$ among all $(W, f)$, where $W$ are all possible manifolds with boundary $A$, and $f : W \to M$ are mappings of $W$ into $M$, which are the identity on $A$, so that the mapping $f_0$, or the film

$X_0 = f_0(W_0)$, the image of the manifold $W_0$ in $M$, possesses the reasonable minimal properties; in particular, so that $\mathrm{vol}_k(X_0) \leqslant \mathrm{vol}_k(X)$, where $X = f(W)$ is any film from the above class (Fig. 11)?

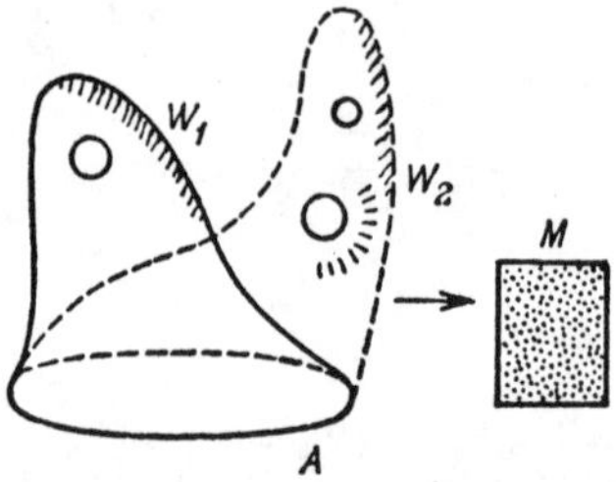


**Fig. 11**

Problem A is thereby that of finding the absolute minimum with respect to all films $X \subset M$ of all possible topological types, which span the given contour $A \subset M$.

By the "reasonable minimal properties" of the image $X = f_0(W_0)$ in $M$ in addition to the inequality $\mathrm{vol}(X_0) \leqslant \mathrm{vol}(X)$, we can, e.g., understand the property of the existence of a subset $Z$ of singular points, nowhere dense in the film $X_0$, so that each point $P \in X_0 \setminus Z$ possesses a neighbourhood $U$ in $M$, for which $(X_0 \setminus Z) \cap U$ consists of smooth submanifolds of $V_\alpha$ of dimensions not exceeding $k$ (of maximally possible dimension of $X_0$), each $V_\alpha$ being a minimal submanifold in the sense of classical differential geometry, i.e., the mean curvature vector $H$ is identically zero.

The second problem is that of finding the absolute minimum in the class of all bordism-variations (see §2.6) of a given closed submanifold $V^k \subset M^n$, which was made more precise.

**Problem B.** Let $(V, g)$ be a pair of $V = V^k$, a compact closed $k$-dimensional manifold, $g : V \to M$ its continuous (or piecewise smooth) mapping into $M^n$, and $X = g(V)$ the image of $V$ in $M$. We will say that a pair $(V', g')$ is a bordism-variation of $(V, g)$ if there exists a compact manifold $Z$ with boundary $\partial Z = V \cup (\pm V')$ and a continuous mapping $F : Z \to M$ such that $F|_V \equiv g$, $F|_{V'} \equiv g'$ (Fig. 12). Can we find such a pair $(V_0, g_0)$ among all $(V, g)$ of the above form, so that the image $X_0 = g_0(V_0)$ possesses reasonable minimal properties; in particular, so that $\mathrm{vol}_k(X_0) \leqslant \mathrm{vol}_k(X)$, where $X = g(V)$ is any film from the indicated class?

Like Problem A, this consists of finding the absolute minimum in the class of all spectral bordism-variations of a given pair $(V, g)$. Formulated thus, the spectral version of the Plateau problem was solved by the author in [155] and [160] (see its precise statement and proof below).

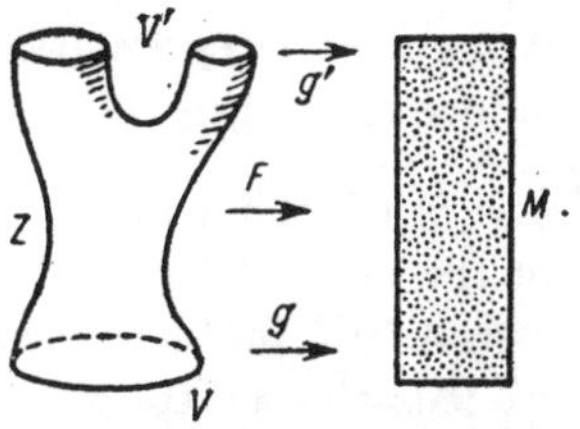


**Fig. 12**

In Problems A and B, the topology type of the varying film may alter, though not arbitrarily. The bordism-variation class naturally arises not only in the problem of finding the absolute minimum, e.g., with respect to all films spanning the given contour, but also in studying level surfaces of smooth functions on a compact manifold. Assume that a smooth function $\alpha$ given on a manifold $M$ is Morse, i.e., all its critical points are nondegenerate, and consider two arbitrary level surfaces $V = \{\alpha = c\}$, $V' = \{\alpha = c'\}$ for two noncritical values $c$ and $c'$ of $\alpha$ (in other words, no critical points of $\alpha$ on $V$ and $V'$). It is evident that the manifold $V'$ is then obtained from $V$ by a bordism-deformation, whereas the corresponding film $Z$ is given as $Z = \{c \leqslant \alpha \leqslant c'\}$ with $\partial Z = V \cup (-V')$.

### 3.2. The Classical Formulations (Finding a Relative Minimum)

Consider the variational problem with fixed boundary $A \subset M$ (where $A$ is a given $(k-1)$-dimensional submanifold without boundary), and also all possible films bounded by $A$ and obtained from each other by continuous deformations. More precisely, we formulate the problem of finding a minimum in the homotopy sense.

**Problem A'.** Can we find a pair $(W, f_0)$ among all $(W, f)$, where $W$ is a certain fixed manifold with boundary $A$ and $f : W \to M$ are all possible continuous (or piecewise smooth) mappings homotopic to a certain fixed mapping $f'$, which are the identity on $A$, i.e., coincident with a fixed homeomorphism of $A$ onto itself, so that the mapping $f_0$ or the film $X_0 = f_0(W)$, the image of $W$ in $M$, possess the minimality properties; in particular, so that $\mathrm{vol}_k(X_0) \leqslant \mathrm{vol}_k(X)$, where $X = f(W)$ is any film from the given homotopy class?

This is the problem of finding a minimum in each homotopy class. It is just in this sense that we speak of a relative minimum in contrast with the absolute minimum in Problem A, which should be found among all the homotopy classes.

**Problem B'.** Can we find a mapping $g_0$ among all $g : V^k \to M^n$, where $V$ is a certain fixed manifold, which are homotopic to a certain original mapping

$f : V \to M$, so that it possesses the minimal property; in particular, so that $\mathrm{vol}_k\, g_0(V) \leqslant \mathrm{vol}_k\, g(V)$, where $g$ is any representative of the above mapping class?

In Problems A′ and B′, we are looking for a minimum in the homotopy class of a given mapping (in the case of A′, with fixed boundary), whereas, in Problems A and B, the variational problem is considered in a wider class of admissible variations, since it is allowed to vary the topological type of the film $W$ (in Problem A) and manifold $V$ (in Problem B).

The four problems do not follow from each other. For example, the existence of the absolute minimum, i.e., the minimum with respect to all homotopy classes in the problem of spanning a fixed contour, does not follow from that of a minimal mapping in each homotopy class; the converse is false, too.

We start with the absolute minimum problem, because it is important for many applications to possess information just about the least film spanning a contour, the topological type being automatically determined by the absolute minimality condition. We fix the terminology, and call problems of type A and A′ those of *spanning a contour*, whereas we call the problems of type B and B′ those of *realization* (e.g., of a nontrivial homotopy class). In the case of realizing a nontrivial class, surfaces without singularities are especially interesting.

We will call the minimal surfaces to be found in these problems *globally minimal* in contrast with the *locally minimal* surfaces characterized by zero mean curvature. The version of Plateau's problem in the homotopy class of multivarifolds was solved by Đào Chông Thi in [95] and [98], and its solution was obtained by further developing the author's methods worked out to minimize the stratified surface-stratified volume (see below).

One possible way to make more precise the idea of minimizing the mapping $f$ in Problems A and A′, or $g$ in Problems B and B′, consists of minimizing the Dirichlet functional considered on these mappings. This method may be successful if the ambient manifold $M$ possesses an additional and sufficiently rigid structure, say, if it is a manifold of negative curvature. Then there exists a mapping $g_0$ minimizing the Dirichlet functional, and Problem B has a solution for the two-dimensional case ($\dim V = 2$) in each homotopy class of the mapping $g : V \to M$ (S. I. Alber [2], [3], [4], J. Eels and L. Lemaire [118]). However, if the ambient manifold is arbitrary, then the earlier methods involving the Dirichlet functional cannot provide for the convergence of minimizing sequences (as shown by simple examples).

### 3.3. Difficulties Arising in the Minimization of the Volume Functional $\mathrm{vol}_k$ for $k > 2$. Appearance on Nonremovable Strata of Small Dimensions

In solving the two-dimensional Plateau problem, we ignore the following effect which becomes essential in greater dimensions. Consider the contour $A = S^1$ represented in Fig. 13, and also the film $X_t = f_t(W)$ striving to occupy in

$\mathbf{R}^3$ a position associated with the least two-dimensional area. It is clear that the film will collapse (coalesce) at one moment, and a segment $P$ joining the upper and lower bases will appear instead of the thin tube $T$. In the two-dimensional case, we can easily get rid of $P$ by continuously mapping $T$ into the two-dimensional disc spanning $A$. The other way is to reject $P$ and to only retain the "two-dimensional part" of the minimal film. In the case $k > 2$, where we take a $(k-1)$-dimensional submanifold $A$ in $M$ as the contour, the appearance of a situation analogous with the above one will make the minimization problem much more complicated. As the $k$-dimensional volume $\mathrm{vol}_k$ of the deforming film $X_t$ tends to minimum, the film $X_t = f_t(W)$ strives to occupy the corresponding "minimal position"; meanwhile, gluing may occur, i.e., the mapping $f_1 : W \to M$, which is homotopic to the original mapping $f$, may decrease the dimension of certain subsets open in $W$ (Fig. 14), thus leading to the appearance of pieces $P$ of dimension $s$, $s \leqslant k-1$, in the image $X_1 = f_1(W)$. In contrast to the two-dimensional case, all these portions $P$, in general, can be neither rejected nor continuously mapped into the "massive" $k$-dimensional part $X^{(k)}$ of the film $X_1$. Recall that we would like to solve the minimum problem in the class of films $X = f(W)$ admitting continuous parametrization by means of manifolds $W$; therefore, for any technique of removing "pieces of small dimension" $P \subset X$, we have to guarantee that the film $\widetilde{X}$ obtained after such restructuring should still admit continuous parametrization, viz., $\widetilde{X} = \tilde{f}(\widetilde{W})$, though, possibly, by means of another manifold $\widetilde{W}$. The rejection of "pieces of small dimensions" $P$ as one of the restructuring variants, i.e., transfer from $X$ to $X^{(k)} = \widetilde{X} = \overline{X \setminus P}$ (the bar meaning closure), or $k$-dimensional part of $X$, does not fulfill the condition, since it is easy to give examples where the "remainder" $\widetilde{X} = X^{(k)}$ does not at all admit, for a fixed contour $A$, any continuous parametrization $\tilde{f} : \widetilde{W} \to X^{(k)}$, where $\tilde{f} : \partial\widetilde{W} \to A$ is a homeomorphism.

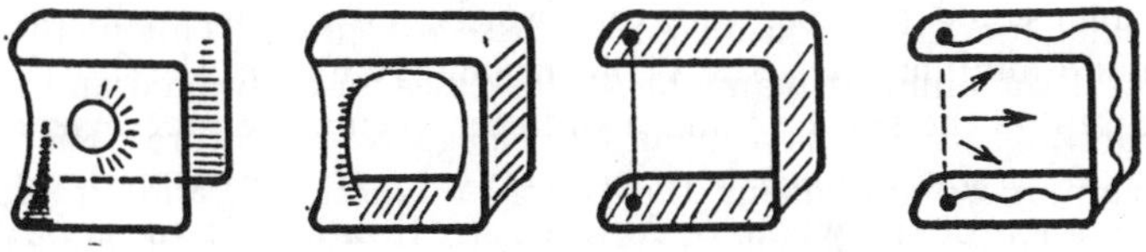

**Fig. 13**

Another restructuring variant may be the construction of a continuous mapping $\lambda : X \to \widetilde{X} = X^{(k)}$ such that $\lambda$ is the identity on the closed subset $X^{(k)} \subset X$, and $\lambda(P) \subset X^{(k)}$, i.e., $\lambda$ carries all the "pieces of small dimensions" $P$ into the "$k$-dimensional part" of the film $X$. In the two-dimensional case, such a mapping $\lambda$ (if $X^{(2)}$ is connected) always exists (e.g., see Fig. 13). However, simple examples demonstrate that it is not so in greater dimensions, it being sufficent to consider the space glued together from the two-dimensional disc $D^2$ and a certain space $X^{(k)}$, $k \geqslant 3$, relative to the mapping $h : S^1 \to X^{(k)}$, of the boundary $S^1$ of $D^2$ into $X^{(k)}$ nonzero element of the fundamental group $\pi_1(X^{(k)})$. It is clear that obstacles to the construction of the required mapping $\lambda : P \to X^{(k)}$ are purely topological, and arise in extending their continuous mapping from the skeleton of dimension $s-1$ to that of dimension $s$.

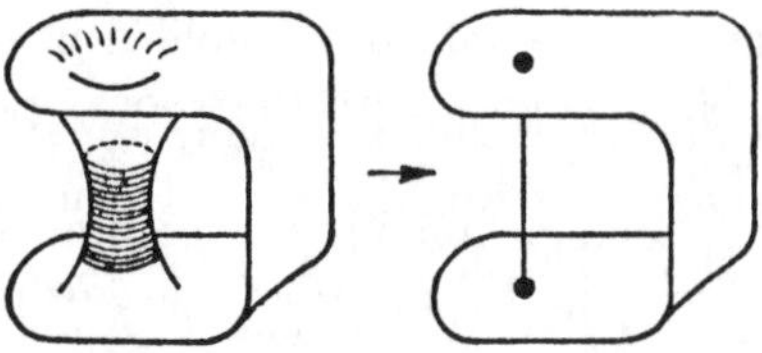

**Fig. 14**

Thus, whether we like it or not, we cannot ignore pieces of small dimensions, arising in the minimization process, and are forced to take them into account if we want to solve the classical Plateau problem, i.e., find a minimum, say, in the homotopy class of a given mapping. It is natural to expect that not only the "$k$-dimensional part" $X^k$ of a minimal film $X_0 = f_0(W)$, e.g., in the homotopy class of the given mapping, should be a $k$-dimensional minimal submanifold in $M$ (possibly, with singularities) but that also all pieces of small dimensions $P$ should be minimal submanifolds of smaller dimensions except, possibly, a certain set of singular points of measure zero. It will be shown below that this is really so. However, due to serious topological difficulties in dealing with pieces of smaller dimensions, we might ignore $P$ for the time being and restrict ourselves only to the $k$-dimensional volume functional $\mathrm{vol}_k$ with the purpose of making the problem simpler, with all the "pieces of small dimension" being immaterial from the standpoint of $\mathrm{vol}_k$, since they have zero $k$-dimensional volume. In other words, we can first solve a narrower problem and minimize only the $k$-dimensional volume of the film $X$, which is concentrated on the set $X^{(k)}$, meanwhile preserving the topological functions of the pieces of small dimensions, because they are necessary for continuous parametrization of the whole of $X$ to exist. However, it turns out that the problem's solution requires more information regarding their behaviour even then (see the details below). Therefore, we again turn to the problem of studying simultaneously both the $k$-dimensional volume functional $\mathrm{vol}_k$ and $\mathrm{vol}_s$, where $s < k$. Hence, the mathematical nature of the multidimensional Plateau problem in dimensions greater than two makes us introduce (a) the *stratified surfaces* $X = X^{(k)} \cup X^{(k-1)} \cup \ldots$, where each subset, or stratum, $X^{(s)}$ is an $s$-dimensional "surface" in $M$, of dimension $s$ at each of its points, and (b) the *stratified volumes* $SV(X) = (\mathrm{vol}_k X^{(k)}, \mathrm{vol}_{k-1} X^{(k-1)}, \ldots)$ represented by a vector with each component equal to the volume of the corresponding stratum in each dimension. Finding a "minimal surface $X$" means proving the existence of a stratified surface whose stratified volume $SV$ is the least in the lexicographic sense. By minimizing the stratified volume vector, we will understand the minimization of its first coordinate, or $k$-dimensional volume, at first, then, fixing it, the minimization of its second coordinate, and afterwards the third coordinate by fixing the minimal values of the first two, and so on.

### 3.4. Formulations of the Plateau Problem in the Language of the Usual Spectral Homology

Due to the above difficulties, the efforts of many a mathematician were made to develop a new language that would remove the influence of the strata of small dimensions (see §3.3). This homology language originated in the 1960s in the famous works of H. Federer, W. Fleming, F. Almgren, and E. Reifenberg (with J. Adams' participation). The Plateau problem was solved in various homology classes in terms of just this language on the basis of the homology group concept. In other words, the existence of absolutely (globally) minimal surfaces was proved in the class of surfaces homologous to each other. The solution of the Plateau problem in the homology, or analogous, terms and formulations was given in the works of H. Federer and W. Fleming [131], [133], [135], [138], F. Almgren [14], [16], E. Reifenberg [351]–[353], C. Morrey [288]–[289], E. de Giorgi [103]–[107], E. Giusti [176], R. Harvey and H. Lawson [192], [194], [196], H. Lawson [238], W. Meeks and S.-T. Yau [269]–[270], and M. Miranda [279]–[281].

Here, we give, e.g., the solution of the homology Plateau problem, obtained by E. Reifenberg for $\mathbb{R}^n$, and then extended by C. Morrey to the case of an arbitrary Riemannian manifold $M^n$.

The problem of spanning a contour, or $(k-1)$-dimensional submanifold $A$ in a Riemannian manifold $M$, was formulated in the language of the usual homology as follows: Denote by $H_{k-1}(A)$ the $(k-1)$-dimensional homology group of a closed manifold $A$, the homology being considered with the integral of periodic coefficients (e.g., see [115]). Let $A \subset X \subset M$, with $X$ being an arbitrary $k$-dimensional "surface" in $M$ such as of subcomplexes or measurable compact sets. Consider the class $\{X\}$ of all $X$ for which the homomorphism $i_* : H_{k-1}(A) \to H_{k-1}(X)$ induced by the embedding $i : A \to X$ annihilates the whole of $H_{k-1}(A)$, i.e., $A$ is the homology boundary of $X$. We also consider the $k$-dimensional volume functional $\mathrm{vol}_k$ on the class $\{X\}$. Let $\lambda_k = \inf_{X\in\{X\}} \mathrm{vol}_k X$. It turns out that then there always exists a minimal surface $X_0 \in \{X\}$ such that $\lambda_k = \mathrm{vol}_k X$. It is the absolute minimum in the class of all surfaces of variable topological type, and with given boundary. Here, we may not enter into the details, referring the reader to the above-cited literature. Note that, within the framework of the described approach, more geometric (see [289], [351]–[353]), and more functional ([12]–[14], [51], [131]–[132], [138], [145], [163]) directions were singled out, and the remarkable existence theorems for the absolute minimum in the homology class, and regularity almost everywhere of minimal surfaces, were proved within the framework of each.

In this approach, it was essential that if $X \supset Y$, where $X, Y$ are compact "surfaces," and $\dim(X \backslash Y) < k$, then $H_k(X) = H_k(Y)$ (and $\mathrm{vol}_k X = \mathrm{vol}_k Y$). In other words, this means that if a manifold $A$ is homologous to zero in the surface $X$, then it is also homologous to zero in $Y$ obtained from $X$ by rejecting pieces $P = X \backslash Y$ of small dimensions; and no problem of nonremovable pieces $P$ (or strata of small dimensions) arises, which characterizes the classical Plateau problem in the language of films parameterized by manifolds. This permitted us to restrict ourselves only to the minimization of one functional $\mathrm{vol}_k$ and to study only the $k$-dimensional piece of $X$.

However, the use of the usual homology to define the concepts "boundary" and "spanning" (see above) moved us farther from the classical Plateau problem in the spirit of J. Douglas and T. Radó, since if $A$ is a submanifold in $M$, and $X_0$ is a minimal surface in the generalized sense and spans $A$ homologically, i.e., $A$ is homologous to zero in $X_0$, then, in general, there exists no such manifold $W$ with boundary $A$ that $X_0 = f(W)$, and $X_0$ may not admit continuous parametrization by a manifold in the sense of §3.1. Therefore, having solved the minimum problem in the homology class, we still cannot say anything regarding the existence of a minimal surface spanning $A$ in the classical sense (see §3.1). The substantial difference between the formulation in the spirit of J. Douglas and T. Radó, and that in the usual homology language, can also be seen from any submanifold $A$ in $\mathbb{R}^n$ being homologous to zero, since $H_k(\mathbb{R}^n) = 0$; meanwhile, there may not exist a single film $X \subset A$ such that $X = f(W)$, i.e., no spanning film (in the homology sense) is parametrized by a manifold.

## 3.5. The Classical Multidimensional Plateau Problem (the Absolute Minimum) and the Language of Bordism Theory

We would like to return to the Plateau problem in the class of films parametrized by manifolds with fixed prescribed boundary $A^{k-1}$. In contrast to all the minimization methods employed earlier, we will minimize not only the $k$-dimensional volume functional, but also the stratified volume $SV$, i.e., the vector made up of all the volumes of $X$ in all its dimensions. To realize this programme, we need a language in which the contour-spanning as well as the realization problem could be formulated precisely. The language should be different from that making use of the usual homology theory (see above). It turns out that the material necessary to introduce such a new language into variational problems is already available. Recall certain topological definitions.

Let $V^{k-1}$ be a closed, oriented manifold. We denote by $-V^{k-1}$ an oriented manifold different from $V^{k-1}$ in its orientation. We call the manifold that is the disjoint union of two closed, compact, and oriented manifolds $V_1$ and $V_2$, preserving their orientation, *the disjoint union* $V_1^{k-1} \cup V_2^{k-1}$.

**Definition 3.5.1.** A closed, oriented manifold $V^{k-1}$ is said to be *null-bordant* and denoted by $V^{k-1} \sim 0$, if there exists a compact, oriented manifold $W^k$

whose boundary $\partial W$ is diffeomorphic to $V^{k-1}$ and preserves the orientation. Two closed, oriented manifolds $V_1^{k-1}$, $V_2^{k-1}$ are said to be *bordant* and denoted by $V_1 \sim V_2$, if their disjoint union $V_1 \cup (-V_2)$ is null-bordant.

In this definition, we consider manifolds regardless of their embedding into any Euclidian space.

**Lemma 3.5.1.** *The bordism relation is that of equivalence on the class of closed, oriented $(k-1)$-dimensional manifolds. The set $\Omega_{k-1}$ of equivalence classes is an Abelian group in which addition is induced by the disjoint union of manifolds.*

**Proof.** We denote by $[V]$ the equivalence class to which the manifold $V$ belongs, and by $\Omega_{k-1}$ the set of all equivalence classes (see Fig. 15). Then the structure of an Abelian group in $\Omega_{k-1}$ is given by $[V_1] + [V_2] = [V_1 \cup V_2]$. The zero element consists of null-bordent manifolds. It is clear that $-[V] = [-V]$. The lemma is thus proved.

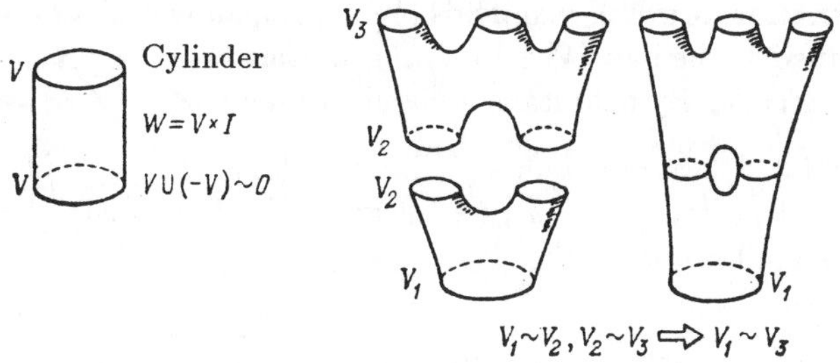


**Fig. 15**

The direct sum of groups $\oplus_{k=1}^{\infty}\Omega_{k-1}$ is usually denoted by $\Omega_*$ (bordism ring), with the ring structure being naturally defined and induced by the direct manifold product (e.g., see [80]). In general, the groups $\Omega_{k-1}$ are different from zero. It can be shown (which is not required here) that the ring $\Omega_* \otimes Q$, where $Q$ is the field of rational numbers, is a polynomial algebra spanned by the bordism classes of the complex projective spaces $CP^{2m}$, $m = 1, 2, \ldots$. It turns out that, along with the oriented bordism theory $\Omega_*$, the nonoriented bordism theory $N_*$ plays an important role in variational problems. For its construction, all closed manifolds are used, and no restrictions of orientability type are imposed, the bordism definition otherwise copying the above for the orientable case. This makes it possible to construct the groups $N_{k-1}$ (analogues of $\Omega_{k-1}$) and the group $N_* = \oplus_{k=1}^{\infty} N_{k-1}$, which is a commutative algebra over the field $Z_2$.

Now, let there be given a space $Y$ and its subspace $Z$.

**Definition 3.5.2.** We call a pair $(V^{k-1}, f)$, where $V^{k-1}$ is a compact, oriented manifold with boundary $\partial V$, and $f$ a continuous mapping $(V, \partial V) \to (Y, Z)$, or $f(V) \subset Y$, $f(\partial V) \subset Z$, an *oriented, singular manifold* of the pair $(Y, Z)$.

An oriented, singular manifold $(V^{k-1}, f)$ of $(Y, Z)$ is said to be null-bordant if there exists a compact, oriented manifold $W^k$, and a continuous mapping $F : W^k \to Y$, such that $V^{k-1}$ is a regular submanifold of the boundary of $W^k$, i.e., $V \subset \partial W$, and the orientation of $V$ coincides with that induced on it by the orientation of $W$, where $F|V = f, F(\partial W \backslash V) \subset Z$ (Fig. 16).

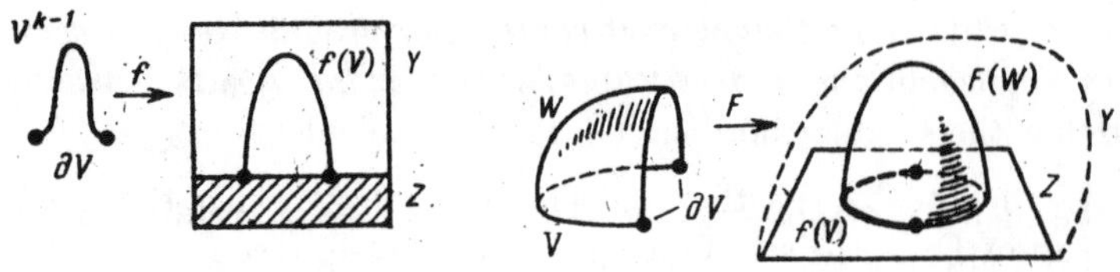


**Fig. 16**

For two singular, oriented manifolds $(V_1^{k-1}, f_1)$, and $(V_2^{k-1}, f_2)$, the disjoint union is defined as the pair $(V_1 \cup V_2, f_1 \cup f_2)$, where $V_1 \cap V_2 = \emptyset, (f_1 \cup f_2)|_{V_1} = f_1, (f_1 \cup f_2)_{V_2} = f_2$. By definition, we will assume that $-(V, f) = (-V, f)$.

**Definition 3.5.3.** Two oriented, singular manifolds $(V_1, f_1)$ and $(V_2, f_2)$ of a pair $(Y, Z)$ will be said to be *bordant* if and only if the disjoint union $(V_1 \cup V_2, f_1 \cup f_2)$ is null-bordant (Fig. 17).

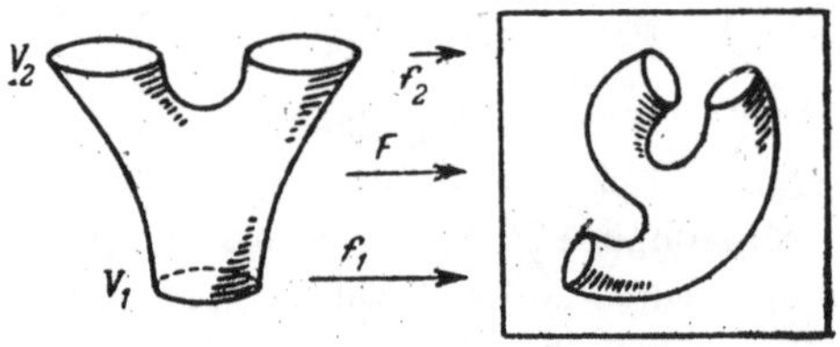


**Fig. 17**

As above, it is easy to verify that the two definitions specify a transitive equivalence relation on the set of all oriented, singular manifolds of $(Y, Z)$. The bordism class of a singular, oriented manifold $(V, f)$ is usually denoted by $[V, f]$ and called a *singular bordism*, whereas the set of all such classes is designated by $\Omega_{k-1}(Y, Z)$. The structure of an Abelian group on $\Omega_{k-1}(Y, Z)$ is induced by disjoint union, i.e., $[V_1, f_1] + [V_2, f_2] = [V_1 \cup V_2, f_1 \cup f_2]$. The class of null-bordant manifolds is the zero element in this group, and $-[V, f] = [-V, f]$. The group $\Omega_p(Y, Z)$ is called the $p$-dimensional oriented bordism group of (Y,Z). We denote the direct sum $\oplus_p \Omega_p(Y, Z)$ by $\Omega_*(Y, Z)$. It can be naturally endowed with the graded module structure over the ring $\Omega_*$ (see above), which obviously can be identified with the ring $\Omega_*(x, \emptyset)$, where $x$ is a point. If we consider the pairs $(V, f)$, where $V$ are compact, closed manifolds, now not necessarily orientable (films $W$ may be nonorientable, too), then the described construction will lead us to the nonoriented bordism group $N_{k-1}(Y, Z)$.

We now consider the spanning problem. Let $A^{k-1}$ be a compact, closed, and oriented submanifold in $M$, $i : A \to X$ an embedding, $A \subset X$, where $X$ is a compact subset of $M$. Then the Plateau problem A (see §3.1) admits the following equivalent re-formulation.

**Problem A.** Can we find a compact set $X_0$ that would possess the minimal properties among all the compact sets $X$ containing $A$ and possessing the property that the singular bordism $(A, i)$ is equivalent to zero in $X$?

Since the identity mapping $e : A \to A$ determines a certain element $\sigma \in \Omega_{k-1}(A)$ (and also $\sigma \in N_{k-1}(A)$ if nonoriented bordisms are considered), the above class of films—the compact sets $X$—is characterized by $i_*(\sigma) = 0$, where $i_* : \Omega_{k-1}(A) \to \Omega_{k-1}(X)$ (resp. $i_* : N_{k-1}(A) \to N_{k-1}(X)$) is the homomorphism induced by the embedding $i : A \to X$.

The second Plateau problem, that of realization, is now formulated as follows.

**Problem B.** Can we find a singular manifold $(V_0, g_0)$ such that the film $X_0 = g_0(V_0)$ possesses the minimal property among all the singular manifolds $(V, g)$, $g : V \to M$, bordant, or equivalent, to a given singular manifold $(V', g')$, $g' : V' \to M$? In other words, can we find $(V_0, g_0)$ such that the film $X_0 = g_0(V_0)$ is minimal in $M$ among all the representatives $(V, g)$ of a given bordism class $\alpha \in \Omega_k(M)$ (or $N_k(M)$)?

Since we consider films of topological type that can vary within the framework of a fixed bordism class, both problems are those of finding the absolute minimum. We now demonstrate that the use of different bordism groups makes it possible to embrace a wide class of geometrical variational problems.

**Example 1.** Consider a minimal two-dimensional film $X_0$ (Fig. 18) homeomorphic to a smooth two-dimensional manifold with boundary $A = S^1$, realizing the absolute minimum of two-dimensional area in the sense of Problem A. The film annihilates the singular manifold $(A, e) \in \Omega_1(A)$ (in the sense of bordisms). Thus, films of this (oriented) type can be tracked by means of the groups $\Omega_*(X, A)$.

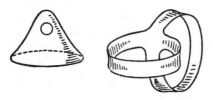

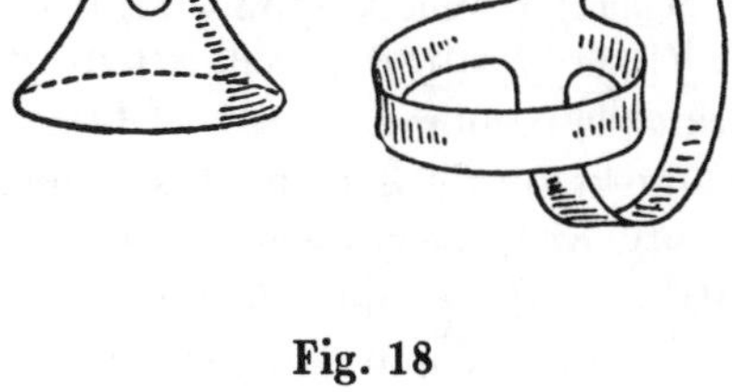

**Fig. 18**

**Example 2.** The minimal film $X_0$, the Möbius band (Fig. 19), can be most naturally tracked by means of nonoriented bordism groups, since it is a nonorientable manifold with boundary $A$.

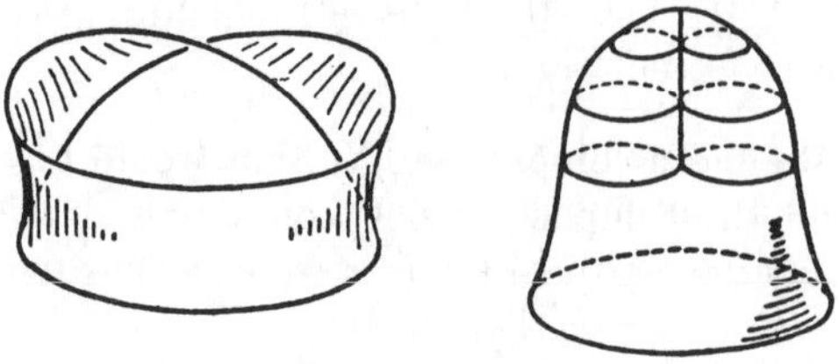

**Fig. 19**

**Example 3.** There exists a class of minimal films $X_0$ for whose description the so-called bordisms " mod $p$" can be most naturally made use of. The triple Möbius band with boundary $A = S^1$ clearly supplies the absolute minimum of the two-dimensional area with a given boundary $A$ (Fig. 20). However, to describe such films, the language of oriented or nonoriented bordisms is insufficient. Therefore, it is useful to consider new groups denoted by $\Omega_*^p$, which we will call *singular bordism groups* mod $p$.

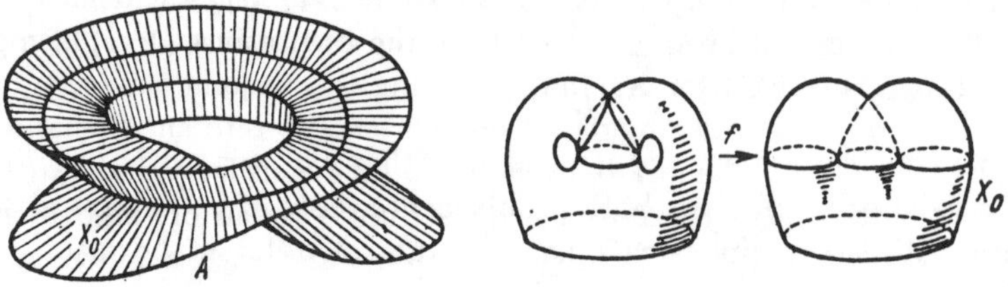


**Fig. 20**

We call the pair $(V, f)$, where $V$ is a $(k-1)$-dimensional compact, oriented manifold with boundary $\partial V$ represented as the disjoint union of $p$ closed, oriented $(k-2)$-dimensional manifolds, viz., $\partial V = B_1 \cup B_2 \cup \cdots \cup B_p$, with the numeration of the manifolds $B_i$ being fixed, the *singular manifold* mod $p$ of the space $Z$, where $p$ is a prime number, *of dimension* $k-1$. Furthermore, for each subscript $i$, we fix a certain diffeomorphism $d_i : B_i \xrightarrow{\cong} B$, where $B$ is the same manifold for each $i$. The mapping $f : V \to Z$ is continuous and possesses the property $f|_{B_i} = f'd_i|_{B_i}$ for each $i$, $1 \leqslant i \leqslant p$, where $f' : B \to Z$ is a certain fixed continuous mapping, and $f|_{B_i}$ denotes the restriction of $f$ to $B_i$. We do not require that these manifolds should be connected. The manifold $B$ may be empty, with the consequence that $V$ is closed.

Two singular manifolds $(V_1; f_1)$ and $(V_2; f_2)$ are said to be *equivalent* mod $p$ if there exists a $k$-dimensional compact, oriented manifold $W$ with boundary $\partial W$ such that

$$\partial W \supset V_1 \cup (-V_2), V_1 \cap V_2 = \emptyset, \partial W = V_1 \cup (-V_2) \cup \left[\bigcup_{i=1}^{p} C_i\right],$$

where each compact, oriented manifold $C_i$ is diffeomorphic to the same manifold $C$ by means of a fixed diffeomorphism $\alpha_i : C_i \overset{\cong}{\to} C$, where $\partial C_i = (B_{1,i}) \cup (-B_{2,i}) (1 \leqslant i \leqslant p)$ and $\partial V_1 = \bigcup_{i=1}^{p} B_{1,i}$, $\partial V_2 = \bigcup_{i=1}^{p} B_{2,i}$, and if, in addition, there exists a continuous mapping $F : W \to Z$ such that $F|_{V_1} = f_1$, $F|_{V_2} = f_2$, $F|_{C_i} = F'_{\alpha_i}|_{C_i}$, where $F' : C \to Z$ is a certain continuous mapping, $C_i \cap (V_1 \cup (-V_2)) = B_{1,i} \cup (-B_{2,i})$ (Fig. 21).

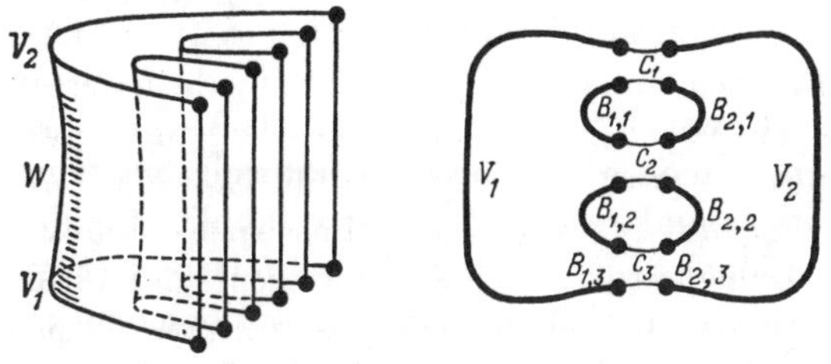


**Fig. 21**

A singular manifold $(V, f)$ mod $p$ is said to be *equivalent to zero* if it is equivalent to $(\emptyset, f_0)$, where $f_0 : \emptyset \to Z$. The operations of replacing the orientation by the opposite, and that of taking the disjoint union, turn the set of equivalence classes of singular manifolds mod $p$ into an Abelian group denoted by $\Omega^p_{k-1}(Z)$. To be null-bordant, mod $p$ means for a closed manifold $V$ that $V \subset \partial W$, where the $k$-dimensional film $W$ is an oriented manifold such that $\partial W = V \cup [\bigcup_{i=1}^{p} T_i]$ (disjoint union), and each component $T_i$ is diffeomorphic to the same manifold $T$ (possibly, $\emptyset$).

Let $Z = x$ (point). It is known that each element $\sigma \in \Omega^p_*(x)$ for $p \neq 2$ is of order $p$. We shall use it in investigating the 2-stability of nontrivial elements $\alpha \in \Omega^p_*(M)$ on a 2-connected manifold $M$.

We now come back to the triple Möbius band. It is clear that a minimal film $X$ spanning $A = S^1$ spans the circle $S^1$ mod 3, i.e., the singular bordism $\sigma = (S^1; e)$ ($e = 1_A$ being the identity mapping, and $\sigma \in \Omega^p_1(S^1)$) vanishes under the homomorphism $i_* : \Omega^3_1(A) \to \Omega^3_1(X_0)$, where $i : A \to X_0$ is an embedding.

Thus, the classical Plateau problem requires the use of the groups $\Omega_{k-1}(A)$, $N_{k-1}(A)$, $\Omega^p_{k-1}(A)$ and their homomorphisms into the groups $\Omega_{k-1}(X)$, $N_{k-1}(X), \Omega^p_{k-1}(X)$, respectively, arising under the embedding $A \to X$.

### 3.6. Spectral Bordism Theory as an Extraordinary Homology Theory

The above bordism groups have originally been defined for cell complexes and polyhedra (simplicial complexes). However, in variational problems, minimal surfaces with complicated singularities arise naturally. Therefore, it is convenient to extend the definition of bordisms to the category of arbitrary compact subsets in a Riemann manifold. This extension is carried out via the standard spectral process with respect to the so-called "nerves of coverings." Eventually, for each compact set, it is possible to compute the bordism groups that we will sometimes call spectral bordisms. The extension of bordism theory to that of spectral bordisms will be described below in greater detail. This process is absolutely analogous to the classical construction of spectral homology (Čech homology). In general, an element of a spectral bordism group is an infinite sequence (spectrum) of smooth manifolds (see the details below). The existence theorems proved by us below for minimal surfaces are related mostly to spectral extraordinary homology and cohomology theories, in particular, to spectral bordisms. If a compact set in question is a "nice" space, e.g., a finite cell complex or pholyhedron, then, certainly, its spectral bordism groups coincide with those of the usual singular bordisms. Hence, in the following, speaking of compact space bordisms, we will mean spectral singular bordisms. For brevity, we will sometimes speak simply of bordisms.The introduction of bordism groups into the statements of multidimensional variational problems is due to the fact that, as will now be shown, the groups $\Omega_*(X,A)$, $N_*(X,A)$, and $\Omega_*^p(X,A)$ satisfy the Eilenberg–Steenrod axioms, thereby forming extraordinary (generalized) homology theories. For definiteness, we consider the groups $\Omega_*(X,A)$, with the argument being completely analogous for the remaining cases. We take pairs of compact spaces $(X,A)$ and continuous mappings $\varphi$. If $\varphi:(X,A)\to(X_1,A_1)$ is a continuous mapping, then the induced natural homomorphism $\varphi_*:\Omega_k(X,A)\to\Omega_k(X_1,A_1)$ is defined, where $\varphi_*[V^k,f]=[V^k,\varphi f]$. Besides, the boundary homomorphism $\partial:\Omega_k(X,A)\to\Omega_{k-1}(A)$, where $\partial[V^k,f]=[\partial V_k, f|_{\partial V^k}]$, is also defined. It is clear that the mapping $\partial$ is defined correctly and is a homomorphism.

**Proposition 3.6.1.** *The triple $(\Omega_*(X,A),\varphi_*,\partial)$ (called a covariant functor in homology theory) satisfies the first six Eilenberg–Steenrod axioms and does not satisfy the seventh, "point," axiom A7 (see [126]). For a one-point space $x$, the group $\Omega_*(x)$ is isomorphic to the oriented bordism group $\Omega_*$ (see §3.5).*

We verify the axioms in the order of [126], with the first three being obvious.

**Lemma 3.6.1.** *If a mapping $i:(X,A)\to(X,A)$ is the identity, then $i_*:\Omega_k(X,A)\to\Omega_k(X_1,A_1)$ is the identity automorphism.*

**Lemma 3.6.2.** *If two mappings $\varphi:(X,A)\to(X_1,A_1)$ and $\psi:(X_1,A_1)\to(X_2,A_2)$ are given, then we have $(\psi\varphi)_*=\psi_*\varphi_*$.*

**Lemma 3.6.3.** *For any mapping $\varphi:(X_1,A)\to(X_1,A_1)$, the diagram is commutative:*

$$\begin{array}{ccc} \Omega_k(X, A) & \xrightarrow{\partial} & \Omega_{k-1}(A) \\ \varphi_* \downarrow & & \downarrow (\varphi \mid A)_* \\ \Omega_k(X_1, A_1) & \xrightarrow{\partial} & \Omega_{k-1}(A_1) \end{array}$$

**Lemma 3.6.4.** *If a mapping $\varphi_0$ is homotopic to $\varphi_1$, where $\varphi_0, \varphi_1 : (X, A) \to (X_1, A_1)$, then $(\varphi_0)_* = (\varphi_1)_*$.*

**Proof.** Let $F : (X \times I, A \times I) \to (X_1, A_1)$ be a homotopy connecting $\varphi_0$ and $\varphi_1$, and $(V^k, f)$ a singular manifold of the pair $(X, A)$. We define the mapping $\psi : V^k \times I \to X_1$ by the formula $\psi(x,t) = F(f(x), t)$. Then $\psi(x, 0) = \varphi_0 f(x)$, $\psi(x, 1) = \varphi_1 f(x)$. Since $V^k \times I$ is a manifold with boundary $\partial(V^k \times I) = (V^k \times \partial I) \cup ((-\partial V^k) \times I)$, the manifold $(V^k \times 1) \cup ((-V^k) \times 0)$ is a regular submanifold of the boundary, and $\psi((-V^k) \times I) \subset A_1$. Hence, $[V^k, \varphi_0 f] = [V^k, \varphi_1 f]$, which completes the proof.

**Lemma 3.6.5.** *For any pair $(X, A)$, the sequence of groups and homomorphisms $\cdots \to \Omega_k(A) \xrightarrow{j_*} \Omega_k(X) \xrightarrow{i_*} \Omega_k(X, A) \xrightarrow{\partial} \Omega_{k-1}(A) \to \ldots$ is exact.*

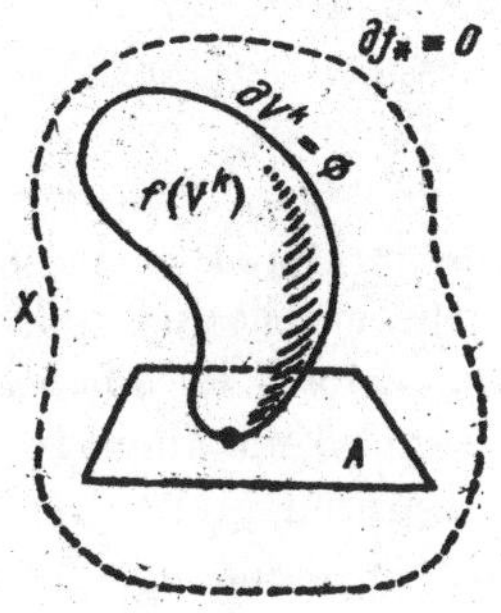

**Fig. 22**

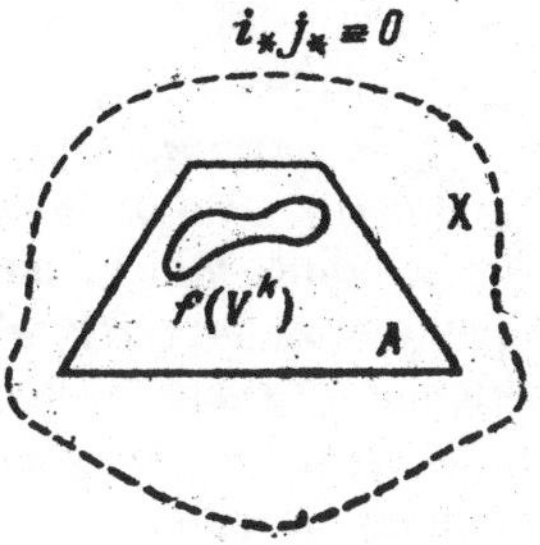

**Fig. 23**

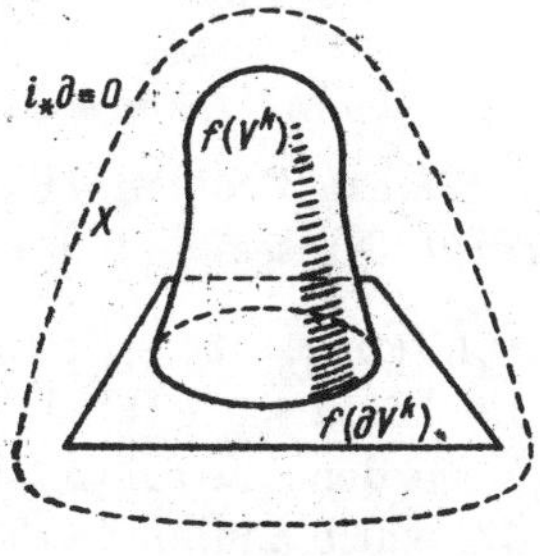

**Fig. 24**

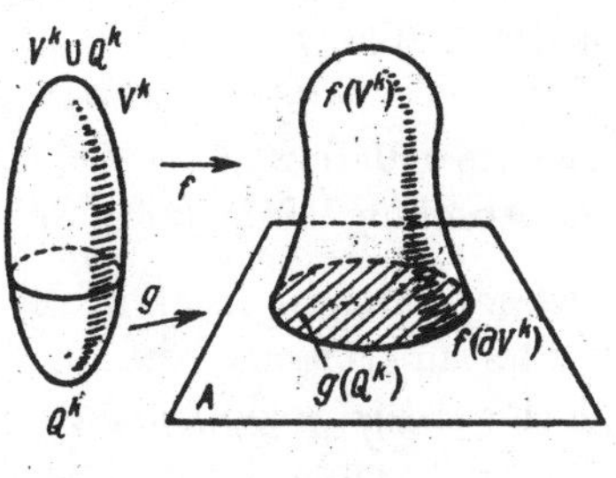

**Fig. 25**

**Proof.** The equality $\partial j_* \equiv 0$ is proved as shown in Fig. 22, and the identity $i_* j_* \equiv 0$ is verified in Fig. 23. For $i_* \partial \equiv 0$, see Fig. 24. Now, let an element $[V^k, f]$ belong to the kernel of the operator $\partial$, which means that there exists a singular manifold $(Q^k, g)$ such that $g(\partial Q^k) = f(\partial V^k), \partial Q^k = \partial V^k, g(Q^k) \subset A$. It is then obvious that $[V^k, f] = j_*[V^k \cup (-Q^k), f \cup g]$ (gluing along the common boundary), i.e.,$[V^k, f] \in \operatorname{Im} j_*$ (Fig. 25). The remaining equalities are verified similarly, which completes the proof.

**Lemma 3.6.6 (Excision Axiom).** *The embedding* $i(X\backslash U, A\backslash U) \subset (X, A)$ *for any* $U$ *such that* $\overline{U} \subset \operatorname{Int} A$ *induces the group isomorphism* $i_* : \Omega_k(X\backslash U, A\backslash U) \xrightarrow{\approx} \Omega_k(X, A)$.

**Proof.** The monomorphy of $i_*$ is obvious. We refer to Fig. 26 for the proof of the epimorphism. Indeed, if $[V^k, f]$ is an element of the group $\Omega_k(X, A)$, $f(\partial V^k) \subset A$, then $[V^k, f] = i_*[Q^k, g]$, where $Q^k \subset V^k$, $Q^k \supset f^{-1}(X \backslash \operatorname{Int} A)$. The lemma is thus proved.

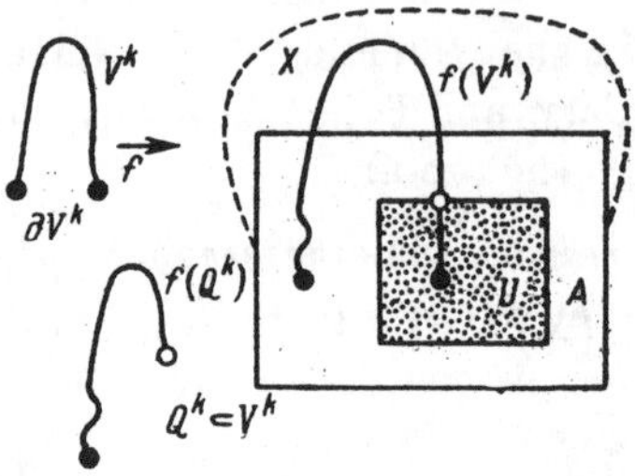


**Fig. 26**

The last, seventh, axiom does not hold, since $\Omega_k(x) \not\equiv 0$ for $k > 0$, with $x$ being a point, which makes bordism theory sharply different from that of the usual homology $H_*$, where the "point axiom" does hold. For variational problems, we shall need a reduced bordism group. Identify the group $\Omega_k$ with $\Omega_k(x)$ for a point $x$, and define the reduced group $\widetilde{\Omega}_k(X)$ as the kernel of the homomorphism $\varepsilon_* : \Omega_k(X) \to \Omega_k(x)$, where $\varepsilon_*$ is induced by the projection of the space $X$ onto $x$. It is evident that $\Omega_k(X) \cong \Omega_k \oplus \widetilde{\Omega}_k(X)$. The oriented, singular manifold $(V^k, f)$ of $X$ is an element of $\widetilde{\Omega}_k(X)$ if and only if $[V^k] = 0$ in $\Omega_k$.

## 3.7. The Formulation of the Solution to the Plateau Problem (Existence of the Absolute Minimum in Spectral Bordism Classes)

Here, we formulate a theorem supplying a solution to the spectral Plateau problem for an arbitrary "contour" $A^{k-1}$ in Euclidian space $\mathbb{R}^n$. Let $A^{k-1}$ be a fixed, closed, and smooth submanifold in $\mathbb{R}^n$. We denote by $A(\sigma)$ the class of all measurable compact sets (films) $X, X \subset \mathbb{R}^n$, which annihilate the bordism $\sigma = [A, e]$, with $e : A \to A$ being the identity mapping. In other words, the singular manifold $(A, i)$, with $i : A \to X$ being an embedding, is null-bordant in $X$, i.e., $i_*(\sigma) = 0$, where $i_*$ is the homomorphism mapping the group of $(k-1)$-dimensional bordisms $A$ into the group of $(k-1)$-dimensional spectral bordisms $X$. We designate by $\operatorname{vol}_k$ either the Riemannian volume of $X$ or the $k$-dimensional Hausdorff measure if $X$ is an arbitrary Hausdorff-measurable compact set.

We now introduce the new concept of stratified surface. We will say that $X \in A(\sigma)$ is a *stratified surface* if $X$ admits a representation $X = X^k \cup X^{k-1} \cup \ldots$, where each subset, or stratum, $X^s$ is an $s$-dimensional surface in $\mathbb{R}^n$, of dimension $s$ at each of its points, and contained in $X \setminus \bigcup_{i=s+1}^{k} X^i$ with $X^s \cap \overline{\left(\bigcup_{i=s+1}^{k} X^i\right)} = \emptyset$. In other words, $X$ is the union of strata of different dimensions, and "dimensionally inhomogeneous," i.e., each of its nonempty strata is outside the union of all strata of greater dimensions.

The concept of stratified surface can be equivalently introduced by means of the spherical density functions $\Psi_s$. Recall that the spherical density $\Psi_s(x)$ of a compact set $X$ at a point $x \in X$ is determined as $\Psi_s(x) = \lim_{a \to 0} \left( \frac{\mathrm{vol}_s(X \cap B^n(x,a))}{\gamma_s a^s} \right)$, where $B^n(x,a)$ is an $n$-dimensional ball in $\mathbb{R}^n$ with centre at $x$ and radius $a$, with $\gamma_s a^s$ being the volume of an Euclidean $s$-dimensional ball of radius $a$.

Then we can define an $s$-dimensional stratum $X^s$ of a surface $X$ as the set of all points $x \in X$ such that $\Psi_s(x) > 0$, and $\Psi_i(x) = 0$ for $i > s$.

And now we present the new concept of the stratified volume of a surface $X$.

We call the vector made up of the volumes of all strata $X^s$ of a surface $X$, i.e., $SV(X) = (\mathrm{vol}_k X^k, \mathrm{vol}_{k-1} X^{k-1}, \ldots)$, the *stratified volume* $SV(X)$ of $X$. We also call a stratum $X^k$ of maximal dimension the *greatest stratum*, and its volume $\mathrm{vol}_k X^k$ its *greatest volume*, whereas the other strata and their volumes will be called *least*. Lexicographic order is naturally introduced into the set of all stratified volumes, viz., a vector $SV(X)$ is "greater" than a vector $SV(Y)$ (we will write $SV(X) > SV(Y)$) if $\mathrm{vol}_i X^i = \mathrm{vol}_i Y^i$ for $k \geqslant i \geqslant s$ and certain $s$, and $\mathrm{vol}_{s+1} X^{s+1} > \mathrm{vol}_{s+1} Y^{s+1}$.

Therefore, we can determine the least vector $SV_{\min} = \inf SV(x)$, where inf is taken with repect to all surfaces $X$ from a given class $A(\sigma)$. The spectral Plateau problem can now be formulated thus: Is the minimum $SV_{\min}$ attained on any surface $X \in A(\sigma)$? It turns out that the answer is positive.

Assume that the volume minimization problem makes sense, i.e., the infimum of stratified volumes of statified surgaces is finite (for the precise formulation, see Chapter 6).

**Theorem 3.7.1.** *Let $A$ be an arbitrary smooth $(k-1)$-dimensional closed submanifold in $\mathbb{R}^n$ and such that the minimization problem makes sense. Then there always exists an absolutely (globally) minimal stratified surface $X_0 = S^k \cup S^{k-1} \cup \ldots$ in the class $A(\sigma)$ of all surfaces with boundary $A$ (see the description above), which is stratified by strata $S^i$, so that its stratified volume $SV(X_0)$ is the least possible in $A(\sigma)$. This surface is the solution to the spectral Plateau problem; in particular, it has the least volume $\mathrm{vol}_k$. Each nonempty stratum $S^i$ of an absolutely minimal surface $X_0$ is, possibly, with the exception of a set of $i$-dimensional measure zero, an analytic minimal $i$-dimensional submanifold in $\mathbb{R}^n$.*

A similar theorem has been proved by the author also for an arbitrary Riemannian manifold $M^n$. The concepts of stratified surface and stratified volume are formulated similarly also for $M^n$.

Besides, Problem B has been solved positively by the author in a similar manner (i.e., the Plateau problem in the class of closed spectral bordant surfaces; see above). Namely, there always exists an absolutely (globally) minimal stratified surface in the class of surfaces $B(\alpha)$ realizing a nonzero element $\alpha$ of the bordism group of a manifold $M$, for which $\mathrm{vol}_k$ is minimal; moreover, its stratified volume is least possible.

We now give the complete statement of the theorem. Assume that the minimization problem makes sense, i.e., that there exist stratified surfaces with finite stratified volumes.

**Theorem 3.7.1*.** *(1) Let $\{X\}_k$ be the class of all compact sets $X$, $A \subset X \subset \mathbb{R}^n$ such that $X \in A(\sigma)$ and $\mathrm{vol}_k X = d_k = \inf(\mathrm{vol}_k Y)$, $Y \in A(\sigma)$. Then the class is nonempty, $d_k > 0$, and each $X \in \{X\}_k$ contains a uniquely determined $k$-dimensional subset $S^k$, i.e., of the dimension $k$ at each of its points, $S^k \subset X \backslash A$ such that $A \cup S^k$ is a compact set in $\mathbb{R}^n$, contains a certain "singular" subset $Z_k$ which may be empty, where $\mathrm{vol}_k(Z_k) = 0$ and $S^k \backslash Z_k$ is a smooth minimal $k$-dimensional submanifold in $\mathbb{R}^n$, with $S^k \backslash Z_k$ being everywhere dense in $S^k$, and $\mathrm{vol}_k(S^k) = \mathrm{vol}_k X = d_k > 0$.*

*(2) Furthermore, if $\{X\}_{k-1}$ is the class of all compact sets $X$, $A \subset X \subset \mathbb{R}^n$ such that $X \in A(\sigma)$, $X \in \{X\}_k$ and $\mathrm{vol}_{k-1}(X \backslash S^k) = d_{k-1} = \inf \mathrm{vol}_{k-1}(Y \backslash S^k)$, $Y \in \{X\}_k$, then the class $X_{k-1}$ is nonempty, and if $d_{k-1} > 0$, each $X \in \{X\}_{k-1}$ contains a uniquely determined $(k-1)$-dimensional subset $S^{k-1}$ of dimension $k-1$ at each of its points, $S^{k-1} \subset X \backslash A \backslash S^k$ such that $A \cup S^k \cup S^{k-1}$ is a compatum in $\mathbb{R}^n$ and contains a certain "singular" subset $Z_{k-1}$ (which may be empty), where $\mathrm{vol}_{k-1}(Z_{k-1}) = 0$ and $S^{k-1} \backslash Z_{k-1}$ is a smooth minimal $(k-1)$-dimensional submanifold in $\mathbb{R}^n$, everywhere dense in $S^{k-1}$ with $\mathrm{vol}_{k-1}(S^{k-1}) = \mathrm{vol}_{k-1}(X \backslash A \backslash S^k) = d_{k-1} > 0$. If, however, $d_{k-1} = 0$, then we put $S^{k-1} = \emptyset$.*

*(3),(4), etc., with respect to decreasing dimension.*

**Corollary 3.7.1.** *In the class $A(\sigma)$, there exists a globally minimal surface $X_0 = A \cup S^k \cup S^{k-1} \cup \ldots$ whose stratified volume $SV(X_0) = (\mathrm{vol}_k S^k, \mathrm{vol}_{k-1} S^{k-1}, \ldots$ is the least among those of all other films $X \in A(\sigma)$, i.e., the film $X_0$ is globally minimal in all dimensions.*

If we confine ourselves only to the first statement of Theorem 3.7.1, then we obtain the existence theorem for a parametrized film $X_0$ which is globally minimal in its greatest dimension $k$. A theorem which is similar to Theorem 3.7.1 and Corollary 3.7.1 also holds for the spectral Plateau Problem B (realization problem). All these theorems are particular cases of the general existence theorem for globally minimal surfaces in all dimensions, which was proved by the author for a wide class of variational problems on a Riemannian manifold (see [154]–[156], [158], [161]–[162]).

For example, let a minimal surface $X_0$ with fixed boundary $A$ be a finite cell complex. All concrete examples of minimal surfaces known today (and whose boundary is a compact minimal manifold) are, in fact, finite cell complexes. It then follows from Therorem 3.7.1 that $X_0$ is of the form $f(W_0)$, where $W_0$ is a certain smooth manifold with boundary $A$ and $f : W_0 \to X_0$ is a continuous mapping. In other words, in this case, $X_0$ is a solution of the classical Plateau problem, i.e., there exists a smooth manifold $W_0$ with boundary $A$, whose continuous image is a minimal surface $X_0$.

## §4 Extraordinary (Co)homology Theories Determined for "Surfaces with Singularities"

In this section, we describe spectral extraordinary homology and cohomology theories, because they turn out to be associated with variational problems, each with its own, admitting a complete solution.

It happens so that, given some extraordinary homology or cohomology theory, we also specify a certain type of boundary conditions which minimal surfaces, with or without boundary, can satisfy. Since the stock of extraordinary (co)homology is enormous, this permits us to consider extremely different forms of boundary conditions. Since, in general, solutions of variational problems possess singularities that may sometimes be rather complicated, these (co)homology theories should be defined on the set of surfaces with singularities. As the widest possible class of such "surfaces," we take that of compact sets in Riemannian manifolds, necessarily including all probable solutions to the variational problems. The present section is of technical nature and may be omitted on first reading; the reader may at once focus his attention on the formulation of variational problems in terms of spectral extraordinary theories.

### 4.1. The Characteristic Properties of (Co)Homology Theories

Let a certain admissible category $K$ be given (see [126]). Suppose that each pair $(X, A) \in K$ and each $q \in Z$ are associated with an Abelian group $h_q^{(q)}(X; A)$. Let each admissible mapping $f : (X, A) \to (Y, B)$ and each $q \in Z$ be associated with a homomorphism $f_*^{(*)}$ defined on $h_q(X, A)$ (resp. $h^q(Y, B)$), and a (co)boundary operator $\partial_q : h_q(X, A) \to h_{q-1}(A, \emptyset)$ (resp. $\delta_q : h^{q-1}(A, \emptyset) \to h^q(X, A)$) defined for $(X, A) \in K$. We require that the triple $(h_*^{(*)}(X, A); f_*^{(*)}; \partial(\delta))$ should satisfy the six Eilenberg–Steenrod axioms A1–A6.

**A1.** If $f \in K$ is the identity mapping, then $f_*^{(*)}$ is the identity homomorphism.

**A2.** We have $(gf)_* = g_* f_*$ (resp. $(gf)^* = f^* g^*$).

**A3.** The (co)boundary operators $\partial(\delta)$ are natural, i.e., they commute with the induced homomorphisms $f_*^{(*)}$.

**A4 (Exactness Axiom).** If $(X, A) \in K$, then the sequence is exact

$$\ldots \to h_q(A, \varnothing) \xrightarrow{i_*} h_q(X, \varnothing) \xrightarrow{j_*} h_q(X, A) \xrightarrow{\partial} h_{q-1}(A, \varnothing) \to \ldots$$
$$(\ldots \leftarrow h^q(A, \varnothing) \xleftarrow{i^*} h^q(X, \varnothing) \xleftarrow{j^*} h^q(X, A) \xleftarrow{\delta} h^{q-1}(A, \varnothing) \leftarrow \ldots).$$

The homomorphisms $i_*^{(*)}$ and $j_*^{(*)}$ are induced by the embeddings.

**A5 (Homotopy Axiom).** If two mappings $f$ and $g$ are homotopic in $K$, then $f_*^{(*)} = g_*^{(*)}$.

In applications, the following reformulation may be useful:

**A5′.** Let $(X, A) \in K$, and $g_i : (X, A) \to (X \times I; A \times I)$ ($i = 0, 1$, whereas $I$ is the unit interval), with $g_0(x) = (x, 0)$, $g_1(x) = (x, 1)$.
Then $(g_0)_*^{(*)} = (g_1)_*^{(*)}$.

**A6 (Excision Axiom).** Let $(X, A) \in K$, $V \subset X$, $U \subset X$ be open subsets such that $\overline{U} \subset V \subset A$ ($\overline{U}$ is the closure of $U$ in $X$), and the embedding $i : (X \backslash U; A \backslash U) \to (X, A)$ be admissible. Then all the homomorphisms $i_*^{(*)}$ are isomorphisms.

The following reformulation of this axiom may be useful:

**A6′.** Let $X = \text{Int}(X_1) \cup \text{Int}(X_2)$, where $X_1$ is closed in $X$, and the embedding $i : (X_1; X_1 \cap X_2) \to (X, X_2)$ admissible in $K$. Then all the homomorphisms $i_*^{(*)}$ are isomorphisms.

**Definition 4.1.1.** We call a functor $h_*^{(*)} = \oplus_q h_q^{(q)}$ satisfying Axioms A1–A6 an *extraordinary (co)homology theory* of the category $K$ with values in the category of Abelian groups $AB$.

To make use of the language of functors $h_*^{(*)}$ in variational problems, we shall need the functors $h = h_*^{(*)}$ satisfying a stronger excision axiom, viz., $\widetilde{\text{A6}}$.

**$\widetilde{\text{A6}}$ (Relative Invariance Axiom).** Let $f : (X, A) \to (Y, B)$ be a relative homeomorphism of $(X, A)$ onto $(Y, B)$, i.e., $X \backslash A \underset{\sim}{\xrightarrow{f}} Y \backslash B$ a homeomorphism. Then all homomorphisms $f_*^{(*)}$ are isomorphisms.

It is clear that if a functor $h$ on $K$ satisfies $\widetilde{\text{A6}}$, then it also satisfies A6=A6′; in general, the converse is not true. We will make use of another variant of the excision axiom.

**A6⁰.** Let $(X, A, B)$ be an admissible triple in $K$, $X = A \cup B$, and $i : (A, A \cap B) \to (X, B)$ an embedding. Then all homomorphisms $i_*^{(*)}$ are isomorphisms.

If a functor $h$ satisfies Axiom $\widetilde{\text{A6}}$ on $K$, then it also satisfies A6⁰, since the embedding $(A, A \cap B) \to (A \cup B, B)$ is a relative homeomorphism. In general, the converse does not hold. If $h$ satisfies A6$^0$ on $K$, then it also satisfies A6. In general, the converse is invalid. Therefore, Axiom A6$^0$ is "placed" between $\widetilde{\text{A6}}$ and A6.

Let $x$ be a one-point topological space, and an embedding $i : x \to X$ admissible in the category $K$. Suppose that a cohomology theory $h_*$ is fixed on $K$.

**Definition 4.1.2.** We call the subgroup $\mathrm{Ker}(i^*) \subset h^q(X)$; $i^* : h^q(X) \to h^q(x)$ a reduced cohomology group $\tilde{h}^q(X)$ of dimension $q$ with respect to a point $x$.

Let $f : X \to x$ be a projection of $X$ onto $x$, and $h_*$ a homology theory.

**Definition 4.1.3.** We call the subgroup $\mathrm{Ker}(f_*) \subset h_q(X)$; $f_* : h_q(X) \to h_q(x)$ a reduced homology group $\tilde{h}_q(X)$ of dimension $q$ with respect to a point $x$.

### 4.2. Extraordinary (Co)Homology Theories for Finite Cell Complexes

Consider an arbitrary spectrum $E$, i.e., the family $E = \{E_n; \varepsilon_n\}$, where $n \in Z$, $E_n$ are topological spaces, and $\varepsilon_n : \Sigma E_n \to E_{n+1}$ are certain mappings ($\Sigma X$ denoting the reduced suspension over the space $X$). $E$ can be described by giving the mappings $\bar{\varepsilon}_n : E_n \to \Omega E_{n+1}$ instead of $\varepsilon_n$ ($\Omega X$ denoting the loop space on $X$). It is sometimes called an $\Omega$-spectrum if all $\bar{\varepsilon}_n$ are homotopy equivalences.

**Example 1.** $E = \{S^n; \varepsilon_n\}$, where $\varepsilon_n : \Sigma S^n \to S^{n+1}$ is the identity mapping, and $S^n$ a sphere of dimension $n$.

**Example 2.** Let $E$ be a certain spectrum, and $X$ an arbitrary $CW$-complex. Put $E'_n = E_n \wedge X$, recalling that $A \wedge B = A \times B \backslash A \vee B$, where $A \vee B$ denotes the wedge of two pointed spaces and forms the coordinate axes in our case, and $\varepsilon'_n : \Sigma E'_n \to E'_{n+1}$, $\varepsilon'_n : \Sigma E'_n = S^1 \wedge E'_n = S^1 \wedge E_n \wedge X \xrightarrow{\varepsilon_n \wedge 1} E_{n+1} \wedge X = E'_{n+1}$ (here, we have made use of the obvious relation $\Sigma A = S^1 \wedge A$). Then, the collection $\{E'_n; \varepsilon'_n\}$ forms a spectrum called the suspension of $E$ if $X = S^1$ and denoted by $E' = E \wedge X$.

**Example 3.** Let $E$ be a spectrum, and $X$ a $CW$-complex. We put $F_n = F(X, E_n)$, where $F(A, B)$ is the space of all continuous mappings of $A$ into $B$. We define the mappings $\varphi_n : \Sigma F_n \to F_{n+1}$ as $\varphi_n(s \wedge f)(x) = \varepsilon_n(s \wedge f(x))$, where $E = \{E_n, \varepsilon_n\}$; $f \in F_n$; $s \in S^1$; $x \in X$; $\Sigma F_n = S^1 \wedge F_n$. We obtain a spectrum $F(X, E) = \{F_n, \varphi_n\}$ called the loop spectrum of $E$ in the case $X = S^1$, and denoted by $\Omega E$.

Recall that a spectrum $X$ is said to be *convergent* if there exists a number $N$ such that the spaces $E_{N+i}$ are $i$-connected, i.e., $\pi_s(E_{N+i}) = 0$ for $s \leqslant i$ and all $i \geqslant 0$. For example, the spectrum of spheres is convergent (see Example 1). Let $E$ be an arbitrary convergent spectrum, and $\alpha_k = (\varepsilon_k)_* \Sigma_*$ be the homomorphism

$$\pi_{n+k}(E_k) \xrightarrow{\Sigma_*} \pi_{n+k+1}(\Sigma E_k) \xrightarrow{(\varepsilon_k)_*} \pi_{n+k+1}(E_{k+1}),$$

where $\Sigma_*$ is the suspension homomorphism, and $\pi_\alpha(A)$ are the homotopy groups of the space $A$. It is clear that the groups $\pi_{n+k}(E_k)$ and homomorphisms $\alpha_k$ form a direct spectrum (stabilizing for fixed $n$ up to $(\varepsilon_k)_*$, because $E$ is convergent), which permits us to define $\pi_n(E) = \varinjlim_{(k)} \{\pi_{n+k}(E_k), \alpha_k\}$. Note that $n$ may be either positive or negative. If $E$ is the spectrum from Example 1, then $\pi_n(E) = \pi_n^S(S^0)$ is an $n$-dimensional stable homotopy group. The groups $\pi_n(E)$ are called the homotopy groups of $E$.

Let $P^2$ be the category of pairs $(X, A)$, where $X$ and $A$ are finite cell complexes, $P$ that of all finite cell complexes, and $P^0$ that of finite cell complexes with base points. We denote the category of Abelian groups by $AB$, and that of compact Abelian groups by $ABC$. Let $E = \{E_n, \varepsilon_n\}$ be a convergent spectrum, $X \in P^0$ an arbitrary $CW$-complex, and $x \in X$ a base point. We put $\tilde{h}_q(X) = \tilde{h}_q(X, E) = \pi_n(E \wedge X)$, where $E \wedge X$ designates the spectrum from Example 2.

It is easy to verify that we have constructed from the category $P^0$ a reduced homology theory $\tilde{h}_*(\cdot, E)$, proceeding from an arbitrary convergent spectrum $E$. We now construct the homology theory $h_*$ on $P^2$ from $\tilde{h}_*$ on $P^0$. Let $(X, A) \in P^2$. We then put $h_q(X, A) = \tilde{h}_q(X/A)$, and if $A = \emptyset$ (which is admissible in $P^2$), then we assume that $X/\emptyset = X \cup \{x\}$ (disjoint union). We can verify that the functor $h_*$ is an extraordinary theory on $P^2$ in the sense of Definition 4.1.1. It is easily proved that $h_*$ on $P^2$ satisfies not only A6, but also A6$^0$.

There exists a one-to-one correspondence between theories $h_*$ on $P^2$ and $\tilde{h}_*$ on $P^0$. We have just constructed a correspondence associating each theory $\tilde{h}_*$ on $P^0$ with a theory $h_*$ on $P^2$, and now we construct the converse correspondence. Let $h_*$ be a theory on $P^2$. If $(X, x) \in P^0$, then we can put $\tilde{h}_q(X) = h_q(X, x)$.

Let $h_*$ be an extraordinary homology theory on $P^2$. Then the groups $h_q(x)$, $q \in Z$ are called the *coefficient groups* of $h_*$. The question arises whether it is possible to represent any theory $h_*$ on $P^2$ as $h_*(\cdot, E)$. The answer is given by the following.

**Proposition 4.2.1 (See [117],[425]).** *For an arbitrary spectrum $E$, the functor $h_*(\cdot, E)$ determines an extraordinary homology theory on $P^2$, satisfying A1–A6$^0$. Conversely, let $\tilde{h}_*$ be a reduced extraordinary homology theory on $P^0$, so that all coefficient groups $\tilde{h}_q(S^0)$ are at most countable. Then there exists an $\Omega$-spectrum $E$ such that there is a unique natural isomorphism $T : \tilde{h}_*(X, E) \cong \tilde{h}_*(X)$ for any $X \in P^0$, with $E$ being uniquely determined up to homotopy equivalence.*

A similar statement is also valid for $h_*$ on $P^2$.

Consider the cohomology case. Let $E = \{E_n, \varepsilon_n\}$ be an arbitrary spectrum,

$x \in P^0$, and $x \in X$ a base point. We construct the spectrum $F(X, E)$ and put $\tilde{h}^q(X) = \tilde{h}^q(X, E) = \pi_{-q}[F(X, E)]$ (see Example 3).

We can verify that we have thereby matched each spectrum $E$ with a cohomology theory on $P^0$. Just as in the case of homology, we can construct $h^*$ on $P^2$, proceeding from $\tilde{h}^*$ on $P^0$. Let $(X, A) \in P^2$. Put $h^q(X, A) = \tilde{h}^q(X/A)$; $h^q(X) = h^q(X, \emptyset) = \tilde{h}^q(X/\emptyset) = \tilde{h}^q(X \cup \{x\})$.

The obtained functor $h^*$ is an extraordinray cohomology theory on $P^2$, satisfying A6 and A6$^0$. Similarly to the cohomology case, there exists a one-to-one correspondence between $\tilde{h}^*$ on $P^0$ and $h^*$ on $P^2$ (see [117], [424]).

### 4.3. The Construction of Extraordinary (Co)Homology Theories for "Surfaces with Singularities" (on Compact Sets)

Until now, we have considered the functors $h$ (meaning $h_*^{(*)}$) satisfying A1–A6, with values in the Abelian group category $AB$. However, they are not suitable for the construction of theories $h$ on the category of compact pairs $U_C$, because the passage to the inverse limit violates the exactness of the homology sequency of a pair.

(a) Let a functor $h_*$ be given on $P^2$ satisfying A1–A6, with values in the category $GFf$ of finite-dimensional vector spaces over a field $F$, i.e., $h_q(X, A) \in GFf$. We will assume that all the homomorphism $f_*$ and $\partial$ are $F$-linear mappings.

(b) Let a functor $h_*$ be given on $P^2$ satisfying A1–A6, with values in the category $ABC$ of topological compact Abelian groups, i.e., $h_q(X, A) \in ABC$. We will assume that all homomorphisms $f_*$ and $\partial$ are continuous.

(c) Let a functor $h^*$ be given on $P^2$ satisfying A1–A6, with values in the category $GR$ of all $R$-modules of over a ring $R$, i.e., $h^q(X, A) \in GR$. We will assume that all homomorphisms $f^*$ and $\delta$ are $R$-linear.

Any cohomology functor $h^*$ in the sense of Definition 4.1.1 satisfies (c), since it can be assumed that $R = Z$ and then $AB = GZ$. Embed $P^2$ into the category $U_C$, and extend the functors $h$ from $P^2$ to the whole of $U_C$. Let a theory $\tilde{h}$ be given on $P^2$ satisfying (a), or (b), or (c). Let $(X, A) \in U_C$, $\mathrm{Cov}^f(X, A)$ be the directed set of all open finite coverings of $(X, A)$, $\alpha \in \mathrm{Cov}^f(X, A)$ an arbitrary finite covering, and $(X_\alpha, A_\alpha)$ the nerve of the covering $\alpha$. Since $(X_\alpha, A_\alpha) \in P^2$, we put $h_\alpha = h(X_\alpha, A_\alpha)$. Then, if $\alpha < \beta$, i.e., if $\beta$ is a refinement of $\alpha$, there arise homomorphisms $\pi_\alpha^\beta : (h_q)_\beta \to (h_q)_\alpha$. We call the systems $\{(h_q)_\alpha, \pi_\alpha^\beta\}$ (resp. $\{(h^q)_\alpha, \pi_\alpha^\beta\}$) *$q$-dimensional (co)homology spectra* of the pair $(X, A)$. The elementary argument demonstrates that a homology (resp. cohomology) spectrum is the inverse (resp. direct) spectrum of the groups $h_\alpha$ over the directed set $\mathrm{Cov}^f(X, A)$.

**Definition 4.3.1.** We call the direct (resp. inverse) limit of a $q$-dimensional (co)homology spectrum a pair $(X, A)$ over $\mathrm{Cov}^f(X, A)$ the *$q$-dimensional extraordinary spectral (co)homology group* of $(X, A) \in U_C$, and still denote it by $h_q^{(q)}(X, A)$.

**Lemma 4.3.1.** *If $h_q(X,A) \in GFf$ (case (a)) and $(X,A) \in P^2$, then $h_q(X,A) \in GF$ for $(X,A) \in U_C$. If $h_q(X,A) \in ABC$ (case (b)) and $(X,A) \in P$, then $h_q(X,A) \in ABC$ for $q \in \mathbf{Z}$, $(X,A) \in U_C$. If $h^q(X,A) \in GR$ (case (c)) and $(X,A) \in P^2$, then $h^q(X,A) \in GR$ for $(X,A) \in U_C$. The category type of the groups h is preserved in the cases (b) and (c) and may alter in the case (a).*

The proof follows from Theorems VIII.3.14 and VIII.4.12 in [126].

Let $f : (X,A) \to (Y,B)$ be a continuous mapping from the category $U_C$, $f_0^{-1} : \mathrm{Cov}^f(Y,B) \to \mathrm{Cov}^f(X,A)$ be the corresponding mapping of the coverings of $(Y,B)$ into those of $(X,A)$, and $f_\alpha : (X_{\alpha'}, A_{\alpha'}) \to (Y_\alpha, B_\alpha)$ that of the nerves of the coverings. Here, $\alpha \in \mathrm{Cov}^f(Y,B)$ and $\alpha' = f_0^{-1}(\alpha)$; therefore $f_\alpha$ is an embedding of $(X_{\alpha'}, A_{\alpha'})$ into $(Y_\alpha, B_\alpha)$. It can be shown that the homomorphisms $(f_\alpha)_*^{(*)}$, together with the mapping $f_0^{-1}$, where $\alpha$ ranges over the whole set $\mathrm{Cov}^f(Y,B)$, form a mapping $\Phi(f)$ of the (co)homology spectrum of $(X,A)$ into the spectrum of $(Y,B)$. The proof consists in verifying the commutativity of the corresponding diagrams, and we omit it. The construction of the spectral groups and induced homomorphisms is thus complete. We now verify A1–A6.

### 4.4. Verifying the Characteristic Properties of the Constructed Theories

**Lemma 4.4.1.** *The theories h constructed in §4.3 satisfy A1 and A2 on the category $U_C$.*

The proof is obvious.

**Lemma 4.4.2.** *The theories h constructed in §4.3 satisfy the homotopy axiom A5=A5′ on the category $U_C$.*

**Proof.** Let $g_i : (X,A) \to (X \times I, A \times I)$ be defined as $g_0(x) = (x,0)$, $g_1(x) = (x,1)$. A covering $\alpha$ of the interval $I = [0,1]$ with open, connected sets $\alpha_i$, where $i = 0,1,2,\dots,n, n > 0$, is said to be *regular* if $0 \in \alpha_0$, $0 \notin \alpha_1$, $1 \in \alpha_n$, $1 \notin \alpha_{n-1}$, $\alpha_i \cap \alpha_{i+1} \neq \emptyset$ for $0 \leqslant i \leqslant n-1$, and $\alpha_i \cap \alpha_j = \emptyset$ for $|i-j| > 1$. Hence, the regular coverings form a cofinal subset in $\mathrm{Cov}^f(I)$ (see Lemma IX.5.4 in [126]). If $\alpha \in \mathrm{Cov}^f(X,A)$, then we denote its indexing pair by $(V_\alpha, V_\alpha^A)$ (see [126]). Assume that $\alpha$ is such that each index $v \in V_\alpha$ is associated with a certain finite, regular covering $\beta^v = \{\beta_i^v\}$ of $I$ with the index set $N^v = \{0,1,\dots,n^v\} = \{i\}$. Since $V_\alpha^A \subset V_\alpha$, it is possible that $v \in V_\alpha^A$. Let $W$ be the set of all pairs $(v,i)$, where $v \in V_\alpha$, $i \in N^v$. In other words, we associate each index $v$ with the set of indices $N^v$. Let $W^A \subset W$ be a subset consisting of all pairs $(v,i)$ such that $v \in V_\alpha^A$. A covering $\gamma \in \mathrm{Cov}^f(X \times I, A \times I)$ with indexing pair $(W, W^A)^\alpha$, defined by the formula $\gamma_{v,i} = \alpha_v \times \beta_i^v$, where $v \in V_\alpha$, $0 \leqslant i \leqslant n^v$, is called a *bricked* covering with base $\alpha$, and the sets $\alpha_v \times \beta_i^v$ are called the *bricks* of the covering $\gamma$. It follows from Lemma IX.5.6 in [126] that bricked coverings form a cofinal subset in $\mathrm{Cov}^f(X \times I, A \times I)$. Let

$\alpha \in \mathrm{Cov}^f(X,A)$, $\gamma = \gamma(\alpha)$, and let $\gamma$ be a bricked covering such that the index set $N^v$ does not depend on $v \in V_\alpha$, which means that $\gamma_{v,i} = \alpha_v \times \beta_i$, where $\beta_0, \beta_1, \ldots, \beta_n$ is a regular covering of $I$, and that $N^v \equiv N = \{0, 1, \ldots, n\}$. We call such coverings *direct* and prove that they form a cofinal subset in $\mathrm{Cov}^f(X \times I, A \times I)$. It suffices to prove the cofinality of direct coverings in the set of direct coverings. Let $\gamma \in \mathrm{Cov}^f(X \times I, A \times I)$ be a covering. It is required to find a direct covering that is a refinement of the former. $\gamma = \gamma(\alpha)$ defines sets $\bar{\gamma}_{v,i} = p(\gamma_{v,i}) = p(\alpha_v \times \beta_i^v)$ on the interval $I$, where $p : (X \times I) \to I$ is the projection onto the factor $I$. Intervals $\beta_i^v$ with endpoints $(x_i^v, y_i^v)$, $0 \leqslant i \leqslant n^v$, thereby arise on $I$. Consider a finite set of points $K = K(\gamma) = \bigcup_i \bigcup_v (x_i^v, y_i^v)$ and order them, starting with the point 0 in ascending order up to 1, i.e., $K = \bigcup_{q=0}^{M} z_q$, where $M = M(\gamma)$. We can assume that $z_\alpha \neq z_\beta$ for $\alpha \neq \beta$ and reject all superfluous points otherwise. Construct a covering $w$ of $I$ with sets $w_q = (z_q, z_{q+2})$, $0 \leqslant q \leqslant M - 2$, $w = \bigcup_q w_q$, generating a covering $\gamma' = \gamma'(\gamma)$ if we take $(\alpha_v \times w_q) \subset X \times I$ as its elements. Then $\alpha_v \times I = \alpha_v \times \left(\bigcup_q w_q\right) = \bigcup_i (\alpha_v \times \beta_i^v)$, i.e., for each fixed $v$, we obtain a covering of the set $\bigcup_i (\alpha_v \times \beta_i^v)$ with new ones $\{\alpha_v \times w_q\}$ which are smaller than $\alpha_v \times \beta_i^v$. It is clear that, for any $q$ and $v$, the set $\alpha_v \times w_q$ is contained in one of the form $\alpha_v \times \beta_i^v$ for a certain $i$; therefore, $\gamma'$ is a refinement of $\gamma$ and direct.

Q.E.D.

It follows that we can make use of only direct coverings in order to define the groups $h_q^{(q)}(X \times I, A \times I)$. At the same time, the nerve $(X \times I, A \times I)_\gamma$ of a direct covering $\gamma$ is homeomorphic to the direct product $(X_\alpha \times I, A_\alpha \times I)$, where $\alpha$ is the base of $\gamma$ and $(X_\alpha, A_\alpha)$ the nerve of the covering $\alpha$. Consider two simplicial mappings $\varphi_1, \varphi_2 : (X_\alpha, A_\alpha) \to (X \times I, A \times I)_\gamma$, where $\gamma = \gamma(\alpha)$, $\alpha$ is the base for the covering $\gamma$, $v \in \alpha$, $\varphi_1(v) = (v, 0)$, $\varphi_2(v) = (v, M-2)$ (recall that $\gamma$ is a direct covering). Since $\varphi_1$ and $\varphi_2$ are homotopic, $\varphi_{1*}^{(*)} = \varphi_{2*}^{(*)}$, where $\varphi_{i*} : h_*(X_\alpha, A_\alpha) \to h_*[(X \times I)_\gamma, (A \times I)_\gamma]$. Denote by $\Pi$ a subset in $\mathrm{Cov}^f(X \times I, A \times I)$, made up of direct stacked coverings, $\gamma \in \Pi$. Consider two coverings $\gamma_0 = g_0^{-1}(\gamma)$ and $\gamma_1 = g_1^{-1}(\gamma)$ of the pair $(X, A)$, and inclusion mappings $g_{i\gamma} : (X_{i\gamma}, A_{i\gamma}) \to [(X \times I)_\gamma, (A \times I)_\gamma]$. Since $\gamma_0 = \gamma_1$, we have $\alpha = \gamma_i$ and $\varphi_i = g_{i\gamma}$, i.e., $g_{0\gamma*}^{(*)} = g_{1\gamma*}^{(*)}$. Passing to the limit with respect to $\gamma \in \Pi$, we obtain $g_{0*}^{(*)} = g_{1*}^{(*)}$.

Q.E.D.

**Lemma 4.4.3.** *The theories $h$ constructed in §4.3 satisfy the excision axiom A6=A6′ on the category $U_C$.*

**Proof.** Consider an embedding $i : (X \backslash U, A \backslash U) \to (X, A)$, where $\overline{U} \subset V \subset A$. If $\alpha \in \mathrm{Cov}^f(X, A)$, then we put $\beta = i^{-1}(\alpha)$, and there arises an embedding of the nerve $(X'_\beta, A'_\beta)$ into the nerve $(X_\alpha, A_\alpha)$, where $(X'_\beta, A'_\beta)$ denotes the nerve of the covering $\beta$ of the pair $(X \backslash U, A \backslash U)$. We can assume that the embedding $\varphi_\alpha : (X'_\beta, A'_\beta) \to (X_\alpha, A_\alpha)$ is a relative homeomorphism (see [126]). Therefore, since the original theories $h$ on $P^2$ satisfy not only A6, but also $A6^0$, the homomorphisms $\varphi_{\alpha *} : h_q(X'_\beta, A'_\beta) \to h_q(X_\alpha, A_\alpha)$ (resp. $\varphi^*_\alpha$) are isomorphisms, which entails the statement of the lemma.

In the following, we shall prove that the $h$ not only satisfy A6, but are even relatively invariant on $U_C$, with the excision axiom in the form $\widetilde{A6}$ being required for applications.

(Co)boundary operators $\partial(\delta)$ are defined as the limits of the corresponding (co)boundary operators from the category over the index set $\mathrm{Cov}^f(X)$.

**Lemma 4.4.4.** *The theories h constructed in §4.3 satisfy the exactness axiom A4 on the category $U_C$.*

The proof is similar to that of exactness of the theories $\check{\mathrm{H}}$ in the classical case (if A7 is fulfilled); therefore, it is performed by the scheme of [126]. We will not dwell on it.

Further, it is easily verified that the theory $h$ constructed by us on the category $P^2$ embedded into $U_C$ coincides with the original $h$.

## 4.5. Additional Properties of Extraordinary Spectral Theories

Consider an inverse spectrum $\{(X_\tau, A_\tau), \pi^{\tau_1}_{\tau_2}\}$, where $(X_\tau, A_\tau) \in U_C$ and $\pi^{\tau_1}_{\tau_2}$ are continuous mappings. Let $(X, A) = \varprojlim\{(X_\tau, A_\tau), \pi^{\tau_1}_{\tau_2}\}$ be the inverse limit of the pairs $(X_\tau, A_\tau)$, and $\omega_\tau : (X, A) \to (X_\tau, A_\tau)$ are the projections of the inverse spectrum $(X, A)$. Then a mapping $\Omega_* : h_*(X, A) \to \varprojlim h_*(X_\tau, A_\tau)$ (resp. $\Omega^* : \varinjlim h^*(X_\tau, A_\tau) \to h^*(X, A)$) arises. Recall that a (co)homology theory $h_*$ on $U_C$ is said to be *continuous* if the functor $h$ commutes with the operation of the passage to the inverse limit.

**Lemma 4.5.1.** *Let h be a spectral (co)homology theory on $U_C$, constructed by means of the theory h on $P^2$. Then the homomorphism $\Omega^{(*)}_*$ is an isomorphism, i.e., h is continuous on $U_C$ in all the cases (a), (b), and (c).*

The proof is similar to that for the theories $H$ satisfying A7 (see Theorem X.3.1 in [126]). We omit it here.

**Lemma 4.5.2.** *The extraordinary spectral and continuous theories h on $U_C$ satisfy $\widetilde{A6}$, i.e., are relatively invariant on $U_C$. Moreover, if $(X, A) \in U_C$ and $p : (X, A) \to (X/A, x)$ is the natural projection, then $p^{(*)}_*$ is an isomorphism.*

**Proof.** It follows from Lemma 4.4.3 and Theorem X.5.4 in [126] that the $h$ are relatively invariant on $U_C$. Let $(X, A) \in U_C$, and $f : X \backslash A \to Y = X \backslash A$ the identity mapping. It then follows from the relative invariance of $h$ and $(Y \cup x, x) = (X/A, x)$ that $p^{(*)}_*$ is an isomorphism, completing the proof.

Thus we have proved:

**Theorem 4.5.1.** *Let a theory $h$ satisfying A1–A5, A6$^0$ be fixed on $P^2$. Then by means of the spectral process, $h$ generates a covariant (contravariant) functor $h$ such that (1) if it satisfies on $U_C$ A1–A3, A5, A6, and (2) if it satisfies (a), or (b), or (c), then it is transformed into a (co)homology theory for which A4 and $\widetilde{A6}$ are fulfilled, which is continuous on $U_C$, and coincides with the original theory $h$ on $P^2 \subset U_C$. In the case (a), we have $h_* \in GF$; in the case (b), $h_* \in ABC$; in the case (c), $h^* \in GR$, where $(X,A) \in U_C$.*

## 4.6. Reduced (Co)Homology Groups on "Surfaces with Singularities"

Let $h$ be a continuous (co)homology theory on $U_C$. We can construct the groups $\tilde{h}_*(X)$ and $\tilde{h}^*(X)$, where $x \in X$ is a fixed point, $i : x \to X$ is an embedding, and $f : X \to x$ is the projection. In the homology case, we then consider the subgroup $G_q(X) = i_* h_q(x)$, whereas, in the cohomology case, $G^q(X) = f^* h^q(x)$. But first we take up the cohomology case. It is clear that if $g : X \to Y$ is continuous and $y = g(x)$, then $g^* \tilde{h}^q(Y) \subset \tilde{h}^q(X)$ and $\mathrm{Ker}(g^*) \subset \tilde{h}^q(Y)$. The subgroup $G^q(Y)$ is isomorphically mapped onto $G^q(X)$ by the homomorphism $g^*$. Since $f_i$ is the identity, $h^q(X)$ decomposes into the direct sum $h^q(X) = \tilde{h}^q(X) \oplus G^q(X)$, with $f$ inducing the isomorphism between $G^q(X)$ and $h^q(x)$.

Consider the homology case. If $g : X \to Y$, $g(x) = y$, then $g_* \tilde{h}_q(X) \subset \tilde{h}_q(Y)$, and $g_*$ isomorphically maps the subgroup $G_q(X)$ onto $G_q(Y)$. It is clear that $h_q(X)$ decomposes into the direct sum $h_q(X) = \tilde{h}_q(X) \oplus G_q(X)$, with $i$ inducing the isomorphism between $h_q(X)$ and $G_q(X)$.

Axiom A7 (point Axiom) requires that the groups $h_q^{(q)}(x)$ should vanish for $q \neq 0$. Then the group $h_0^{(0)}(x) = G$ is the coefficient group of the theory $h$. If $h$ satisfies A7, then it means that the index $q$ acquires a distinctly geometrical meaning: It is the geometric dimension of the simplexes (or cells) forming $q$-dimensional (co)chains.

# §5 The Coboundary and Boundary of a Pair of Spaces (*X*,*A*)

We introduce the concept of the (co)boundary of a pair of spaces $(X, A)$ for the general case of an arbitrary extraordinary (co)homology theory on the category of compact pairs $U_C$ (see the construction in §4), which will permit us in the following to embrace an extremely wide class of new variational problems, including the classical Plateau problem reformulated above in the language of bordism theory. We start with the cohomology case, with the $K$-functor being an important example of an extraordinary cohomology theory, since the stock of cohomology functors in whose terms variational problems are completely solvable turns out to be quite large.

### 5.1. The Coboundary of a Pair $(X,A)$

Let $h^*$ be a continuous, relatively invariant cohomology theory on $U_C$; in particular, we can assume that $h^* = H^*$, i.e., it is the usual cohomology theory satisfying A7. We can take the $K$-functor as another example.

**Definition 5.1.1.** Let $(X,A) \in U_C$, and $x \in A$ a fixed point. We call the set of all elements $\alpha \in \tilde{h}^{k-1}(A)$ such that $\alpha \notin \operatorname{Im}(\tilde{i}^*)$, where $i : A \to X$ is an embedding, and the homomorphism $\tilde{i}^* : \tilde{h}^{k-1}(X) \to \tilde{h}^{k-1}(A)$ is induced by the embedding, the *coboundary* $\nabla^k(X,A)$ of the pair $(X,A)$ in the dimension $k$ with respect to $x$. In general, $k$ is not related to the topological dimension of $X$. We then put $\nabla^*(X,A) = \bigcup_{k\in Z} \nabla^k(X,A)$.

If we consider the reduced sequence of the pair $h^k(X,A) \xleftarrow{\tilde{\delta}} \tilde{h}^{k-1}(A) \xleftarrow{\tilde{i}^*} \tilde{h}^{k-1}(X)$, then it becomes clear that $\alpha \in \nabla^k(X,A)$ if and only if $\tilde{\delta}(\alpha) \neq 0$, i.e., $\nabla^k(X,A) = \tilde{h}^{k-1}(X,A)\backslash \operatorname{Im}(\tilde{i}^*) = \tilde{h}^{k-1}(A)\backslash \operatorname{Ker}(\tilde{\delta})$; in particular, $\nabla^k(X,A)$ is not a subgroup of the group $\tilde{h}^{k-1}(A)$, and this is the essential difference between the cohomology variant of the theory under construction and its homology analogue, since the boundary $\Delta_k(X,A)$ to be defined below will be a subgroup of the group $\tilde{h}_{k-1}(A)$.

The above notion of coboundary is prompted by the intuitive idea of the geometric boundary of a film $X$ when it makes sense to speak of this boundary. For example, if $X = CA$ (cone over $A$), then, obviously, $\nabla^*(CA,A) = \bigcup_k [\tilde{h}^{k-1}(A)\backslash 0]$, i.e., $CA$ completely spans $A$. Another example: Let $X$ be a $k$-dimensional manifold with a closed $(k-1)$-dimensional manifold as its boundary $A$, and $h^* = H^*$, the usual cohomology theory. Then $\nabla^k(X,A) = H^{k-1}(A)\backslash 0$, which again corresponds to intuition. Meanwhile, extraordinary theories provide us with ample opportunities to choose spanning films $X$ not necessarily of, e.g., dimension $k$ for the coboundary $\nabla^k(X,A)$ to be nonempty.

### 5.2. The Boundary of a Pair $(X,A)$

Let an extraordinary homology theory $h_*$ be given on $U_C$, satisfy A1–A6, and be continuous on $U_C$. For example, we can take bordism theory.

**Definition 5.2.1.** Let $(X,A) \in U_C$, and $x \in A$ a fixed point. We call a subgroup $\operatorname{Ker}(\tilde{i}_*) \cap \tilde{h}_{k-1}(A)$, where $i : A \to X$ is an embedding and the homomorphism $\tilde{i}_* : \tilde{h}_{k-1}(A) \to \tilde{h}_{k-1}(X)$ induced by it, the *boundary* $\Delta_k(X,A)$ of the pair $(X,A)$ in dimension $k$ with respect to $x$. We then put $\Delta_*(X,A) = \bigcup_{k\in Z} \Delta_k(X,A)$.

From the exact reduced sequence of the pair $h_k(X,A) \xrightarrow{\tilde{\partial}} \tilde{h}_{k-1}(A) \xrightarrow{\tilde{i}_*} \tilde{h}_{k-1}(X)$, it follows that $\Delta_k(X,A) = \operatorname{Im}(\tilde{\partial})$. In contrast to the cohomology case, the boundary $\Delta_k(X,A)$ is a subgroup of $\tilde{h}_{k-1}(A)$. If $h_* = H_*$ is the usual homology theory with compact coefficients, then Definition 5.2.1 turns into the one formulated and investigated for the variational problems by Adams and Reifenberg, who did not consider the cohomology boundary concept [351]. As well as in the cohomology case, the notion of boundary is linked to the intuitive idea of geometric boundary. If $X = CA$ (cone over $A$), then $\Delta_k(CA, A) = \tilde{h}_{k-1}(A)$.

Meanwhile, the boundary thus formulated (e.g., in the case $h_* = H^*$) does not permit us to translate certain geometric cases into the language of algebra, viz., if $A = S^1$ (circle), and $X$ is the connected sum of the double and triple Möbius bands (Fig. 27), then, from the standpoint of visual geometry, $A$ is the boundary of $X$, though the boundary $\Delta_2(X, A)$ is zero: As a matter of fact, $A$ is a retract of $X$, in which case the film $X$ contains a one-dimensional subset of singular points, which is homeomorphic to $S^1$. $A$ is a retract of $X$, but not a deformation retract.

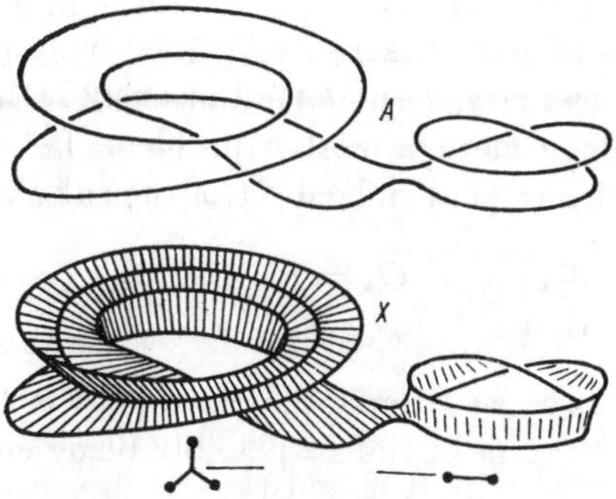


Cross section

**Fig. 27**

This example by J. Adams demonstrates that the concept of algebraic (co)boundary is not comprehensive. The interesting geometric problem is the visual construction of a retraction of a surface $X$ onto its boundary $A$ (see Fig. 27), carried out by T. N. Fomenko in [171] and [455] on the basis of a new proof of the classical Hopf retraction theorem (see its description also in [170]).

There also exist examples of soap films $X$, or two-dimensional minimal surfaces, with boundary $A = S^1$, for which $S^1$ is a deformation retract of $X$; therefore, $\Delta_2(X, A) = 0$ (see Poston [343]).

## §6 Determination of Classes of Admissible Variations of Surfaces in Terms of (Co)Boundary of Pair $(X, A)$

### 6.1. The Variational Classes $h(A, L, L')$ and $h(A, \tilde{L})$

Let $M^n$ be a Riemannian manifold without boundary (not necessarily compact), $A \subset M$ a fixed compact set, $x \in A$ a fixed point, $h$ a continuous, relatively invariant (co)homology theory on $U_C$, $(X,A) \subset M$ a compact pair, and $i : A \to X$, $j : X \to M$ embeddings. We first consider the homology case. Suppose that $L = \{L_p\}$ is a fixed set of subgroups $L_p \subset \tilde{h}_p(A)$, where $p \in Z$ and $L' = \{L'_q\}$ is a fixed set of subgroups $L'_q \subset \tilde{h}_q(M)$.

**Definition 6.1.1.** We denote by $\mathcal{O}_* = h_*(A, L, L')$ the class of all compact sets $X$, $A \subset X \subset M$, such that (1) $L \subset \mathrm{Ker}(\tilde{i}_*) = \Delta_*(X, A)$, (2) $L' \subset \mathrm{Im}(\tilde{j}_*)$.

Consider the cohomology case. Let $L = \{L_p\}$ be a fixed set of subsets $L_p \subset \tilde{h}^p_x(A) \setminus 0$, and $L' = \{L'_q\}$ a fixed set of subsets $L'_q \subset \tilde{h}^q_x(M) \setminus 0$.

**Definition 6.1.2.** We denote by $\mathcal{O}^* = h^*(A, L, L')$ the class of all compact sets $X$, $A \subset X \subset M$, such that (1) $L \subset \tilde{h}^*(A) \setminus \mathrm{Im}(\tilde{i})^* = \nabla^*(X, A)$, and (2) $L' \subset \tilde{h}^*(M) \setminus \mathrm{Ker}(\tilde{j}^*)$.

The class $\mathcal{O}$ consists of all compact sets $X \subset M$ spanning the (co)homological "holes" $L$ in the boundary $A$ and simultaneously realizing those $L'$ in the manifold $M$. We now consider the relative problem. Let $\alpha : (X, A) \to (M, A)$ be an embedding, and $\tilde{L} = \{\tilde{L}_r\}$ a fixed set of the subgroups $\tilde{L}_r \subset h_r(M, A)$.

**Definition 6.1.3.** We denote by $\tilde{\mathcal{O}}_* = h_*(A, \tilde{L})$ the class of all compact sets $X$, $A \subset X \subset M$, such that $\tilde{L} \subset \mathrm{Im}(\tilde{\alpha}_*)$.

In the cohomology case, we assume that $\tilde{L} = \{\tilde{L}_r\}$ is a fixed set of subsets $L_r \subset \tilde{h}^r(M, A) \setminus 0$, with some of the $L_r$ possibly being empty.

**Definition 6.1.4.** We denote by $\tilde{\mathcal{O}}^* = h^*(A, \tilde{L})$ the class of all compact sets $X$, $A \subset X \subset M$, such that $\tilde{L} \subset \tilde{h}^*(M, A) \setminus \mathrm{Ker}(\tilde{\alpha}^*)$.

Generally, $\mathcal{O}$ and $\tilde{\mathcal{O}}$ are weakly related to each other, though, if $A = x$, we can assume that $L = 0(\emptyset)$, $L' = \tilde{L}$, and then $h(x, 0(\emptyset), L') = h(x, \tilde{L})$. This particular case applies to realizing classes, i.e., such that each of their elements realizes an absolute (co)cycle in the manifold. However, if $A \neq x$, then even if $h = H$, the knowledge of all compact sets in $\mathcal{O}$ does not permit us to describe those from $\tilde{\mathcal{O}}$, and vice versa.

Let $h = H$, i.e., let Axiom A7 be fulfilled, and let a set $L$ be selected in $\mathcal{O}$ so that $L_p \neq 0(\emptyset)$ only if $p = k - 1$, and a set $L'$ so that $L'_q \neq 0(\emptyset)$ only if $q = k$. Then the value $k$ acquires a geometric meaning, viz., we can single out a nonempty subclass of the compact sets in $\mathcal{O}$ such that the dimension of each is $k$.

We now concentrate our attention on the classes $\mathcal{O}$, since the study of $\tilde{\mathcal{O}}$ on $M$ turns out to be reduced to $\mathcal{O}$ on $M/A$, which is provided for by the invariance of theory $h$, i.e., if $\alpha : (X, A) \to (M, A)$ is an embedding, then the homomorphism $\alpha_* : h_*(X, A) \to h_*(M, A)$ coincides with $\alpha'_* : \tilde{h}_*(X/A) \to \tilde{h}_*(M/A)$ (resp. homomorphism $(\alpha')^*$ in the cohomology case). In other words, we can consider $\mathcal{O}$ on the space $M/A$, a manifold everywhere except on point $x$, and not $\tilde{\mathcal{O}}$ on the manifold $M$.

Consider the two limit cases of the classes $h(x, 0(\emptyset), L')$ and $h(A, L, 0(\emptyset))$. To simplify the notation, we will also write 0 instead of $\emptyset$ in the cohomology case and consider the classes $h(x, 0, L')$ and $h(A, L, 0)$. In the first case, $\mathcal{O}$ consists only of compact sets $X$ realizing the "holes" in $M$ without any additional boundary condition, since $A = x$; we call such compact sets *realizing* compact sets (Plateau Problem B; see §3). In the second case, $\mathcal{O}$ consists of compact sets $X$ spanning the "holes" in the compact set $A$, and $M$ plays the role of the underlying space; we call such compact sets *spanning* compact sets (Plateau Problem A).

## 6.2. The Stability of Variational Classes

The concept of $p$-stability is introduced due to the fact repeatedly stressed above, viz., that the dimension $k$ of elements $\alpha$ of the group $h_k^{(k)}(X)$ is weakly related to the geometric one of the supports of $\alpha$ in the case of an extraordinary theory $h$. The geometry of $\alpha$ is "spread" over all dimensions $s$ such that $s \leqslant |k|$. Since our analytic constructions will require a certain freedom in treating cells of small dimensions (more often than not, 1 and 2) in compact sets $X$, it is desirable that the operation of rejecting cells of small dimensions should not influence the realizing and spanning properties of $X$, which leads us to the concept of $p$-stability.

**Definition 6.2.1.** We call a finite simplicial complex $Z$, embedded into a smooth manifold $M^n$, *smooth* if (1) each open complex $\Delta^i \subset Z$ of dimension $i$ is a smooth submanifold in $M$, (2) if there exists a positive number $\varepsilon = \varepsilon(Z)$ such that, for any point $P \in Z$ in the ball $B^n(P, \varepsilon)$ with centre in $P$ and radius $\varepsilon$, we can introduce the local coordinates $x^1, \dots, x^n$ such that the intersection $Z \cap B(P, \varepsilon)$ is a linear simplicial complex in the ball $B(P, \varepsilon)$ in these coordinates.

**Definition 6.2.2.** Let a compact set $A$ be fixed in $M^n$, so that the class $h(A, L, L')$ (resp. $h(A, \tilde{L})$) is nonempty. The class $h(A, L, L')$ (resp. $h(A, \tilde{L})$) and the set $(L, L')$ (resp. $\tilde{L}$) are said to be *p-stable*, where $p$ is an integer, $1 \leqslant p \leqslant n-2$, if it follows from a compact set $X$ belonging to $h(A, L, L')$ (resp. $h(A, \tilde{L})$) that any subcompact set $Y \subset X$, with $X \setminus Y$ being a smooth simplicial subcomplex in $M$ of dimension not greater than $p$, also belongs to $h(A, L, L')$ (resp. $h(A, \tilde{L})$).

Consider by way of example the usual homology theory $H$ on $U_C$. Let the set $(L, L')$ possess the property that $L_{p-1} \neq 0$ only if $p = k$, and $L'_q \neq 0$ only when $q = k$. If $X \in H(A, L, L') = h(A, L_{k-1}, L'_k)$ and $Y \subset X$, $\dim \overline{(X \setminus Y)} \leqslant k-1$, then, obviously, $Y \in H(A, L_{k-1}, L'_k)$, i.e., any class $\mathcal{O}$ or $\tilde{\mathcal{O}}$ of the above form is $(k-1)$-stable in the sense of our definition. It is just because of this fact that we have considered until now variational problems in the classes of the usual homologies $H$ always in one geometric dimension coinciding with that of supports $X$ (see [135], [289]).

In the following, 2-stable classes will be most important; therefore, the following problem is of interest, viz., what are the topological conditions to be imposed on the arguments $\{A, L, L'\}$, so that the corresponding classes $h(A, L, L')$ may be 2-stable? For simplicity, we put $A = x$ (then $L = 0$) and consider the stability of the realizing classes $h(x, 0, L')$. It turns out that *any* class $h(x, 0, L')$ is *2-stable* on the *2-connected* manifold $M$, i.e., $\pi_1(M) = \pi_2(M) = 0$. In particular, for $M^n = \mathbf{R}^n$, all the variational classes are 2-stable. We will prove this theorem later.

## §7 Solution of the Plateau Problem (Finding Globally Minimal Surfaces (Absolute Minimum) in the Variational Classes $h(A,L,L')$ and $h(A,\tilde{L})$)

### 7.1. The Formulation of the Problem

The statement of variational problems, given below, and their solution, i.e., the proof that globally minimal surfaces exist, is due to the author. Consider a compact, smooth, and closed Riemannian manifold $M$. Let $h$ be a certain extraordinary (co)homology theory on $U_C$, continuous, and relatively invariant on the category of compact pairs, $A$ a fixed compact set in $M$. Then the variational classes $h(A, L, L')$, $h(A, \tilde{L})$ are defined (see §6), in each of which the problem of finding a globally minimal surface arises. For each $X \in \mathcal{O}$ or $\tilde{\mathcal{O}}$ we construct its stratification $X = A \cup S^k \cup S^{k-1} \cup \ldots$, where $S^k$ is the maximal subset in $X \setminus A$, of dimension $k$ in each of its points, $S^{k-1}$ that in $X \setminus A \setminus S^k$, of dimension $k-1$ in each of its points, etc. We call the $S^i$ *strata*; if they are measurable, then the stratified volume $SV(X) = (\mathrm{vol}_k S^k, \mathrm{vol}_{k-1} S^{k-1}, \ldots)$ represented as a vector with $k$ coordinates is defined. By varying the "surface" $X$ in the class of admissible variations $\mathcal{O}$ or $\tilde{\mathcal{O}}$, we perform the same operation with the surface-stratified volume vector, and the problem is then to find a surface with the least stratified volume. We understand the least vector $SV = (d_k, d_{k-1}, \ldots)$ in the following sense. We first try to minimize the first coordinate of $SV$, i.e., we seek a surface (compact set) $X_k$ in $\mathcal{O}$ for which $\mathrm{vol}_k(S^k) = \mathrm{vol}_k(X \setminus A) = d_k = \inf_{Y \in \mathcal{O}} \mathrm{vol}_k(Y \setminus A)$. If such surfaces $X_k$ do exist, then we start minimizing the second coordinate of the stratified volume vector, viz., we look for such a surface $X_{k-1}$ in the class of $X_k$ with the minimal first coordinate $(\mathrm{vol}_k(X_k \setminus A) = d_k)$, for which

$$\mathrm{vol}_{k-1}(X_{k-1} \setminus A \setminus S^k) = d_{k-1} = \inf_{\mathrm{vol}_k(X \setminus A = d_k)} \mathrm{vol}_{k-1}(X_k \setminus A \setminus S^k),$$

and so on. We minimize the next coordinate of the stratified volume vector each time on condition that all the previous ones should be already minimized and fixed, and the minimization is carried out in the class of films with already minimal previous volumes. If the process is determined correctly (and it is just what we are going to prove), then it stops on a surface whose stratified volume is globally minimal in the class of all stratified surfaces from the given variational class $\mathcal{O}(\widetilde{\mathcal{O}})$.

Our method for minimizing the volume $SV$ applies not only to spectral bordism variational classes, but also to any of the form $\mathcal{O}(\widetilde{\mathcal{O}})$. With this approach, each extraordinary (co)homology theory $h$ and each triple $(A, L, L')$ (or pair$(A, \tilde{L})$), determines a subset $\mathcal{O}$ (or $\widetilde{\mathcal{O}}$) in the space $\theta(M)$ of all compact sets in the manifold $M$; therefore, each $h$ determines its own kind of "boundary conditions" whose stock is extremely large due to a great variety of the theories $h$. We give an example of the "contravariant" variational problem determined by the $K$-functor. Let a stable nontrivial vector bundle $\xi$ be given on $M$. Consider the class of all compact sets $X \subset M$ such that the restriction of $\xi$ on $X$ is still stable nontrivial, i.e., $X$ is the support of $\xi$ (Fig. 28). The question arises: Is it possible to find a globally minimal surface $X_0$ (in the sense of stratified volume) among all such surfaces $X$? In what follows, we will show that the answer is positive.

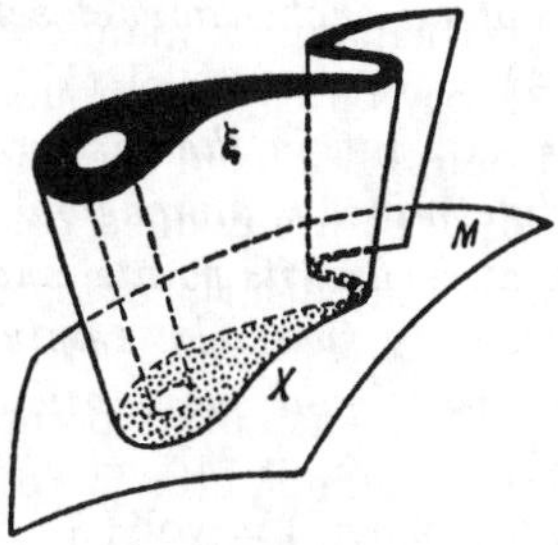


**Fig. 28**

**Theorem 7.1.1.** *Let $A \subset M^n$ be an arbitrary compact set in a smooth Riemannian manifold $M$, $\tilde{h}$ any extraordinary (co)homology theory satisfying the above conditions (in particular, we can take spectral bordisms, stable homotopy groups, $K$-functor, etc.), and $\mathcal{O}$ (or $\widetilde{\mathcal{O}}$) the 2-stable class of surfaces $X$ in $M$, for which $A$ is the boundary in the sense of $\tilde{h}$.*

*Similarly, we can consider the class of surfaces $X$ realizing some nonzero subgroup in $\tilde{h}(M)$, where $M$ is compact and closed. Then there always exists an absolutely (globally) minimal stratified surface in each of them, solving the Plateau problem, which means that its greatest volume and stratified volumes are the least possible. Each of its strata is almost everywhere a minimal submanifold in $M$.*

We also give a more detailed statement of the theorem.

### 7.2. The Basic Existence Theorem for Globally Minimal Surfaces. Solution of the Plateau Problem

Assume that the volume minimization problem makes sense, i.e., the infimum of $s$-dimensional volumes of stratified surfaces is finite for any $s$ (in the class $\mathcal{O}$ or $\widetilde{\mathcal{O}}$). For the precise formulation, see Chapter 6.

**Theorem 7.2.1.** *Let $M^n$ be a compact, closed Riemannian manifold of the class $C^r$, $r \geqslant 4$, $A \subset M^n$ a fixed compact set, and $x \in A$ a fixed point. Let $\tilde{h}$ be a reduced, continuous, and relatively invariant extraordinary (co)homology theory on the category of compact pairs $U_C$. Consider an arbitrary nonempty and 2-stable class $\mathcal{O} = h(A, L, L')$ or $\widetilde{\mathcal{O}} = h(A, \tilde{L})$ (e.g., for a spectral bordism theory and 2-connected manifolds, i.e., $\pi_1(M) = \pi_2(M) = 0$, any nonempty class $\mathcal{O}$ or $\widetilde{\mathcal{O}}$ is 2-stable). Let $k$ be the least of integers $s$, $s < n$, for which $d_s = d_s(h(A, L, L')) < \infty$ (resp. $\tilde{d}_s = \tilde{d}_s(h(A, \tilde{L})) < \infty$). Assume that $3 \leqslant k \leqslant n - 1$ and that the minimization problem makes sense. Then the following statements are valid.*

*(1) If $\{X\}_k$ is the class of all such compact sets $X$, $A \subset X \subset M$, that $X \in \mathcal{O}(\widetilde{\mathcal{O}})$ and $\mathrm{vol}_k(X \setminus A) = d_k = \inf \mathrm{vol}_k(Y \setminus A)$, $Y \in \mathcal{O}(\widetilde{\mathcal{O}})$, then we assert that $\{X\}_k \neq \emptyset$, $d_k < \infty$, and in the case where $d_k > 0$, each $X$ from $\{X\}_k$ contains a uniquely determined $k$-dimensional subset $S^k \subset X \setminus A$, $S^k = S^k(X)$, i.e., of dimension $k$ at each of its points, such that $A \cup S^k$ is a compact set in $M$; $S^k$ contains a subset $Z_k$ (possibly, empty), where $\mathrm{vol}_k(Z_k) = 0$ and $S^k \setminus Z_k$ is a topological $k$-dimensional submanifold in $M$ without boundary and everywhere dense in $S^k$, i.e., $Z_k$ is the set of all $k$-dimensional singular points of the surface $X$, with $\mathrm{vol}_k(S^k) = \mathrm{vol}_k(X \setminus A) = d_k > 0$. However, if $d_k = 0$, then we put $S^k = 0$.*

*(2) Further, if $\{X\}_{k-1} \subset \{X\}_k$ is the class of the compact sets $X$, $A \subset X \subset M$, such that $X \in \mathcal{O}(\widetilde{\mathcal{O}})$, $X \in \{X\}_k$ and $\mathrm{vol}_k(X \setminus A \setminus S^k) = d_{k-1} = \inf \mathrm{vol}_{k-1}(Y \setminus A \setminus S^k)$, $Y \in \{X\}_k$, then we assert that $\{X\}_{k-1} \neq \emptyset$, $d_{k-1} < \infty$ and, in the case where $d_{k-1} > 0$, each $X \in \{X\}_{k-1}$ contains a uniquely determined $(k-1)$-dimensional subset $S^{k-1} \subset X \setminus A \setminus S^k$, $S^{k-1} = S^{k-1}(X)$ of dimension $k-1$ at each of its points, so that $A \cup S^k \cup S^{k-1}$ is a compact set $M$. The set $S^{k-1}$ contains a subset $Z_{k-1}$ (possibly, empty), where $\mathrm{vol}_{k-1}(Z_{k-1}) = 0$ and $S^{k-1} \setminus Z_{k-1}$ is a topological $(k-1)$-dimensional submanifold in $M$ without boundary and everywhere dense in $S^{k-1}$, i.e., the set $Z_{k-1}$ is that of all $(k-1)$-dimensional singular points of $X$, with $\mathrm{vol}_{k-1}(S^{k-1}) = \mathrm{vol}_{k-1}(X \setminus A \setminus S^k) = d_{k-1} > 0$. However, if $d_{k-1} = 0$, then we put $S^{k-1} = \emptyset$.*

*(3), etc.*

...

*$(k-2)$. Finally, if $\{X\}_3 \subset \{X\}_4$ is the class of all compact sets $X$, $A \subset X \subset M$, such that $X \in \mathcal{O}$ $(\widetilde{\mathcal{O}})$, $X \in \{X\}_4$, and $\mathrm{vol}_3\left(X \setminus A \setminus \bigcup_{i=4}^k S^i\right) = d_3 = \inf \mathrm{vol}_3\left(Y \setminus A \setminus \bigcup_{i=4}^k S^i\right)$, $y \in \{X\}_4$, then we assert that $\{X\}_3 \neq \emptyset$,*

*$d_3 < \infty$, and in the case where $d_3 > 0$, each $X \in \{X\}_3$ contains a uniquely determined three-dimensional subset $S^3 \subset X \setminus A \setminus \bigcup_{i=4}^k S^i$, $S^3 = S^3(X)$ of dimension 3 at each of its points, so that $A \cup \left(\bigcup_{i=3}^k S^i\right)$ is a compact set in $M$, with $S^i \cap S^j = \emptyset$ if $i \neq j$; $S^3$ contains a subset $Z_3$, where $\mathrm{vol}_3(Z_3) = 0$ and $S^3 \setminus Z_3$ is a topological three-dimensional submanifold in $M$ without boundary and everywhere dense in $S^3$, i.e., $Z_3$ is the set of all three-dimensional singular points of the surface $X$, with $\mathrm{vol}_3(S^3) = \mathrm{vol}_3\left(X \setminus A \setminus \bigcup_{i=4}^k S^i\right) = d_3 > 0$. However, if $d_3 = 0$, then we put $S^3 = \emptyset$. Due to the 2-stability of the class $O(\widetilde{O})$, the relation $\{X\}_3 = \{X\}_2 = \{X\}_1$ holds. Further, each subset $K^i = S^i \setminus Z^i$, $K^i \subset X$, $X \in \{X\}_3$, is actually a submanifold of the class $C^{r-2}$ in the manifold $M$; however, if $M$ is a smooth (or analytic) manifold, then the submanifolds $K^i$ for $3 \leqslant i \leqslant k$ are also smooth (resp. analytic) in $M$, each being $i$-dimensional and minimal, i.e., the mean curvature is equal to zero. If the class $O(\widetilde{O})$ is $q$-stable, where $q \geqslant 3$, then all the sets $S^i$ are empty for $3 \leqslant i \leqslant q$. If $X = A \cup \left(\bigcup_{i=3}^k S^i\right)$ and $S^{i_0} \neq \emptyset$ for a certain $i_0$, then all the compact sets $\widetilde{X}_\alpha = A \cup \left(\bigcup_{i=\alpha}^k S^i\right)$, where $i_0 + 1 \leqslant \alpha \leqslant k$, do not belong to $O(\widetilde{O})$, i.e., no nonempty sets $S^{i_0}$ can be removed from $X \in \{X\}_3$ without violating the latter's topological properties. For each dimension $i$, the standard spherical density function $\Psi_i(x, S^i)$ defined on $S^i$ possesses the property $\Psi_i(x, S^i) \geqslant 1$ for any $x \in S^i$, and $\Psi_i(x, S^i) = 1$ if and only if $x \in K^i = S^i \setminus Z^i$, i.e., $x$ is a regular point of the surface $S^i$.*

**Remark.** Recall that if $d_s(h, A, L, L') = \infty$ for any $s < n$, then the minimum problem in $O$ becomes senseless due to the triviality of the answer, i.e., the condition $s < n$ is not a restriction (see the statement of the theorem).

As pointed out in §3, we can solve the minimization problem for only one volume $\mathrm{vol}_k$ greatest in dimension, without being interested (from the metric standpoint) in pieces of smaller dimensions. This can be also solved by Theorem 7.2.1, for which it suffices to restrict ourselves to only its first item.

**Corollary 7.2.1.** *With the assumptions of Theorem 7.2.1, there always exists a globally minimal surface $X_0$ in each class $O(\widetilde{O})$ whose stratified volume $SV = (d_k, d_{k-1}, \dots)$ is the least in all dimensions and which possesses uniquely determined stratification $X_0 = A \cup S^k \cup S^{k-1} \cup \dots$, where each subset $S^i$ (possibly, except for a set of measure zero, consisting of singular points) is a smooth minimal submanifold of a smooth manifold $M$, i.e., the mean curvature is identically zero. Meanwhile, $d_i = \mathrm{vol}_i(S^i)$.*

Consider the most interesting special cases:

1. If we take spectral bordism theory as extraordinary homology theory, then we obtain a solution to the Plateau problem in the class of "spectral" surfaces spanning a given "contour" $A$ in $M$ (in particular, see Theorem 3.7.1). As the principal bordism theories, we can take $\Omega_*$ (oriented bordisms), $N_*$

(nonoriented bordisms), and $\Omega_*^p$ (bordisms mod $p$). These groups determine extraordinary homology theories on the category $P^2$ (see above). Since, in general, minimal surfaces do possess singularities that are sometimes rather complicated, we should extend these theories from $P^2$ to the category of compact pairs $U_C$, which we do in accordance with the scheme of §4. Because $\Omega_*^p$ and $N_*$ assume values in the category of compact topological groups $ABC$, their extension to $U_C$ does not encounter obstructions (see §4). We have to be more accurate with respect to $\Omega_*$. Since $\Omega_*$ on $P^2$ does not satisfy the conditions (a) and (b) (see §4.3), we have to consider the groups $\Omega_* \otimes_{\mathbf{Z}} Q_p = {}^p\Omega_*$, where $Q_p$ is the group of $p$-adic integers ($Q_p$ being a flat module, $\Omega_*$ form an exact homology theory). Since any bordism can be determined by selecting a suitable $p$, we do not restrict ourselves from the geometric standpoint. The ${}^p\Omega_*$ form a homology theory with values in $ABC$; therefore, it can be extended from $P^2$ to $U_C$, the category of "surfaces with singularities." Thus, let $A^{k-1} \subset M^n$ be a closed manifold, and $h_*$ one of ${}^p\Omega_*$, $N_*$, $\Omega_*^p$. The manifold $A$ determines an element $\sigma = [A, e] \in h_{k-1}(A)$, where $e : A \to A$ is the identity mapping. Suppose that $L$ is a subgroup in $\tilde{h}_{k-1}(A)$, generated by $\sigma$, $x \in A$.

**Corollary 7.2.2.** *Assume that the class $h_*(A, L, 0)$ is nonempty and 2-stable (e.g., the manifold $M$ is 2-connected). Then there exists a globally minimal surface $X_0$ annihilating $\sigma$ (see Theorem 7.2.1), which is (possibly, with singularities) a solution of the Plateau problem in the class of all films spanning $A$ and admitting continuous parametrization via spectra of manifolds, i.e., a solution of the spanning spectral Problem A. If we consider the second limit case, or the class $h_*(x, 0, L')$, then we obtain the existence of a globally minimal surface $X_0$ in the class of all those which realize the given bordism $\{\sigma'\} = L' \subset \tilde{h}_k(M)$, a solution of the realization spectral Problem B. Besides, in the case of Problem A, the inequality $d_k > 0$ always holds.*

In the "contour" spanning problem for a minimal film $X_0$ with complicated singularities, there generally exists a spectrum of manifolds $\{W_\alpha\}$ with boundary $\partial W_\alpha = A$, which spans $A$ in $X_0$ in the sense of §4.2 and §4.3. However, if $X_0$ is a cell complex (and the most typical singularities encountered in applications always correspond just to this case; see below), then all $W_\alpha$ are homeomorphic to the same $W_0$, which is just what parametrizes $X_0$, i.e., $X_0 = f_0(W_0)$, $\partial W_0 = A$.

2. If we take the usual theory $H_*$ (satisfying A7) as a homology theory, then Theorem 7.2.1 entails the results obtained in [289], [351], and [352].

If we take the $K$-functor as a cohomology theory, then we obtain an existence theorem for a minimal surface in the class of surfaces such that the restriction to them of a stable nontrivial bundle $\xi$ on $M$ is still nontrivial.

If we consider stable homotopy groups $\pi_*^S$ as an extraordinary homology theory, then we obtain from Theorem 7.2.1 an existence theorem for a globally minimal surface in each spectral homotopy class $\sigma \in \pi_*^S(M)$ of stable homotopy spectral mappings of the sphere into $M$. In particular, there exists a global minimum in the stable homotopy (relative) spectral class of discs spanning a fixed sphere in $M$.

### 7.3. A Rough Outline of the Existence Theorem

Here, we briefly describe the principal steps in the proof of the existence of a minimal surface in the classes $\mathcal{O}(\widetilde{\mathcal{O}}$, whereas the details are given in Chapter 6.

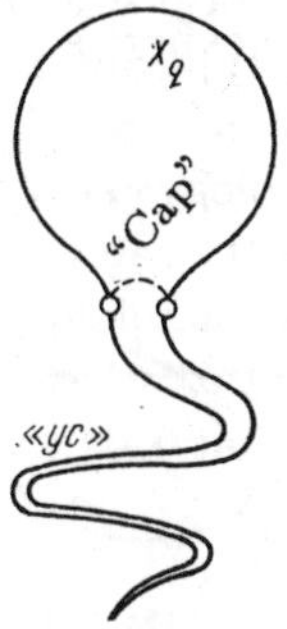


**Fig. 29**

**Step 1.** We first minimize the greatest volume $\mathrm{vol}_k$, for which we consider a minimizing sequence of compact sets $X_q \in \mathcal{O}$ such that $\mathrm{vol}_k(X_q \setminus A) \to d_k = \inf_{Y \in \mathcal{O}} \mathrm{vol}_k(Y \setminus A)$. It is desirable to derive the minimal surface $X_0$ as the limit. Meanwhile, it is clear that the original sequence should be treated accordingly in order that its limit may be determined sensibly, since, though the volumes $\mathrm{vol}_k(X_q \setminus A)$ tend to a minimum, the surfaces $X_q$ themselves can “sweep out” practically the whole of the manifold $M$ on account of the thin “whiskers” whose measure approaches zero, and the whiskers themselves fill still a greater and greater domain in $M$. Therefore, as a top priority, we should get rid of them, cutting the whiskers at the base and replacing them by “caps” of small volume and diameter (Fig. 29). In doing so, it is important to smooth the “whiskers” so that the compact set $\widetilde{X}_q$ obtained should again belong to the same variational class $\mathcal{O}$ as the original compact set $X_q$. It turns out that such an operation called by us the *S-smoothing* can be defined correctly. Having $S$-smoothed all the surfaces $X_q$, we obtain a new minimizing sequence $\{\widetilde{X}_q\}$ for which still $\mathrm{vol}_k(\widetilde{X}_q \setminus A) \to d_k$, but the $\widetilde{X}_q$ are of considerably more “regular” form than the original ones.

**Step 2.** The passage to the limit is carried out as follows. Each $\widetilde{X}_q$ is associated with the spherical density function $\Psi_k(P, \widetilde{X}_q)$ defined by $\Psi_k(P, \widetilde{X}_q) = \lim_{\varepsilon \to 0} \frac{\mathrm{vol}_k(\widetilde{X}_q \cap B^n(P,\varepsilon))}{\gamma_k(\varepsilon)}$, where the point $P \in M$, $B^n(P, \varepsilon)$ is an $n$-dimensional ball in $M$ with centre at $P$ and radius $\varepsilon$, and $\gamma_k(\varepsilon)$ is the $k$-dimensional volume of the standard Euclidean ball of radius $\varepsilon$ and dimension $k$. If $P \notin \widetilde{X}_q$, then it is obvious that $\Psi_k(P, \widetilde{X}_q) = 0$; however, if $P \in \widetilde{X}_q$, then $\Psi_k(P, \widetilde{X}_q) \geqslant 0$ (Fig. 30). It is clear that $\Psi_k(P, \widetilde{X}_q)$ measures the deviation of $\widetilde{X}_q$ at $P$ from the $k$-dimensional disc. If $\widetilde{X}_q$ is a smooth submanifold in $M$ in the vicinity of $P \in \widetilde{X}_q$, then, obviously, $\Psi_k(P, \widetilde{X}_q) = 1$, since the intersection $B^n(P, \varepsilon) \cap \widetilde{X}_q$ for small $\varepsilon$ is little different from the $k$-dimensional smooth disc of radius $\varepsilon$. Thus, the sequence of $\widetilde{X}_q$ determines that of the functions $\Psi_k(P, \widetilde{X}_q)$. It turns out that we can define the limit function $\Psi_k(P) = \lim_q \Psi_k(P, \widetilde{X}_q)$ when we take the support of $\Psi_k(P)$, i.e., the set of points for which $\Psi_k(P) > 0$, as the limit surface in dimension $k$ (Reifenberg's idea).

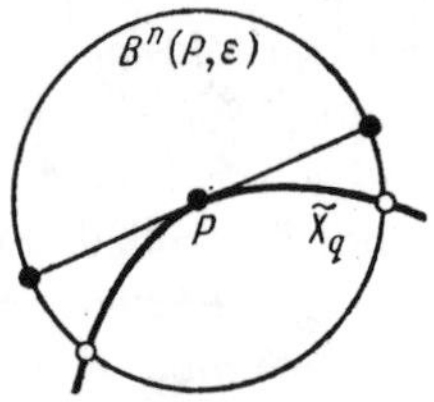


**Fig. 30**

**Step 3.** It is proved that the set $S^k = \{P : \Psi_k(P) > 0\}$ is a compact set. It is clear that, in general, $S^k$ does not belong to the class $\mathcal{O}$, thus being different from the usual homology. We then repeat the above procedure of the passage to the limit in all the lower dimensions prior to $k$, which is one of the most complicated stages of the proof. As the result of the construction, we obtain a stratified set $X_0 = A \cup S^k \cup S^{k-1} \cup \ldots$.

**Step 4.** It is proved that $X_0$ is a compact set in $M$.

**Step 5.** It is proved that the variational classes $\mathcal{O}(\widetilde{\mathcal{O}})$ are closed with respect to the passage to the limit, i.e., if $X_0 = \lim_q \widetilde{X}_q$, where $\widetilde{X}_q \in \mathcal{O}(\widetilde{\mathcal{O}})$, then $X_0 \in \mathcal{O}(\widetilde{\mathcal{O}})$.

**Step 6.** It is proved that the stratified volume of the surface $X_0$ is the least in $\mathcal{O}(\widetilde{\mathcal{O}})$.

**Step 7.** In each dimension $s \leqslant k$, it is proved that the function $\Psi_s(P, X_0) \geqslant 1$ on the whole of $S^s$, and $\Psi_s(P, X_0) = 1$ on an open subset in $S^s$, which is everywhere dense in it. Then the points from $S^s$, for which $\Psi_s > 1$, turn out to be singular in $S^s$ and fill a set $Z^s$ with zero $s$-dimensional volume.

**Step 8.** It is proved that $S^s \setminus Z_s$ is a smooth minimal submanifold in $M$.

## §8 Solution of the Problem of Finding Globally Minimal Surfaces in Each Homotopy Class of Multivarifolds

In §3, we formulated Problems A′ and B′ of finding a minimal surface in each homotopy class. For example, films of constant topological type, which are homotopic to each other, are considered by the spanning Problem A′ in contrast to Problem A, in which the absolute minimum is sought with respect to all homotopy classes. As it turns out, the concepts of stratified volume and stratified minimal surface, which were introduced by the author, make it possible to solve also Problems A′ and B′ after convenient reformulation in the functional varifold language. The solution (in terms of multivarifolds) is due to Đào Chông Thi who formulated Problem A′ in the following form: Can among all multivarifolds $g : W^k \to M^n$ homotopic to the original multivarifold $f$ and such that $g|_{\partial W} = f|_{\partial W}$ a multivarifold $g_0$ be found which would minimize the greatest $k$-dimensional volume? With such a restatement, the problem of minimizing the greatest volume $\mathrm{vol}_k$ is solved, whereas the behaviour of volumes of smaller dimensions is not controlled. Further, Đào Chông Thi solved Plateau's problem by establishing the existence of a locally Lipschitz mapping $g_0 : W^k \to M^n$ in terms of currents, which minimizes the $k$-dimensional volume functional in the class of all locally Lipschitz mappings $g : W \to M$ such that $g|_{\partial W} = f|_{\partial W}$ (the problem of finding the absolute minimum with respect to all homotopy classes of multivarifolds) [88]–[100]. As well as in the preceding cases, the greatest volume $\mathrm{vol}_k$ is minimized in terms of currents. The solution of these problems for the above statements became possible after Đào Chông Thi had introduced the concept of multivarifolds, functional analogues of stratified surfaces $X = A \cup S^k \cup S^{k-1} \cup \ldots$. This reformulation also generalizes the functional language worked out in [131], [132], and [138] (for details see [88]–[100]; for the detailed Đào Chông Thi theory, see [458]; and for further results,see Lê Hông Vân's works [244] and [245].

CHAPTER 3

# Explicit Calculation of Least Volumes (Absolute Minimum) of Topologically Nontrivial Minimal Surfaces

## §9 Exhaustion Functions and Minimal Surfaces

### 9.1. Certain Classical Problems

We now describe a number of problems leading to the one whose solution is going to be discussed in the present chapter (for one important special case). In many branches of multidimensional variational calculus, algebraic geometry, and complex analysis, the following situation often arises:

(a) Given a smooth manifold $M^n$ and its exhaustion (e.g., in the sense of [180]) by $n$-dimensional domains $B_r^n$ depending on the real parameter $r$ and expanding with its growth, i.e,. $B_{r_1}^n \supset B_{r_2}^n$ for $r_1 > r_2$ , they should exhaust (cover) the entire manifold or an open and everywhere dense domain in it.

(b) A $k$-dimensional globally minimal surface is given in $M^n$ (in the following, we will call it a *GM-surface*), e.g., $X^k$, a complex algebraic submanifold or complex analytic subset in a Kählerian manifold.

(c)The question is posed: How can the volume function $\mathrm{vol}_k(X^k \cap B_r^n) = \Psi(X^k, r)$ depending on $r$ be given a lower estimate?

The particular cases of this general problem have been studied in many works. For example, to solve the functional problems considered in [355], the inequality $(2r)^{2n-2} \leqslant \mathrm{vol}_{2n-2}(X^{2n-2} \cap I_r^{2n})$ is necessary, where $I_r^{2n}$ is the $2n$-dimensional standard cube in $\mathbf{C}^n(z^1, \ldots, z^n) \cong \mathbf{R}^{2n}$, $I_r^{2n} = \{|\operatorname{Re} z^i| \leqslant r, |\operatorname{Im} z^i| \leqslant r\} = B_r^{2n}$, and $X^{2n-2} = \{f = 0\}$, i.e, it coincides with the null level surface of the complex analytic function $f(z^1, \ldots, z^n)$, the surface $X^{2n-2}$ passing through the origin $0 \in \mathbf{C}^n$, $(2r)^{2n-2} = \mathrm{vol}_{2n-2}\, I_r^{2n-2}$. In other words, the volume of the part of $X^{2n-2}$, contained in the cube, should be given a lower estimate by the $(2n-2)$-dimensional volume of the minimal plane section. Such a section is just the standard cube $I_r^{2n-2}$ of the same dimension as the surface itself. For $n = 2$, this result has been proved in [223], and its corollaries in [355]. For $n > 2$, the statement remains undecided.

If $M = \mathbb{R}^n$ and $B_r^n = D_r^n = \{x \in \mathbb{R}^n, |x| \leqslant r\}$ is the ball of radius $r$ and centre at the point 0, then the classical result holds: The estimate $\mathrm{vol}_k(X^k \cap D_r^n) \geqslant \mathrm{vol}_k D_r^k = \gamma_k r^k$ is valid for $k = n-1, n-2$, where $X^k$ is a GM-surface in $\mathbb{R}^n$, passing through 0, and $\gamma_k$ is the $k$-dimensional volume of the unit standard $k$-dimensional ball. A similar lower estimate $\mathrm{vol}_k(X^k \cap D_r^n) \geqslant \mathrm{vol}_k D_r^k$ for a minimal surface of an arbitrary codimension $X^k \subset \mathbb{R}^n$ has been proved in [156],[162] (see also the classical Lelong estimates for $k = n-1$, $M = \mathbb{R}^n$, where $D_r^n$ is a ball of radius $r$ [180]).

We now formulate the general **Conjecture A.**

Let $X^k \subset \mathbb{R}^n$ be a GM-surface without boundary, i.e., "stretching to infinity" in $\mathbb{R}^n$, which passes though 0, and $B^n \subset \mathbb{R}^n$ a central symmetric convex domain with piecewise smooth boundary $\partial B^n$, with $0 \in B^n$ being the central point for $B^n$. Then the inequality

$$\mathrm{vol}_k(X^k \cap B^n) \geqslant \min_{\mathbb{R}^k} \mathrm{vol}_k(\mathbb{R}^k \cap B^n) = \mathrm{vol}_k(\mathbb{R}_0^k \cap B^n)$$

holds, where the minimum is taken over all $k$-dimensional plane sections of $B^n$ by planes $\mathbb{R}^k$ passing through 0, and $\mathbb{R}_0^k \cap B^n$ is the minimal plane section in the sense of intersection volume; certainly, there can be many least sections of the type. The case of strictly convex domains $B^n$ which are centrosymmetric with respect to 0 are of special interest.

**The complex version of Conjecture A.** Let $X^k \subset \mathbb{C}^n$ be a complex analytic surface (hence, globally minimal; see §2.6) without boundary, of complex dimension $k$, and $0 \in X^k$, $B^n \subset \mathbb{C}^n$ a convex symmetric domain in $\mathbb{C}^n$ with piecewise smooth boundary. Then the inequality $\mathrm{vol}_{2k}(X \cap B) \geqslant \min_{\mathbb{C}^k} \mathrm{vol}_{2k}(\mathbb{C}^k \cap B)$ holds, where $\mathbb{C}^k \cap B$ are $2k$-dimensional plane sections of the domain $B$ (over $\mathbb{R}$) with the $k$-dimensional complex planes $\mathbb{C}^k$ passing through 0.

If $B$ is nonconvex, then it is easy to see that the above estimate is generally invalid. In fact, consider a ball $D_r^n$ of radius $r$ in $\mathbb{R}^n$. Let $X^k \subset D_r^n$ be an arbitrary GM-surface passing through the centre 0 and different from a plane. Further, let $D_\varepsilon^n$ be a ball of sufficiently small radius $\varepsilon$. As a nonconvex domain $B$, we take $D_r^n \backslash U_s$, where $U_s$ is a tubular neighbourhood of small radius $s$ of the surface $X^k \cap (D_r^n \backslash D_\varepsilon^n)$ (Fig. 31). Put $X^k = \widetilde{X}^k \cap B$ and let the value $s$ tend to zero. Then $B$ tends to $D_r^n \backslash \widetilde{X}^k$. It is clear that we have $\mathrm{vol}_k(\widetilde{X}^k \cap D_\varepsilon^n) \cong \mathrm{vol}_k X^k$ for sufficiently small $s$. Meanwhile, for any $k$- dimensional plane section $T^k = \mathbb{R}^k \cap B$, we have $\mathrm{vol}_k T^k \cong \mathrm{vol}_k D_r^k$. By selecting $\varepsilon$ sufficiently small, we can make the inequality $\mathrm{vol}_k D_r^k \geqslant \mathrm{vol}_k X^k$ hold.

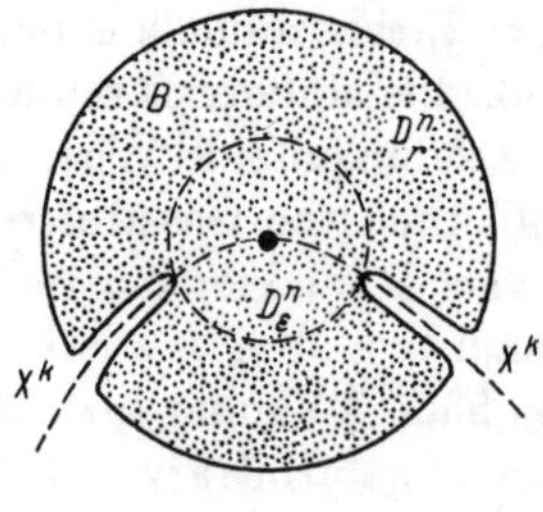


**Fig. 31**

In the case where $M$ is a complex analytic manifold and $X \subset M$ is a divisor of complex codimension one, problem (c) in §9.1 becomes that of finding a lower estimate of the volume function $\Psi(X,r) = \mathrm{vol}(X \cap \{f \leqslant r\})$, where $\{x \in M | f(x) \leqslant r\} = \{f \leqslant r\}$, and $f$ is the exhaustion function (see the definitions in [180]). Let $M = \mathbf{C}^n$, $f(x) = |x|^2$. The divisor $D$ is algebraic if and only if the volume function is asymptotically (i.e., for large $r$) of the form $\Psi(X,r) \sim O(1) \cdot r^{2n-2}$ (see [180]). Similarly, a series of results related to the solution of the Bernshtein problem (e.g., see the survey in [132]) also contains a solution of one particular case of problem (c) in §9.1, viz., it turns out that, for large $r$, the function $\Psi(X^{n-1}, r)$ grows as $r^{n-1}$, where $X^{n-1}$ is and arbitrary locally minimal surface uniquely projectable on the hyperplane $\mathbf{R}^{n-1} \subset \mathbf{R}^n$, and $\Psi(X^{n-1}, r) \sim c \cdot r^{n-1}$, where $c$ is the density (or multiplicity) of $X^{n-1}$ at the point $0 \in X^{n-1}$.

Below, we shall prove a general theorem on the lower estimation of the volume function $\Psi(X,r)$ for gobally minimal (GM) surfaces in $M$, whose particular cases are some earlier results describing the behaviour of the volume function. Besides, the theorem establishes a certain universal point of view regarding the listed different approaches to the description of the global properties of the volume function on GM-surfaces.

### 9.2. Bordisms and Exhaustion Functions

Let $M^n$ be a smooth, compact, orientable, and connected manifold with boundary $\partial M = M_1^{n-1} \cup M_2^{n-1}$, where $M_i^{n-1}$ are smooth, connected, and orientable manifolds, i.e., $M_1$ and $M_2$ are bordant. We will also consider such films $M$ that $M_1 = \emptyset$. Let $f : M \to \mathbf{R}$ be a Morse function on $M$, with critical points not on the boundary $\partial M$, and $f|_{M_1} = 0$, $f|_{M_2} = 1$, $0 \leqslant f(x) \leqslant 1$. We will also consider such functions $f$ whose critical points do not contain either local maxima or minima if $M_1 \neq \emptyset$ $M_2 \neq \emptyset$ (see [277] for their existence). If $M_1 = \emptyset$, then we assume that the level set $\{f = 0\}$consists of one point, of a nondegenerate minimum for $f$. We denote the level hypersurface $f^{-1}(r) \subset M$ by $F_r = \{f = r\}$; in the case where $M_1 = \emptyset$, the hypersurfaces $F_\varepsilon$ are spheres $S_\varepsilon^{n-1}$ for sufficiently small $\varepsilon > 0$.

Consider the subsets $B_r^n = \{x \in M | f(x) \leqslant r\}$ filling the whole of the manifold $M$, exhausting it, as they gradually expand when $r$ varies from 0 to 1 in the domain $B_r$. If $r \in [0,1]$ is a noncritical value for the function $f$, then the boundary $F_r = \partial B_r$ is a submanifold. When $r$ passes through a critical value $\tilde{r}$, the manifold $F_r$ is subject to the Morse surgery at all the critical points placed outside $F_{\tilde{r}}$. We call $f$ the *exhaustion function*, $B_r$ a *wave*, and its boundary $F_r$ a *wavefront*.

Consider a smooth vector field $v$ on $M$, with all singular points isolated, nondegenerate, and not on $M_1 \cup M_2$.

Assume that

(a) $v(f) > 0$ at all nonsingular points of $v$;

(b) the set of critical points of the function $f$ is contained in that of $v$, e.g., we can take $\operatorname{grad} f$ as $v$; due to (a), the vector of $v$ points into the domain $\{f > r\}$ at each nonsingular point $x \in F_r$;

(c) the indices of all the singular points of $v$ (or the dimension of the separatrix disc filled with the integral curves of the field, which pass through the singular points) are different from zero and $n$.

We call a field $v$ satisfying conditions (a)–(c) *f-monotonic*.

### 9.3. GM-Surfaces

Let $M$ be a Riemannian manifold with boundary. Then the $k$-dimensional Hausdorff measure $\operatorname{vol}_k X$ is defined for any $k$-measurable subset $X^k \subset M^n$. If $X$ is a submanifold (possibly, with singularities), then $\operatorname{vol}_k X$ coincides with its $k$-dimensional Riemannian volume. Recall the definition of global minimality as applied to the present case. Since $\partial M = M_1 \cup M_2$, for any $X \subset M$, we have $X \cap \partial M = (X \cap M_1) \cup (X \cap M_2)$. Let $\partial_1 X = X \cap M_1$, $\partial_2 X = X \cap M_2$, $\partial X = \partial_1 X \cup \partial_2 X$. Let $A = A^{k-1} \subset M_2$ be a fixed $(k-1)$ dimensional compact set (subcomplex), $H_{k-1}^{(k-1)}(A, G)$ the $(k-1)$-dimensional group of a (co)homology with coefficients in $G$, $H_{k-1}^{(k-1)}(A) \neq 0$, $L \subset H_{k-1}^{(k-1)}(A)$, and $L \neq 0$ a fixed subgroup (or subset in $H^{k-1}(A)$) (see §5). Consider the class $\mathcal{O}(L)$ of all compact sets (subcomplexes) $X \subset M$ such that (1) $\dim X = k$, $\Psi(x, X) \geqslant 1$ for any point $x \in X \backslash \partial X$, where $\Psi_k(x, X)$ denotes the spherical density of subsets in a Riemannian manifold (see §7.1.3 or [158]); (2) A $\supset \partial_2 X$; (3) under the embedding $i : A \to \partial_2 X \to X/\partial_1 X$, we have $i_*(L) = 0$, where $i_* : H_{k-1}(A) \to H_{k-1}(X/\partial_1 X)$ or $L \subset \operatorname{Im} i^*$ with $i^* : H^{k-1}(X/\partial_1 X) \to H^{k-1}(A)$. Let $\mathcal{O}(L) \neq 0$. If $d_k(L) = \inf_{X \in \mathcal{O}(L)} \operatorname{vol}_k X$, then there exists a GM-surface $X_0 \in \mathcal{O}(L)$ such that $\operatorname{vol}_k(X_0) = d_k(L)$ (see §7.2). The class $\mathcal{O}(L)$ consists of all surfaces $X$ spanning a fixed subgroup (subset) $L$, but different surfaces $X$, $X' \in \mathcal{O}(L)$ must not necessarily be (co)homologous in $H_k^{(k)}(M/M_1)$.

All the results below are also valid for an arbitrary extraordinary (co)homology theory $h$. The following properties of a GM-surface $X_0$ (see §7) will be employed:

(1) $X_0 \supset Z$, where $\mathrm{vol}_k\, Z = 0$, $X_0 \setminus Z$ is an open subset everywhere dense in $X_0$, and an analytic submanifold of dimension $k$; for the density function $\Psi_k(x, X_0 \backslash Z) \equiv 1$, $x \in X_0 \backslash Z$; (2) $\Psi_k(x, X_0) \geqslant 1$ for $x \in X_0 \backslash \partial X_0$. Therefore, we will assume that $\Psi_k(x, X_0) \geqslant 1$ hold on a GM-surface (see §30).

Let the surface $X_0/\partial_1 X_0$ pass through a singular point $* = \pi M_1$ in the quotient $M/M_1$, where $\pi : M \to M/M_1$ is a factorization map, i.e., $\partial_1 X_0 \neq \emptyset$ (Fig. 32). The case of an arbitrary triple $(M, f, X_0)$ can be reduced to the one with $\partial_1 X_0 \neq \emptyset$. We need to consider the film $M' = \{x \in M | \rho \leqslant f(x) \leqslant 1\} \subset M$, $\rho = \inf_{x \in X_0} f(x)$. Then $M' = M_1' \cup M_2$, $M_1' = F_\rho$, $X_0 \cap M_1' \neq \emptyset$. Consider an example that will be useful for the following.

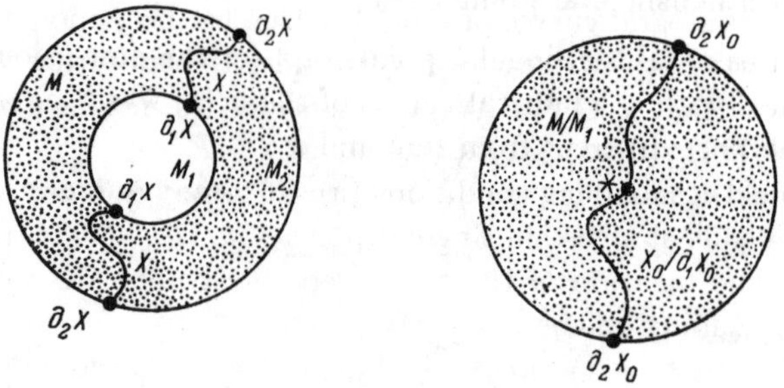


**Fig. 32**

Let $M = D^n(b) \backslash D^n(a)$, where $a < b$, $D^n(s)$ is the Euclidean ball in $\mathbb{R}^n$ with centre at 0 and radius $s$. Then the GM-surfaces $X_0 \subset M$ satisfying the condition $X_0 \cap M_1 \neq \emptyset$, where $M_1 = S_a^{n-1}$, $M_2 = S_b^{n-1}$ are spheres, exist for a convenient choice of $A \subset M_2$ due to the existence of a parallel displacement in $\mathbb{R}^n$. The case $a \to 0$ is important, where we consider a GM-surface passing through the point $0 \in D^n(b)$ as $X_0$.

### 9.4. Formulation of the Problem of a Lower Estimate of the Minimal Surface Volume Function

Let $M^n \supset X^k$, $* = X/\partial_1 X$, $X$ a GM-surface, and an exhaustion function $f$ on $M$ be given. We construct the function $\Psi(X, f, r) = \mathrm{vol}_k(X \cap B_r^n)$, $0 \leqslant r \leqslant 1$. It is clear that $\Psi$ is a nondecreasing function with respect to $r$. The general problem is to supply an exact lower estimate of $\Psi$ in terms of the Riemannian metric on a manifold irrespective of the topological type of a GM-surface $X \subset M$. We understand the "exactness" in the sense that the required estimate should turn into equality for sufficiently concise series of concrete triples $(M, f, X)$. The involvement of GM-surfaces in all these problems can be accounted for by a number of reasons.

In particular,

(a) complex analytic and algebraic surfaces are GM-surfaces in $\mathbb{C}^n$ (which is important for many applications; see above);

(b) it turns out that certain important results regarding the behaviour of complex algebraic and complex analytic surfaces can be proved with the use of their global minimality only.

We shall have to base ourselves on the existence theorem (and regularity almost everywhere) for GM-surfaces in a given (co)homology class, i.e., on a sufficiently nontrivial fact.

## §10 Definition and Simplest Properties of the Deformation Coefficient of a Vector Field

Let a triple $(M, f, X)$ and an $f$-monotonic field $v$ (e.g., $v = \operatorname{grad} f$) be given. Denote by $\gamma$ the integral curves of $v$ on $M$. Due to the $f$-monotonicity of $v$, almost all $\gamma$ starting on $M_1$ reach $M_2$. Consider the wavefront $F_r$ of a wave $B_r$, and suppose that $x \in F_r$. We also consider the field $-v$ and issue an integral curve $\gamma(\tau)$, $\gamma(0) = x$, from each point $x \in F_r$.

Two variants are possible: (a) $\gamma(\tau)$ starts at $x$ and ends on $M_1$; (b) $\gamma(\tau)$ starts at $x$ and ends at a certain singular point of $v$, i.e., $\gamma(\tau)$ is the separatrix of zero of $v$. The measure of the set of separatrix curves is zero, with the "typical curve" ending on the boundary of $M_1$.

Let $H_x(v)$ be the $(n-1)$-dimensional hyperplane orthogonal to the vector $v \neq 0$ at $x$, and $\Pi_x^{k-1} \subset H_x(v)$ an arbitrary $(k-1)$-dimensional plane in $H_x(v)$. Consider an exponential mapping along the geodesics $\exp_x : T_xM \to M$. Suppose $S_\varepsilon^{k-1} = S_\varepsilon = \exp_\varepsilon D^{k-1}(\varepsilon)$, where $D^{k-1}(\varepsilon) \subset \Pi^{k-1}$ is a ball of small radius $\varepsilon$ in the plane $\Pi^{k-1}$, with centre at the point 0 and $x = \exp_x(0)$. Then $S_\varepsilon$ may be regarded as a $(k-1)$-dimensional ball of radius $\varepsilon$ with centre at $x$. We let a curve $\gamma_y(\tau)$ of the field $-v$ emanate from each point $y \in S_\varepsilon$, and extend it until $\gamma_y(\tau)$ reaches $M_1$ or ends at a critical point of $-v$. The totality of all the curves $\{\gamma_y(\tau) | y \in S_\varepsilon\}$ forms a tube $CS_\varepsilon$, which is a $CW$-complex of dimension $k$ due to the $f$-monotonicity of $v$. $CS_\varepsilon$ is almost everywhere a smooth $k$-dimensional submanifold with boundary in the manifold $M$ in the sense of $\operatorname{vol}_k$ (Fig. 33).

Put $\kappa_k(v, x, \Pi_x^{k-1}) = \lim_{\varepsilon \to 0} (\operatorname{vol}_k CS_\varepsilon)/(\operatorname{vol}_{k-1} S_\varepsilon)$. Let $\kappa_k(v, x) = \sup_{\Pi_x^{k-1}} \kappa_k(v, x, \Pi_x^{k-1})$. We call the function $\kappa(v, x)$ the $k$-dimensional deformation coefficient of the field $v$. In concrete situations, the deformation coefficient of $v$ is normally easy to calculate.

**Example 1.** Let $M = D^n(R) \backslash D^n(0)$ (see example above, where $a = 0$), and $M = \{x \in \mathbf{R}^n, 0 \leqslant |x| \leqslant R\}$. We put $f(x) = |x|$ and $v = \operatorname{grad} f$. Then $\kappa_k(v, x) = r/k$, where $r = |x|$, and the coefficient $\kappa_k$ depends only on $r$, not depending on $x \in F_r$, whereas the curves $\gamma(\tau)$ coincide with the radii of the ball $D^n(r)$.

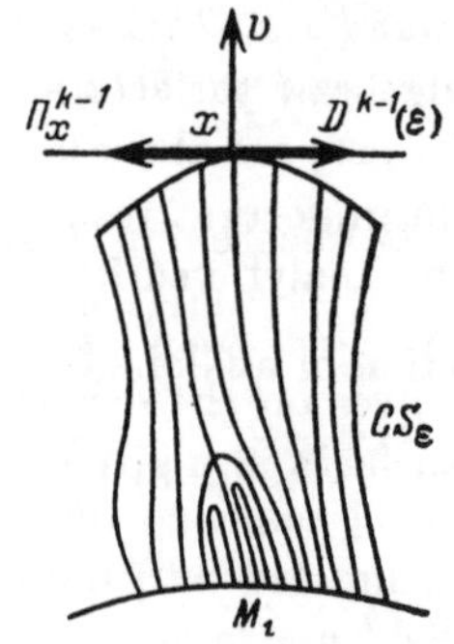


**Fig. 33**

**Example 2.** Let $M = D^n(R)\setminus D^n(0)$, where $D^n(R)$ is a ball of geodesic radius $R$, embedded into the sphere $S^n(q)$ of radius $q$. Then $R = q \cdot \varphi_0$, where $\varphi_0$ is the angle of the spherical sector determining $M$. If $f(x) = r =$ length of meridian from $N$ to $x$, $v = \operatorname{grad} f$, then $B_r^n$ =ball of radius $r, r = \varphi \cdot q$, and

$$\varkappa_k(v, x) = q \cdot \frac{\int_0^{r/q} \sin^{k-1} t \, dt}{\sin^{k-1}(r/q)}.$$

The coefficient $\kappa_k$ in $\mathbb{R}P^n$ in the space of constant negative curvature is calculated absolutely in the same manner, the trigonometric functions being replaced by hyperbolic ones. An explicit formula for $\kappa_k$ in $\mathbb{C}P^n$ endowed with the standard invariant metric can be easily obtained from the representation of $\mathbb{C}P^n$ as the quotient $U_{n+1}/U_1 \times U_n$.

## §11 Formulation of the Basic Theorem for the Lower Estimate of the Minimal Surface Volume Function

**Theorem 11.2.1 (Special case)** *Let $f$ be a smooth function on a manifold $M^n$, so that $0 \leqslant f \leqslant 1$, $f(x_0) = 0$, $f(\partial M) = 1$, and it is a Morse function on $M\setminus(x_0 \cup \partial M)$, with only critical saddle points whose indices do not exceed a certain fixed value $k-2$, where $k < n$, and $X^k$ is a globally minimal surface in $M^n$, passing through the point $x_0$, and possessing a nontrivial boundary $\partial X$ wholly contained in the boundary $\partial M$, i.e., from the class of surfaces $\mathcal{O}(L)$ (see above). Then for the volume function $\Psi(r) = \Psi(X, f, r)$ of this surface, we have: $\Psi(r) \geqslant q(r) \cdot l$, where the constant $l = \lim_{a\to 0} \frac{\Psi(a)}{q(a)}$ does not depend on $r$ and is determined only by the structure of the minimal surface $X$ in the vicinity of $x_0$, whereas the function $q(r)$ is of the form $q(r) = \exp \int [\max_{x\in\{f=r\}}(\kappa_k(x)|\operatorname{grad} f(x)|)]^{-1} dr$, where $\kappa_k(x) = \kappa_k(\operatorname{grad} f, x)$. In particular, $\operatorname{vol}_k X = \Psi(1) \geqslant l \cdot q(1)$. Thus, the behaviour of the volume function $\Psi(r)$ on the minimal surface $X_0$ is determined by that at the initial* *moment of time $r = 0$, or around $x_0$. This estimate turns into equality for many concrete minimal surfaces and we obtain an exact expression for the minimal surface volume.*

Below, we give the proof of the theorem in a more general form.

### 11.1. Functions of the Interaction of a Globally Minimal Surface with a Wavefront

Let $x \in F_r$, $T_xF_r$ be the tangent plane to the wavefront $F_r$, and $n(x)$ the unit exterior normal to $T_xF_r$, directed into the domain $\{f \geqslant r\}$. If $v(x)$ is the value of an $f$-monotonic field $v$ at a nonsingular point $x$, then the angle $\alpha$ between the normal $m$ and the vector $v$ is determined. Though $\alpha$ is not defined at singular points, this will not alter the further argument.

Consider in $M$ a GM-surface $X \in \mathcal{O}(L)$. Since $\partial_1 X \neq \emptyset$, for almost all values $r \in [0,1]$, the surfaces $X$ and $F_r$ intersect transversally at regular points of $X$; in particular, $\mathrm{vol}_{k-1}(X \cap F_r) < \infty$. Let $x \in X \cap F_r$. We put $V_x^{k-1} = T_xX \cap T_xF_r$. Since $V_x^{k-1} \subset T_xX$, the normal $m \in T_xX$, $m \perp V_x^{k-1}$ is defined. It is exterior with respect to $F_r$ (Fig. 34). Let $\beta$ be the angle between $n$ and $m$. Since $v$ is an $f$-monotonic field, we have $v \notin V_x^{k-1}$, and the $k$-dimensional plane $R_x^k$ spanned by the plane $V_x^{k-1}$ is determined uniquely (as well as the vector of $V$). Let $l$ be the exterior normal in the plane $R_x^k$ to $V_x^{k-1}$ (Fig. 34); $l$ is on the same side of $T_xF_r$ as $v, n$, and $m$. Suppose $\varphi$ is the angle between $l$ and $v$; $\alpha$, $\beta$, and $\varphi$ are almost everywhere smooth functions on $X$ of $x \in X$.

### 11.2. Formulation of the Basic Volume Estimation Theorem

**Theorem 11.2.1 (General case).** *Let $f$ be a Morse function on $M^n$, $0 \leqslant f(x) \leqslant 1$, $\partial M = M_1 \cup M_2$, $f|_{M_1} = 0$, $f|_{M_2} = 1$, $v$ be an $f$-monotonic field on $M$, all critical points of $f$ be singular for the field $v$, all singular points $x_\lambda$ of the field are nondegenerate, $0 < \lambda = \mathrm{ind}\, x_\lambda \leqslant k-2$ (where $k$ is an integer, $k < n$), and $X^k \subset M^n$ a globally minimal surface $X \in \mathcal{O}(L)$, $\partial_1 X \neq \emptyset$, $L \neq 0$. Then $\Psi(X,f,r) \geqslant \lim_{a\to 0} \frac{\Psi(X,f,a)}{h(a)} \cdot h(r)$, where the constant $l' = \lim_{a \to 0} \frac{\Psi(X,f,a)}{h(a)}$ does not depend on $r$ and is determined only by the GM-surface $X$. The function $h(r)$ is of the form*

$$h(r) = \exp \int \frac{dr}{\max_{x \in F_r}[\kappa_k(v,x)|\operatorname{grad} f(x)| \cos\varphi(x) \cos\beta(x)]}.$$

*Thus, the behaviour of the volume function $\Psi(X,f,r)$ is determined by that at the moment $a = 0$, i.e., on the boundary of $M_1$, and the geometry of the manifold $M$. This estimate is exact in the sense that there exist sufficiently extensive series of quadruples $(M,X,f,v)$ for which inequality turns into equality, and in which cases we obtain an exact and explicit expression for $\Psi(X,f,r)$*

*on a GM-surface in terms of the field deformation coefficient; in particular, this permits us to calculate the volume of this GM-surface precisely.*

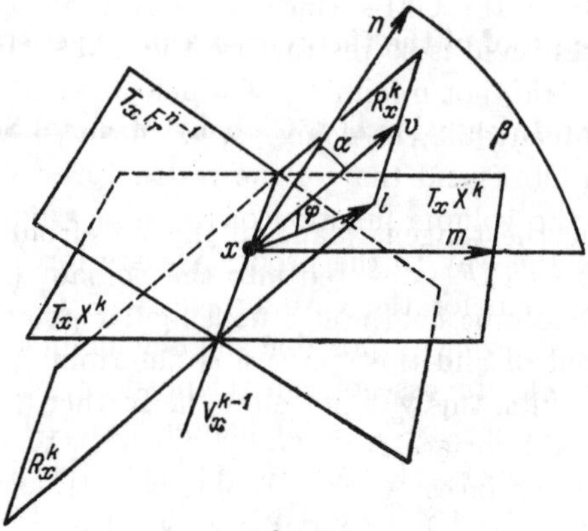


**Fig. 34**

The function $h(r)$ determined by the indefinite integral is thus given up to a constant; however, this does not affect the estimate because of the expression $[h(a)]^{-1}$ involved in the formula for $l'$, which is just what makes up for the indeterminacy.

**Corollary 11.2.1.** *With the data of Theorem 11.2.1, we have* $\Psi(X,f,r) \geqslant \lim_{a\to 0} \frac{\Psi(X,f,a)}{q(a)} \cdot q(r)$, *where*

$$q(r) = \exp \int \frac{dr}{\max\limits_{x\in F_r}[\kappa_k(v,x)|\operatorname{grad} f(x)|]}.$$

*There exist sufficiently extensive series of quadruples* $(M,X,f,v)$ *for which the above inequality turns into equality, thus supplying explicit formulas for the volumes of such GM-surfaces.*

Since $h(r) \geqslant q(r)$ for all $r$, the estimate in Corollary 11.2.1 is rougher than in Theorem 11.2.1; however, as shown by numerous applications, it is sufficient if we need a nontrivial statement regarding the behaviour of the density function $\Psi_k(x_0, X)$. Theorem 11.2.1 is a corollary of the following deeper statement on the behaviour of $\Psi(X,f,r)$.

**Theorem 11.2.2.** *Let a quadruple* $(M,X,f,v)$ *satisfy all the conditions of Theorem 11.2.1. Then the piecewise continuous function* $\frac{\Psi(X,f,r)}{h(r)}$ *is not decreasing with respect to* $r$, *i.e.,* $\frac{\Psi(X,f,r_1)}{h(_1)} \leqslant \frac{\Psi(X,f,r_2)}{h(r_2)}$ *for* $r_1 < r_2$. *Similarly, we have* $\frac{\Psi(X,f,r_1)}{q(r_1)} \leqslant \frac{\Psi(X,f,r_2)}{q(r_2)}$ *for* $r_1 < r_2$.

Theorem 11.2.1 and Corollary 11.2.1 are both obtained from Theorem 11.2.2 when $r_1 \to 0$, $r_2 \to 0$.

## §12 Proof of the Basic Volume Estimation Theorem

We start with the topological part of the problem. Note that we do not know very much about the GM-surface $X$, only that $X \in \mathcal{O}(L)$, $L \neq 0$, $\partial_1 X \neq 0$. Our nearest goal is to restructure the GM-surface by replacing it with another (in general, not minimal), of which we shall know considerably more than of $X$. Meanwhile, in doing so, we have to establish a relation between the volume of $X$ and that of the restructured surface $X'$ to make a conclusion about the volume function on $X$ after studying $X'$. Note one of the difficulties on the way to the proof. We are forced to base ourselves on the existence theorem for the GM-surface $X \in \mathcal{O}(L)$ which is not, in general, a submanifold owing to possible singular points. Consider the level surface $F_r = \{f = r\}$. In case of general position, $A_r^{k-1} = F_r \cap X$ is a $(k-1)$-dimensional surface in $F_r$ (see Chapter 6 and [156],[158],[161]). We introduce the function $\widetilde{\varphi}(r) = \int_0^r \mathrm{vol}_{k-1}(X \cap F_t)dt$. Then $\widetilde{\varphi}(r)$ is continuous with respect to $r$ (see Chapter 6 and [161],[162]) in contrast to the generally discontinuous function $\Psi(X, f, r)$. It will be proved in Chapter 6 that almost for all $r \in [0,1]$, $\mathrm{vol}_{k-1} A_r < \infty$, $\mathrm{vol}\, A_r = \widetilde{\varphi}'_r = \mathrm{vol}_{k-1}(X \cap F_r)$ hold. Consider the $k$-dimensional tube $CA_r = \bigcup_{x \in A_r} \gamma_x(r)$, where $0 \leqslant r \leqslant T'_x$ (see §10), which is the union of all the integral curves of the field $-v$, emanating from the points $A_r$. In case of general position, $CA_r$ is a $k$-dimensional surface in the sense of §7, placed in the domain $\{f \leqslant r\}$. Construct a new surface $X' = [X \backslash (X \cap \{f \leqslant r\})] \cup CA_r$, i.e., replace part of $X$ in $\{f \leqslant r\}$ by $CA_r$.

**Lemma 12.1.** *The surface $X'$ continues to span $L$ in the group $H_{k-1}^{(k-1)}(A)$ under the embedding $A \to X'/X' \cap M_1 = X'/\partial_1 X'$. In other words, $X' \in \mathcal{O}(L)$, $L \neq 0$, which means that the above surgery of $X$ does not take us outside the variational class $\mathcal{O}(L)$.*

**Proof.** For definiteness, we consider the homology case. We prove that, under the embedding $i_r : A \to CA_r$, the relation $i_{r*} H_{k-1}(A_r) = 0$ holds in $H_{k-1}(CA_r)$. We assume first that the $k$-dimensional surface $CA_r$ contains no singular points of the field $v$. Then $CA_r \backslash (CA_r \cap M_1)$ is homeomorphic to the direct product $I \times A_r$ due to the $f$-monotonicity of $v$, and it is evident that the embedding $i : A_r \to CA_r / CA_r \cap M_1$ completely annihilates the group $H_{k-1}(A_r)$, since the cycle $[A_r]$ is homologous to the union of those in the boundary of $M_1$. Consider the general case. Let $x_\lambda \in \{f < r\}$ be singular points, and $\mathrm{ind}\, x_\lambda = \lambda$. By the data of the theorem, $0 < \lambda \leqslant k-2$. Suppose $r_0 = \max f(x_\lambda)$. Then $0 < r_0 < r$. There are no singular points of $v$ in the interval $(r_0, r)$; therefore, $A_r$ is homeomorphic to $A_s$ when $s \to r_0$, $s > r_0$. In particular, $A_r$ and $A_s$ are homologous in $M$. We prove that $A_r$ is homologous to $A_s$ for $s = r_0 - \varepsilon$, where $\varepsilon > 0$ is sufficiently small. Consider the surgery of $A_s$, $s > r_0$, when $s$ passes through the critical value $r_0$, for which it suffices to study it in the vicinity of one singular point $x_\lambda$, $f(x_\lambda) = r_0$. Since $x_\lambda$ is a nondegenerate point, a sufficiently small neighbourhood $U(x_\lambda)$ is diffeomorphic to the direct product

$D^\lambda \times D^{n-\lambda}$. All the curves $\gamma(\tau)$ for the field $-v$, which make up the disc $D^{n-\lambda}$, end at $x_\lambda$, where those making up $D^\lambda$ start. Let $A_s^U = A_s \cap U(x_\lambda)$. It is clear that $A_s^U \backslash (A_s^U \cap D^{n-\lambda})$ is homeomorphic $A_b^U \backslash (A_b^U \cap D^\lambda)$ for $s > r_0$ and $b < r_0$. The homeomorphism is established along the integral curves $\gamma$ not passing through $x_\lambda$. The intersection $A_s^U \cap D^{n-\lambda}$, where $s > r_0$, undergoes surgery and turns into $A_b^U \cap D^\lambda$, where $b < r_0$. Since $\lambda \leqslant k-2$, the relation $\dim(A_b^U \cap D^\lambda) \leqslant k-2$ holds, and because $A_b^U = \overline{[A_b^U \backslash (A_b^U \cap D^\lambda)]} \cup (A_b^U \cup D^\lambda)$, we have $\partial_{k-1}^* A_b^U = \partial_{k-1}^* \overline{[A_b^U \backslash (A_b^U \cap D^\lambda)]} = A_b^U \backslash (A_b^U \cap D^\lambda)$ due to the equality $\partial_{k-2}^*(A_b^U \cap D^\lambda) = 0$, $\partial_{k-1}^*$ denoting the boundary operator of taking the $(k-1)$-dimensional fundamental cycle. Meanwhile, we have used the relations $(\partial_{k-2} A_b^U) \cap D^\lambda = 0$, $\dim(\partial_{k-2} A_b^U) \cap D^\lambda \leqslant k-3$, where $\partial$ is the boundary operator. It follows that $A_b^U$ is homologous to $A_s^U$ in $U(x_\lambda)$ due to the $k$-dimensional film $CA_t \cap U(x_\lambda) \cap \{s \geqslant (t = f(x)) \geqslant b\}$. Note that, for $\lambda = k-1$, this statement is, in general, invalid. For $\lambda \leqslant k-2$, $A_b(b = r_0 - \varepsilon)$ is homologous to $A_s(s = r_0 + \varepsilon)$. We repeat the above surgery procedure on each critical level containing the singular points of the field $v$ until we attain the surface $M_1 = \partial_1 M$. Thus, $A_r$ is homologous to $\lim_{b \to 0} A_b \subset M_1$; therefore, under the embedding $j_r : A_r \to CA_r/\partial_1 CA_r$, we have $(j_r)_* H_{k-1}(A_r) = 0$, and the surgery operation $X \to X'$ is an $S$-surgery in the sense of Chapter 6 (see also [128],[130]). Hence, by Theorem 3.2 from [161] (see also Chapter 6), we obtain $X' \in \mathcal{O}(L)$, $L \neq 0$.

Q.E.D.

**Lemma 12.2.** *Let $X'$ be the surface constructed in Lemma 12.1. Then the inequalities* $\mathrm{vol}_k X' \geqslant \mathrm{vol}_k X$ *and* $\mathrm{vol}_k(X \cap \{f \leqslant r\}) \leqslant \mathrm{vol}_k CA_r$ *hold.*

**Proof.** Since, by Lemma 12.1, $X' \in \mathcal{O}(L)$, $L \neq 0$, we have $\mathrm{vol}_k X' \leqslant \mathrm{vol}_k X = \inf_{Y \in \mathcal{O}(L)} \mathrm{vol}_k(Y)$. The second inequality follows from $X \cap \{f \geqslant r\} = X' \cap \{f \geqslant r\}$. The Lemma is thus proved.

We have made use of the fact that $X$ is a GM-surface in $M$, i.e., referred to the existence theorem for GM-surfaces in the class $\mathcal{O}(L)$. If $X$ were a "prelimit" surface, then the inequality $\mathrm{vol}_k X' \geqslant \mathrm{vol}_k X$ would not be guaranteed. We are aware of the topological behaviour of the new surface $X' \cap \{f \leqslant r\} = CA_r$ along the field $v$ : $CA_r = \cup \gamma_x(\tau)$, $x \in A_r$, which is just what permits us to estimate $\mathrm{vol}_k X'$.

Consider the hypersurface $f_r$. Let $F_r \cap X = A_r$. Denote by $d\Lambda^{k-1}$ the exterior $(k-1)$-dimensional form (possibly with singularities) of $(k-1)$-dimensional volume on $A_r$. Due to the general properties of GM-surfaces (see §7), we can assume in case of general position that $d\,\Lambda^{k-1}$ is the usual $(k-1)$-form on an open submanifold $T$ of complete measure in $A_r$.

**Lemma 12.3.** *For almost all $r$, $0 < r < 1$, the inequality is valid:*

$$\frac{d\Psi(X, f, r)}{dr} = \int_{A_r} \frac{1}{|\operatorname{grad} f| \cos\beta} d\Lambda^{k-1}.$$

**Proof.** Due to the general properties of GM-surfaces, we can assume that

$$\Psi'_r = \lim_{\Delta r \to 0} \frac{1}{\Delta r} (\Psi(r + \Delta r) - \Psi(r)) = \lim_{\Delta r \to r} \frac{\Delta \Psi}{\Delta r}$$

holds for almost all $r$, where $\Psi(X, f, r)$ is denoted by $\Psi(r)$. Let partition $A_r$ into the union of disjoint infinitesimal spherical volumes $d\sigma(V_x^{k-1})$, where $V_x^{k-1}$ is the tangent plane to the ball $d\sigma(V_x^{k-1}) \subset A_r$ and $x \in T$ is a regular point of $A_r$. Let $\mathrm{vol}_{k-1}\, d\sigma(V_x^{k-1}) = d\Lambda^{k-1}(V_x^{k-1})$ be the $(k-1)$-dimensional measure of the elementary ball $d\sigma(V_x^{k-1})$. Then $\mathrm{vol}_{k-1} A_r = \int_{A_r} d\Lambda^{k-1}(V_x^{k-1})$. Consider two nearby hypersurfaces $F_r$ and $F_{r+\Delta r}$. Suppose that $\Delta X = X \cap \{r \leqslant f(x) \leqslant r + \Delta r\}$ is an infinitesimal layer between $F_r$ and $F_{r+\Delta r}$. Then $\Delta\Psi = \mathrm{vol}_k(\Delta X)$. For each ball $d\sigma(V_x^{k-1})$, we construct an infinitesimal cylinder whose base is $d\sigma(V_x^{k-1})$ and whose height is the segment $xR$ of the normal $m$, cut out on $m$ by $F_r$ and $F_{r+\Delta r}$ (Fig. 35). Then $\Delta X$ is

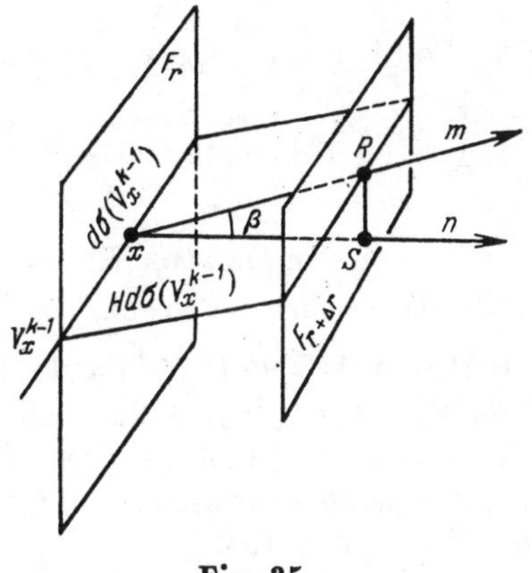


**Fig. 35**

partitioned into the union of disjoint cylinders of the form $H d\sigma(V_x^{k-1})$, i.e., $\Delta\Psi = \mathrm{vol}_k(\Delta X) = \sum \mathrm{vol}_k\, H d\sigma(V_x^{k-1})$. We find $\mathrm{vol}_k\, H d\sigma(V_x^{k-1})$. Since $m$ is orthogonal to $d\sigma(V_x^{k-1})$ at the point $x$, and because of the infinitesimality of the entire argument above, we have replaced all the infinitesimal surfaces embedded into $M$ by their diffeomorphic images in the tangent planes, the segment $xR$ is orthogonal to $d\sigma(V_x^{k-1})$. Thus,

$$\mathrm{vol}_k\, H\, d\sigma(V_x^{k-1}) = |xR|\, \mathrm{vol}_{k-1}\, d\sigma(V_x^{k-1}) = |xR|\, d\Lambda^{k-1}(V_x^{k-1}).$$

Due to the definition of $\beta$, we have $|xR| = \frac{|xS|}{\cos\beta}$, where $xS$ is directed along the normal to $F_r$ and determined uniquely by the increment $\Delta r = \Delta f(x)$. Hence, $\Delta r = |\mathrm{grad}\, f(x)| \cdot |xS|$ up to infinitesimals of higher orders, for

$$\lim \frac{\Delta r}{|xS|} = \langle \operatorname{grad} f,\ n\rangle = \left\langle \operatorname{grad} f,\ \frac{\operatorname{grad} f}{|\operatorname{grad} f|}\right\rangle = |\operatorname{grad} f|,$$

where $\langle\, ,\, \rangle$ denotes the scalar product. We have

$$\operatorname{vol}_k H\, d\sigma(V_x^{k-1}) = \frac{d\Lambda^{k-1}(V_x^{k-1})\cdot \Delta r}{\cos\beta\cdot|\operatorname{grad} f|},$$

which gives

$$\lim_{\Delta r\to 0} \frac{\Delta\Psi}{\Delta r} = \lim_{\Delta r\to 0} \frac{1}{\Delta r}\cdot \operatorname{vol}_k(\Delta X)$$

$$= \lim_{\Delta r\to 0} \frac{1}{\Delta r}\sum_{A_r} \operatorname{vol}_k H\, d\sigma(V_x^{k-1}) = \int_{A_r} \frac{d\Lambda^{k-1}(V_x^{k-1})}{\cos\beta\cdot|\operatorname{grad} f|}.$$

Q.E.D.

**Lemma 12.4.** *For almost all $r$, $0 < r < 1$, the equality holds:*

$$\operatorname{vol}_k CA_r = \int_{A_r} \cos\varphi\cdot\varkappa_k(v,\ x,\ \Pi_x^{k-1})\, d\Lambda^{k-1}(V_x^{k-1}),$$

*where the plane $\Pi_x^{k-1}$ is contained in the plane $R_x^k = V_x^{k-1} + v$ and orthogonal to the vector $v \in R_x^k$ (Fig. 34).*

**Proof.** Consider the partition $A_r = d\sigma(V_x^{k-1})$ again (see Lemma 12.3), and construct the tube $Cd\sigma(V_x^{k-1}) = \bigcup \gamma_y(\tau)$ for each ball $d\sigma(V_x^{k-1})$, where $y \in d\sigma(V_x^{k-1})$. Then $CA_r = \bigcup Cd\sigma(V_x^{k-1})$; therefore, $\operatorname{vol}_k CA_r = \sum \operatorname{vol}_k Cd\sigma(V_x^{k-1})$, i.e., it remains to consider $\operatorname{vol}_k Cd\sigma(V_x^{k-1})$ as a function of points $x \in A_r$. It suffices to prove that

$$\operatorname{vol}_k C\, d\sigma(V_x^{k-1}) = \cos\varphi\cdot\varkappa_k(v,\ x,\ \Pi_x^{k-1})\cdot d\Lambda^{k-1}(V_x^{k-1}).$$

Up to the infinitesimals of higher orders, we have

$$\operatorname{vol}_k Cd\sigma(V_x^{k-1}) = \operatorname{vol}_k Cd\sigma(\Pi_x^{k-1}),$$

where $d\sigma(\Pi_x^{k-1})$ denotes the ball tangent to $\Pi_x^{k-1}$ at the point $x$ (Fig. 36). We are making use of the fact that the curves $\gamma_y(\tau)$, $y \in d\sigma(V_x^{k-1})$ may be regarded as parallel in the vicinity of $x$ if $v \neq 0$. By the definition of the deformation coefficient of the vector field $v$ (see §10), we have

$$\operatorname{vol}_k C\, d\sigma(\Pi_x^{k-1}) = \varkappa_k(v,\ x,\ \Pi_x^{k-1})\operatorname{vol}_{k-1} d\sigma(\Pi_x^{k-1}).$$

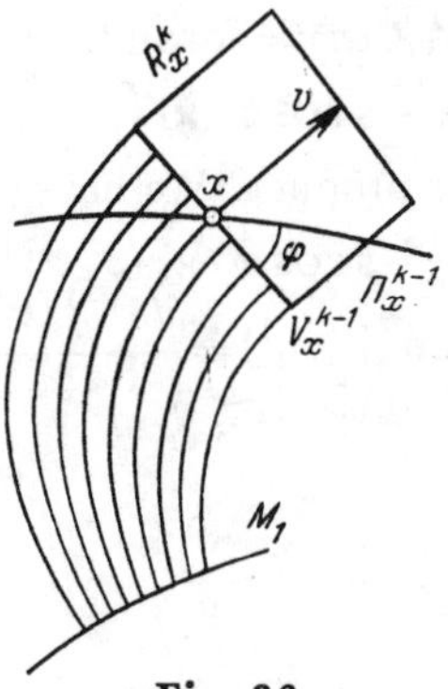


**Fig. 36**

But since the angle between $V_x^{k-1}$ and $\Pi_x^{k-1}$ in the plane $R_x^k$ equals $\varphi$, or that between $v$ and l (Fig. 34), $\mathrm{vol}_{k-1}\, d\sigma(\Pi_x^{k-1}) = \mathrm{vol}_{k-1}\, d\sigma(V_x^{k-1})\cos\varphi$. Thus,

$$\mathrm{vol}_k\, C\, d\sigma(\Pi_x^{k-1}) = \varkappa_k(v,\ x,\ \Pi_x^{k-1})\cos\varphi\cdot \mathrm{vol}_{k-1}\, d\sigma(V_x^{k-1}).$$

Finally,

$$\begin{aligned}\mathrm{vol}_k\, CA_r = \sum \mathrm{vol}_k\, C\, d\sigma_x(V_x^{k-1}) &= \sum \mathrm{vol}_k\, C\, d\sigma_x(\Pi_x^{k-1})\\ &= \int_{A_r} \varkappa_k(v,\ x,\ \Pi_x^{k-1})\cos\varphi\cdot d\Lambda^{k-1}(V_x^{k-1}).\end{aligned}$$

Q.E.D.

**Lemma 12.5.** *The inequality*

$$\mathrm{vol}_k\, CA_r \leqslant \int_{A_r} \cos\varphi\cdot\varkappa_k(v,\ x)\, d\Lambda^{k-1}(V_x^{k-1})$$

*holds.*

**Proof.** Since $\kappa_k(v,x) = \max_{\Pi_x^{k-1}} \kappa_k(v,x,\Pi_x^{k-1})$ and $\cos\varphi \geqslant 0$, the statement follows from Lemma 12.4.

**Lemma 12.6.** *Let a function $s = s(r)$, $0 \leqslant r \leqslant 1$, be such that*

$$F(v,\ f,\ x,\ X) = \frac{1}{\cos\beta\cdot|\mathrm{grad}\, f|} - s(r)\cos\varphi\cdot\varkappa_k(v,\ x)$$

*is nonnegative on each level surface $F_r$, $0 \leqslant r \leqslant 1$. Then the inequality $\frac{d}{dr}\Psi(X,f,r) \geqslant s(r)\Psi(X,f,r)$ holds.*

**Proof.** Consider the difference $\Psi_r' - s\Psi$. Then, due to Lemmas 12.2 and 12.4,

$$\Psi_r' - s\Psi \geqslant \Psi_r' - s\,\mathrm{vol}_k CA_r =$$
$$= \int_{A_r} \left[\frac{1}{\cos\beta \cdot |\operatorname{grad} f|} - s\cos\varphi \cdot \varkappa_k(v,\ x,\ \Pi_x^{k-1})\right] d\Lambda^{k-1}(V_x^{k-1})$$
$$\geqslant \int_{A_r} \left[\frac{1}{\cos\beta \cdot |\operatorname{grad} f|} - s\cos\varphi \cdot \varkappa_k(v,\ x)\right] d\Lambda^{k-1}(V_x^{k-1})$$
$$= \int_{A_r} F(v,\ f,\ x,\ X)\, d\Lambda_x^{k-1} \geqslant 0.$$

Q.E.D.

**Lemma 12.7.** *Put*

$$s_m(r) = \left\{\max_{x\in F_r} [\varkappa_k(v,\ x) \cdot |\operatorname{grad} f| \cdot \cos\beta \cdot \cos\varphi]\right\}^{-1}.$$

*Hence,* $\Psi_r' - s_m\Psi \geqslant 0$.

**Proof.** It is clear that, due to the definition of $s_m$, the inequality $F(v, f, x, X) \geqslant 0$ holds; therefore, the required statement follows from Lemma 12.6.

It is verified similarly that we can take the function

$$\tilde{s}_m = \left(\max_{x\in F_r}[\kappa_k(v,x)|\operatorname{grad} f|]\right)^{-1}$$

as $s$. Then

$$(\cos\beta \cdot |\operatorname{grad} f|)^{-1} - \tilde{s}_m \cdot \cos\varphi \cdot \varkappa_k(v,\ x)$$
$$= (\cos\beta \cdot |\operatorname{grad} f|)^{-1} \cdot (1 - [\cos\beta \cdot \cos\varphi \cdot \varkappa_k |\operatorname{grad} f|]$$
$$\times [\max(\varkappa_k |\operatorname{grad} f|)]^{-1}) \geqslant 0.$$

Therefore, $\Psi_r' - \tilde{s}_m\Psi \geqslant 0$, which entails Corollary 11.2.1.

The role of Lemma 12.7 is in the elimination of the last parameter unknown to us, viz., the boundary $A_r$ determined by the surface $X$, which, together with the surgery $X \to X'$ (see Lemma 12.2), permits us to remove the influence of the unknown topological structure of $X$. Meanwhile, we stay in the same topological class of GM-surfaces. Smoothing a surface by means of its surgery in many cases turns out to be the identity transformation; therefore, the obtained estimate often happens to be an exact equation.

**Proof of Theorem 11.2.2.** Due to Lemma 12.7, we have $\Psi_r'(X, f, r) \geqslant s_m(r) \cdot \Psi(X, f, r)$. Consider the required inequality $\left(\frac{\Psi}{h}\right)_r' \geqslant 0$; hence, $\frac{\Psi' h - h'\Psi}{h^2} \geqslant 0$, which is equivalent to $\Psi' \geqslant \frac{h'}{h}\Psi$. Consider also the differential equation $\frac{h'}{h} = s_m$. Then $\exp\int s_m(r)dr = h(r)$. To prove Corollary 11.2.1, we have to take the function $\tilde{s}_m$ instead of $s_m$. Thus, we have found a function $h$ for which $\left(\frac{\Psi}{h}\right)'_r \geqslant 0$. If $\Psi$ were differentiable, then the theorem would already follow. However, in general, $\Psi$ is discontinuous; therefore, the monotonicity of $\frac{\Psi}{h}$ requires additional proof.

**Lemma 12.8.** *The functions $\frac{\Psi}{h}$ and $\frac{\Psi}{q}$ are not increasing with respect to $r$.*

**Proof.** We define a continuous function $\int \Psi(t)dt = \chi(r)$, and put $\omega = \Psi - \chi$. Then $\Psi = \chi + \omega$, where $\omega$ is the singular part of $\Psi$. It is clear that $\omega'_r = 0$ almost everywhere on the interval $(0, 1)$. Further, applying the argument of [162] to the proof of Lemma 10.2 (see also Chapter 6), we obtain $d\omega(r) \geqslant 0$, which entails the statement of Theorem 11.2.2.

## §13 Certain Geometric Consequences

### 13.1. On the Least Volume of Globally Minimal Surfaces Passing Through the Centre of a Ball in Euclidean Space

The estimate obtained in Theorem 11.2.1 is of the form $\Psi \geqslant c(M_1, X) \cdot h(r)$, where the constant $c(M_1, X)$ is determined by the GM-surface $X$ and the boundary of $M_1$. In many cases, it can, in turn, be given a lower estimate by another and more simple constant not depending on $X$. We illustrate this theorem in the case of minimal surfaces of arbitrary codimension in $\mathbb{R}^n$. As the domain we take a ball with its centre in the origin and consider GM-surfaces passing through the origin. Recall that the spherical density function $\Psi_k$ on GM-surfaces is not less than unity at each point.

**Corollary 13.1.1.** *Let $X^k \subset D_r^n \subset \mathbb{R}^n$ be a GM-surface passing through the centre $O$ of the ball $D_r^n$ of radius $r$.*

*Then*

$$\Psi(X, r) = \mathrm{vol}_k(X \cap D_r^n) \geqslant \Psi_k(O, X)\gamma_k r^k \geqslant \gamma_k r^k = \mathrm{vol}_k D_r^k,$$

*where $\gamma_k$ is the $k$-dimensional volume of the standard $k$-dimensional ball of radius 1, $\Psi_k(O, X) \geqslant 1$ is the spherical density function of the minimal surface $X$ at $O$, and $\Psi_k(O, X) = 1$ if and only if $O$ is a regular point of $X$. Since $D_r^k$ is the standard plane $k$-section of the domain $D_r^m$, Conjecture A is valid (see §9), i.e., the volume of any minimal surface passing through the centre of the ball, with boundary on its boundary, is not less than that of the standard central plane section of the ball of the same dimension.*

**Proof.** Put $M = D_r^n$. Then $M_2 = S_r^{n-1}$, $M_1 = \emptyset$. We take $f = |x|$ as $f$, having no critical points on $M \backslash O$. Put $v = \operatorname{grad} f$. Then $|v| = 1$ on $M \backslash O$. Due to Example 1 from §10, we have $\kappa_k(v, x) = r/k$, where $r = |x|$. Apply Corollary 11.2.1. We obtain

$$\mathrm{vol}_k(X\cap D_r^n)\geqslant\left(\lim_{a\to 0}\frac{\Psi(X,f,a)}{\gamma_k q(a)}\right)q(r)\gamma_k,$$

where

$$q(r)=\exp\int_0^r\frac{dr}{\max\limits_{x\in F_r}[\varkappa_k(v,x)\,|\operatorname{grad}f(x)|]}=\exp\int_0^r\frac{dr}{\max\limits_{x\in F_r}(r/k)},$$

because $|\operatorname{grad}f|\equiv 1$. Thus, $q(r)=\exp\int_0^r\frac{kdr}{r}=r^k$. Hence,

$$\lim_{a\to 0}\left(\frac{\Psi(X,f,a)}{\gamma_k a^k}\right)=\lim_{a\to 0}\left(\frac{\mathrm{vol}_k(X\cap D_a^n)}{\gamma_k a^k}\right)=\Psi_k(O,X).$$

Therefore,

$$\mathrm{vol}_k(X\cap D_r^n)\geqslant\Psi_k(O,X)\gamma_k q(r)=\Psi_k(O,X)\gamma_k r^k$$
$$=\Psi_k(O,X)\cdot\mathrm{vol}_k\cdot D_r^k\geqslant\mathrm{vol}_k D_r^k.$$

Q.E.D.

**Corollary 13.1.2.** *Let $X^k\subset D_r^n\subset\mathbb{R}^n$ be a GM-surface passing through the origin, and $O$ the centre of the ball $D_r^n$. Then $\mathrm{vol}_{k-1}S_r^{k-1}\leqslant\mathrm{vol}_{k-1}(X^k\cap\partial D_r^n)$, where $S_r^{k-1}=\mathbb{R}^k\cap\partial D_r^n$ is the standard $(k-1)$-dimensional equator in the sphere $S_r^{n-1}=\partial D_r^n$. In particular, let a complex algebraic surface of degree $p$ be given, i.e., $X\subset D_r^n$, where $n=2m$, $D^{2m}\subset\mathbb{R}^{2m}\cong(\mathbb{C}^m)^{\mathbb{R}}$. Consider the intersection $X^k\cap S_r^{2m-1}=\bigcup_{i=1}^N Y_i^{k-1}$, where, in case of general position, the number of components is finite, $N<\infty$, $Y_i^{k-1}$ are $(k-1)$-dimensional connected, smooth, compact and closed submanifolds in the sphere $S_r^{2m-1}$. Then, among the components $Y_i^{k-1}$ there is $Y_{i_0}^{k-1}$ with "large" volume, viz., there exists a constant $c(k,p,n)$ not depending on the radius $r$ and the surface $X^k\subset\mathbb{C}^m$, so that for a certain subscript $i_0$, the inequality $c(k,p,n)\cdot r^{k-1}\leqslant\mathrm{vol}_{k-1}Y_{i_0}^{k-1}$ holds. In some particular cases, inequality turns into equality.*

**Proof.** Consider the function $\Psi(X,f,r)$, $f=|x|$. Let $A_r=X\cap S_r^{n-1}$. Then, due to Lemma 12.2, $\Psi(X,f,r)\leqslant\mathrm{vol}_k CA_r$, where $CA_r$ is a cone over $A_r$ with vertex at $O$, made up of the radii. Since $v=\operatorname{grad}f$, we have $\kappa_k=\frac{r}{k}$, $r=|x|$, i.e., $\mathrm{vol}_k CA_r=\frac{r}{k}\cdot\mathrm{vol}_{k-1}A_r$ (see §10). By Corollary 13.1.1, we obtain $\mathrm{vol}_k(X\cap D_r^n\geqslant\mathrm{vol}_k D_r^k$. It is clear that $\frac{r}{k}\cdot\mathrm{vol}_{k-1}S_r^{k-1}=\mathrm{vol}_k D_r^k$, and we have $\mathrm{vol}_{k-1}S_r^{k-1}\leqslant\mathrm{vol}_{k-1}A_r$, which proves the first half of Corollary 13.1.2. The second statement follows from the fact that, as is generally known, the number $N$ of the connected components $Y_i^{k-1}$ of the intersection $X\cap S_r^{2m-1}$ in case of general position is bounded above for a fixed degree $p$ by a constant

$c_1(k,p,n)$ irrespective of the choice of the radius $r$ and the surface $X$ of the given degree.

**Hypothesis B.** Let $B^n \subset \mathbf{R}^n$ be a convex, bounded domain in $\mathbf{R}^n$, which is centrosymmetric about a point $O \in B^n$, and with piecewise smooth boundary. Then there exists a constant $c(B^n,k)$ depending only on $B^n$ and $k < n$, so that, for any GM-surface $X^k \subset \mathbf{R}^n$ without boundary, passing through the point $O$, $\Psi_k(O,X) \geqslant 1$, $q \in X^k$, and $X^k$ being globally minimal in $\mathbf{R}^n$ with respect to its boundary $A^{k-1} = X^k \cap \partial B^n$, there exists at least one "sufficiently large" connected component $Y_{i_0}^{k-1}$ of the boundary of $A$, i.e., there exists a component $Y_{i_0}$ among $Y_i^{k-1}$, $A = \bigcup_{i=1}^N Y_i^{k-1}$, for which $\mathrm{vol}_{k-1} Y_{i_0} \geqslant c(B^n,k)$.

Recently, Lê Hông̃ Vân has proved Conjecture A for a sufficiently wide class of domains in $\mathbf{R}^n$. Namely, let $B^n$ be a rectangular parallelepiped or ellipsoid with centre in $O \in \mathbf{R}^n$. Then the area of the intersection of any two-dimensional simply-connected, globally minimal surface passing through the centre of $B^n$ with it is always not less than that of the minimal two-dimensional central plane section [243].

## 13.2. On the Least Volume of Globally Minimal Surfaces Passing through a Fixed Point in a Manifold

**Corollary 13.2.1.** *Let a quadruple $(M^n, f, v, X)$ satisfy all the conditions of Theorem 11.2.1, $M_1 = \emptyset$, and let $x_0 \in M$ be a fixed point (Fig. 37). Suppose that the Morse function $f$ has no minimum or maximum points on $M \backslash x_0$ (saddle points being permitted), attains an absolute minimum on $M$ at $x_0$, and $f(x_0) = 0$. Assume the $f$ is of the form $f(x) = |x|$ in certain local coordinates in the vicinity of $x_0$, where $|x|$ is the length of the vector $x_0x$ in the metric on $M$, and that the integral curves of the field $v$ in the neighbourhood of $x_0$ coincide with those of* $\operatorname{grad} f$ *(which, e.g., occurs for $v = \operatorname{grad} f$ on $M$). Let $X^k \subset M$ be a globally minimal surface in $M$, passing through $x_0$ (see the conditions of Theorem 11.2.1). Then we have the following lower estimate on the volume function $\Psi(X,f,r)$ on the surface $X$, viz.,*

$$\mathrm{vol}_k(X \cap \{f \leqslant r\}) = \Psi(X, f, r) \geqslant \Psi_k(x_0, X) \cdot \gamma_k \cdot q(r) \geqslant \gamma_k \cdot q(r),$$

*where $\gamma_k$ is the $k$-dimensional volume of the $k$-dimensional Euclidean ball of radius 1, $\Psi_k(x_0,X)$ is the density function of $X$ at $x_o = 0$, and*

$$q(r) = \exp \int_0^r \left[\max_{x \in F_r} (\varkappa_k(v, x) \cdot |\operatorname{grad} f(x)|)\right]^{-1} dr.$$

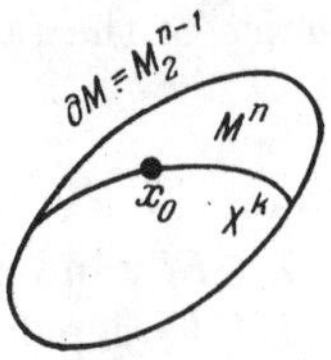


**Fig. 37**

**Proof.** We obtain from Corollary 11.2.1 that $\Psi(X,f,r) \geqslant \left(\lim_{a\to 0} \frac{\Psi(X,f,a)}{\gamma_k q(a)}\right) \cdot q(r)\cdot\gamma_k$. It remains to show that $c = \lim_{a\to 0} \frac{\Psi(X,f,a)}{\gamma_k q(a)} = \Psi_k(x_0, X)$. Consider a sufficiently small neighbourhood $V$ of $x_0$. Then, due to the assumptions and behaviour of $v$ and $f$ in $V = V(x_0)$, we can assume the computation of the constant $c$ to be performed in the neighbourhood of $O \in X^k$ in $\mathbf{R}^n$, with $f(x) = |x|$. Since $F_a \cong S_a^{n-1}$ for small $a$, the field $v$ is orthogonal to $F_a$; therefore, $\kappa_k(v,x) = \frac{a}{k}$, $a = |x|$, and, because of 13.1.1,

$$\lim_{a\to 0} \frac{\Psi(X,\ f,\ a)}{\gamma_k q(a)} = \lim_{a\to 0} \frac{\operatorname{vol}(X \cap \{|x| \leqslant a\})}{\gamma_k a^k} = \Psi_k(x_0,\ X).$$

Q.E.D.

## 13.3. On the Least Volume of Globally Minimal Surfaces Formed by the Integral Curves of a Field $v$

Consider the important case of a GM-surface $X^k \subset M^n$ being made up of the integral curves $\gamma_x(t)$ of $v$, passing through all points $x \in A = \partial_2 X \subset M_2$.

**Corollary 13.3.1.** *Let* $X = \bigcup_{x\in A} \gamma_x(t)$. *Then, with the conditions of Theorem 11.2.1,*

$$\Psi(X,\ f,\ r) \geqslant \lim_{a\to 0}\left(\frac{1}{\bar{h}(a)}\cdot\Psi(X,\ f,\ a)\right)\bar{h}(r),$$

*where* $\bar{h}(r) = \exp\int_{r_0}^{r} [\max_{x\in F_r}(\kappa_k|\operatorname{grad} f|\cos\alpha)]^{-1}\,dr$. *If* $v = \operatorname{grad} f$, *then* $\cos\alpha = 1$ *(in the subsequent sections, see examples of GM-surfaces of the form* $X = \bigcup \gamma_x(t)$, *for which the estimate turns into equality, thereby expressing their volumes).*

Let $M_1 = \emptyset$, $x_0 \in M^n$ be a fixed point, a surface $X^k$ pass through $x_0$, and $f(a) = |a|$, where the point $a$ belongs to a sufficiently small neighbourhood $U(x_0)$, and $v = \operatorname{grad} f$ on $U(x_0)\backslash x_0$. Combining Corollary 13.3.1 with 13.2.1, we then obtain $\Psi(X,f,r) \geqslant \Psi(x_0,X)\cdot\gamma_k\cdot\bar{h}(r)$, where $\bar{h}(r) \equiv q(r)$ if $v = \operatorname{grad} f$ on the whole of $M^n$.

## §14 Nullity of Riemannian, Compact, and Closed Manifolds. Geodesic Nullity and Least Volumes of Globally Minimal Surfaces of Realizing Type

### 14.1. The Definition of the Nullity of a Manifold

Let $M$ be a closed, connected, and compact Riemannian manifold, and $x_0$ an arbitrary point on $M$. Then $M$ can be represented as a cellular decomposition containing only one $n$-dimensional cell $D^n$ of maximal dimension, homeomorphic to an $n$-dimensional disc whose boundary is glued to a certain cell complex $C$ of dimension not exceeding $n-1$, $M^n = D^n \cup C$, where $\partial D^n \to C$. The subcomplex $C$ is not determined uniquely. Then $M \backslash (x_0 \cup C)$ is homeomorphic to the disc $D^n$ without $x_0$. Further, consider on $M$ a continuous function $f$ which is smooth on $M^n \backslash (x_0 \cup C)$, has precisely one minimum at $x_0$, $f(x_0) = 0$, takes the maximal value 1 on $C$, and has no critical points on the complement of $x_0 \cup C$, i.e., on $D^n \backslash x_0$, $0 \leqslant f \leqslant 1$. We call such a function *centred* at $x_0$, and denote the set of all functions centred at $x_0$ by $F(x_0)$. Meanwhile, only $x_0$ is fixed, whereas the subcomplex $C$ varies (Fig. 38). As before, we denote by $v$ an $f$-monotonic vector field on $D^n \backslash x_0$ and assume that $v$, as well as the function $f$, has no singular point on the "annulus" $D^n \backslash x_0$. Taking the triple $(D^n \backslash x_0, f, v)$, we get into the situation considered in the previous sections and can define the function

$$q_{x_0}(r) = \exp \int_0^r \left( \max_{x \in F_r} [\varkappa_k(v, x) \, |\operatorname{grad} f(x)|] \right)^{-1} dr,$$

depending on $x_0$, $f$, and $v$.

**Fig. 38**

**Definition 14.1.1.** We call $\Omega(x_0, f, v) = \gamma_k q_{x_0}(1) \lim_{a \to 0} \left( \frac{a^k}{q_{x_0}(a)} \right)$ the *k-nullity function*, where $q_{x_0}(1) = \lim_{r \to 1} q_{x_0}(r)$. At each point $x_0 \in M$, we consider the value $\Omega_k(x_0) = \sup_{v, f \in F(x_0)} \Omega(x_0, f, v)$ and call $\Omega_k = \inf_{x_0 \in M} \Omega_k(x_0)$ the *k-nullity of the manifold M*.

The value constructed depends only on $M$ and the number $k$, it being clear that $\Omega_k > 0$. For example, if $v = \operatorname{grad} f$ and the function $f$ in a small

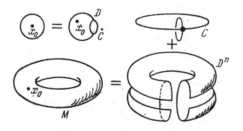

neighbourhood is of the form $f(x) = |x|$, where $x_0$ is taken as the origin in the neighbourhood, then we have $\lim_{a \to 0} \frac{a^k}{q_{x_0}(a)} = 1$ due to Corollary 13.2.1, and also $\Omega_k(x_0, f, v) = \gamma_k \cdot q_{x_0}(1)$.

The concept of geodesic diffeomorphism is formulated by us for the case of an arbitrary smooth, compact Riemannian manifold $M^n$. Let $x_0 \in M^n$ be a fixed point, and $\exp_{x_0} : T_{x_0}M \to M$ a mapping determined by the pencil of geodesics emanating from a point $P$, i.e., for a point $y \in T_{x_0}M$, its image in $M$ is $\gamma(t)$ under the mapping $\exp_{x_0}$, where $t$ is the length of the vector $Oy$ in $T_{x_0}M$, joining $O$ to $y$, and $\dot{\gamma}(0) = \frac{Oy}{t}$. It is generally known that, for small $\varepsilon$, the mapping $\exp_{x_0}$ establishes a diffeomorphism between the disc $D^n(O,\varepsilon)$ embedded into $T_{x_0}M$, with centre at $O$ and radius $\varepsilon$, and its image in $M$, which we denote by $Q^n(x_0,\varepsilon)$. Consider all those values of $t$ for which $\exp_{x_0}$ sets a diffeomorphism between $D^n(O,t)$ and $Q^n(x_0,t)$. Let $R(x_0) = \sup(t)$, i.e., $\exp_{x_0}$ ceases being a diffeomorphism for $t > R(x_0)$. In particular, this means that if $\bar{D}^n(O, R(x_0))$ is a closed disc in $T_{x_0}M$, then $\exp_{x_0}$ is not a diffeomorphism on it, and identifies certain points on the boundary of $\bar{D}^n(O, R(x_0))$. It is also clear that, for all $t < R(x_0)$, the closure $\bar{Q}^n(x_0,t)$ of an open disc in $M$ is homeomorphic to a closed disc in $M$. Thus, the value $R(x_0)$ is the maximal radius of an open geodesic disc $Q^n(x_0,t)$ with centre at $x_0$, which can be inscribed in $M$. We will regard as the radii of the disc all the geodesics emanating from $x_0$, and which are the images under $\exp_{x_0}$ of rays from O in the tangent plane $T_{x_0}M$. For each point $x \in Q^n(x_0, R(x_0))$, there exists precisely one radius joining it to $x_0$. The above diffeomorphism of $D^n(O, R(x_0))$ on $Q^n(x_0, R(x_0))$ is said to be *geodesic*. It is obvious that the disc $\bar{Q}^h(x_0, R(x_0))$ must not necessarily coincide with $M$; in the general case, the complement to $M$ is nonempty. Meanwhile, for certain symmetric spaces $M$ (e.g., of rank 1), this disc completely exhausts by itself the whole manifold.

Consider a compact Riemannian manifold $M^n$, a fixed point $x_0 \in M$, a geodesic diffeomorphism and its corresponding disc $Q^n(x_0, R(x_0))$ consisting of the pencil of geodesic radii emanating from $x_0$. We introduce the natural parameter $r$ on them, which varies from 0 to $R(x_0)$. Let $f(x) = r$ be a smooth function of $Q^n(x_0, R(x_0))\backslash x_0$, where $x = \gamma(r)$, i.e., the value of this function at $x$ equals the distance to $x_0$, measured along the unique radius, or geodesic, joining them. It is obvious that it has no critical points on $Q(x_0, R(x_0))\backslash x_0$. As the $f$-monotonic vector field $v$, we take $\operatorname{grad} f$. Clearly, $|v| \equiv 1$ on $Q(x_0, R(x_0))\backslash x_0$. Consider now the triple $(Q(x_0, R(x_0))\backslash x_0, f, v)$. Then, following §10, we can calculate the deformation coefficient $\kappa_k(v,x)$ for $v$. Proceeding as we do generally for the $k$-nullity of $M$, we determine its geodesic $k$-nullity as follows.

**Definition 14.1.2.** We call

$$\Omega_k^0(x_0) = \gamma_k q_{x_0}(R(x_0)) \cdot \lim_{a \to 0} \frac{a^k}{q_{x_0}(a)}$$

the *geodesic k-nullity function*, where

$$q_{x_0}(r) = \exp \int_0^r \left( \max_{x \in \{f=r\}} \varkappa_k(v, x) \right)^{-1} dr.$$

Remember that $|\operatorname{grad} f| \equiv 1$. Put $\kappa_k(x_0, r) = \max_{x \in \{f=r\}} \kappa_k(v, x)$. We call $\Omega_k^0 = \inf_{x_0 \in M} \Omega_k^0(x_0)$ the geodesic *k-nullity* of the manifold $M$.

Since $v$ is of the form $\operatorname{grad} f$, and $|\operatorname{grad} f| \equiv 1$, we get

$$\lim_{a \to 0} q_{x_0}(a) = \lim_{a \to 0} \exp \left[ \exp \int_0^a \frac{k\,dr}{r} \right] = \lim_{a \to 0} (a^k).$$

We have thereby proved the following lemma.

**Lemma 14.1.1.** *For* $\Omega_k^0(x_0)$,

$$\Omega_k^0(x_0) = \gamma_k q_{x_0}(R(x_0)).$$

## 14.2. The Theorem on the Relation of Nullity with the Least Volumes of Surfaces of Realizing Type

For simplicity, we consider the usual (co)homology theory $H_*^{(*)}$, and an arbitrary variational class $H(x, 0, L')$ introduced in §6. We denote the latter just by $\mathcal{O}(L')$. Recall that it is made up by the surfaces (compact sets) $X$ in $M$, realizing in the homology case the subgroup $L'$ of the group $H_k(M)$, whereas in the cohomology case, the subset $L'$ of $H^k(M)$. Though all the results below are also valid for an arbitrary theory $h_*^{(*)}$ (e.g., for a bordism theory) and an arbitrary class $\mathcal{O}(\widetilde{\mathcal{O}})$, we concentrate our attention on the latter simple case. Geometrically, the globally minimal surface $X$ from $\mathcal{O}(L')$ represents the smallest support of $L$ in $M$, smallest in volume. The question arises as to the limits within which the volume of the support of $L'$ in a fixed manifold $M$ can vary. The upper estimate is supplied by the volume of any concrete representative from $\mathcal{O}(L')$, and the lower by a nontrivial problem which we shall solve with the use of the concept of the nullity of $M$.

Let a dimension $k$ be in the interval $3 \leqslant k \leqslant n-1$ (see the existence theorem for GM-surfaces, §7).

**Theorem 14.2.1.** *Let $X_0^k \subset M^n$ be a globally minimal surface realizing a subgroup (subset) $L' \neq 0$, i.e., $X_0 \in \mathcal{O}(L')$. Then the inequality $\operatorname{vol}_k(X_0) \geqslant (\sup_{x_0 \in X_0} \Psi_k(x_0, X_0))\, \Omega^k \geqslant \Omega^k > 0$ is valid. The value $\Omega_k(M)$ thereby turns out to be a universal constant supplying a lower estimate on the k-dimensional volume of any closed minimal surface realizing nontrivial cycles ((co)cycles) in the manifold M. In general, this estimate cannot be improved, i.e., there exist* *large series of examples where it is attained on concrete minimal surfaces. A similar inequality holds also for the geodesic nullity of* $\Omega_k^0$, *viz.,* $\mathrm{vol}_k X_0^k \geqslant (\sup_{x_0\in X_0}\Psi_k(x_0,X_0))\,\Omega_k^0 \geqslant \Omega_k^0 > 0$. *It is clear that* $\Omega_k^0 \leqslant \Omega_k$.

**Proof.** Consider a decomposition $M = D^n \cup C$, $x_0 \in X_0$, where $X_0$ is a fixed minimal surface from the class $\mathcal{O}(L')$. Consider an arbitrary function $f$ centred at $x_0$ (see §14.1). Let $v$ be an $f$-monotonic field without singularities on $D^n\backslash x_0$. Since all the conditions of Theorem 11.2.1 are fulfilled, we derive the inequality

$$\begin{aligned}\mathrm{vol}_k X_0 &\geqslant \mathrm{vol}_k(X_0\cap\{f\leqslant r\}) = \Psi(X_0, f, r)\\ &\geqslant \lim_{a\to 0}\left(\frac{\Psi(X_0, f, a)}{q_{x_0}(a)}\right)\cdot q_{x_0}(r) = \lim_{a\to 0}\left(\frac{\Psi(X, f, a)}{\gamma_k a^k}\right)\cdot \lim_{a\to 0}\left(\frac{a^k}{q_{x_0}(a)}\right)\cdot \gamma_k q_{x_0}(r).\end{aligned}$$

Hence, letting $r$ tend to unity, we obtain

$$\mathrm{vol}_k X_0 \geqslant \Psi_k(x_0, X_0)\cdot\gamma_k q_{x_0}(1)\lim_{a\to 0}\left(\frac{a^k}{q_{x_0}(a)}\right) = \Psi_k(x_0, X_0)\,\Omega_k(x_0, f, v).$$

Then

$$\mathrm{vol}_k X_0 \geqslant \Psi_k(x_0, X)\,\Omega_k(x_0).$$

Further

$$\begin{aligned}\mathrm{vol}_k X_0 &\geqslant \sup_{x_0\in X_0}[\Psi_k(x_0, X_0)\,\Omega_k(x_0)]\\ &\geqslant \Big(\sup_{x_0\in X_0}\Psi_k(x_0, X_0)\Big)\inf_{x_0\in M}\Omega_k(x_0) = \Big(\sup_{x_0\in X_0}\Psi_k(x_0, X_0)\Big)\,\Omega_k,\end{aligned}$$

which completes the proof of the theorem. Recall that $\Psi_k(x_0, X_0) \geqslant 1$ on $X_0$.

As we show below, there exist such surfaces $X_0$ that $\mathrm{vol}_k X_0 = \Omega_k > 0$. It turns out that there are no singular points at all on them.

**Proposition 14.2.1.** *Let* $X_0 \in \mathcal{O}(L')$ *be a minimal surface for which the inequality in Theorem 14.2.1 turns into equality, i.e.,* $\mathrm{vol}_k X_0 = \Omega_k$ *or* $\mathrm{vol}_k X_0 = \Omega_k^0$. *Then* $X_0$ *is a smooth, compact, and closed submanifold without singularities in* $M$.

**Proof.** We derive from Theorem 14.2.1 that

$$\Omega_k = \mathrm{vol}_k X_0 \geqslant \left(\sup_{x_0\in X_0}\Psi_k(x_0, X_0)\right)\Omega_k,$$

i.e., $\Psi_k(x_0, X_0) \equiv 1$ on the whole of $X_0$. Due to the basic Theorem 7.2.1, it follows that all the points of $X_0$ are regular, and $X_0$ is a smooth submanifold in $M$. Q.E.D.

Furthermore, it turns out that the above class of minimal surfaces made up by the integral curves of the vector field $v$ can also be characterized in terms of the geodesic $k$-nullity function for the manifold $M$.

**Proposition 14.2.2.** *Let $X_0 \in \mathcal{O}(L')$ be a minimal surface, at whose point $x_0$, $\operatorname{vol}_k X_0 = \Omega_k(x_0, f, v)$ holds, with $f$ being a certain function from the class $F(X_0)$ or $\operatorname{vol}_k X_0 = \Omega_k^0(x_0)$. Then $X_0$ is a smooth submanifold in a certain neighbourhood of $x_0$; moreover, $X_0$ wholly consists of integral curves of the vector field $v$ there (which emanate from $x_0$).*

**Remark.** The condition of Proposition 14.2.2 is different from that of Proposition 14.2.1, since

$$\Omega_k = \inf_{x_0 \in M} \sup_{v, f \in F_{x_0}} \Omega(x_0, f, v).$$

**Proof.** It follows from the proof of Theorem 14.2.1 that

$$\Omega_k(x_0, f, v) = \operatorname{vol}_k X_0 \geqslant \Psi_k(x_0, X) \cdot \Omega_k(x_0, f, v),$$

i.e., $\Psi_k(x_0, X_0) = 1$ at $x_0$. Therefore, by Theorem 7.2.1, $X_0$ in a certain neighbourhood of $x_0$ is a smooth minimal submanifold of $M$. Since

$$\Psi(X_0, f, 1) \geqslant \Omega_k(x_0, f, v) = \gamma_k q_{x_0}(1) \lim_{a \to 0}\left(\frac{a^k}{q_{x_0}(a)}\right)$$
$$= \operatorname{vol}_k X_0 \geqslant \Psi(X_0, f, 1),$$

we have

$$\Psi(X_0, f, 1) = q_{x_0}(1) \lim_{a \to 0}\left(\frac{\Psi(X_0, f, a)}{q_{x_0}(a)}\right),$$

because $\lim_{a \to 0} \Psi(X_0, f, a) = \lim_{a \to 0}(\gamma_k a^k)$ in the neighbourhood of $x_0$. We derive that

$$\lim_{r \to 1} \frac{\Psi(X_0, f, r)}{q_{x_0}(r)} = \lim_{a \to 0} \frac{\Psi(X_0, f, a)}{q_{x_0}(a)}.$$

Therefore, as it is nondecreasing on $(0,1)$, the function $\frac{\Psi(X_0,f,r)}{q_{x_0}(r)}$ is, in fact, constant. Thus, the derivative of $\Psi/q$ is identically zero, and in the proof of Lemma 12.6, we have $\Psi' - s\Psi \equiv 0$, i.e., $\Psi' - s\Psi = \Psi' - s \operatorname{vol}_k CA_r$. Hence, for any $r$, $\Psi = \operatorname{vol}_k CA_r$, which means that, for sufficiently small $r$, when the level surface $F_r$ can be regarded as a sphere, two films $CA_r$ and $X_0 \cap B_r$ with the same $k$-dimensional volume for each $r$ span the "contour" $A_r$ contained in the sphere $F_r$. Because the film $X_0$ is minimal, $CA_r$ is also minimal, but then, due to the uniqueness theorem holding in this situation, $X_0 \cap B_r = CA_r$, which completes the proof.

This will soon be applied to completely describing globally minimal surfaces of realizing type in symmetric spaces of rank 1. In conclusion, we explain the geometric meaning of a nullity $\Omega_k$. Since the interpretation is not important for the following, we do not dwell on it for long. We can calculate the nullity $\Omega_k(x_0, f, v)$ in terms of differential geometric invariants of the Riemannian manifold $M$ (e.g., its curvature). It turns out that the maximal deformation coefficient of the vector field $v$ is realized along such trajectories of the field that possess maximal curvature, and along which the $k$-dimensional pencil $\xi$ of nearby integral curves is deformed most. Then $\Omega_k(M)$ is the $k$-dimensional volume of the body obtained by the revolution around a certain plane $\Pi^k$ of $\xi$, which should rotate, always remaining tangent to the plane. The obtained manifold with boundary does not, in general, coincide with the original manifold $M$. However, in many important cases, e.g., for symmetric spaces of rank 1, we can assume that this coincidence does take place in some sense.

### 14.3. Proof of the Reifenberg Conjecture Regarding the Existence of a Universal Upper Estimate of the "Complexity" on the Singular Points of Minimal Surfaces of Realizing Type

**Definition 14.3.1.** We will say that a singular point $x_0$ of a minimal surface $X_0^k$ is of the "first type" if there exists a neighbourhood in $X_0^k$, the union of a number $r(x_0) \geqslant 3$ of $k$-dimensional half-discs (Fig. 39).

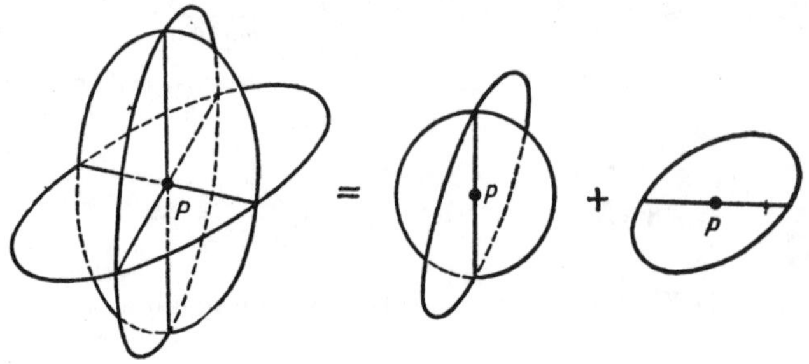


**Fig. 39**

Certain half-discs can form whole ones. If $r(x_0) = 2$, then $x_0$ is a regular point. Singular points of the first type are second less complex on minimal surfaces after regular ones.

**Extended Reifenberg Conjecture.** *Let $M$ be a compact, smooth and closed Riemannian manifold, and $X_0 \in \mathcal{O}(L')$ a minimal surface of realizing type. There therefore exists a constant $F(M,k) = F$ depending only on the dimension $k$ and manifold $M$, so that $r(x_0) \leqslant F$ for any singular point of the first type on any minimal surface $X_0 \in \mathcal{O}(L')$.*

Reifenberg has formulated the conjecture for minimal surfaces of spanning type in Euclidean space, whose boundary is a smooth $(k-1)$-dimensional submanifold $A$. Many an observation is in its favour. For example, if $k = 2$, then, in fact, on each one-dimensional singular line, precisely three sheets of the film converge at the angle of 120° in three-dimensional space (Fig. 40). The quadruple point represented is clearly decomposable into two triple points under the surface minimization. Meanwhile, it should be noted that points of the first type do not at all exhaust all types of singular points of minimal surfaces. We shall soon get acquainted with the singular points of "cone-type," which may be of considerably more complicated structure, and are cones over rather complex manifolds. Therefore, it is pertinent to pose the problem of the upper estimation of the "complexity" of a singular point irrespective of its type. As shows, e.g., Theorem 7.2.1., the most important characteristic of a singularity is its spherical density function $\Psi_k(x_0, X)$; hence, it is natural to solve the problem of an upper estimate immediately.

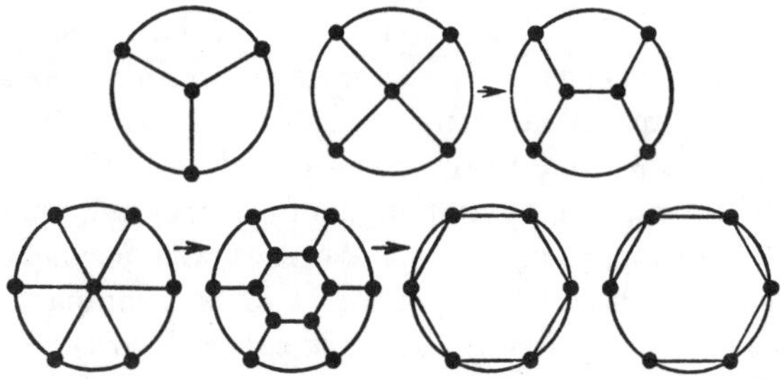

**Fig. 40**

**Proposition 14.3.1.** *Let $M$ be a smooth, closed and compact Riemannian manifold, and $X_0 \in \mathcal{O}(L')$ an arbitrary minimal surface of realizing type in $M$. Then there exists a constant $F = F(M,k)$ depending only on the value $k$ and the manifold $M$, so that, for any $X_0 \in \mathcal{O}(L')$, where $3 \leqslant k \leqslant n-1$ at any point $x_0 \in X_0$, we have $\Psi_k(x_0, X_0) \leqslant F(M,k)$. $F(M,k)$ can be calculated explicitly, and the estimate is generally optimal. In particular, for $X_0 \in \mathcal{O}(L')$, the extended Reifenberg conjecture is valid.*

**Proof.** It follows from Theorem 14.2.1 that $\Psi_k(x_0, X_0) \leqslant \frac{\mathrm{vol}_k(X_0)}{\Omega_k}$, where $\Omega_k > 0$ is the nullity of $M$. Since $M$ is compact, there exists a finite number of minimal surfaces $X_1, \ldots, X_q$ (see Theorem 7.2.1) such that the surface $\widetilde{X} = \bigcup_{i=1}^q X_i$, their union, realizes any subgroup (subset) $L'$ of the group $H_k^{(k)}(M)$. Since $\mathrm{vol}_k(\widetilde{X}) \leqslant \mathrm{vol}_k\left(\bigcup_{i=1}^q X_i\right) < \infty$, we obtain $\Psi_k(x_0, X_0) \leqslant \frac{\mathrm{vol}_k(\widetilde{X})}{\Omega_k}$. Thus, we derive a universal upper estimate and may set $F(M,k) = \frac{\mathrm{vol}_k(\widetilde{X})}{\Omega_k}$. It follows from the construction of $F$ and the examples below that it is generally optimal, i.e., there exists an extensive series of examples of minimal surfaces for which it turns into equality. It is also clear that if a singular point $x_0 \in X_0$ is of the first type (see Definition 14.3.1), then $\Psi_k(x_0, X_0) = \frac{1}{2}\tau(X_0)$.

Q.E.D.

The Reifenberg conjecture can be proved for surfaces of realizing type even without involving Theorem 14.2.1, but then we should apply the procedure to be worked out below in proving the basic minimal surface existence theorem. The second variant of the proof will, however, not be required, since it does not supply an exact estimate.

## §15 Certain Topological Corollaries. Concrete Series of Examples of Globally Minimal Surfaces of Nontrivial Topological Type

### 15.1. Globally Minimal Surfaces Realizing Nontrivial (Co)Cycles in Symmetric Spaces

The considerable difficulty of variational problems of the above type accounts for comparatively few concrete examples of globally minimal surfaces until now, also due to serious difficulties in proving the global minimality of some concrete topological surface. It is remarkable that the above methods based on the constructions, by means of which the general solution existence Theorem 7.2.1 is proved, permit us not only to deduce existence theorems, but also to prove the global minimality of concrete surfaces. We shall demonstrate the efficiency of our method of studying and constructing minimal surfaces in some nontrivial topological situations. We start with a relatively simple case where the underlying Riemannian manifold $M$ is a compact and symmetric space of rank 1 (see the description of such spaces below).

**Theorem 15.1.1.** *(1) Let* $M^{2n} = \mathbb{C}P^n$ *be a complex projective space,* $n > 1$, *and* $\mathbb{C}P^k = X_0^{2k}$, $1 \leqslant k \leqslant n-1$, *standardly embedded complex projective subspaces (whose description is given below), each of which realizes a generator of the* $2k$*-dimensional cohomology group of the ring* $H^*(\mathbb{C}P^n, \mathbb{Z}) = \mathbb{Z}[x_2]/(x_2^{n+1})$, *where* $\dim x_\alpha = \alpha$.

*(2) Let* $M^{4n} = \mathbb{Q}P^n$ *be a quarternion projective space,* $n \geqslant 1$, *and* $\mathbb{Q}P^k = X_0^{4k}$, $1 \leqslant k \leqslant n-1$, *standardly embedded quarternion projective subspaces (whose embedding is given below), each of which realizes a generator of the* $4k$*-dimensional cohomology group of the ring* $H^*(\mathbb{Q}P^n; \mathbb{Z}) = \mathbb{Z}[x_4]/\left(x_4^{n+1}\right)$.

*(3) Let* $M^n = \mathbb{R}P^n$ *be a real projective space,* $n \geqslant 2$, *and* $\mathbb{R}P^k = X_0^k$, $1 \leqslant k \leqslant n-1$, *standardly embedded real projective spaces, each of which realizes a generator of the* $k$*-dimensional cohomology group of the ring* $H^*(\mathbb{R}P^n, \mathbb{Z}_2) = \mathbb{Z}_2[x_1]/\left(x_1^{n+1}\right)$.

*(4) Let* $M^{16} = F_4/\mathrm{Spin}(9)$ *be a symmetric space containing a standardly embedded sphere* $S^8 = X_0^8$ *(whose embedding is described below), realizing a generator of the eight-dimensional cohomology group of the ring* $H^*(M^{16}, \mathbb{Z}) = \mathbb{Z}[x_8]/(x_8^3)$.

*Then each of the above submanifolds* $X_0^p$ *is not only totally geodesic in* $M$ *but is also a globally minimal surface, with* $\mathrm{vol}_p X_0^p = \Omega_p^0(M)$, *i.e., it has the least volume among all minimal surfaces of the same dimension, realizing nontrivial (co)cycles. Moreover, any globally minimal surface* $X_0^p \subset M$ *realizing some nontrivial co-cycle in dimension* $p$ *coincides up to isometry in* $M$

*with the above submanifold $X_0^p$ realizing a generator of the cohomology group of the same dimension. In particular, for each p, there can be only one globally minimal surface $X_0^p$ in M. The above list of manifolds M exhausts all symmetric spaces of rank 1.*

**Theorem 15.1.2.** *(1) Let $M = G^{\mathbf{C}}_{p+q,q}$, $1 \leqslant q \leqslant p$, be a complex Grassmannian manifold of q-dimensional planes in (p+q)-dimensional space, and $\mathbf{C}P^s$, $1 \leqslant s \leqslant p$, a set of standardly embedded totally geodesic submanifolds (whose embedding is described below), each of which is diffeomorphic to complex projective space and realizes an element $x_2^s$ of the cohomology ring $H^*(M, Z)$ isomorphic to the polynomial ring $\mathbf{Z}[x_2, x_4, \ldots, x_{2q}]$ up to dimension 2p. All these submanifolds are then globally minimal surfaces in $G^{\mathbf{C}}_{p+q,q}$, and, moreover, are of the least volume $\mathrm{vol}_{2s}(\mathbf{C}P^s) = \Omega^0_{2s}(M)$. Besides, dimensions $2, 4, 6, \ldots, 2p$ are unique in which a minimal surface $X_0^k \subset M$ exists, so that $\mathrm{vol}_k X_0 = \Omega^0_k(M)$. In other words, $\mathrm{vol}_k X_0 > \Omega^0_k(M)$ for $k \neq (2, 4, 6, \ldots, 2p)$ and any surface $X_0 \in O(L'), L' \neq 0, L' \subset H^k(M; \mathbf{Z})$.*

*(2) Let $M = G^{\mathbf{Q}}_{p+q,q}$, $1 \leqslant q \leqslant p$, be a quaternion Grassmannian manifold, and $\mathbf{Q}P^s$, $1 \leqslant s \leqslant p$, a set of standardly embedded totally geodesic submanifolds, each of which is diffeomorphic to a quarternion projective space and realizes the element $x_4^s$ of the cohomology ring $H^*(M, \mathbf{Z})$ isomorphic to the polynomial ring $\mathbf{Z}[x_4, x_8, x_{12}, \ldots, x_{4p}]$ up to dimension $4p+2$. All these submanifolds are then globally minimal surfaces in $G^{\mathbf{Q}}_{p+q,q}$, and moreover, of the least volume $\mathrm{vol}_{4s}(\mathbf{Q}P^s) = \Omega^0_{4s}(M)$. Besides, the values $k = (4, 8, 12, \ldots, 4p)$ are unique dimensions in which there exists a minimal surface $X_0^k \subset M$ such that $\mathrm{vol}_k(X_0^k) = \Omega^0_k(M)$.*

The proofs will be given in the next sections.

## 15.2. Compact Symmetric Spaces and Explicit Form of a Geodesic Diffeomorphism

We now consider symmetric spaces. Recall that a compact manifold $M$ is said to be *symmetric* if it is representable as a homogeneous space $M = \widetilde{G}/\widetilde{H}$, where $\widetilde{G}$ and $\widetilde{H}$ are compact Lie groups, $\widetilde{H} \subset \widetilde{G}$, and an involutive automorphism $\sigma, \sigma^2 = 1$ is defined on $\widetilde{G}$, so that $\widetilde{H} = \{g \in \widetilde{G} | \sigma(g) = g\}$. For simplicity, we will consider only simply-connected and irreducible symmetric spaces not representable as direct products of other symmetric spaces.

We shall need some simple facts from symmetric-space theory, and list them without proofs, referring the reader, e.g., to [46] and [87].

**Definition 15.2.1.** Let a point $e \in M$. Consider all geodesics $\gamma$ emanating from $e$, and distinguish on each the first point conjugate to $e$ along this geodesic. Due to the compactness of the manifold $M$, the first conjugate point necessarily exists on each $\gamma$. The set of all such points will be designated by $C(e)$.

**Definition 15.2.2.** A minimum point for a point $e$ along a geodesic $\gamma$ is such a point $x$ on the curve $\gamma$ that its line segment from $e$ to $x$ is minimal and minimizes the length of the arc between the ends, no greater line segment being minimal. The set $M(e)$ of all the minimum points for $e$ is called the *locus of minimum points* (or *cuts*) for $e$.

It is generally known, e.g., for a simply-connected, compact, and symmetric space $M$, that the equality $M(e) = C(e)$ always holds. It is clear that $\dim C(e) \leqslant n-1$, $n = \dim M$; therefore, the manifold $M$ is representable as the decomposition $M = C(e) \cup (M \backslash C(e))$, where (if $M(e) = C(e)$), the manifold $M \backslash C(e)$ is homeomorphic to a disc of dimension $n$.

Let $M = \widetilde{G}/\widetilde{H}$, where $\widetilde{G}$ and $\widetilde{H}$ are compact Lie groups. Since $M$ is simply-connected, we can assume that $\widetilde{G}$ is simply-connected, too. Let $G$ and $H$ be the Lie algebras of $\widetilde{G}$ and $\widetilde{H}$, respectively. It is generally known that $M$ can be embedded into $\widetilde{G}$ as a Cartan model $V \subset M$, where $V = \{g\sigma(g^{-1})\}$ is a totally geodesic submanifold, and $\sigma$ is an involutive automorphism of $\widetilde{G}$. Then $\widetilde{H} = \{g \in \widetilde{G} | \sigma(g) = g\}$ ($\widetilde{H}$ being connected for $\widetilde{G}$ is simply-connected), $V = g\{\widetilde{G} \in \sigma(g) = g^{-1}\}$, and the projection $p : \widetilde{G} \to V$, $p(g) = g\sigma(g^{-1})$ determines the principal fibre bundle $\widetilde{G} \xrightarrow{\widetilde{H}} M \cong V$. The natural orthogonal decomposition $G = B + H$ arises in $G$, where $H = T_e(\widetilde{H})$, $B = T_e(V)$, $V = \exp(B)$, and $e$ is the unit element in $\widetilde{G}$. Let $P \subset B$ be a Cartan subalgebra, i.e., the maximal Abelian subspace in $B$. Then $G$ admits the following decomposition with respect to the adjoint representation $\mathrm{ad}_P$ on $G$, viz.,

$$G = Z_0 + P + \sum V_{\hat{\alpha}},$$

where $Z_0 = C(P) \cap H$, $C(P)$ is the centralizer of $P$ in $G$, $\hat{\alpha}$ are linear functionals on $P$, and each subspace $V_{\hat{\alpha}} \subset G$ is a two-dimensional subspace over the field of real numbers, invariant under $\mathrm{ad}_P$. With respect to a certain convenient basis for the plane $V_{\hat{\alpha}}$, the operators $\mathrm{ad}_P$ are of the form

$$\mathrm{ad}_p(G)|_{V_{\hat{\alpha}}} = \begin{pmatrix} 0 & 2\pi\hat{\alpha} \\ -2\pi\hat{\alpha} & 0 \end{pmatrix},$$

where $\hat{\alpha}$ is a functional on $P$, and $\hat{\alpha}(p) \in \mathbf{R}$ if $p \in P$; $\hat{\alpha}$ are called the roots of the symmetric space $M$. Since there may be coincident roots among $\hat{\alpha}$ (up to sign), we introduce the subspaces $\widetilde{V}_{\hat{\alpha}} = \sum_{\hat{\beta}=\pm\hat{\alpha}} V_{\hat{\beta}}$; in particular, $\widetilde{V}_{\hat{\alpha}} \equiv \widetilde{V}_{-\hat{\alpha}}$. It is clear that $\widetilde{V}_{\hat{\alpha}} = \left(\widetilde{V}_{\hat{\alpha}} \cap H\right) + \left(\widetilde{V}_{\hat{\alpha}} \cap B\right)$. Let $P' \subset G$ be a Cartan subalgebra in $G$. It is known that we can then assume $P' \supset P$, and if $\{\hat{\theta}_i\}$ is the complete set of roots of $G$ (functionals on $P'$), then all $\hat{\alpha}$ on $P$ are the restriction on $P \subset P'$ of those $\hat{\theta}_i$ which are other than zero on $P$, the multiplicity of each $\hat{\alpha}$ being equal to the number of $\hat{\theta}_i$ whose restriction on $P$ it is. We call the set of all points $x \in P$ such that $\hat{\alpha}(x) \equiv 0 \pmod 1$ (i.e., $\hat{\alpha}(x)$ is an integer) for a certain $\hat{\alpha}$ the *diagram* D $(G; H)$ of an orthogonal involutive Lie algebra $G = H + B$, $\sigma(H) \equiv H$, $\sigma(B) = -B$, corresponding to a simply-connected symmetric space $V \cong M$.

By $h_{\hat{\alpha}}$, we denote an element of the algebra $P$ such that the vector $(0, h_{\hat{\alpha}})$ is orthogonal to the hyperplane $\hat{\alpha}^{-1}(0)$, and in addition, $\hat{\alpha}(h_{\hat{\alpha}}) = 2$. The integral lattice generated by $\{h_{\hat{\alpha}}\}$ is called the *lattice of the pair* $(G, H)$. The points of the manifold $M$, conjugate to $e$, are completely described in terms of $D(G; H)$.

**Proposition 15.2.1 (see [87]).** *Let $x \in P \subset B$. A point $\exp(x) \in V$ is conjugate to a point $e \in V$ along a geodesic $\gamma^*(t) = \exp(tx)$ if and only if $x \in D(G; H)$ and $\hat{\alpha}(x) \neq 0$, i.e., $\hat{\alpha}(x) \equiv 0 \pmod 1$, whereas $\hat{\alpha}(x) \neq 0$ for a certain root $\hat{\alpha}$. If $\widetilde{V}_{\hat{\alpha}} = \sum_{\hat{\beta}=\pm\hat{\alpha}} V_{\hat{\beta}}$, then the kernel of the differential of the mapping $\exp_e : B \to V$ at the point $x \in P \subset B$, $\exp_e x = e^x$ is a linear subspace in $B$, obtained by a Euclidean parallel displacement into $x$ of the subspace $\sum_{\hat{\alpha}} \widetilde{V}_{\hat{\alpha}} \cap B$, where summation is over all $\hat{\alpha}$ such that $\hat{\alpha}(x) \equiv 0 \pmod 1$, $\hat{\alpha}(x) \neq 0$. The multiplicity of $e^x$ as the conjugate point along the geodesic $\exp(tx)$ equals $\dim[C(e^x) \cap \widetilde{H}]/[C(x) \cap \widetilde{H}]$, where $C(x)$ and $C(e^x)$ are the centralizers of the elements $x$ and $e^x$.*

Since the orthogonal Killing $\langle x, y\rangle$ metric is fixed in the algebra $G$, all the roots $\hat{\alpha}$ on $P$ can be realized by the vectors $\alpha$ in the subalgebra $P$: Namely, consider a vector $\alpha \in P$ such that $\hat{\alpha}(x) \equiv \langle x, \alpha\rangle$ for any $x \in P$, with $\alpha$ existing and being determined uniquely. Then we have $\hat{\alpha}(\alpha) = |\alpha|^2$, where $|\alpha|^2 = \langle \alpha, \alpha\rangle$. Since we are interested in the set of the first conjugate points $C(e) = M(e)$, we consider only those vectors $x \in P$ for which $\hat{\alpha}(x) = 1$ for a certain $\hat{\alpha}$, and then $C(e) = \{e^x\}$, where the $x$ are chosen in the above way. The ends of all $x \in P$ lie in the hyperplanes $\Pi_{\alpha'}$ orthogonal to $\alpha$ and passing through the ends of $\alpha' = \alpha/|\alpha|^2$ because $\hat{\alpha}(\alpha') = 1$. It is clear that $|\alpha'| = 1/|\alpha|$ and $\alpha' = h_{\hat{\alpha}}/2$, i.e., all $\alpha'$ are completely reconstructible from the lattice of the pair $(G; H)$. Meanwhile, we consider the collection of all roots $\alpha \in P$, without restricting ourselves only to simple ones. We call $\alpha'$ the *s-root* associated with $\alpha$. Note that an s-root $\alpha'$ is, in fact, a root of the algebra $P$ if and only if $|\alpha| = |\alpha'| = 1$. Consider an open set $\widetilde{K}_1 \subset P$ consisting of all points lying on the rays emanating from 0 and ending on the first plane $\Pi_{\alpha'}$ encountered, where $\alpha'$ ranges over all s-roots of $P \subset B$. Note that none of the $\Pi_{\alpha'}$ passes through 0. It is clear that $\widetilde{K}_1$ is diffeomorphic to an open ball $B^l$, where $l = \dim P =$ rank of $V$. Put $\widetilde{K}_t = t \cdot \widetilde{K}_1$, where $t$ is the similarity coefficient, $0 \leqslant t \leqslant 1$, $\widetilde{F}_t = \partial \widetilde{K}_t$. It is evident that $\widetilde{F}_t$ is homeomorphic to the sphere $S^{l-1}$. Consider the adjoint action of the group $\widetilde{H}$ on $B$. Then the orbits $\widetilde{H}(x)$ orthogonal to the plane $P$ at $x$ are of all $x \in \widetilde{K}_t$. Put $\widetilde{B}_t = \bigcup_{x \in K_t} \widetilde{H}(x)$, $\widetilde{S}_t = \bigcup_{x \in F_t} \widetilde{H}(x)$. Then $\widetilde{S}_t = \partial \widetilde{B}_t$, with $\widetilde{B}_t$ and $\widetilde{S}_t$ being symmetric with respect to the origin $O$. Consider $B_t = \exp_e(\widetilde{B}_t)$ and $S_t = \exp_e(\widetilde{S}_t)$, $S_t = \partial B_t$, $F_t = \exp_e(\widetilde{F}_t)$; $K_t = \exp_e(\widetilde{K}_t)$; $\partial K_t = F_t$.

**Lemma 15.2.** *Let $M^n \cong V$ be a compact, simply connected and symmetric space. Then $S^1 = C(e) = M(e)$, $B_t$ is diffeomorphic to the open disc $D^n$ for $0 < t \leqslant 1$ with centre at the point $e \in M$, and $S_t$ is homeomorphic to the sphere $S^{n-1}$ for $0 < t < 1$.*

The proof is obvious.

The set $S_1$ is not homeomorphic to the sphere and coincides with the penultimate skeleton of the manifold $M$ in its decomposition into cells, $\dim S_1 \leqslant n-1$.

Thus, we have constructed a diffeomorphism $D^n \to B_1 = B(e) \subset M$, where the radii of the ball $B_1$ are geodesics in $M$, i.e., this is a geodesic diffeomorphism. Further, the closure $B_1 = M$, and $M \backslash \partial B_1 = M \backslash S_1 \cong B_1$, since the complement of the locus of cuts of a simply-connected manifold is diffeomorphic to the open disc $D^n$ (see [46]). Hence, inflating the ball $B_t$ from $e$, we finally exhaust the whole of $M$ by the closure of $B_1$.

### 15.3. Explicit Computation of the Deformation Coefficient of a Radial Vector Field on a Symmetric Space

Consider a simply-connected symmetric space $M^n$. Let $e \in M$ be a fixed point that can be identified with the unit element of a group $\widetilde{G}$ in realizing $M$ in it. Consider a geodesic diffeomorphism and the corresponding maximal open disc $Q^n(e, R(e))$. We realize $M$ in the motion group $\widetilde{G}$ as a Cartan model $V$. Suppose that $B \subset G$, and the ball $\widetilde{B}_t \subset \widetilde{B}_1 \subset B$. We inscribe the standard ball $\widetilde{Q}_t$ of greatest radius and centre at $O$ into the ball $\widetilde{B}_t$, certainly, of generally variable radius. It then becomes clear that the radius $r = r(t)$ of $\widetilde{Q}_t$ equals $t|\alpha_0'|$, where $\alpha_0$ is a root (not necessarily simple) of greatest length, i.e., the $s$-root $\alpha_0'$ is the shortest among all others of the algebra $P \subset B$ (Fig. 41).

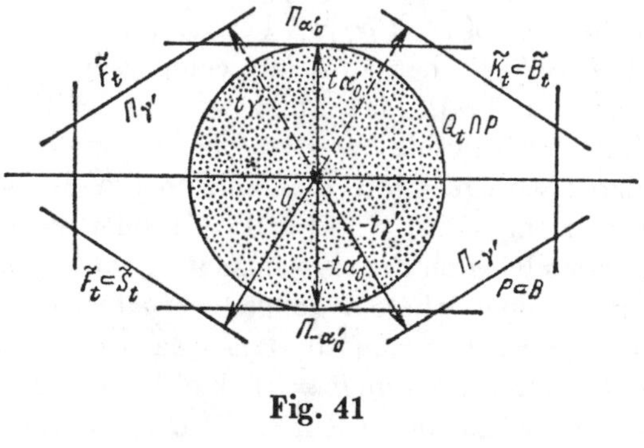


**Fig. 41**

Note that there may be several such $\alpha_0'$. We project the whole construction into the manifold $M$. Put $Q_t = \exp_e(\widetilde{Q}_t)$. Then $Q_t$ is an open ball in $M$ of constant radius $r = r(t) = t|\alpha_0'|$, filled with the geodesic radii emanating from $e$, and $\partial Q_t$ is a sphere smoothly embedded into $M$ for $0 < t < 1$. (We shall sometimes write $Q(r)$ instead of $Q_t$.) In the notation of the previous items, we have $Q(e, R(e)) = Q_1$, $R(e) = |\alpha_0'|$. Put $R_t = \partial Q_t = \partial Q(r)$. Then the sphere $R_t$ is the diffeomorphic image of the standard sphere $\widetilde{R}_t = \partial \widetilde{Q}_t \subset B$ for $t < 1$, or $r < |\alpha_0'|$. With the following in mind, we fix the sphere $\widetilde{R}_1 \subset B$ of radius $|\alpha_0'|$. Note that the mapping $\exp_e$ is not, in general, a diffeomorphism on $R_1$.

Consider an arbitrary pair $(x, \Pi_x^{k-1})$, where $x \in \partial Q(r)$, $r < |\alpha_0'|$. Then $(x', \Pi_{x'}^{k-1})$ is determined uniquely, with $x' \in \widetilde{Q}(r) \subset B$, $\exp_e(x') = x$, $d(\exp_e)$ $('\Pi_{x'}^{k-1}) = \Pi_x^{k-1}$, and $d(\exp_e)$ denoting the differential of $\exp_e$ at the point $x' \in B$. Consider a ray $(O, x')$ in $B$, and carry out a parallel displacement of the plane $'\Pi_{x'}^{k-1}$ along the ray into a point $y$ at which the extension of the ray meets $\widetilde{R}_1 \subset B$. We designate the obtained plane by $f_t(\Pi_x^{k-1})$. By the definition of a geodesic diffeomorphism, the velocity vector $\dot{\gamma}$ at a point $x \in \partial Q(r)$ is orthogonal to $\partial Q(r)$, i.e., $\Pi_x^{k-1} \subset T_x^{n-1}(R_t)$. But then, due to the properties of exp, we have $'\Pi_{x'}^{k-1} \subset T_x^{n-1}(\widetilde{R}_t)$, or $f_t(\Pi_x^{k-1}) \subset T_y(\widetilde{R}_1)$, where $y$ denotes the end of $(O, x')$ meeting $\widetilde{R}_1 \subset B$. We will denote by $\gamma = \gamma(r)$ a geodesic $\gamma$ parametrized by arc length $r$, and the same geodesic parametrized by $t (0 \leqslant t \leqslant 1)$ by $\gamma^* = \gamma^*(t)$. We are interested in the coefficient

$$\varkappa_k(r) = \max_{x \in \partial Q(r)} (\varkappa_k(v, x) \,|\, \operatorname{grad} f(x)\,|) = \max_{x \in \partial Q(r)} \varkappa_k(v, x)$$

involved in stating all basic theorems from §11–15. Since $\kappa(r)$ is unaltered under $e$-preserving automorphisms of the space $M$, and

$$\varkappa_k(r) = \varkappa_k(v, x, \Pi_x^{k-1}), \quad x = \gamma(r) \quad (r = r_\gamma = t\,|\gamma|)$$

for a certain pair $(x, \Pi_x^{k-1})$ (see above), it suffices to study the behaviour of the coefficient $\kappa_k(v, x, \Pi_x^{k-1})$ only for the points $x' \in P \subset B$. Due to the choice of the ball $Q(|\alpha_0'|)$, the whole geodesic $\gamma^*(t)$, $0 \leqslant t \leqslant 1$, $\gamma(r) = x$, then wholly belongs to the torus $T = \exp(T)$, and $x = \exp(x') \in T$.

**Remark.** $\kappa_k(r) < \infty$ if $r < |\alpha_0'|$, since the transformation of $\kappa_k(r_0)$ into infinity for a certain $r_0$ would mean the existence of conjugate points in $Q(|\alpha_0'|)$, which is impossible due to its choice. Recall that $M(e) = C(e)$; therefore no two geodesic radii intersect in $Q(|\alpha_0'|)$. Recall also that $\gamma^*(1) = \exp_e(\tilde{y})$, $\gamma = |(O, \tilde{y})|$, $y \in \widetilde{F}_1$. Let $x \in T$ be a regular element. Then the tangent plane $T_x(R_t)$ admits the orthogonal decomposition $T_x(R_t) = A_x + A_x' + A_x''$, where $A_x = T_x(T \cap R_t)$, $(A_x' + A_x'')$ is the tangent plane to the orbit $\widetilde{H}(x)$ of $x$ under the adjoint action of $\widetilde{H}$ on $M$, $\widetilde{H}(x)$ being orthogonal to the torus $T$ at $x$. We now describe the planes $A_x'$ and $A_x''$. $\widetilde{H}(x)$ contains an orbit $\widetilde{H}'(x) = C\gamma^*(1)/C(x)$, where $\gamma(r) = x$, $\gamma^*(1) \in F_1 = \partial K_1$, $\gamma(0) = e$, $C(x)$ is the centralizer of $x$ in the group $\widetilde{H}$, and $C\gamma^*(1)$ is the centralizer of $\gamma^*(1)$ in $\widetilde{H}$. It is clear that $C\gamma^*(1)$ coincides with the centralizer in $\widetilde{H}$ of the whole geodesic $\gamma(r')$, $0 \leqslant r' \leqslant |\alpha_0'|$, since $Q(|\alpha_0'|)$ contains no points conjugate of $e$, from which it follows, due to the compactness of the symmetric space $M$, that the plane $T_x\widetilde{H}'(x)$ is completely filled with the vectors $W(x)$ of the Jacobi fields $W$ vanishing only at the two points $e$ and $\gamma^*(1)$, i.e., $\dim \widetilde{H}'(x)$ equals the multiplicity of $\gamma^*(1)$ with respect to $e$ along $\gamma^*(t)$. Put $A'_x = T_x\widetilde{H}'(x)$ and take the orthogonal complement to $A'_x$ in $T_x\widetilde{H}(x)$ as $A''_x$. Since, by assumption, $x$ is regular, $\dim T + \dim \widetilde{H}(x) = n$. Let $x \in T$ be singular, i.e., $\dim T + \dim \widetilde{H}(x) < n$. Construct the decomposition $A_x + A'_x + A''_x$. As well as for a regular element, we put $A_x = T_x(T \cap R_t)$, $\dim A = l - 1$, and consider the orbits $\widetilde{H}(x)$ and $H'(x) \subset \widetilde{H}(x)$ (see the definition above). We again put $A'_x = T_x\widetilde{H}'(x) \subset T_x\widetilde{H}(x)$, and determine $A''_x$ as the orthogonal complement of the sum $A_x + A'_x$ in $T_x(R_t)$. It is clear that $(A'_x + A''_x) \supset T_xH(x)$ for a singular element; however, $A'_x + A''_x$ does not any more coincide with $T_xH(x)$ in contrast to the regular case. Note that, for an arbitrary $x \in T$, $\dim A_x = l - 1$, $\dim A'_x = \nu =$ multiplicity of the point $\gamma^*(1)$ conjugate of $e$ along $\gamma^*(t)$, and, if $\gamma^*(t) = \exp_e(t\alpha')$, then $\nu$ is the multiplicity of the root $\alpha$ in the algebraic sense, which follows from Proposition 15.2.1, i.e., $\nu = \dim A'_x = \dim(\sum_{\hat{\alpha}} \widetilde{V}_{\hat{\alpha}} \cap B)$, where $\hat{\alpha}(a) \equiv 0 \pmod 1$, $\hat{\alpha}(a) \not\equiv 0$. Put $A_y = f_t(A_x)$, $A'_y = f_t(A'_x)$, $A''_y = f_t(A''_x)$, where $y \in R_1 \subset B$, and is the end of the ray $(O, x')$. Note that the decomposition $T_y(\widetilde{R}_1) = A_y + A'_y + A''_y$ depends only on the geodesic $\gamma(r) = \gamma^*(t)$, $r = t|\gamma|$, and does not depend on the position of $x$ on $\gamma^*(t)$. In other words, it is completely determined by the point $y \in \widetilde{R}_1$, because the rays $(O, y)$ and $\gamma^*(t)$ determine each other (Fig. 42).

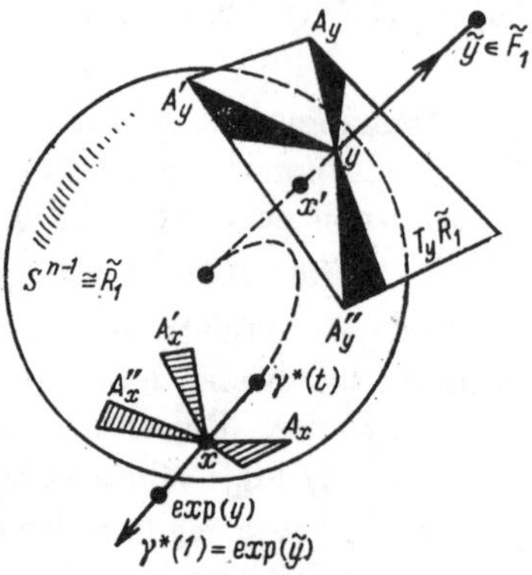


**Fig. 42**

**Lemma 15.3.1.** *Let $M$ be a compact, simply-connected, and symmetric space, $e \in M^n$, $2 \leqslant k \leqslant n-1$. Consider the above canonical geodesic diffeomorphism with centre at the point $e$, and a geodesic $\gamma^*(t) = \exp(t\alpha'_0)$, where $0 \leqslant t \leqslant 1$, $\alpha'_0$ any of the s-roots of the algebra $P \subset B$ of least length, $\gamma(r) = \gamma^*(t)$, $r = t|\alpha'_0|$. We fix the decomposition $T_x(R_t) = A_x + A'_x + A''_x$ at each point*

*$x = \gamma(r)$. Let $T_y(R_1) = A_y + A'_y + A''_y$ be the corresponding decomposition at the point $y \in \widetilde{R}_1$, not depending on $r$. We consider a plane $\Pi_x^{k-1}$ at each $x = \gamma(r)$, so that the plane $f_t(\Pi_x^{k-1}) = \Pi_y^{k-1}$ is the same one fixed in $T_y(\widetilde{R}_1)$, not depending on $r = t|\alpha'_0|$, i.e., the family of $\{\Pi_x^{k-1}\}$ along the geodesic $\gamma(r)$ consists of parallel planes. We choose the plane $\Pi_y^{k-1}$ in accordance with the following three cases and value of $k$, viz.,*

*(1) if $2 \leqslant k \leqslant \nu+1$, where $\nu$ is the multiplicity of the root $\alpha_0$ corresponding to the s-root, $\alpha'_0$, then we take a $(k-1)$-dimensional plane such that $\Pi_y^{k-1} \subseteq A'_y$, $\dim A'_y = \nu$ as $\Pi_y^{k-1}$;*

*(2) if $\nu_2 \leqslant k \leqslant n-l+1$, where $l = \dim P =$ rank of $M$, then we take a $(k-1)$-dimensional plane such that $A'_y \subset \Pi_y^{k-1} \subseteq (A'_y + A''_y)$, $\dim(A'_y + A''_y) = n - l$ as $\Pi_y^{k-1}$; its position is described in [448];*

*(3) if $n-l+1 < k \leqslant n-1$ (this case being possible only for $l \geqslant 3$), then we take a $(k-1)$-dimensional plane such that $A'_y + A''_y \subset \Pi_y^{k-1}$ as $\Pi_y^{k-1}$; its position is described in [448].*

*We then assert that the coefficient $\kappa_k(v, x, \Pi_x^{k-1})$, where $x = \exp(t\alpha'_0)$ and $\Pi_x^{k-1}$ have been chosen in the above way, is the maximal coefficient $\kappa_k(r)$.*

*Conversely, let $\kappa_k(r)$ be the maximal coefficient. Then there always exists a pair $(x, \Pi_x^{k-1})$ such that $\kappa_k(v, x, \Pi_x^{k-1}) = \kappa_k(r)$, where $x = \gamma(r)$ and $\gamma$ is the geodesic radius. It turns out that $\gamma(r) = \gamma^*(t) = \exp(t\alpha'_0)$, $0 \leqslant t \leqslant 1$, where $\alpha'_0$ is one of the s-roots $\alpha'$ of least length, $r = t|\alpha'_0|$, and we can assume that $\Pi_x^{k-1}$ is determined only by $\alpha'_0$, whereas the plane $f_t(\Pi_x^{k-1}) \subset T_y(\widetilde{R}_1)$ does not depend on $x$, i.e., does not alter its position when $r$ varies from 0 to $|\alpha'_0|$. Besides, the position of $f_t(\Pi_x^{k-1})$ is described in accordance with cases (1)–(3).*

**Proof.** Consider a sufficiently narrow $k$-dimensional pencil of geodesics $f_{\alpha'_0}$ emanating from $e$, which has the geodesic $\exp(t\alpha'_0)$ as its axis, and a sufficiently small ball $\Delta_{\alpha'_0}$ of dimension $k-1$, placed in $\exp_x(\Pi_x^{k-1})$ with centre at $x$ as the base, where $\Pi_x^{k-1}$ is chosen in accordance with the conditions of the lemma. In contrast, we consider another pencil $f_d$ of geodesics with axis $\exp(td)$, $d \in P \subset B$, $d \in \widetilde{F}_1$, $\alpha'_0 \in \widetilde{F}_1 \cap \widetilde{Q}(|\alpha'_0|)$, having a small ball $\Delta_d$ of the same radius as for $\Delta_{\alpha'_0}$ as the base, and placed at the point $\gamma_d(r)$, where $d = \dot{\gamma}_d$ and $\gamma_{\alpha'_0}(r) = x$, $\alpha'_0 = \dot{\gamma}_{\alpha'_0}$. We need to compare the $k$-dimensional volumes $\mathrm{vol}_k(f_{\alpha'_0})$ and $\mathrm{vol}_k(f_d)$ of $f_{\alpha'_0}$ and $f_d$.

Since $M \cong V \subset \widetilde{G}$, $\widetilde{G}$ can be embedded into a unitary group $U(N)$, and $U(N)$ into a sphere of constant radius in Euclidean space $\mathbf{R}^\varphi$, where $\varphi = 2N^2$, of $N \times N$-matrices. Then all metric calculations can be carried out in the standard Euclidean metric inducing the Killing metric both on $U(N)$ and $\widetilde{G}$. Meanwhile, we shall use the fact that the inner automorphisms of the group $\widetilde{G}$ are orthogonal rotations in $\mathbf{R}^\varphi$.

Three types of directions in accordance with the decomposition $T_y(\widetilde{R}_1) = A_y + A'_y + A''_y$, with $y$ being the point where the vector $d$ meets the sphere $R_1$, $y = y(d)$, are naturally distinguished from the geodesic ones of $f_d$ (as well as $f_{\alpha'_0}$). Lifted to the space $B$, $f_d$ turns into a narrow cone of the rays of $\tilde{f}_d$ stretching from the point $O$ to a sufficiently small ball on $\widetilde{R}_1$, and then on $\widetilde{S}_1$. The directions to the points of the plane $A_y$, which are sufficiently close to $y$, are Euclidean, i.e., projected by exp into $M$, they get into the torus $T$. The directions to the points of the plane $A'_y$ are conjugate (sinusoidal), i.e., exp applied, they change into geodesics infinitesimally close to $\exp(td)$, intersecting it at a finite point $\exp(d) = \exp(\tilde{y}_d) \in \partial B_1$. The directions to the points of the plane $A''_y$ if the element $x = \gamma_d(r)$ is regular are interpreted as orbital, i.e., exp applied, they turn into geodesics infinitesimally close to $\exp(td)$, and obtained from the latter under small rotations shifting the point $\exp(d) \in S_1 = \partial B_1$. In other words, they are wholly contained in the union of projections of planes orthogonal to $T_x\widetilde{H}'(x)$ in $T_x\widetilde{H}(x)$ with respect to all $x = \gamma_d(r)$, i.e., each direction of this type is associated with a point from $\widetilde{H}(x)/\widetilde{H}'(x)$. However, if $x = \gamma_d(r)$ is singular, then $A''_x \supset T_x[\widetilde{H}(x)/\widetilde{H}'(x)]$, $A''_x$ not exhausted by it wholly; therefore, not all directions along $A''_y$ are orbital. Nevertheless, since the regular elements are everywhere dense in $T$, and $\kappa_k(r) < \infty$ in $Q(|\alpha'_0|)$, it suffices to consider only those of the torus $T$ for our purpose. (Note that, since regular elements are dense, any direction along $A''_y$ can be approximated by orbital ones as accurately as we please.) Hence, from the metric standpoint, we can regard all the directions along $A''_y$ as orbital.

Thus, there arises a decomposition in the ball $\Delta_d$ (as well as in $\Delta_{\alpha'_0}$) consistent with the above of the pencil $f_d$. We will formally write the fact as

$$\Delta_d = s_d + s'_d + s''_d,$$

where

$$s_d = \Delta_d \cap \exp_e(A_y),\ s'_d = \Delta_d \cap \exp_e(A'_y),\ s''_d = \Delta_d \cap \exp_e(A''_y).$$

Put $s(d) = s = \dim(s_d)$, $s'(d) = s' = \dim(s'_d)$, $s''(d) = s'' = \dim(s''_d)$. Then $k-1 = s + s' + s''$. Consider the first case. Since we have $\widetilde{\Pi}^{k-1}_y \subset A'_y$ in the pencil $\tilde{f}_{\alpha'_0}$, all the trajectories of the pencil $f_{\alpha'_0}$ are conjugate, and $(k-1) = s'(\alpha'_0)$, $s = s'' = 0$, $f(s'_{\alpha'_0}) = f_{\alpha'_0}$. Consider an arbitrary pencil $f_d$, where $d \neq \alpha'_0$, and assume that $|d| > |\alpha'_0|$, recalling that $d \in \widetilde{F}_1$. Let $\Delta_d = s_d + s'_d + s''_d$ be the canonical decomposition (see above). Since $\mathrm{vol}_{k-1}(\Delta_d) = \mathrm{vol}_{k-1}(\Delta_{\alpha'_0})$ due to the choice of the discs, $\mathrm{vol}_{s+1} f(s_d) < \mathrm{vol}_{s+1} f(s_1, \alpha'_0)$, where $k-1 = s'(\alpha'_0) = s_1 + s_2 + s_3$, $s_1 = s = s(d)$, $s_2 = s' = s'(d)$, $s_3 = s''(d)$, i.e., $\Delta_{\alpha'_0}$ is decomposed into the sum of three conjugate pencils associated with the values $s_1$, $s_2$, and $s_3$. The above inequality follows from the Euclideaness of the pencil $f(s_d)$. Since $|d| > |\alpha'_0|$, $\mathrm{vol}_{s'+1} f(s'_d) < \mathrm{vol}_{s'+1} f(s_{2,\alpha'_0})$, because the conjugate point $\exp(d)$ is farther from $e$ than the conjugate point $\exp(\alpha'_0)$. The inequality $\mathrm{vol}_{s''+1} f(s''_d) < \mathrm{vol}_{s''+1} f(s_{3,\alpha'_0})$ is verified similarly. Thus, in the case where $|d| > |\alpha'_0|$, we have $\mathrm{vol}_k f_d < \mathrm{vol}_k f_{\alpha'_0}$, which is just what was required. Now, let $|d| - |\alpha_0'|$. Then $d$ is an $s$-root, and due to the transitivity of the Weyl group on roots of the same length, we can put $d = \alpha_0'$, in which case $\mathrm{vol}_{s'+1} f(s_d') = \mathrm{vol}_{s'+1} f(s_{2,\alpha_0'})$, though two subpencils $f(s_d')$ and $f(s_{2,\alpha_0'})$ now having the common axis $\exp(t\alpha_0')$ may be noncoincident. However, for the remaining two pencils, if at least one of them is nonempty, the inequalities $\mathrm{vol}_{s+1} f(s_d) < \mathrm{vol}_{s+1} f(s_{1,\alpha_0'})$ (if $s_1 \neq 0$) and $\mathrm{vol}_{s''+1} f(s_d'') < \mathrm{vol}_{s''+1} f(s_{3,\alpha_0'})$ (if $s_3 \neq 0$) hold, i.e., if one of the values $s_1$ or $s_3$ is different from zero, then $\mathrm{vol}_k f_d < \mathrm{vol}_k f_{\alpha_0'}$. Nevertheless, if $s_1 = s_3 = 0$, then the whole pencil $f_d$ is filled with conjugate directions; it therefore extends along the domain of maximal coefficient action.

Before considering case (2), we shall give the following auxiliary statements. Consider an $n$-dimensional disc of radius $R\varphi_0$, where $\varphi_0$ is the spherical angle, on the sphere $S^n$ of radius $R$. Then the volume of this ball is $V_n|_0^{\varphi_0} = R^n\alpha(n-1)\int_0^{\varphi_0}\sin^{n-1}(\varphi)d\varphi$, where $\alpha(n-1)$ is the $(n-1)$-dimensional volume of the unit sphere $S^{n-1}$. For example, if $n = 2$, then $V_2|_0^{\varphi_0} = 4\pi R^2\sin^2\left(\frac{\varphi_0}{2}\right)$, and the coefficient $\kappa_2(r)$ equals $R\tan(r/2R)$.

Consider an arbitrary vector $d \in F_1$. Let all three types of the two-dimensional coefficients be represented along it, viz., Euclidean, conjugate, and orbital. We denote them by $\kappa_2^e(r,\Pi)$, $\kappa_2^s(r,\Pi)$, and $\kappa_2^0(r,\Pi)$, respectively, where $\Pi$ is the straight line orthogonal to $d$ and determining the inclination of the two-dimensional element of area. It is then easy to see that, for $r \leqslant |\alpha_0'|$, the strict inequalities $\kappa_2^e(r,\Pi) < \kappa_2^0(r,\Pi) < \kappa_2^s(r,\Pi)$ hold (recall that $|\alpha_0'| \leqslant |d|$). Thus, we consider case (2), where orbital directions are added to the conjugate ones of the pencil $f_d$, $\nu$, or the multiplicity of the root $\alpha_0$, in number. We assume at first that $|d| > |\alpha_0'|$, $k-1 = s + s' + s''$ for $f_d$, $s_1 = s$, $s_2 = s'$, $s_3 = s''$, and $k - 1 = s(\alpha_0') + s'(\alpha_0') + s''(\alpha_0')$ for $f_{\alpha_0'}$. Then $s(\alpha_0') = 0$, $s'(\alpha_0') = \nu$, i.e., $k - 1 = \nu + s''(\alpha_0')$ for $f_{\alpha_0'}$. The following two cases (a) $s'(d) \leqslant s'(\alpha_0')$ and (b) $s'(d) > s'(\alpha_0')$ are possible. In (a), we distinguish a subpencil $\tilde{f}_d$ from $f_d$, so that $\tilde{f}_d \supset f(s_d')$ and $\dim\tilde{f}_d = \dim f(s'_{\alpha_0'}) = s'(\alpha_0') = \nu$. Then, due to (1) and $|d| > |\alpha_0'|$, we have $\mathrm{vol}_{\nu+1}\tilde{f}_d < \mathrm{vol}_{\nu+1} f(s'_{\alpha_0'})$. Since $s'(d) \leqslant \nu$, the quotient pencil $f_d/\tilde{f}_d$ contains no conjugate directions, whereas the quotient pencil $f_{\alpha_0'}/f(s'_{\alpha_0'})$ consists of only orbital trajectories. We now compare these pencils, and replace all Euclidean directions, if any, in $f_d/\tilde{f}_d$ by orbital ones. We can thereby only increase the volume of $f_d/\tilde{f}_d$. Thus, there is only one orbital direction in $[f_d/f_d]'$ for each orbital direction in $f_{\alpha_0'}/f(s'_{\alpha_0'})$. Since $|d| > |\alpha_0'|$, the curvature of $f_{\alpha_0'}/f(s'_{\alpha_0'})$ is greater than that of $[f_d/\tilde{f}_d]'$, which entails the inequality $\mathrm{vol}_{k-\nu-1} f_{\alpha_0'}/f(s'_{\alpha_0'}) > \mathrm{vol}_{k-\nu-1}[f_d/\tilde{f}_d]' \geqslant \mathrm{vol}_{k-\nu-1} f_d/\tilde{f}_d$, and we therefore obtain that $\mathrm{vol}_k f_{\alpha_0'} > \mathrm{vol}_k f_d$.

Consider case (b), i.e., $s'(d) > s'(\alpha_0')$. There exists such an open neighbourhood $U$ of the $s$-root $\alpha_0'$ that, for any vector $d$ belonging to $U$ and ending on $\widetilde{F}_1$, the inequality $s'(d) \leqslant s'(\alpha_0')$ holds; therefore, if $s'(d) > s'(\alpha_0')$, then $d \notin U$.

This remark is based on the fact that the multiplicity of $s'(d)$ is determined by the hyperplanes which $d$ meets, and if it is close to $\alpha'_0$, then it meets the same plane as $\alpha'_0$. We construct two subpencils $f' \subset f_{\alpha'_0}$, $\dim f' = s'(\alpha'_0) - 1$, $f' \subset f(s'_{\alpha'_0})$ and $\tilde{f}_d \subset f_d$, where $\tilde{f}_d \subset f(s'_d)$ and $\dim f_d = \dim f'$, i.e., both $f'$ and $f_d$ are made up of conjugate directions. Since $|d| > |\alpha'_0|$, we have $\mathrm{vol}_{s'(\alpha'_0)} f' > \mathrm{vol}_{s'(\alpha'_0)} \tilde{f}_d$ due to item (1). We increase the volume of $f_d$ by replacing all its Euclidean and orbital trajectories by conjugate ones and denote the obtained pencil by $f'_d$. There is now only one conjugate trajectory $\gamma_0$ in $f'_{\alpha'_0}/f'$, and the others are orbital. Compare $f_{\alpha'_0}/f'$ and $f'_d/\tilde{f}_d$, where the latter consists of only conjugate directions. We add another orbital direction from the former to $\gamma_0 \subset f_{\alpha'_0}/f'$ forming the two-dimensional element of area $\sigma_2$ along with the trajectory $\exp(t\alpha'_0)$, and obtain a certain three-dimensional pencil $\omega_3$. We also consider in $f'_d$ a three-dimensional pencil $\tau_3$ consisting of two conjugate directions in $f'_d/\tilde{f}_d$, and also of the axis of the pencil $\exp(td) \subset \tilde{f}_d$. Straightforward computation shows that $\mathrm{vol}_3(\tau_3) < \mathrm{vol}_3(\omega_3)$ (recalling that $|d| > |\alpha'_0|$). This argument paves the way for induction on the number of orbital directions in $f_{\alpha'_0}$. It turns out that the coefficient along $\omega_i$ made up of $\sigma_2$ and $i-2$ orbital directions in $f_{\alpha'_0}$, $i = \dim \omega_i$, and containing one conjugate direction is greater than that of $\tau_i$, where $\tau_i \subset f'_d$, $\tau_i$ consists of $i-1$ conjugate directions in $f'_d/\tilde{f}_d$, and also of the trajectory $\exp(td) \subset \tilde{f}_d$, which can be easily seen by direct computation.

Since $f_{\alpha'_0}/f' \cup \gamma_0$ now consists of only orbital directions, we shall eventually exhaust the whole of $f_{\alpha'_0}$ by adding one orbital direction at a time, which completes the proof. Thus, the case $|d| > |\alpha'_0|$ is investigated in full.

Suppose now that, in situation (2), we have $|d| = |\alpha'_0|$. Making use of the Weyl group, we can assume that $d - \alpha'_0$. Then the inequality $s'(d) \leqslant s'(\alpha'_0)$ always holds, since, from the data of item (2), $s'(\alpha'_0) = \nu$, and $\nu$ is the maximal dimension of the space of Jacobi fields along $\exp(t\alpha'_0)$. We apply the scheme of the above argument and distinguish a subpencil $\tilde{f}_d$ from $f_d$, $\dim \tilde{f}_d = \nu + 1 = s'(\alpha'_0) + 1$, $\tilde{f}_d \supseteq f(s'_d)$. Due to item (1), we have $\mathrm{vol}_{\nu+1}(\tilde{f}_d) < \mathrm{vol}_{\nu+1}[f(s'_{\alpha'_0})]$. Further, the transition from $\tilde{f}_d$ to $f_d$ is carried out by adding orbital and Euclidean directions. Since there is an orbital direction in $f_{\alpha'_0}/f(s'_{\alpha'_0})$ for each Euclidean in $f_d/\tilde{f}_d$, adding Euclidean directions to $\tilde{f}_d$ continues to increase the difference in the coefficients already available. When all the Euclidean directions are exhausted, we will add orbital directions from both pencils, which preserves the difference in coefficients, already accumulated, since $d = \alpha'_0$.

However, if there have been no Euclidean directions at all in $f_d$, then we have $\mathrm{vol}_k(f_d) < \mathrm{vol}_k(f_{\alpha'_0})$ for $s'(d) < s'(\alpha'_0)$. If $s'(d) = s'(\alpha'_0)$, then $f_d$ coincides with the maximal pencil described by the lemma, though it does not necessarily coincide with $f_{\alpha'_0}$.

We now come to case (3). It is clear that, for any $d \in \widetilde{F}_1$, $f_d$ cannot contain more than $n-l$ orbital and conjugate directions; therefore, if $n-l+1<k\leqslant n-1$, then $s'(d)+s''(d)\leqslant n-l=s'(\alpha'_0)+s''(\alpha'_0)$. Consider two subpencils $\tau=f(s'_{\alpha'_0}\cup s''_{\alpha'_0})\subset f_{\alpha'_0}$, $\dim(\tau)=s'(\alpha'_0)+s''(\alpha'_0)$ and $\tilde{f}_d=f(s'_d\cup s''_d\cup\xi)$, where $\dim\tilde{f}_d=s'(\alpha'_0)+s''(\alpha'_0)=s'(d)+s''(d)+\dim\xi$. Then, due to item (2), we have $\mathrm{vol}_{n-l+1}(\tau)>\mathrm{vol}_{n-l+1}\tilde{f}_d$ if at least one of the inequalities $s'(d)<s'(\alpha'_0)$, $s''(d)<s''(\alpha'_0)$ holds, i.e., if $\dim\xi>0$. Since both quotient pencils $f_{\alpha'_0}/\tau$ and $f_d/\tilde{f}_d$ consist now of only Euclidean directions, the strict inequality also holds in the transition from $\tau$ and $\tilde{f}_d$ to $f_{\alpha'_0}$ and $f_d$, respectively; hence, $\mathrm{vol}_k(f_{\alpha'_0})>\mathrm{vol}_k(f_d)$, which was required. However, if $\dim\xi=0$, then $f_d$ is the maximal pencil, because $s'(d)=s'(\alpha'_0)=\nu$, $s''(d)=s''(\alpha'_0)=n-l-\nu$.

In conclusion, we note that, as $r\to|\alpha'_0|$, $\kappa_k(r)\to\infty$ for any $k$, and that if $\kappa_k(v,x,\Pi_x^{k-1})$ is not maximal, then, for $r>|\alpha'_0|$, we have $\kappa_k(v,x,\Pi_x^{k-1})<\infty$.

Q.E.D.

### 15.4. An Explicit Formula for the Symmetric Space Geodesic Nullity

Thus, if $M$ is a compact, symmetric (simply-connected) space, then $\Omega_k^0(e)=\omega_k|\alpha'_0|^k$, where $|\alpha'_0|=|\alpha_0|^{-1}$ and $\alpha_0$ is any of the roots of maximal length in $M$, and we can derive an explicit analytic expression for the coefficient $\omega_k$ from Lemma 15.3.1 (which is not required). Note that the variation of $\omega_k$ undergoes two essential break points as dimension $k$ increases, viz., for the values $k=\nu+1$ and $k=n-l+1$ (see Lemma 15.3.1).

Consider all possible motions $g\in\tilde{G}$ of $M$. We can send a point $e\in M$ to any point $P\in M$ by a convenient motion $g$. Since $g$ are metric-preserving, $\Omega_k^0(e)\equiv\Omega_k^0(P)$, where $P=g(e)$. The latter general value obviously coincides with $\Omega_k^0$, the geodesic $k$-nullity of $M$.

**Proposition 15.4.1.** *Let $M^n$ be a compact, homogeneous Riemannian space (not necessarily symmetric) endowed with the invariant metric, and $\Omega_k^0$ the geodesic $k$-nullity of $M$. Suppose that a surface $X_0\in\mathcal{O}(L')$, $L'\neq 0$, is a globally minimal surface, $\dim X_0=k$, and $\mathrm{vol}_k(X_0)=\Omega_k^0$. Then $X_0$ is a compact, closed and totally geodesic submanifold in $M$.*

**Proof.** Let $P\in X_0$ be an arbitrary point. Then $\Omega_k^0=\Omega_k^0(P)$, where $g(e)=P$. By Proposition 14.2.2, there exists a neighbourhood $U=U(P)$ such that $U=\exp_p T_p(X_0)$ and diffeomorphic to the open disc $D^k$. Since this is true for any point $P\in X_0$, and the integral curves of the field $v$ are geodesics emanating from $P$, $X_0$ is a totally geodesic submanifold.

Q.E.D.

### 15.5. Globally Minimal Surfaces of Least Volume ($\mathrm{vol}_k X_0=\Omega_k^0$) in Symmetric Spaces are Symmetric Spaces of Rank 1

For the present, we only know of minimal surfaces $X_0$ with the property $\mathrm{vol}_k X_0=\Omega_k^0$ (if these exist at all) that they are totally geodesic submanifolds in $M$ (see Proposition 15.4.1). Theorem 15.5.1 gives a full answer to the above question in the case where $M$ is a symmetric and compact space.

**Definition 15.5.1.** Let $V^k \subset M^n$ be a totally geodesic submanifold, $Q(e, R(e)) = Q(e)$ the maximal open geodesic ball in $M$ with centre at a point $e$ and radius $R(e)$, $x \in V^k \cap Q(e)$, and $\gamma(r)$ a geodesic from $e$ to $x$. We will say that $V^k$ *extends* along the maximal coefficient $\kappa_k(e,r)$ of the manifold $M$ if $\kappa_k(e,r) \equiv \kappa_k(v, x, \Pi_x^{k-1})$, where $x$ is an arbitrary point from $V^k \cap Q(e)$, $\Pi_x^{k-1} \subset T_x(V^k)$, and the plane $\Pi_x^{k-1}$ is orthogonal to the velocity vector $\dot{\gamma} \in T_x(V^k)$ (Fig. 43).

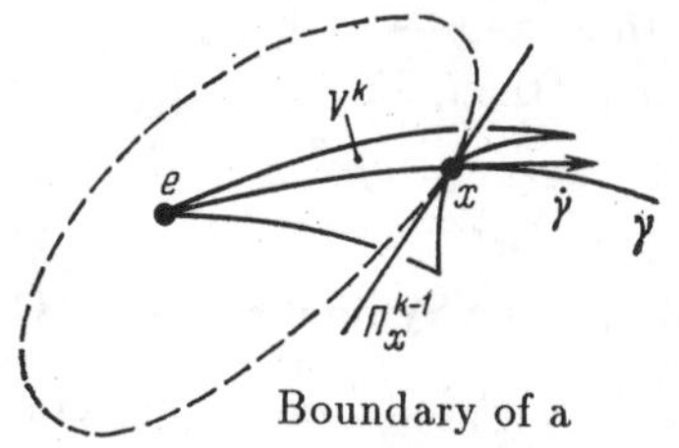


**Fig. 43**

**Theorem 15.5.1.** *Let $M^n$ be a compact, simply-connected, and symmetric space, and $X_0 \subset M$ a $k$-dimensional minimal surface of realizing type, i.e., $X_0 \in \mathcal{O}(L')$, $L' \neq 0$, $3 \leqslant k \leqslant n-1$. Assume that* $\mathrm{vol}_k(X_0) = \Omega_k^0(M)$, *where $\Omega_k^0$ is the geodesic nullity of the manifold $M$. Then $X_0$ is a simply-connected, compact, and totally geodesic submanifold in $M$ (and, therefore, $X_0$ is a symmetric space) of rank 1, i.e., the manifold $X_0$ is diffeomorphic to one of the manifolds (1) $S^k$; (2) $\mathbb{C}P^{k/2}$; (3) $\mathbb{Q}P^{k/4}$; (4) $F_4/\mathrm{Spin}(9)$. Besides, the embedding $X_0 \to M$ is such that on applying a convenient motion of $X_0$ in $M$, the unique root $\alpha_0$ of the symmetric space $X_0$ coincides with one of the roots of greatest length, i.e., with the one associated with a certain minimal s-root, of the symmetric space $M$, and the manifold $X_0$ wholly extends along the maximal coefficient $\kappa_k(e,r)$ in accordance with the description of this zone in Lemma 15.3.1. Conversely, let $X_0 \subset M$ be a $k$-dimensional simply-connected, totally geodesic submanifold of rank 1, whose unique root $\alpha_0$ coincides with a certain root of maximal length in the space $M$. Besides, we suppose that $X_0$ extends along the maximal coefficient $\kappa_k(e,r)$ in accordance with Lemma 15.3.1, and that $X_0 \in \mathcal{O}(L')$, i.e., that it realizes a certain nontrivial subgroup or subset in $H_k^{(k)}(M)$. Then $X_0$ is a globally minimal surface in $M$, and also* $\mathrm{vol}_k(X_0) = \Omega_k(M) = \Omega_k^0(M)$.

**Proof.** Let $e \in X_0 \subset M_0$ be a fixed point. Then, due to Proposition 15.4.1, the surface $X_0$ is a totally geodesic submanifold in $M$. It follows from the equality $\mathrm{vol}_k(X_0) = \Omega_k^0$ that

$$\Omega_k^0 = \mathrm{vol}_k X_0 \geqslant \Psi(X_0, f, R(e)) \geqslant \Psi_k(e, X_0) \cdot \lim_{a \to 0} \left( \frac{a^k}{q_e(a)} \right) \cdot \gamma_k q_e(R(e)) = \Psi_k(e, X_0) \cdot \Omega_k^0(x_0) = \Psi_k(e, X_0)\, \Omega_k^0 \geqslant \Omega_k^0.$$

Hence, $\Psi_k(e, X_0) = 1$, $\frac{\Psi(X_0,f,R(e))}{q_e(R(e))} = \lim_{a\to 0} \frac{\Psi(X_0,f,a)}{q_e(a)}$, i.e., the function $\frac{\Psi(X_0,f,r)}{q_e(r)}$ is constant. Therefore, $\left(\frac{\Psi}{q}\right)' = 0$, $\Psi = \frac{q}{q'}\Psi'$, but, because $\frac{q}{q'} = \max_{x\in\{f=r\}} \kappa_k(v, x) = \kappa_k(e, r)$, $\Psi(r) = \kappa_k(e,r)\Psi'(r)$. Since the velocity vector $\dot{\gamma}$ is orthogonal to the wavefront $\partial Q(r)$ for a geodesic diffeomorphism, it follows that $X_0$ extends along the maximal coefficient $\kappa_k(e,r)$, for otherwise we would obtain that $\Psi(r_0) < \kappa_k(e, r_0)\Psi'(r_0)$ for a certain $r_0$. We derive from Lemma 15.3.1 that, for any geodesic $\gamma \subset X_0$ emanating from $e$, $\kappa_k(v, x, \Pi_x^{k-1}) \equiv \kappa_k(e, r)$, $x = \gamma(r)$, $\Pi_x^{k-1} \subset T_x(X_0)$, and the plane $f_t(\Pi_x^{k-1})$ does not depend on $x$, its position being described by Lemma 15.3.1. It also follows that each $\gamma \subset X_0$ can be made coincident by a convenient rotation with the geodesic $\exp(t\alpha_0')$, where $\alpha_0'$ is a certain minimal $s$-root in $M$. Since there is a finite number of $\alpha_0'$ and all of them are isolated, all $\gamma \subset X_0$ passing through $e \in X_0$ can be made coincident with the same geodesic $\exp(t\alpha_0') \subset X_0$. It follows that the motion group of the space $X_0$ is transitive on the vectors of the plane $T_e(X_0)$; therefore, the rank of $X_0$ is 1. We now prove that $X_0$ is simply-connected. In fact, we assume that $\pi_1(X_0) \neq 0$. Consider the ball $Q[R(e)]$. Since $\mathrm{vol}_k(X_0) = \Omega_k^0$, we have that $X_0 \subset \bar{Q}[R(e)]$. Let $\gamma(r)$ be a geodesic in $X_0$ emanating from $e$ and ending on $\partial Q[R(e)] = R_1$. Since $k > 2$, the universal covering $\widetilde{X}_0$ over a manifold $X_0$ is diffeomorphic to one of the manifolds (1) $S^k$, (2) $\mathbb{C}P^{k/2}$, (3) $\mathbb{Q}P^{k/4}$, (4) $F_4/\mathrm{Spin}(9)$, which entails, along with $\pi_1(X_0) \neq 0$, the existence of such a geodesic $\gamma_0 \subset X_0$ that the first conjugate point on it is $e$ after a complete rotation along the closed geodesic $\gamma_0$. Recall that $X_0$ is obtained from $\widetilde{X}_0$ by factorization with respect to a certain finite subgroup of the motion group. Let $M(e, X_0)$ be the set of minimum points for $X_0$. Then $M(e, X_0) = M(e, M) \cap \bar{Q}(|\alpha_0'|) = M(e, X_0) \cap R_1$. Further, since $M$ is simply-connected, we obtain that $M(e, M) = C(e, M)$, and that each point $\varphi \in M(e, X_0)$ must necessarily be conjugate of $e$. Indeed, if some point is not conjugate of $e$, then the submanifold $X_0$ cannot be served by the maximal coefficient $\kappa_k(e, r)$ (recalling that, due to Lemma 15.3.1, the pencil $f_{\alpha_0'}$ associated with $\kappa_k(e, r)$ should contain at least one conjugate direction), which contradicts the existence of $\gamma_0$, the first conjugate point along which is $e$ itself, and which proves the first part of the theorem.

Conversely, let $X_0 \subset M$ be a totally geodesic submanifold of rank 1, embedded into $M$ in accordance with the requirements of the theorem. Retracing our steps and making use of the scheme of the proof of Theorem 14.2.1, we obtain a chain of equalities $\Psi(X_0, f, |\alpha_0'|) = \mathrm{vol}_k X_0 \Omega_k^0$, since $X_0 \subset \bar{Q}(|\alpha_0'|)$. Because $X_0 \in \mathcal{O}(L')$, $X_0$ is a globally minimal surface.

Q.E.D.

We have resorted to the fact that $X_0$ is simply-connected, referring to its being perfectly served by the maximal coefficient, which is impossible for a non-simply-connected, symmetric space $X_0$, as follows from the first part of the theorem. Note that Theorem 15.5.1 reduces the problem of finding minimal surfaces with the property $\mathrm{vol}_k(X_0) = \Omega_k^0$ in the case of a symmetric space $M$ to the study of totally geodesic submanifolds of rank 1, which is now an algebraic problem that can be solved quite easily.

**Remark.** Let $M^n$ be a simply-connected, compact, and symmetric space of rank 1, and $\xi > 0$ the greatest dimension of a simply-connected, compact, and totally geodesic submanifold of rank 1 in $M$, whose root $\alpha_0$ coincides with a certain root of maximal length in $M$. It then follows from Theorem 15.5.1 that $\xi \leqslant n - l + 1$, and for any minimal surface $X_0^k \in \mathcal{O}(L')$, where $\xi + 1 \leqslant k \leqslant n - 1$, the strict inequality $\mathrm{vol}_k(X_0) > \Omega_k^0(M)$ holds.

## 15.6. Proof of the Classification Theorem for Surfaces of Least Volume in Certain Classical Symmetric Spaces

We now turn to the proofs of Theorems 15.1.1 and 15.1.2. The last statement of Theorem 15.1.1 should be understood as follows. If $L_k' = \{\sigma\}$ and $\widetilde{L}_k' = \{p\sigma\}$ are subgroups generated by the elements $\sigma$ and $p\sigma$, respectively, then the minimal compact set $\widetilde{X}_0$ realizing $\widetilde{L}_k'$ is different from the minimal surface $X_0$ realizing $L_k'$ if and only if $\mathrm{vol}_k(\widetilde{X}_0) < \mathrm{vol}_k(X_0)$, for otherwise $X_0$ serves in the sense of minimality not only $L_k'$, but also $\widetilde{L}_k'$. In particular, it follows from Theorem 15.1.1 that an increase in the multiplicity of a (co)homology class in symmetric spaces of rank 1 at any rate does not increase the volume of the support.

**Proof of Theorem 15.1.1.** Consider the case where $M$ is simply-connected. (1) $M^{2n} = \mathbb{C}P^n$, $\nu = \nu(\alpha_0) = 1$ ($\nu$ being the multiplicity of the root $\alpha_0$), and therefore, $\kappa_k(e, r)$ is linked to the pencil $f_{\alpha_0'}$ with precisely one conjugate direction, and $k - 2$ is orbital. Since $X_0^k = \mathbb{C}P^{k/2}$, $\dim X_0^k - \dim(\mathbb{C}P^{\frac{k}{2}-1}) = 2$; therefore, $X_0$ extends along the maximal coefficient (see Definition 15.5.1), and statement (1) of Theorem 15.1.1 follows from Theorem 15.5.1. In case (2), $M^{4n} = \mathbb{Q}P^n$, $\nu = \nu(\alpha_0) = 3$, $\dim(X_0^k) - \dim(\mathbb{Q}P^{\frac{k}{4}-1}) = 4$, which entails the required statement as in case (1). In case (3), $\nu(\alpha_0) = 7$, $\dim(X_0^8) - 0 = 8$, which is just what is required in order to apply Theorem 15.5.1. In case (4), $M^n = \mathbb{R}P^n$, and $\pi_1(M^n) \neq 0$; therefore, the above technique cannot be applied formally, but we have $M(e, M) = \mathbb{R}P^{n-1}$, though $M(e) \neq C(e) = e$, and can we put $Q[R(e)] = \mathbb{R}P^n \backslash \mathbb{R}P^{n-1}$, repeating the entire argument from the proof of Theorem 15.5.1 with $\nu(\alpha_0) = 0$. We proceed further as in cases (1)–(3). Conversely, let $\widetilde{X}_0^k \subset M^n$ be an arbitrary minimal surface of the realizing class $\mathcal{O}$. Then there is a minimal surface $X_0^k$ in the same dimension $k$, coinciding with one of the elements of the above filtration. Since $X_0^k$ realizes the whole group $H_k^{(k)}(M)$, $\mathrm{vol}_k(\widetilde{X}_0^k) \geqslant \mathrm{vol}_k(\widetilde{X}_0^k)$. Because, by Theorem

14.2.1, $\mathrm{vol}_k(\widetilde{X}_0^k) \geqslant \Omega_k(M) \geqslant \Omega_k^0(M)$, we have $\mathrm{vol}_k(X_0^k) = \mathrm{vol}_k(\widetilde{X}_0^k) = \Omega_k^0$, and then, by Theorem 15.5.1, $\widetilde{X}_0^k$ should be obtained from the submanifold $X_0$ by a certain isometry of the manifold $M$. The theorem is thus proved.

So, if the rank of $M$ is unity, then the values $\Omega_k^0$ are always attained and completely describe minimal surfaces in $M$.

Let us consider the more complex situation of Theorem 15.1.2. Let $M^n = \mathrm{SU}(p+q)/S[U(p) \times U(q)]$, $p \geqslant q \geqslant 1$, be a simply-connected, complex Grassmannian manifold $G^{\mathbf{C}}_{p+q,q}$. When $q = 1$, we have $M^n = \mathbf{C}P^{n/2}$. The manifold $M^n$ is modelled in the group $\mathrm{SU}(p+q)$ as a Cartan model $V \cong M^n = \exp B$, where $B = \{b\}$, $b = \begin{pmatrix} 0 & Z \\ -\bar{Z}^T & 0 \end{pmatrix}$, and $Z$ is a complex matrix with $q$ rows and $p$ columns. If we denote the standard basis for the algebra $\mathrm{su}(p+q)$ by $E_{ij}$, then the maximal Abelian subspace in the plane $B$ is determined as $\sum_{i=1}^{q} \mathbf{R}(E_{i,q+i} - E_{q+i,i})$, with $\mathbf{R}$ being the field of real numbers. Put $\tilde{\alpha} \in P$, $\tilde{\alpha} = 2\pi(E_{1,q+1} - E_{q+1,1})$, and distinguish from $B$ the plane $C_s$:

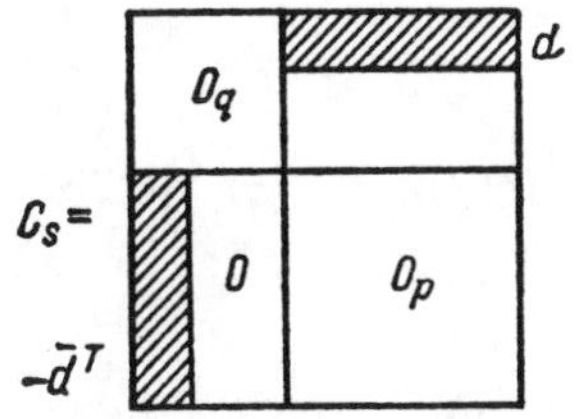


where $d$ is a $p$-vector of the form $d = (\varphi_1, \varphi_2, \ldots, \varphi_s, 0, \ldots, 0)$, with $s$ being fixed, $1 \leqslant s \leqslant p$, and $0_q$ and $0_p$ denoting two null square matrices of orders $q$ and $p$, respectively. Then $\tilde{\alpha} \in C_s$, and $\exp(C_s) \subset \exp(B) = G^{\mathbf{C}}_{p+q,q}$ is a totally geodesic submanifold in $G^{\mathbf{C}}_{p+q,q}$, diffeomorphic to $\mathbf{C}P^s$, and which can be written as $\mathbf{C}P^s = \{l\}$ under a convenient motion on $M^n$, where $l = A \oplus I_{p+q-s-1}$, $I_\alpha$ is the unit matrix of order $\alpha$, $A = (a_{ij})$, $1 \leqslant i, j \leqslant s+1$; $a_{11} = -1 + 2|z_0|^2$; $a_{1,r+1} = -\bar{a}_{r+1,1} = 2\bar{z}_0 \cdot z_r$, $1 \leqslant r \leqslant s$; $a_{ii} = 1 - 2|z_{i-1}|^2$, $2 \leqslant i \leqslant s+1$; $a_{ji} = \bar{a}_{ji} = 2z_{i-1}\bar{z}_{j-1}$, $2 \leqslant i \leqslant s+1$, $2 \leqslant j \leqslant s+1$, $i \neq j$; $\sum_{i=0}^{s} |z_i|^2 = 1$.

**Proof of the first part of Theorem 15.1.2.** Consider $\mathbf{C}P^s \subset M$. Let $\gamma^*(t) = \exp(t\tilde{\alpha})$ be a geodesic in $\mathbf{C}P^s$. Then the first conjugate point on $\gamma^*(t)$ arises for $t = \frac{1}{2}$. Put $\tilde{\alpha}' = \frac{\tilde{\alpha}}{2}$. Then the orbit of the point $\exp(\tilde{\alpha}')$ in $\exp(C_s) = \mathbf{C}P^s$ is a submanifold diffeomorphic to $\mathbf{C}P^{s-1}$ formed by the midpoints of closed geodesics $\gamma^*(t)$, $0 \leqslant 1$, i.e., the vector $\tilde{\alpha}'$ is a unique $s$-root of the space $\mathbf{C}P^s$. Thus, the structure of the pencil $f_{\tilde{\alpha}'}$ in $\mathbf{C}P^s$ is such that $f_{\tilde{\alpha}'}$ contains one conjugate and $2s-2$ orbital directions. It is clear that $\partial Q(|\alpha_0'|) \cap \mathbf{C}P^s = \mathbf{C}P^{s-1}$, where $|\alpha_0'|$ is the length of the minimal $s$-root in $M$. By a convenient rotation of the submanifold $\mathbf{C}P^s$ in $M$, we can change the vector $\tilde{\alpha}'$ into $\alpha' = \pi i(E_{11} - E_{22})$. Consider the scalar product $\langle X, Y \rangle = -(2\pi)^{-2} \operatorname{Re}\operatorname{Spur}(X \cdot Y)$

in the Lie algebra su$(p+q)$. It is generally known that the roots of su$(p+q)$ in the canonical basis are of the form $\{\lambda_i \pm \lambda_j\}$; $i \neq j$. With respect to canonical coordinates on the torus $T^{p+q-1}$, we have $\alpha_{ij} = 2\pi i(E_{ii} - E_{jj})$, $i \neq j$, in su$(p+q)$, the simple roots being the vectors $\alpha_{i,i+1}$. Since $|\alpha_{ij}|^2 = 2$, the corresponding $s$-roots are of the form $\alpha'_{ij} = \pi i(E_{ii} - E_{jj})$. It is clear that the image of $\tilde{\alpha}'$ after the rotation (see above) coincides with the $s$-root $\alpha'_{12} = \pi i(E_{11} - E_{22})$. Since all $s$-roots in su$(p+q)$ are of the same length, we obtain that $\alpha'$ of $\mathbf{C}P^s$ after a convenient rotation coincides with minimal $\alpha'_{12}$ of su$(p+q)$. We now calculate the multiplicity of the first conjugate point $\exp(\alpha')$ along $\exp(t\alpha')$ in the group SU$(p+q)$. It is clear that

$$\nu = \dim [C(\exp(\alpha'))/C(\alpha')] = \dim [C(\exp(\tilde{\alpha}'))/C(\tilde{\alpha}')] = \dim \{SU(2)/S[U(1)\times U(1)]\} = 2.$$

The coefficient associated with $f_{\tilde{\alpha}'}$ in $\mathbf{C}P^s \subset$ SU$(p+q)$ is thereby not maximal in SU$(p+q)$; however, it is maximal in the submanifold $\exp B \cong G^{\mathbf{C}}_{p+q,q}$.

It is easy to calculate that

$$\nu_1 = \dim [C(\exp(\tilde{\alpha}')) \cap S(U(p)\times U(q))]/[C(\tilde{\alpha}') \cap S(U(p)\times U(q))] = \dim S^1 = 1,$$

therefore, $f_{\tilde{\alpha}'} \subset \mathbf{C}P^s$ in the manifold $\exp(B)$ is associated with the maximal coefficient $\kappa_k(e,r)$ in $\exp(B)$. Since all submanifolds $\mathbf{C}P^s$, $1 \leqslant s \leqslant p$, realize the additive generators of the groups $H^{2s}(M,\mathbf{Z})$, due to Theorem 15.5.1, all $\mathbf{C}P^s$ are minimal surfaces in $\exp B \cong G^{\mathbf{C}}_{p=q,q}$, and besides, $\mathrm{vol}_{2s}(\mathbf{C}P^s) = \Omega^0_{2s}$. It remains to be proved that, with $k \neq \{2,4,6,\ldots,2p\}$, we have $\mathrm{vol}_k(X_0) > \Omega^0_k$ for any surface $X_0 \in \mathcal{O}(L')$. Because $H^*(M,\mathbf{Z}) = \mathbf{Z}[x_2,x_4,\ldots x_{2q}]$ up to dimension $2p$, for $k \leqslant 2p$ and any coefficient group, nontrivial minimal surfaces may be only in even dimensions, and we return to the above situation. Thus, suppose that $k \geqslant 2p+2$. Since $\mathbf{C}P^p$ is a symmetric space of rank 1, maximal in dimension, and can be obtained in the group SU$(p+q)$ by rotating the geodesic $\exp(t\tilde{\alpha}')$, by the remark after Theorem 15.5.1, the value $\xi$ equals $2p$; therefore, $\mathrm{vol}_k(X_0) > \Omega^0_k(M)$ for any $k$, $2p+2 \leqslant k \leqslant n-1$, and the first part of Theorem 15.1.2 is now proved.

We now consider the quarternion case, which will be a little more delicate as to the technique of the proof.

Let

$$M^n = G^{\mathbf{Q}}_{p+q,\,q} = \mathrm{Sp}(p+q)/\mathrm{Sp}(p)\times\mathrm{Sp}(q), \qquad p \geqslant q \geqslant 1,$$

be a quarternion Grassmannian manifold. Then the ring $H^*(M;\mathbf{Z})$ is isomorphic (up to dimension $4p+2$) to the polynomial ring $\mathbf{Z}[y_4,y_8,\ldots,y_{4q}]$.

Consider the standard embeddings $\mathbf{C}P^r = G^{\mathbf{Q}}_{r+1,1}$ in $G^{\mathbf{Q}}_{p+q,q}$, $1 \leqslant r \leqslant p$, realizing the elements $y_4, y_4^2, y_4^3, \ldots, y_4^p$ of the ring $H^*(M;\mathbf{Z})$. Let the group $\mathrm{Sp}(p+q)$ be standardly embedded into $\mathrm{SU}(2p-2q)$, $V \cong G^{\mathbf{Q}}_{p+q,q} \subset \mathrm{Sp}(p+q)$ be the Cartan model of the manifold $M$, and $G = B + H$ the corresponding decomposition of the algebra $\mathrm{sp}(p+q)$, where the plane $H$ is made up of matrices of the form $\begin{pmatrix} A & D \\ -\bar{D} & \bar{A} \end{pmatrix}$, $A = X_{11} \oplus X_{22}$, $D = X_{13} \oplus X_{24}, X_{11} \in u(p)$, $X_{22} \in u(p)$, $X_{13}$ being a symmetric matrix of order $q \times q$, and $X_{24}$ a symmetric matrix of order $p \times p$. The plane $B$ consists of matrices $\begin{pmatrix} M & N \\ -\bar{N} & \bar{M} \end{pmatrix}$, where $M = \begin{pmatrix} 0 & Y_{12} \\ -\bar{Y}^T_{12} & 0 \end{pmatrix}$, $N = \begin{pmatrix} 0 & Y_{14} \\ Y^T_{14} & 0 \end{pmatrix}$. The maximal torus $P \subset B$ is obtained if we set $Y_{14} = 0$, $Y_{12} = \sum_{i=1}^q \mathbf{R}\cdot E_{ii}$. The Cartan subalgebra in $\mathrm{sp}(p+q)$ consists of elements of the form $Z \oplus Z$, where $Z \in u(p+q)$. We distinguish from $B$ a subspace $C_r$ consisting of all such elements that $(Y_{12})_{1\beta} = \varphi_\beta$, $1 \leqslant \beta \leqslant r$; $(Y_{12})_{1\beta} = 0$, $r < \beta \leqslant p$; $(Y_{12})_{ij} = 0$, $i > 1$; $(Y_{14})_{1\beta} = \psi_\beta$, $1 \leqslant \beta \leqslant r$; $(Y_{14})_{1\beta} = 0$, $r < \beta \leqslant p$; $(Y_{14})_{ij} = 0$, $i > 1$; $\varphi_\beta$, $\psi_\beta \in \mathbf{C}$. It is clear that $\exp(C_r)$ is a totally geodesic submanifold in $V$, diffeomorphic to $\mathbf{Q}P^r$, the indicated embedding $\mathbf{Q}P \xrightarrow{r} V$ being topologically standard. (Note that $\mathbf{Q}P^r$ can be given explicitly by formulas similar to those describing $\mathbf{C}P^s$ in $G^{\mathbf{C}}_{p+q,q}$.)

**Proof of the second item of Theorem 15.1.2.** Fix in $\mathrm{sp}(p+q)$ the scalar product $\langle X, Y\rangle = -(8\pi^2)^{-1} \cdot \mathrm{Re}\,\mathrm{Spur}(\bar{X} \cdot Y)$. Let $e_i = 2\pi A_{ii}$, where $A_{ss} = i(E_{ss} - E_{p+q+s,p+q+s})$. Then $|e_i| = 1$. The system of all roots of $\mathrm{sp}(p+q)$ is of the form $\{e_i \pm e_j\}$, and those of maximal length are $\{2e_i\}$. If we project all the roots of $\mathrm{sp}(p+q)$ onto $B'$, where

$$B' = \mathrm{Ad}_C(B), \quad C = D \oplus \bar{D}, \; D \in u(p+q),$$
$$D = (1/\sqrt{2})M \oplus I_{p-q}, \quad M = \begin{pmatrix} I_q & -iI_q \\ I_q & iI_q \end{pmatrix},$$

it becomes clear that $CPC^{-1} \subset P'$, where $P'$ is the Cartan subalgebra in $\mathrm{sp}(p+q)$. The plane $CPC^{-1}$ is spanned by the vectors $\{e_i - e_{i+q}\}$, $1 \leqslant i \leqslant q$. Straightforward calculation, which we omit, shows that the vectors $\{\pm(e_i - e_{q+i})\}$, $1 \leqslant i \leqslant q$, are roots of greatest length of the space $\exp B$. Put $\alpha_i = e_i - e_{i+q}$, $\alpha_i' = \frac{1}{2}\alpha_i$ and $\mu_i' = C^{-1}\alpha_i' C$. We only investigate the $s$-root $\mu_1'$, because the other $s$-roots $\mu_i'$ are obtained from it under the action of the Weyl group. It is obvious that $\exp(\mu_1') = K \oplus K$, where $K = (-I_1) \oplus (I_{q-1}) \oplus (-I_1) \oplus (I_{p-1})$. It is required to calculate the dimension $C[\exp(\mu_1')]/C(\mu_1')$, which will give us the multiplicity of the point $\exp(\mu_1')$ in $\mathrm{SU}(2p+2q)$. It turns out that the pencil $f_{\mu_1'}$ is not associated with the maximal coefficient in $\mathrm{SU}(2p+2q)$; however, it is easy to see by straightforward calculation that

$$[C(\exp(\mu_1')) \cap (\mathrm{Sp}(p) \times \mathrm{Sp}(q))]/[C(\mu_1') \cap (\mathrm{Sp}(p) \times \mathrm{Sp}(q))] \cong S^3,$$

where the subgroup $S^3 \cong \mathrm{SU}(2)$ is embedded into $S^3 \times S^3$ as the diagonal $\Delta$. Thus, the multiplicity $\nu$ of the point $\exp(\mu_1')$ in the manifold $V$ equals 3.

Since $\dim \mathbb{Q}P^r - \dim \mathbb{Q}P^{r-1} = 4$, by Lemma 15.3.1, the submanifold $\exp(C_r)$ contains 3 conjugate directions along the geodesic $\exp(t\mu_1')$, i.e., $f_{\mu_1'}$ is associated with the maximal coefficient $\kappa_k(e,r)$ in $M$. Because this pencil generates a submanifold $\mathbb{Q}P^r$ under a rotation of the vector around $e$, all $\mathbb{Q}P^r \subset M$ are globally minimal surfaces. All the other statements are proved similarly to the first item.

Q.E.D.

The real case of a one-connected and non-simply-connected Grassmannian manifold is investigated exactly in the same manner however, we are not going to dwell on it.

Theorems 15.1.1 and 15.1.2 describe situations where the value $\Omega_k^0(M)$ is attained for many $k$. By comparison, we study several cases where the $\Omega_k^0$ are reached in a bounded number of instances, or not attained at all.

Let $M^n = \widetilde{G}$ be a compact, simply-connected Lie group, and $\alpha_0$ an arbitrary root of maximal length in $M$ (there possibly being several such roots). It then always generates the subgroup $SU(2)$ diffeomorphic to the sphere $S^3$ and realizing a free generator of the group $H_3^{(3)}(\widetilde{G};\mathbb{Z})$. (We recall that $H_1^{(1)}(\widetilde{G};\mathbb{Z}) = H_2^{(2)}(\widetilde{G};\mathbb{Z}) = 0$.) We denote this submanifold by $S_0^3$.

**Proposition 15.6.1.** *Let $M^n = \widetilde{G}$ be a compact, simply-connected Lie group, $S_0^3$ the above subgroup, and $X_0 \in \mathcal{O}(L')$, $3 \leqslant k \leqslant n-1$, an arbitrary minimal surface in $\widetilde{G}$. Then the equality $\mathrm{vol}_k(X_0) = \Omega_k^0(\widetilde{G})$ holds only for $k = 3$; $\mathrm{vol}_3(X_0) = \Omega_3^0$ holds only for the submanifold $X_0 = S_0^3$ up to motions of $\widetilde{G}$; in particular, $S_0^3$ is a globally minimal surface in $\widetilde{G}$.*

**Proof.** Let $X_0$ be a minimal surface in $\widetilde{G}$, so that $\mathrm{vol}(X_0) = \Omega_k^0$. By Theorem 15.5.1, $X_0$ is then a submanifold diffeomorphic to one of the spaces $S^k$, $\mathbb{C}P^{k/2}$, $\mathbb{Q}P^{k/4}$, $F_4/\mathrm{Spin}(9)$, $X_0$ being generated by an $s$-root $\alpha_0'$ of least length in $\widetilde{G}$, in which case the multiplicity of $\exp(\alpha_0')$ as of a conjugate point equals 2. In fact, the multiplicity of an arbitrary root $\hat{\alpha}$ is $\nu(\hat{\alpha}) = \sum \dim \widetilde{V}_{\hat{\beta}}$, where $\hat{\beta}(\alpha_0') \equiv 0 \pmod 1$, and $\hat{\beta}(\alpha') \neq 0$ (see Proposition 15.2.1). Because $\dim_{\mathbb{R}}(V_{\hat{\alpha}}) = 2$ and the addend $\dim V_{\hat{\alpha}}$ is involved in the above sum ($\hat{\alpha}(\alpha') = 1$), we have $\nu(\hat{\alpha}) \geqslant 2$. Since $\alpha_0'$ is the least $s$-root, there is no other root $\hat{\beta} \neq \hat{\alpha}_0$ such that $\hat{\beta}(\alpha_0') \equiv 0 \pmod 1$, for we would otherwise obtain that $|\beta'| < |\alpha_0'|$, which is impossible. Thus, $\nu(\alpha_0') = 2$ holds in $\widetilde{G}$. Since $k \geqslant 3$, the pencil $f_{\alpha_0'}$ associated with the coefficient $\kappa_k(e,r)$ must contain two conjugate directions (see Lemma 15.2.1). They being already contained in the subgroup $S_0^3$ and, by Lemma 15.3.1 and Theorem 15.5.1, the whole of $X_0$ being served by $\kappa_k(e,r)$ and being a totally geodesic submanifold, we have $X_0 \supset S_0^3$. $S_0^3$ is known to realize a free generator in $H_3^{(3)}(\widetilde{G},\mathbb{Z})$; hence, certain of the above symmetric spaces of rank 1, coinciding with $X_0$, must contain an element of infinite order in their three-dimensional (co)homology group, which is only possible if $k = 3$ and $X_0 = S^3$.

The minimality of $S_0^3$ follows from Theorem 15.5.1, thus completing the proof of the proposition.

The values $\Omega_k^0$ are attained on no symmetric spaces for any $k$. For example, consider the symmetric space $M^n = \mathrm{SU}(2m)/\mathrm{Sp}(m)$ of rank $m-1$. It is known that $H^*(M;\mathbf{Z}) \cong \Lambda(x_3, x_9, \ldots, x_{4m-3})$ (see [53]). Let $m \geqslant 2$, and $X_0 \in \mathcal{O}(L')$ be a globally minimal surface. Then the strict inequality $\mathrm{vol}_k(X_0) > \Omega_k^0(M)$ holds. Let $M^n = \mathrm{SU}(2m+1)/\mathrm{SO}(2m+1)$, (another example) and $X_0 \in \mathcal{O}(L')$. Then always $\mathrm{vol}_k(X_0) > \Omega_k^0$. The proofs are as of Proposition 15.6.1.

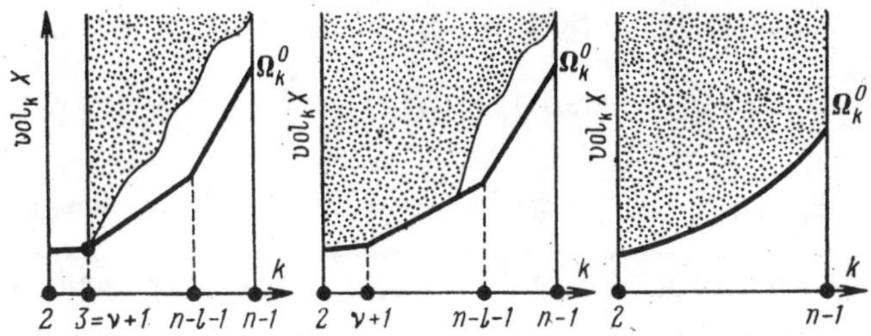


a) $M = \widetilde{G}$ b) $M = \widetilde{G}/\widetilde{H}$ c) $M = \widetilde{G}/\widetilde{H}$ Symmetric space of rank 1

**Fig. 44**

Theorem 15.5.1 permits us to describe completely the geodesic $k$-nullity $\Omega_k^0$ of any compact, symmetric space (e.g., we can base ourselves on their list of these spaces in [198]). We restrict ourselves to the above examples. The study of a considerable number of concrete symmetric spaces permitted us to describe the qualitative behaviour of the function $\Omega_k^0$ in the following manner. We mark off dimensions $k$, $2 \leqslant k \leqslant n-1$ on the horizontal axis, and volumes of closed surfaces of realizing type on the vertical axis, i.e., $X \in \mathcal{O}(L')$, $L' \neq 0$, $L' \subset H_k^{(k)}(M)$ (Fig. 44). As it turns out, the three cases of (a) Lie groups, (b) symmetric spaces other than Lie groups or symmetric spaces of rank 1, and (c) symmetric spaces of rank 1 can be distinguished. In (a) and (b), the behaviour of $\Omega_k^0$ is characterized by the availability of two break points $k_1 = \nu + 1$, $k_2 = n - l - 1$, where $\nu$ is the multiplicity of a root $\alpha_0$ of maximal length (see above) and $l$ is the rank of $M$. In (a), $\Omega_k^0$ osculates the volumes of closed topologically nontrivial surfaces only at one point if $k = 3$, whereas in (b), such osculations increase, there being sufficiently many minimal surfaces with the same volume as of $\Omega_k^0$, and in (c), $\Omega_k^0$ completely describes all minimal surfaces of least volume.

Further examples of globally minimal surfaces realizing nontrivial (co) cycles in symmetric spaces have been obtained by Đao Chông Thi (see [89], [92]). We only give some of them here.

**Theorem 15.6.2.** *The totally geodesic submanifolds in compact, symmetric* *spaces M listed below are globally minimal surfaces of realizing type in their (co)homology classes, viz.,*

(1) $M = U(n)/(U(m)\times U(n-m))$, $n\geqslant 2$, $1\leqslant m\leqslant [n/2]$;
$X_0 = U(p)/(U(q)\times U(p-q))$, $p<n$, $q<m$,

*where $U(p)$, $U(q)$, $U(p-q)$ are subgroups standardly embedded into the corresponding Lie groups $U(n)$, $U(p)$, $U(m)$;*

(2) $M = SO(n)/(SO(2)\times SO(n-2))$, $n\geqslant 3$;
$X_0 = SO(p)/(SO(2)\times SO(p-2))$, $3\leqslant p\leqslant n$,

*where $SO(p)$ and $SO(p-2)$ are standardly embedded into $SO(n)$ and $SO(n-2)$, respectively;*

(3) $M = Sp(2n)/U(n)$, $n\geqslant 2$; $X_0 = Sp(2q)/U(q)$, $2\leqslant q<n$;

*with the embeddings $Sp(2q)\subset Sp(2n)$, $U(q)\subset U(n)$ being standard;*

(4) $M = SO(2n)/U(n)$, $n\geqslant 2$; $X_0 = SO(2q)/U(q)$, $2\leqslant q<n$;

*with the embeddings $SO(2q)\subset SO(2n)$, $U(q)\subset U(n)$ being standard;*

(5) *Let $X_0$ be an arbitrary closed quarternion submanifold of real dimension 4 in a Kählerian, quarternion-symmetric, connected, and compact space $M$. Then it is a globally minimal surface in its (co)homology class.*

This theorem is a corollary to the general result obtained by Ɖao Chông Thi in order to describe currents of minimal mass (see the terminology for the case of symmetric spaces in [135]; see also [88], [89], [92], [94], [243]–[245], [451], and [452].

CHAPTER 4

# LOCALLY MINIMAL CLOSED SURFACES REALIZING NONTRIVIAL (CO)CYCLES AND ELEMENTS OF SYMMETRIC SPACE HOMOTOPY GROUPS

## §16 Problem Formulation. Totally Geodesic Submanifolds in Lie Groups

In the previous sections, we mostly studied globally minimal surfaces and discovered, meanwhile, that those of least volume, say, among surfaces realizing nontrivial (co)cycles in a given dimension, often turn out to be totally geodesic submanifolds. On the other hand, totally geodesic submanifolds are always locally minimal, i.e., they are extremals of the volume functional (possibly "saddles," and not absolute minimal; see §2). Therefore, the problem of describing totally geodesic submanifolds realizing nontrivial (co)cycles in symmetric spaces as in one of the classes of Riemannian manifolds, which is most important in applications, is of special interest. We will completely solve it in the present chapter and then classify all those cases where nontrivial elements of the homotopy groups $\pi_i(M) \otimes \mathbb{Q}$, where $M$ is a symmetric space and $\mathbb{Q}$ are rational numbers, are realized by totally geodesic spheres, i.e., we describe all those homotopy classes of symmetric spaces in which there is a representative given by the standard sphere or a locally minimal surface of maximally simple form. It turns out that this classification reveals close links between the properties of compact Lie groups and the geometry of locally minimal surfaces.

As was shown in §2, each totally geodesic submanifold is locally minimal. It turns out that if the ambient manifold is a Lie group, then totally geodesic submanifolds can be described especially simply (from the local standpoint). Denote the commutator operation in a Lie algebra by $[X, Y]$. Since we are interested in compact Lie groups and compact algebras, we can assume that they are realized as matrix groups, and then the commutator of two elements $X$ and $Y$ is of the form $XY - YX$, where $XY$ is the product of two matrices.

**Definition 16.1.** Let $G$ be a Lie algebra of the field of real numbers, and $B$ a subspace in $G$ called a *triple Lie system* if, for any three elements $X, Y$, and $Z$ belonging to $B$, the element $[[X,Y]Z]$ also belongs to $B$.

We denote the canonical mapping of $G$ into a group $\mathfrak{G}$ by exp. Let $V$ be a totally geodesic submanifold in $\mathfrak{G}$. Then we can assume that it contains the unit element of the group $e$. It is clear that $T_e(V) \subset G$.

**Proposition 16.1 (see [71]).** *Let $\mathfrak{G}$ be a connected, compact Lie group, and $V$ a totally geodesic submanifold in it. Then the subspace $B = T_eV$ is a triple Lie system, whereas $\exp(B) = V$. Conversely, if $B \subset G$ is a triple system, then the submanifold $V = \exp B$ is totally geodesic in $\mathfrak{G}$.*

Any Lie subgroup of positive dimension is a totally geodesic submanifold in $\mathfrak{G}$.

## §17 Necessary Results Concerning the Topological Structure of Compact Lie Groups and Symmetric Spaces

### 17.1. Cohomology Algebras of Compact Lie Groups

Here, we give certain facts regarding the homology and cohomology of compact groups, which we need in the following.

The totality of all simple, compact Lie groups consists of four large series and five special group classes. Classes of locally isomorphic, compact, connected, and simple Lie groups are usually denoted by the symbols $A_n (n \geqslant 1), B_n (n \geqslant 2), C_n (n \geqslant 3), D_n (n \geqslant 4)$, classical series and $G_2, F_4, E_6, E_7, E_8$, special structures. The classical series are defined for all $n \geqslant 1$, but the above restrictions on the group ranks are accounted for by isomorphisms of groups of small dimensions from different series, viz., $A_1 = B_1 = C_1$, $B_2 = C_2$, $A_3 = D_3$, $D_2 = A_1 + A_1$. We designate the centre of a group $\tilde{\mathfrak{G}}$ by $Z(\tilde{\mathfrak{G}})$. Each class of locally isomorphic groups contains a unique simply-connected compact group $\tilde{\mathfrak{G}}$. For $A_n, B_n, C_n$, these simply-connected groups are $\mathrm{SU}(n+1), \mathrm{Spin}(2n+1), \mathrm{Sp}(n)$, where $Z(\mathrm{SU}(n+1)) = \mathbf{Z}_{n+1}, Z(\mathrm{Spin}(2n+1)) = \mathbf{Z}_2, Z(\mathrm{Sp}(n)) = \mathbf{Z}_2$. The complete quotient groups $\tilde{\mathfrak{G}}/Z(\tilde{\mathfrak{G}})$ are the projective transformation groups $\mathrm{PU}(n+1), \mathrm{SO}(2n+1), \mathrm{PSp}(n)$. The simply-connected group $\tilde{\mathfrak{G}}$ of class $D_n$ is $\mathrm{Spin}(2n)$, $Z(\mathrm{Spin}(2n)) = \mathbf{Z}_4$ if $n = 2\alpha+1$ and $Z(\mathrm{Spin}(2n)) = \mathbf{Z}_2 \oplus \mathbf{Z}_2$ if $n = 2\alpha$. The complete quotient group $\mathrm{Spin}(4\alpha+2)/Z$ is the projective orthogonal group $\mathrm{PSO}(2n), n = 2\alpha + 1$. Besides, for $n = 2\alpha + 1$, there exists another group $\mathrm{Spin}(2n)/\mathbf{Z}_2$, where $\mathbf{Z}_2 \subset \mathbf{Z}_4$, which is isomorphic to $\mathrm{SO}(2n)$. If $n$ is even, then, in addition to $\mathrm{SO}(2n)$, there exist another two "semispinor" groups homeomorphic to each other for any $n = 2\alpha$; in particular, for $n = 8$, they are isomorphic to the group $\mathrm{SO}(8)$. The simply-connected representatives in the classes $G_2, F_4, E_6, E_7, E_8$ are groups with cyclic centres $0, 0, \mathbf{Z}_3, \mathbf{Z}_2, 0$, respectively. Below, we list the real cohomology rings for simple, compact Lie groups, viz.,

$$H^*(\mathrm{SU}(n+1)) = \Lambda(x_3, x_5, x_7, \ldots, x_{2n+1});$$
$$H^*(\mathrm{Spin}(2n+1)) = \Lambda(x_3, x_7, x_{11}, \ldots, x_{4n-1});$$
$$H^*(\mathrm{Sp}(n)) = \Lambda(x_3, x_7, x_{11}, \ldots, x_{4n-1});$$
$$H^*(\mathrm{Spin}(2n)) = \Lambda(x_3, x_7, x_{11}, \ldots, x_{4n-5}, \bar{x}_{2n-1});$$
$$H^*(G_2) = \Lambda(x_3, x_{11});$$
$$H^*(F_4) = \Lambda(x_3, x_{11}, x_{15}, x_{23});$$
$$H^*(E_6) = \Lambda(x_3, x_9, x_{11}, x_{15}, x_{17}, x_{23});$$
$$H^*(E_7) = \Lambda(x_3, x_{11}, x_{15}, x_{19}, x_{23}, x_{27}, x_{35});$$
$$H^*(E_8) = \Lambda(x_3, x_{15}, x_{23}, x_{27}, x_{35}, x_{39}, x_{47}, x_{59}).$$

Let $f : \mathfrak{G} \times \mathfrak{G} \to \mathfrak{G}$ be a continuous mapping defined by group multiplication. The homomorphism $f^* : H^*(\mathfrak{G}; \mathbf{R}) \to H^*(\mathfrak{G} \times \mathfrak{G}; \mathbf{R}) = H^*(\mathfrak{G}) \otimes H^*(\mathfrak{G})$ arises. An element $x \in H^*(\mathfrak{G}; \mathbf{R})$ is said to be *primitive* if $f^*(x) = 1 \otimes x + x \otimes 1$. It is known that the generators $\{x_{2k_i-1}\}$ in $H^*(\mathfrak{G}; \mathbf{R})$ may be chosen to be primitive. In the following, we will assume that the generators are primitive.

he groups $\mathrm{SU}(n)$ and $\mathrm{Sp}(n)$ are torsion-free, and $H^*(\mathrm{SU}(n); \mathbf{Z}) = \Lambda(x_3, x_5, \ldots, x_{2n-1}), H^*(\mathrm{Sp}(n); \mathbf{Z}) = \Lambda(x_3, x_7, \ldots, x_{4n-1})$.

Multiplication $f$ in a group defines the homomorphism $f_* : H_*(\mathfrak{G} \times \mathfrak{G}) \to H_*(\mathfrak{G})$. If we take a field as the coefficient group, then we obtain the homomorphism $f_* : H_*(\mathfrak{G}) \otimes H_*(\mathfrak{G}) \to H_*(\mathfrak{G})$. If two elements $a$ and $b$ belong to $H_*(\mathfrak{G}; C)$, then the element $f_*(a \otimes b) \in H_*(\mathfrak{G}; C)$ is called their *Pontryagin product*. Thus, $H_*(\mathfrak{G}; C)$ turns into a ring in which the multiplication is associative, distributive, and possesses a unit element, but is not necessarily anticommutative. However, if $H^*((\mathfrak{G}; C)$ possesses a simple system of primitive generators $\{x_i\}(1 \leqslant i \leqslant m)$, then $H_*(\mathfrak{G}; C) = \Lambda(y_1, y_2, \ldots, y_m)$, where $\deg y_i = \deg x_i, i = 1, 2, \ldots, m$; the converse statement also holds. As the coefficient ring, we take the field of real numbers. Let $\mathfrak{G}$ be a compact group. Then $H^*(\mathfrak{G}; \mathbf{R})$ admits a simple system of primitive generators, and the duality of $H_*(\mathfrak{G}; \mathbf{R})$ and $H^*(\mathfrak{G}; \mathbf{R})$ is valid. The homomorphisms $f^*$ and $f_*$ are adjoint to each other, with the generators $y_i$ being dual of $x_i$, and vice versa, i.e., $f^*$ and the Pontryagin product completely determine each other.

## 17.2. Subgroups Totally Nonhomologous to Zero

Consider a connected, compact Lie group $\mathfrak{G}$. Let $V$ be a compact, simply-connected and totally geodesic submanifold in $\mathfrak{G}$. The problem that we solve in the first part of the present chapter is investigating conditions for $V$ to realize a nontrivial homology cycle in $H_*(\mathfrak{G}; C)$, i.e., for the element $i_*[V]$ to be different from zero, $[V]$ denoting the fundamental class of $V$. We first consider the particular case where that totally geodesic submanifold $V$ is a subgroup. Suppose that $i : V = \mathfrak{H} \to \mathfrak{G}$ is an embedding. Then the homorphism $i^* : H^*(\mathfrak{G}; C) \to H^*(\mathfrak{H} : C)$ arises.

**Definition 17.2.1.** A subgroup $\mathfrak{H}$ is said to be *totally nonhomologous to zero* in a group $\mathfrak{G}$ for a coefficient group $C$ if the homomorphism $i^*$ is an epimorphism.

**Proposition 17.2.1.** *A connected, compact subgroup $\mathfrak{H}$ in a compact Lie group $\mathfrak{G}$ realizes a nontrivial cycle in $H^*(\mathfrak{G};\mathbf{R})$ if and only if $\mathfrak{H}$ is totally nonhomoogous to zero in $\mathfrak{G}$ for real coefficients.*

**Proof.** Let $i_*[\mathfrak{H}] \neq 0$ in $H_*(\mathfrak{G};\mathbf{R})$. The group $H^*(\mathfrak{H};\mathbf{R})$ can be treated as the space of linear functionals on $H_*(\mathfrak{H};\mathbf{R})$. Since $i_*[\mathfrak{H}] = i_*(y_1 \cdot y_2 \dots y_r) = i_*(y_1)i_*(y_2)\dots i_*(y_r) \neq 0$, where $y_1, y_2, \dots, y_r$ are multiplicative generators in $H_*(\mathfrak{H},\mathbf{R})$, we have $i_*(y_\alpha) \neq 0$ for $1 \leqslant \alpha \leqslant r$. Consider in $H_*(\mathfrak{H};\mathbf{R})$ the linear subspace of irreducible elements (alias primitive), which we shall denote by $P(\mathfrak{H})$; $y_1, y_2, \dots y_r$ are the additive basis for this subgroup. It is easy to see that $P(\mathfrak{H})$ is mapped monomorphically by the homomorphism $i_*$. Since $i_*(P(\mathfrak{H})) \subset P(\mathfrak{G})$, the image of the algebra $H_*(\mathfrak{H};\mathbf{R})$ is the exterior subalgebra in $H_*(\mathfrak{G};\mathbf{R})$, generated by the subspace $i_*(P(\mathfrak{H}))$, from which it follows that $i_*$ is a monomorphism on $H_*(\mathfrak{H};\mathbf{R})$.

Q.E.D.

Proposition 17.2.1 reduces our problem in the case where $V = \mathfrak{H}$ to the study of $\mathfrak{H}$ on the total nonhomologousness to zero. We now describe one known construction permitting us to answer the question regarding the total nonhomologousness to zero of a subgroup $\mathfrak{H}$ in $\mathfrak{G}$ in terms of Lie algebras.

Let $\mathfrak{G} \to E_{\mathfrak{G}} \to B_{\mathfrak{G}}$ be the universal fibre space for the group $\mathfrak{G}$. Consider the embedding of $\mathfrak{H}$ into $\mathfrak{G}$. Then there arises a projection $\rho$ of the space $B_{\mathfrak{H}}$ onto the space $B_{\mathfrak{G}}$, inducing the homomorphism $\rho^*(\mathfrak{H},\mathfrak{G}) : H^*(B_{\mathfrak{G}}) \to H^*(B_{\mathfrak{H}})$.

**Proposition 17.2.2 (A. Borel; see [53]).** *Let $\mathfrak{H}$ be a connected, compact subgroup in $\mathfrak{G}$. $\mathfrak{h}$ is totally nonhomologous to zero for real coefficients if and only if $\rho^*(\mathfrak{H},\mathfrak{G})$ is an epimorphism.*

In the case of real coefficients, $\rho^*$ can be computed efficiently. Consider a group $\mathfrak{G}$ of rank $R$. Let $T^R$ be the maximal torus of $\mathfrak{G}$. The tangent space $T_e(T^R) = t^R$, a Cartan subalgebra, is a covering group for $T^R$. Let $t \in T^R$, and $N(t)$ its centralizer. We distinguish from $t^R$ the set $S'$ of all elements $t'$ such that $\exp t' \in S$, where $S$ is the set of all singular elements in $T^R$. $S'$ is called the *diagram* of $\mathfrak{G}$ and contains the "unit lattice," a collection of all such elements $t' = t^R$ that $\exp t' = e$. It is clear that the elements of the Weyl group $W(\mathfrak{G}) = N(t)/T$ transforms both the lattice and diagram into themselves.

**Proposition 17.2.3 (see [53]).** *If a group $\mathfrak{G}$ is connected, then the ring of $I_G(R)$ of polynomials on the Cartan subalgebra, which are invariant with respect to the action of the group $W^*(\mathfrak{G};\mathbf{R})$, is a free algebra with $R$ generators $P_{k_1}, \dots, P_{k_R}$, where $R$ is the rank of $\mathfrak{G}$, and $\deg P_{k_i} = k_i, 1 \leqslant i \leqslant R, P(\mathfrak{G},t) = (1+t^{2k_1-1})\dots(1+t^{2k_R-1})$, with $P(\mathfrak{G},t)$ denoting the Poincaré polynomial.*

Consider the embedding $i : \mathfrak{H} \to \mathfrak{G}$. The explicit form of the generating polynomials $P_{k_i}$ in the ring $I_G$ is given in [108]. Thus, a *compact connected subgroup* $\mathfrak{H}$ realizes a nontrivial cocycle in $H^*(\mathfrak{G}; \mathbf{R})$ if and only if the homomorphism $I^* : I_G \to I_H$ is an epimorphism.

Application of this statement to actual computations is considerably eased by an elaborate technique for the study of rings $I_G$ (due to Ðoàn Quỳnh, O. V. Manturov, and I. Z. Rozenknop, see [108], [257], and [359]; whereas concrete computations of the homomorphism $\rho^*_{\mathbf{R}}(\mathfrak{H}; \mathfrak{G})$ have been carried out, in particular, in [108] and [149]).

We now survey all basic cases of total homologousness to zero, but first describe a large class of subgroups $\mathfrak{H}$ irreducibly acting on the tangent space to the homogeneous space $\mathfrak{G}/\mathfrak{H}$. We determine the action of $\mathfrak{H}$ on $\mathfrak{G}$ by $h(g) = hgh^{-1}$, where $h \in \mathfrak{H}, g \in \mathfrak{G}$, generating a linear representation of $\mathfrak{H}$ on the linear space $G$. A linear subspace $H \subset G$ is invariant with respect to the action of $\mathfrak{H}$ on $G$; therefore, the orthogonal complement $B$ (the groups being compact) to $H$ is also invariant. $H$ and $B$ are invariant relative to $\mathrm{ad}_{\mathrm{H}}$, which entails for the decomposition $G = H + B$ that (1) $[H, H] \subset H$, (2) $[H, B] \subset B$, (3) $\langle X, Y \rangle = 0$, where $X \in H, Y \in B$.

Consider the class $\Psi$ of all thepairs $(\mathfrak{G}, \mathfrak{H})$ such that $\mathfrak{G}$ is a simple, connected, and compact Lie group and that the compact connected subgroup $\mathfrak{H}$ irreducibly acts on the tangent space to $V = \mathfrak{G}/\mathfrak{H}$. We enter only those pairs $(\mathfrak{G}, \mathfrak{H})$ in the subclass $\Psi_1$ for which $[B, B] \subset H$, i.e., $V = \mathfrak{G}/\mathfrak{H}$ is a symmetric space. All the other pairs from $\Psi$ will be assumed to belong to a subclass $\Psi_2$. All $(\mathfrak{G}, \mathfrak{H}) \in \Psi_1$ are described by E. Cartan in [71], whereas all $(\mathfrak{G}, \mathfrak{H}) \in \Psi_2$ by O. V. Manturov in [257]; the complete description of the embeddings $i : \mathfrak{H} \to \mathfrak{G}$ (where $(\mathfrak{G}, \mathfrak{H}) \in \Psi_1$) from the homology standpoint can be found in M. Takeuchi's work [404]. It turns out that the unique pairs $(\mathfrak{G}, \mathfrak{H}) \in \Psi_1$, where the manifold $\mathfrak{G}/\mathfrak{H}$ is not a Lie group for which the subgroup $\mathfrak{H}$ is totally nonhomologous to zero for real coefficients, are the following (the embeddings of $\mathfrak{H}$ into $\mathfrak{G}$ being described below), viz., (1) $(\mathrm{SU}(2m+1), \mathrm{SO}(2m+1)), m \geqslant 1$, (2) $(\mathrm{SU}(2m), \mathrm{Sp}(m)), m \geqslant 2$, (3) $(\mathrm{SO}(2l), \mathrm{SO}(2l-1)), l \geqslant 4$, and (4) $(E_6, F_4)$.

The class $\Psi_2$ was investigated by Ðoàn Quỳhn from the homology standpoint in [108].

It turns out that $(\mathfrak{G}, \mathfrak{H}) \in \Psi_2$, where $\mathfrak{H}$ is totally nonhomologous to zero for real coefficients, are only the following (see the embeddings in [108]), viz.,

(1) $(A_{N-1}, A_{n-1}), n \geqslant 3, N = \frac{n(n+1)}{2}$, (2) $(A_{15}, D_5)$, (3) $(A_{26}, E_6)$, (4) $(B_N, B_n), n \geqslant 3$, if $n$ is odd, $N = \frac{n(2n+1)-1}{2}, (D_N, B_n), n \geqslant 3$, if $n$ is even, $N = \frac{n(2n+1)}{2}$, (5) $(B_N, B_n), n \geqslant 2$, if $n$ is odd, $N = \frac{n(2n+3)-1}{2}$, $(D_N, B_n), n \geqslant 2$, if $n$ is even, $N = \frac{n(2n+3)}{2}$, (6) $(B_N, C_n), n \geqslant 2$, if $n$ is odd, $N = \frac{n(2n+1)-1}{2}, (D_N, C_n), n \geqslant 2$, if $n$ is even, $N = \frac{n(2n+1)}{2}$, (7) $(D_{21}, C_4)$, (8) $(D_4, B_3)$, (9) $(B_2, A_1)$, (10) $(D_{124}, E_8)$, (11) $(B_{66}, E_7)$, (12) $(D_{13}, F_4)$, (13) $(D_{26}, F_4)$, (14) $(B_3, G_2)$, (15) $(D_7, G_2)$, (16) $(C_7, C_3)$, (17) $(G_2, A_1)$, (18) $(E_6, G_2)$, and (19) $(G_2, A_1)$.

According to H. Samelson's work [366], the real Poincaré polynomial of the space $V = \mathfrak{G}/\mathfrak{H}$, where the subgroup $\mathfrak{H}$ is totally nonhomologous to zero in the group $\mathfrak{G}$, is of the form $P(V,t) = \frac{P(\mathfrak{G},t)}{P(\mathfrak{H},t)}$.

### 17.3. Pontryagin Cycles in Compact Lie Groups

The problem of realizing cycles in Lie groups and homogenous spaces is intimately connected to finding the simplest cell decomposition of a manifold. Important results were obtained in [129], where the cellular structure of real and complex Grassmannian manifolds was investigated. In [341], L. S. Pontryagin described the rings $H^*(\mathfrak{G};\mathbf{R})$ and the structure of integral homology of classical compact, simple Lie groups by having constructed the basic cycles explicitly. Forming a basis for cohomology in the groups $U(n)$ and $\mathrm{Sp}(n)$, these cycles are not submanifolds.

Consider the sphere $S^{n-1}$ whose points will be given by the Euclidean coordinates $x_1,\dots,x_n, x_1^2+\dots+x_n^2 = 1$. We associate each point $x \in S^{n-1}$ with $T_n(x) \in \mathrm{SO}(n), T_n(x) = \|t_{ij}\|E_{1,n-1}, t_{ij} = \delta_{ij} - 2x_ix_j$, where $E_{1,n-1} = (-E_1)\oplus E_{n-1}$ and $E_\alpha$ is the identity operator. It is clear that the mapping $T_n$ defines the embedding of the projective space $\mathbf{R}\,\mathrm{P}^{n-1}$ into the group $\mathrm{SO}(n)$. If $n$ is even, then the intersection number for the cycle $X_{n-1} = \mathbf{R}\,\mathrm{P}^{n-1}$ and the subgroup $\mathrm{SO}(n-1)$ equals $+1$. It follows that the submanifold $X_{n-1}$ realizes a nontrivial cycle in $H_*(\mathrm{SO}(n);\mathbf{R})$. If $n$ is odd, then we can construct a cycle $U_{n-3}$ also realizing a nontrivial element in $H^*(\mathrm{SO}(n))$. Consider two chains $X_{n-2}$ and $X_{n-1}$ and construct the chain $U_{2n-3} = X_{n-2}\cdot X_{n-1}$ (Pontryagin product) which is a cocycle, whereas the intersection number for $U_{2n-3}$ and the subgroup $\mathrm{SO}(n-2)$ is $+1$.

Consider the standard embedding $i : \mathrm{SO}(2n-1) \to \mathrm{SO}(2n)$. Then

$$H_*(\mathrm{SO}(2n);\mathbf{R}) = \Lambda(x_3,\dots,x_{4n-5})\otimes\Lambda(y_{2n-1}),$$

and the generator of the subring $H_*(S^{2n-1};\mathbf{R})$ is realized by $X_{2n-1} = \mathbf{R}\,\mathrm{P}^{2n-1}$. It follows easily that the generators of the homology ring $H_*(\mathrm{SO}(n);\mathbf{R})$ admit the following geometric description.

A generator $x_3 \in H_*(\mathrm{SO}(2n+1);\mathbf{R})$ is realized in $\mathrm{SO}(2n+1)$ by the submanifold $X_3 = \mathbf{R}\,\mathrm{P}^3$. All the other generators $x_7, x_{11},\dots,x_{4n-1}$ are realized by the cycles $U_7,\dots,U_{4n-1}, U_{4p-1} = X_{2p-1}\cdot X_{2p}$. Note that $\mathbf{R}\,\mathrm{P}^3 = U_3 = X_1\cdot X_2 = S^1\cdot\mathbf{R}\,\mathrm{P}^2$ (which can be directly seen). In $H_*(\mathrm{SO}(2n);\mathbf{R})$, the generators $x_3,\dots,x_{4n-5},y_{2n-1}$ are realized by the cycles $\mathbf{R}\,\mathrm{P}^3, U_7,\dots,U_{4n-5},\mathbf{R}\,\mathrm{P}^{2n-1}$. The above canonical sets of generators are an additive basis for the linear space of primitive elements.

We now distinguish from these realizations locally minimal, totally geodesic submanifolds. It is evident that the cycles $U_{2m-1}$ in the groups $\mathrm{SO}(n)$ for $2m-1 \neq 3$ are not submanifolds; however, $U_3$ is a subgroup and, hence, a totally geodesic submanifold. Consider the cycle $X_{2n-1}$ associated with the generator $y_{2n-1} \in H_*(\mathrm{SO}(2n);\mathbf{R})$.

**Proposition 17.3.1.** *A submanifold $X_{n-1} \subset \mathrm{SO}(n)$ is totally geodesic in $\mathrm{SO}(n), n \geqslant 3$.*

**Proof.** It is clear that the tangent space $T_e(X_{n-1})$ in the Lie algebra so$(n)$ consists of matrices of the form

$$\left\| \begin{matrix} 0 & a_2 & a_3 \dots a_n \\ -a_2 & & \\ \vdots & & 0 \\ -a_n & & \end{matrix} \right\|,$$

where $a_i$ are arbitrary real numbers. We see by direct verification that $T_e(X_{n-1})$ is a triple system in so$(n)$. It is easily verified that $X_{n-1} = \exp(T_e(X_{n-1}))$, which entails the required statement.

Thus, unique cycles among the Pontryagin ones realizable by totally geodesic submanifolds in a group are $U_3 = \mathbf{R}P^3$ in $\mathrm{SO}(n)$ for $n \geqslant 3$ and $X_{2n-1} = \mathbf{R}P^{2n-1}$ in $\mathrm{SO}(2n)$ for $n \geqslant 2$. It turns out that these cycles belong to a series of the same type of locally minimal cycles which we shall describe below.

### 17.4. Necessary Results Concerning Symmetric Spaces

**Definition 17.4.1.** A connected Riemannian manifold $V$ is called a *symmetric space* if, for each point $p \in V$, there exists an involutive isometry $s_p : V \to V$ other than the identity for which $p$ is an isolated fixed point.

A connected, compact Lie group with invariant Riemannian metric is turned into a symmetric space. Symmetric spaces admit a representation in the form $V = \mathfrak{G}/\mathfrak{H}$, where $\mathfrak{G}$ is a connected Lie group with involutive automorphism $\sigma$ whose fixed-point set coincides with the connected component of the unit element of the subgroup $\mathfrak{H}$. Denote by $I(V)$ the set of all isometries of a symmetric space. Then it is a Lie group. If $\mathfrak{G} = I_o(V)$ designates the connected component of the unit element in $I(V)$, then the compact, simply-connected, and symmetric space is representable in the form $V = \mathfrak{G}/\mathfrak{H}$, where $\mathfrak{H}$ is the stability subgroup, $\mathfrak{H} = \{g \in \mathfrak{G} | \sigma(g) = g\}$, and $\sigma$ is an involution. Denote by $\theta(d\sigma)_e$ the corresponding involutive automorphism of the Lie algebra $G = T_e\mathfrak{G}$. Then the subalgebra $H = T_e\mathfrak{H} \subset G$ is that of fixed points for the automorphism $\theta : G \to G$. It is compactly embedded. We can assume that $H \cap Z(G) = 0$, where $Z(G)$ is the centre of $G$, and $G = B + H$, where $B = \{X \in G | \theta(X) = -X\}$.

A symmetric space $V = \mathfrak{G}/\mathfrak{H}$ is said to be *irreducible* if the adjoint representation $\mathrm{ad}_H$ on the plane $B$, generated by the relation $[H, B] \subset B$, is irreducible.

**Proposition 17.4.1 (see [198]).** *A compact, simply-connected, and symmetric Riemannian space $V$ decomposes into the direct product $V_1 \times \cdots \times V_s$ of irreducible, compact, and symmetric spaces.*

E. Cartan completely classified compact, irreducible, and symmetric spaces, dividing them into Types I and II (see [198]).

**Proposition 17.4.2.** *Compact, simply-connected, irreducible, and symmetric Riemannian spaces of Type I are listed in Table 1.*

Table 1 (A)

| *Type* | | *Space V* | $G$ | $H$ | *Rank* |
|---|---|---|---|---|---|
| $^0A_{2m}$I | $m\geqslant 1$ | SU$(2m+1)$/SO $(2m+1)$ | $A_{2m}$ | $B_m$ | $2m$ |
| $A_{2m-1}$I | $m\geqslant 2$ | SU $(2m)$/SO $(2m)$ | $A_{2m-1}$ | $D_m$ | $2m-1$ |
| $^0A_{2m-1}$II | $m\geqslant 2$ | SU $(2m)$/Sp $(m)$ | $A_{2m-1}$ | $C_m$ | $m-1$ |
| $^*A_l$III$_m$ | $l\geqslant 1$, $1\leqslant m\leqslant \leqslant\left[\frac{l+1}{2}\right]$ | $U(l+1)/U(m)\times U(l+ +l-m)$ | $A_l$ | $T^1+ +A_{m-1}+ +A_{l-m}$ | $m$ |
| $^*B_l$I$_2$ | $l\geqslant 2$ | SO $(2l+1)$/SO $(2)\times \times$ SO $(2l-1)$ | $B_l$ | $T^1+B_{l-1}$ | 2 |
| $B_l$I$_{2m}$ | $l\geqslant 2$, $2\leqslant m\leqslant \leqslant l-1$ | SO $(2l+1)$/SO $(2m)\times \times$ SO $(2l+1-2m)$ | $B_l$ | $D_m+ +B_{l-m}$ | min $(2m, 2l+1- -2m)$ |
| $^*C_l$I | $l\geqslant 3$ | Sp $(l)/U(l)$ | $C_l$ | $T^1+A_{l-1}$ | $l$ |
| $B_l$II | $l\geqslant 3$ | SO $(2l+1)$/SO $(2l)$ | $B_l$ | $D_l$ | 1 |
| $C_l$II$_m$ | $l\geqslant 3$, $1\leqslant m\leqslant \leqslant\left[\frac{l}{2}\right]$ | Sp $(l)$/Sp $(m)\times$ Sp $(l-m)$ | $C_l$ | $C_m+C_{l-m}$ | $m$ |
| $^*D_l$I$_2$ | $l\geqslant 4$ | SO $(2l)$/SO $(2)\times$ SO $(l-2)$ | $D_l$ | $T^1+D_{l-1}$ | 2 |
| $D_l$I$_{2m}$ | $l\geqslant 4$, $2\leqslant m\leqslant \leqslant\left[\frac{l}{2}\right]$ | SO $(2l)$/SO $(2m)\times \times$ SO $(2l-2m)$ | $D_l$ | $D_m+ +D_{l-m}$ | $2m$ |
| $D_l$I$_{2m+1}$ | $l\geqslant 4$, $1\leqslant m\leqslant \leqslant\left[\frac{l-1}{2}\right]$ | SO $(2l)$/SO $(2m+1)\times \times$ SO $(2l-2m-1)$ | $D_l$ | $B_m+ +B_{l-m-1}$ | $2m+1$ |

Table 1 (A) *(continued)*

| *Type* | | *Space V* | *G* | *H* | *Rank* |
|---|---|---|---|---|---|
| ${}^0D_l$II | $l \geq 4$ | $SO(2l)/SO(2l-1)$ | $D_l$ | $B_{l-1}$ | 1 |
| ${}^*D_l$III | $l \geq 4$ | $SO(2l)/U(l)$ | $D_l$ | $T^1+A_{l-1}$ | $\left[\frac{l}{2}\right]$ |
| $E_6$I | | $E_6/PS_p(4)$ | $E_6$ | $C_4$ | 6 |
| $E_6$II | | $E_6/SU(2)\cdot SU(6)$ | $E_6$ | $A_1+A_5$ | 4 |
| ${}^*E_6$ III | | $AdE_6/T^1\cdot Spin(10)$ | $E_6$ | $T^1+D_5$ | 2 |
| ${}^0E_6$IV | | $E_6/F_4$ | $E_6$ | $F_4$ | 2 |
| $E_7$V | | $E_7/SU^*(8)$ | $E_7$ | $A_7$ | 7 |
| $E_7$VI | | $E_7/SU(2)\cdot Spin(12)$ | $E_7$ | $A_1+D_6$ | 4 |
| ${}^*E_7$VII | | $AdE_7/T^1\cdot E_6$ | $E_7$ | $T^1+E_6$ | 3 |
| $E_8$VIII | | $E_8/SO(16)$ | $E_8$ | $D_8$ | 8 |
| $E_8$IX | | $E_8/SU(2)\cdot E_7$ | $E_8$ | $A_1+E_7$ | 4 |
| $F_4$I | | $F_4/SU(2)\cdot Sp(3)$ | $F_4$ | $A_1+C_3$ | 4 |
| $F_4$II | | $F_4/Spin(9)$ | $F_4$ | $B_4$ | 1 |
| $G_2$I | | $G_2/SO(4)$ | $G_2$ | $A_1+A_1$ | 2 |

Table 1 (B)

| *Type* | $\chi(V)$ | *Poincaré Polynomial* $P(V,t)$ |
|---|---|---|
| ${}^0A_{2m}$I | 0 | $(1+t^5)(1+t^9)(1+t^{13})\dots(1+t^{4m+1})$ |
| $A_{2m-1}$I | 0 | $(1+t^5)(1+t^9)(1+t^{13})\dots(1+t^{4m-3})(1+t^{2m})$ |
| ${}^0A_{2m-1}$II | 0 | $(1+t^5)(1+t^9)(1+t^{13})\dots(1+t^{4m-3})$ |
| ${}^*A_l$III$_m$ | $C_{l+1}^m$ | $(1-t^{2(l-m+2)})(1-t^{2(l-m+3)})\dots(1-t^{2(l+1)})/(1-t^2)\times$ $\times(1-t^4)\dots(1-t^{2m})$ |
| ${}^*B_l$I$_2$ | $2l$ | $1+t^2+t^4+\dots+t^{4l-2}$ |

**Table 1 (B)** *(continued)*

| *Type* | $\chi(V)$ | *Poincaré Polynomial* $P(V,t)$ |
|---|---|---|
| $B_l\mathrm{I}_{2m}$ | $2C_l^m$ | $(1-t^{4(l-m+1)})(1-t^{4(l-m+2)})\dots(1-t^{4l})/(1-t^4)\times$ $\times(1-t^8)\dots(1-t^{4(m-1)})(1-t^{2m})$ |
| $^*C_l\mathrm{I}$ | $2^l$ | $(1+t^2)(1+t^4)(1+t^6)\dots(1+t^{2l})$ |
| $B_l\mathrm{II}$ | $2$ | $1+t^{2l}$ |
| $C_l\mathrm{II}_m$ | $C_l^m$ | $(1-t^{4(l-m+1)})(1-t^{4(l-m+2)})\dots(1-t^{4l})/(1-t^4)\times$ $\times(1-t^8)\dots(1-t^{4m})$ |
| $^*D_l\mathrm{I}_2$ | $2l$ | $(1+t^2+t^4+\dots+t^{2l-2})(1+t^{2l-2})$ |
| $D_l\mathrm{I}_{2m}$ | $2C_l^m$ | $\dfrac{(1-t^{4(l-m)})(1-t^{4(l-m+1)})\dots(1-t^{4(l-1)})}{(1-t^4)(1-t^8)\dots(1-t^{4(m-1)})}\times$ $\times\dfrac{1-t^{2l}}{(1-t^{2m})(1-t^{2(l-m)})}$ |
| $D_l\mathrm{I}_{2m+1}$ | $0$ | $(1-t^{4(l-m)})(1-t^{4(l-m+1)})\dots(1-t^{4(l-1)})(1+t^{2l-1})/$ $/(1-t^4)(1-t^8)\dots(1-t^{4m})$ |
| $^0D_l\mathrm{II}$ | $0$ | $1+t^{2l-1}$ |
| $^*D_l\mathrm{III}$ | $2^{l-1}$ | $(1+t^2)(1+t^4)(1+t^6)\dots(1+t^{2l-2})$ |
| $E_6\mathrm{I}$ | $0$ | $(1+t^8+t^{16})(1+t^9)(1+t^{17})$ |
| $E_6\mathrm{II}$ | $36$ | $(1+t^4+t^8+\dots+t^{20})(1+t^6+t^{12})(1+t^8)$ |
| $^*E_6\mathrm{III}$ | $27$ | $(1+t^2+t^4+\dots+t^{16})(1+t^8+t^{16})$ |
| $^0E_6\mathrm{IV}$ | $0$ | $(1+t^9)(1+t^{17})$ |
| $E_7\mathrm{V}$ | $72$ | $(1+t^6+t^{12}+\dots+t^{30})(1+t^8+t^{16})(1+t^{10})(1+t^{14})$ |
| $E_7\mathrm{VI}$ | $21$ | $(1+t^4+t^8+\dots+t^{24})(1+t^8+t^{16})(1+t^{12}+t^{24})$ |
| $^*E_7\mathrm{VII}$ | $56$ | $(1+t^2+t^4+\dots+t^{26})(1+t^{10})(1+t^{18})$ |
| $E_8\mathrm{VIII}$ | $135$ | $(1+t^8+t^{16}+t^{24}+t^{32})(1+t^{12}+t^{24})(1+t^{16}+t^{32})\times$ $\times(1+t^{20}+t^{40})$ |
| $E_8\mathrm{IX}$ | $120$ | $(1+t^4+t^8+\dots+t^{36})(1+t^{12}+t^{24}+t^{36})(1+t^{20}+t^{40})$ |

Table 1 (B) *(continued)*

| *Type* | $\chi(V)$ | *Poincaré Polynomial* $P(V,t)$ |
|---|---|---|
| $F_4$I | 12 | $(1+t^4+t^8+\ldots+t^{20})(1+t^8)$ |
| $F_4$II | 3 | $1+t^8+t^{16}$ |
| $G_2$I | 3 | $1+t^4+t^8$ |

**Proposition 17.4.3.** *Compact, irreducible, and symmetric Riemannian spaces of Type II are compact, connected, and simple Lie groups endowed with the bi-invariant Riemannian metric.*

**Remark (see Table 1)**

(1) The symbol $*$ on the left denotes Hermitian symmetric spaces, and the symbol $^0$ on the left also denotes spaces whose stability subgroups are totally nonhomologous to zero in the underlying group $\mathfrak{G}$.

(2) In the third column, the representation of the manifold $V$ in the form $\mathfrak{G}/\mathfrak{H}$ does not necessarily coincide with $\mathfrak{G}'/\mathfrak{H}'$, where $\mathfrak{G}' = I_0(V)$. Further, $SU^*(8) = SU(8)/\{\pm 1\}$. The stability subgroups in $E_6$II, $^*E_6$III, $E_7$VI, $^*E_7$VII, $E_8$IX, and $F_4$I are not direct products, the intersections of the factors are nonempty and are cyclic subgroups of orders 2, 4, 2, 3, 2, and 2, respectively.

The restrictions on the indices in the second column pursue the goal of evading coincidence between spaces of different classes. The isomorphisms in small dimensions are as follows

*1)* $A_1\mathrm{I} = A_1\mathrm{III}_1 = C_1\mathrm{I} = B_1\mathrm{I}_2$; *2)* $C_2\mathrm{I} = B_2\mathrm{I}_2$; *3)* $C_2\mathrm{II}_1 = B_2\mathrm{I}_4$; *4)* $A_3\mathrm{I} = D_3\mathrm{I}_3$; *5)* $A_3\mathrm{II} = D_3\mathrm{I}_1$; *6)* $A_3\mathrm{III}_2 = D_3\mathrm{I}_2$; *7)* $A_3\mathrm{III}_1 = D_3\mathrm{III}$; *8)* $D_4\mathrm{I}_2 = D_4\mathrm{III}$.

Though the algebra $D_2 = A_1 + A_1$ is not simple, the spaces $D_2$ III and $BD\mathrm{I}(SO(p+q)/SO(p)\times SO(q))$ for $p+q=4$ can be defined, which gives another three isomorphisms, viz., (9) $D_2I_1 = SO(4)/SO(1)\times SO(3) = SO(4)/SO(3) = SU(2)$, (10) $D_2I_2 = SO(4)/SO(2)\times SO(2) = A_1\mathrm{I}\times A_1\mathrm{I}$, (11) $D_2\mathrm{III} = SO(4)/U(2) = A_1\mathrm{I}$.

We are not going to describe involutions for all spaces of Type I: The required ones are specified whenever neccessary.

If an irreducible symmetric space is determined by a semisimple and simply-connected group $\mathfrak{G}$ (see above) and an involutorial automorphism $\theta$ acting in the Lie algebra $G$, then $G$ is a maximal-connected isometry group of the space (see [71], [198]).

## §18 Lie Groups Containing a Totally Geodesic Submanifold Necessarily Contain Its Isometry Group

Let $\mathfrak{G}$ be a connected, compact Lie group, and $V \subset \mathfrak{G}$ a compact, one-connected and totally geodesic submanifold. We need to clarify when the cycle $i_*[V]$ is other than zero in the group $H_*(\mathfrak{G})$. The case where $V$ is a subgroup has been completely investigated in §17.2. Since any totally geodesic submanifold $V$ in $\mathfrak{G}$ is a symmetric space, one of the ways to solve the problem may be that we should consider all embeddings of compact, symmetric spaces into Lie groups and find out which realize nontrivial cycles.

**Theorem 18.1.** *Let $V$ be a compact, simply-connected, and totally geodesic submanifold in a compact Lie group $\mathfrak{G}$, and $\{\mathfrak{G}_\alpha\}$ the collection of all the subgroups of $\mathfrak{G}$, containing $V$. Then $V$ decomposes into the direct product $V = K \times V_{m+1} \times \cdots \times V_s = K \times V'$, where $k$ is a compact group of $\mathfrak{G}$, and each $V_i (m+1 \leqslant i \leqslant s)$ is a totally geodesic submanifold in $\mathfrak{G}$, without being its subgroup. Besides, the decomposition possesses the property that if $A(V) = \bigcap_\alpha \mathfrak{G}_\alpha$, then the subgroup $A(V)$ is compact and semisimple, whereas the universal covering group $\tilde{A}(V)$ is isomorphic to the direct product of groups $\tilde{A}(V) \cong K \times \tilde{I}_0(V_{m+1}) \times \cdots \times \tilde{I}_0(V_s)$.*

**Proof.** We assume that the unit element of the group belongs to $V$. If $V = K$, then the theorem is valid; therefore, we can assume in the following that $V$ is not a subgroup of $\mathfrak{G}$. Denote by $B$ the tangent space $T_e(V)$ to $V$ at a point $e$. Then $B$ is a triple system, and $V = \exp B$. If $[B, B]$ is the commutant of the plane $B$, then $[B, B]$ is a subalgebra in $G$. Consider the subalgebra $G_1 = B + [B, B]$,"+" denoting the sum of the spaces. Denote by $\mathfrak{G}_1$ the subgroup in $\mathfrak{G}$ whose Lie algebra is $G_1$. It is clear that $\mathfrak{G}_1 = A(V)$.

Consider the space $H = B \cap [B, B]$ in $G_1$, where $H$ is the ideal in $G_1$. Let $\bar{\mathfrak{G}}_1$ be the subgroup which is the closure of the subgroup $\mathfrak{G}_1$ in $\mathfrak{G}$, and $\bar{G}_1$ is the Lie algebra of $\bar{\mathfrak{G}}_1$. Then $\bar{G}_1 \supset G_1$. Due to the compactness of $\mathfrak{G}$, $\bar{\mathfrak{G}}_1$ is also compact; therefore, $\bar{\mathfrak{G}}_1$ is a compact subalgebra of the algebra $G$. It is obvious that $G_1$ is the ideal in $\bar{G}_1$. Since each compact algebra $A$ is the direct sum $Z \oplus [A, A]$, where $Z$ is the centre in $A$ and the ideal $[A, A]$ is compact and semisimple $\bar{G}_1 = Z \oplus G_2, G_2 = [\bar{G}_1, \bar{G}_1]$. Consider a simply-connected group $\tilde{\bar{\mathfrak{G}}}_1$ with the Lie algebra $\bar{\mathfrak{G}}_1$. It decomposes into the direct product of two groups $\tilde{R} \times \tilde{\mathfrak{G}}_2$, where $\tilde{R}$ and $\tilde{\mathfrak{G}}_2$ are simply-connected, $\tilde{R}$ is diffeomorphic to Euclidean space (centre of $\tilde{\bar{\mathfrak{G}}}_1$), and $\tilde{\mathfrak{G}}_2$ is compact and semisimple. Since $V$ is simply-connected, it is "lifted" to $\tilde{\bar{\mathfrak{G}}}_1$ "without turning"; therefore, we obtain in $\tilde{\bar{\mathfrak{G}}}_1$ a compact, one-connected, and totally geodesic submanifold $V$ homeomorphic to $V$. Let $\pi_1$ and $\pi_2$ be the projections of the direct sum $Z \oplus G_2$ onto the addends $Z$ and $G_2$, respectively. Denote the corresponding projections $\tilde{R} \xleftarrow{p_1} \tilde{\bar{\mathfrak{G}}}_1 \xrightarrow{p_2} \tilde{\mathfrak{G}}_2$ by $p_1$ and $p_2$. Since $Z$ and $G_2$ are ideals in $\bar{G}_1$, $\exp(\pi_1 B)$ and $\exp(\pi_2 B)$ are totally geodesic submanifolds in $\tilde{R}$ and $\tilde{\mathfrak{G}}_2$.

It is evident that $\exp(\pi_1(B))$ is compact, and $\exp(\pi_1 B) = B_1(\widetilde{V}), \exp(\pi_2) = p_2(\widetilde{V})$. Since $\widetilde{R}$ is diffeomorphic to Euclidean space, $p_1(\widetilde{V}) = e$ in the group $\widetilde{R}$, i.e., the submanifold $\widetilde{V}$ is wholly contained in $\widetilde{\mathfrak{G}}_2$, implying that $B \subset G_2$, and $[B, B] \subset G_2$, or $B + [B, B] = G_1 \subset G_2$. Because $G_1$ is an ideal in $\bar{G}_1$, $G_1$ is also an ideal in $G_2$. Then there exists an ideal $N$ in $G_2$, so that $G_2 = N \oplus G_1$ (meaning the direct sum), i.e., $H$ is an ideal in $G_2$; therefore, the algebras $G_1$ and $H$ are compact and semisimple. Then the compact, semisimple group $\widetilde{\mathfrak{G}}_2$ decomposes into the direct product of groups $\widetilde{N} \times \widetilde{\mathfrak{G}}_1; \widetilde{N} = \exp(N)$, where $\widetilde{N}$ and $\widetilde{\mathfrak{G}}_1$ are compact, simply-connected, and semisimple groups, and $\widetilde{\mathfrak{G}}_1$ is a finite-covering of $\mathfrak{G}$. Thus, the subgroup $\mathfrak{G}_1$ is compact and semisimple. Since $H$ is an ideal in $G_1$, there arises the direct decomposition of the algebra $G_1$ into the sum of two ideals $H \oplus I$, which gives its decomposition into the direct product $\widetilde{K} \times \approx{I}$, where $\widetilde{K}$ and $\approx{I}$ are compact, semisimple and simply-connected. Due to $H \subset B$, there arises the decomposition of $B$ into the direct sum of two subspaces $H + (B \cap I) = H + H_1$, with $[X, Y] = 0$ for any $X \in H, Y \in H_1$, respectively, and $[B, B] = H \oplus H_2, H_1 \subset I, H_2 \subset I, H_2$ being an ideal in $[B, B]$. All these decompositions are orthogonal in the Cartan metric. The ideal $I$ decomposes into the direct sum of subspaces $H_1 + H_2; H_1 \cap H_2 = 0$. Note that the planes $H_1$ and $H_2$ must not necessarily be orthogonal to each other. Because $\widetilde{\mathfrak{G}}_1 = \widetilde{K} \times \approx{I}$ and $\widetilde{K} \subset \widetilde{V}$, $\widetilde{V}$ admits a decomposition into the direct product $\widetilde{K} \times \widetilde{V}'$, $\widetilde{V}' = \exp H_1 \subset \widetilde{\mathfrak{G}}_1$. Since $I$ is an ideal in $G_1$, $H_1$ is a triple system in $I$, whereas $H_2$ is a subalgebra in $I$, i.e., $\widetilde{K}$ and $\widetilde{V}'$ are totally geodesic submanifolds in $\widetilde{\mathfrak{G}}_1$; therefore, in $\widetilde{V}$, too, $\widetilde{V} = \widetilde{K} \times \widetilde{V}'$ signifies the decomposition of $\widetilde{V}$ into the direct product of symmetric spaces $\widetilde{K}$ and $\widetilde{V}'$. Consider a group $\approx{I}$ with the Lie algebra $I$ and a totally geodesic submanifold $\widetilde{V}'$ with the tangent plane $H_1, \widetilde{V}' \subset \approx{I}$. It is easy to see that $H_2 = [H_1, H_1]$; therefore, $I = H_1 + [H_1, H_1], H_1 \cap [H_1, H_1] = 0$.

We define an involutive automorphism $\theta$ on $I$ by setting $\theta(X) = X$ if $X \in [H_1, H_1]$, and $\theta(X) = -X$ if $X \in H_1$. Since $I$ is a compact, semisimple Lie algebra, we see by direct computation that the subalgebra $H_2 \subset I$ contains no nonzero ideals of $I$ and, therefore, no ideals of $G_1$. It follows (see [198]) that the algebra $(I, \theta)$ admits a decomposition into irreducible ideals $(I_j, \theta_j) : I = \oplus_j I_j$ ($\oplus$ meaning the direct sum), $I_j$ being pairwise orthogonal (in particular, $[X_i, X_j] = 0$ if $X_i \in I_i, X_j \in I_j, i \neq j$) and invariant with respect to $\theta$, where $(I_j, \theta|_{I_j})$ are irreducible algebras.

The group $\approx{I}$ decomposes into the direct product $\approx{I}_{m+1} \times \cdots \times \approx{I}_s$, where all $\approx{I}_k, m < k \leqslant s$ are simple, compact, and simply-connected groups; hence, the space $\widetilde{V}'$ decomposes into the product $\widetilde{V}_{m+1} \times \cdots \times \widetilde{V}_s$, where $\widetilde{V}_k$ are compact, irreducible, and symmetric spaces, with $\widetilde{V}_k = \exp B_k, m < k \leqslant s$, where $B_k = H_1 \cap I_k$. Passing from the group $\widetilde{\mathfrak{G}}_1$ to $\mathfrak{G}_1$ by factorizing the former with respect to a certain discrete subgroup of the centre, we see that the submanifold $\widetilde{V}$ is diffeomorphically projected into $\mathfrak{G}_1$ i.e., $V$ is representable in the form $K \times V_{m+1} \times \cdots \times V_s(K \approx \widetilde{K}, V_k \approx \widetilde{V}_k)$, where $K, V_k$ are totally geodesic submanifolds, and $[H, B_k] = 0, [B_k, B_n] = 0, k \neq n, H = T_e$.

We now turn our attention again to $\widetilde{\mathfrak{G}}_1$. The subgroup $\approx{I} \subset \widetilde{\mathfrak{G}}_1$ is simply-connected; therefore, the automorphism $\theta$ is uniquely extendable to the automorphism $\sigma = e^\theta$, and $\widetilde{V}' = \widetilde{\widetilde{I}}/\widetilde{\mathfrak{H}}_2$, where $\widetilde{\mathfrak{H}}_2 = \exp H_2$ is the isotropy subgroup of the manifold $\widetilde{V}'$, or the fixed- point set for $\sigma$. Since the subalgebra $H_2$ does not contain any nonzero ideals of the algebra $I$, $\widetilde{\mathfrak{H}}_2$ operates effectively on $T_e(\widetilde{V}')$, i.e., $(I, \theta)$ determines a Riemannian symmetric pair $(\widetilde{\widetilde{I}}, \widetilde{\mathfrak{H}}_2)$ operating effectively on $\widetilde{V}'$, and then $\widetilde{\widetilde{I}} = \widetilde{I}_0(\widetilde{V}'), \widetilde{\widetilde{I}}_k = \widetilde{I}_0(\widetilde{V}_k)$, or $\widetilde{\mathfrak{G}}_1 = \widetilde{K} \times \widetilde{I}_0(V_{m+1}) \times \cdots \times \widetilde{I}_0(V_s)$, since $\widetilde{V}_k$ is diffeomorphic to $V_k$.

Q.E.D.

None of the spaces $V_k$ is a subgroup in $\mathfrak{G}$, though they can be diffeomorphic to a Lie group. Theorem 18.1 implies that the embedding $V \to \mathfrak{G}$ can be extended to a homomorphism of the group $K \times \widetilde{I}_0(V')$ into the group $\mathfrak{G}$. Note that the group $\widetilde{A}(V) = K \times \widetilde{I}_0(V')$ is not isomorphic to the universal covering group of the isometry group $I_0(V)$, since the latter is isomorphic to the group $K \times K \times \widetilde{I}_0(V')$, i.e., $\widetilde{A}(V) = \widetilde{I}_0(V)/K$, where $K$ is the maximal subgroup in the decomposition of the symmetric space into irreducible components in $\mathfrak{G}$. If $V = V'$, then $\widetilde{A}(V) = \widetilde{I}_0(V)$.

## §19 Reduction of the Problem of the Description of (Co)Cycles Realizable by Totally Geodesic Submanifolds to the Problem of the Description of (Co)Homological Properties of Cartan Models

The following important corollary follows from Theorem 18.1.

**Corollary 19.1.** *Let $V$ be a compact, simply-connected, and totally geodesic submanifold in a compact group $\mathfrak{G} = A(V)$, and $V = V_1 \times V_2 \times \cdots \times V_m \times V_{m+1} \times \cdots \times V_s$ the decomposition of $V$ into irreducible components in $\mathfrak{G}$ (see Theorem 18.1). Then $V$ realizes a nontrivial cycle in $H_*(\mathfrak{G}; \mathbf{R})$ if and only if all the submanifolds $V_i, 1 \leqslant i \leqslant s$, realize nontrivial cycles in $H_*(\mathfrak{G}; \mathbf{R})$, it being assumed that no $V_i$ is a point.*

**Proof.** Let $i_k$ be an embedding of $V_k$ into $\mathfrak{G}$, and $i$ that of $V$ into $\mathfrak{G}$. Consider the elements $i_{k_*}[V_k]$ in $H_*(\mathfrak{G}; \mathbf{R})$. Since a structure of the Pontryagin product is defined in $H_*(\mathfrak{G}; \mathbf{R})$, the element $u = i_{1*}[V_1] \ldots i_{s*}[V_s]$ realized in $\mathfrak{G}$ as a cycle $U = V_1 V_2 \ldots V_s$ is defined, i.e., $U = \{x_1, \ldots, x_s\}, x_k \in V_k$. We prove that $u = i_*[V]$ and first show that $V = V_1 V_2 \ldots V_s$. It was established in Theorem 18.1 that $[B_k, B_n] = 0$ for $k \neq n$, and $[H, B_k] = 0, m < k \leqslant s$, which entails the required relation.

Q.E.D.

Corollary 19.1 reduces the original problem to that of an arbitrary connected, compact Lie group $\mathfrak{G}$ and a compact, simply-connected, totally geodesic, and irreducible submanifold, and it is required to clarify the condition on which it realizes a nontrivial real cycle. It has been pointed out above that if $N$ is a subgroup, then the problem can be completely solved by the investigation of the homomorphism $\rho^*(V;\mathfrak{G})$. Moreover, the one-connectedness of $V$ is not assumed. Hence, we can wholly concentrate our attention on the study of the embedding of the submanifold $V'$ into $\mathfrak{G}$, where $V' = V_{m+1}\times\cdots\times V_s$ (see Theorem 18.1). The embedding $i: V \to \mathfrak{G}$ is the composition $i = i_2 i_1$, where $i_1 : V \to A(V), i_2 : A(V) \to \mathfrak{G}$. Since $i_* = i_{2*}i_{1*}$, we first of all study the homomorphism $i_{1*}$. This problem is completely solved below when we find the element $i_{1*}[V] \in H_*(A(V);\mathbf{R})$ for each irriducible, compact, and symmetric space $V$. Since the embedding $i: V \to \tilde{A}(V)$ is unique up to conjugacy, the pair $(A(V), V)$ can be regarded as an elementary cell which can be monomorphically mapped into compact Lie groups. If $i_{1*}[V] = 0$, then $i_*[V] = 0$. However, if $i_{1*}[V] \neq 0$, this does not mean that also $i_*[V] \neq 0$; therefore, in the case where $i_{2*}[A(V)] = 0$, an additional study of the homomorphism $i_{2*}$ on the element $i_{1*}[V]$ is required. It is clear that this can be done to the full in each concrete case, too.

Thus, the irreducible, simply-connected, and compact space $V$ is embedded into the group $A(V)$ as a totally geodesic submanifold, with $A(V)$ generally not being simply-connected. It is easy to see that $V$ realizes a nontrivial cycle in $A(V)$ if and only if the nontrivial cycle is realized by the submanifold $\tilde{V}$ diffeomorphic to $V$ in the covering $\tilde{A}(V)$. It suffices to investigate embeddings of $V$ into the simply-connected group $\tilde{I}_0(V) = \tilde{A}(V)$.

The embeddings can be described canonically with the use of the available involutive automorphism $\theta$ determining the symmetric space $V$. The embedding of the irreducible, symmetric space $V$ into $\tilde{A}(V)$ generates $\theta$, $\theta(B) = -B, \theta(H) = H$, in the algebra $G_1$. Due to the one-connectedness of $\tilde{A}(V)$, $\theta$ is extendable to the involutive automorphism of the whole group. Then the fixed-point set for the automorphism $\sigma$ is connected (see [198]), in which case this set coincides with the subgroup $\mathfrak{H}$. It is obvious that the totally geodesic submanifold $V \subset \tilde{A}(V)$ consists of those elements for which $\sigma(g) = g^{-1}$.

Conversely, it turns out that any irreducible, symmetric space $V$ admits an embedding into the group $\tilde{I}_0(V)$ as a totally geodesic submanifold.

**Proposition 19.1 (E. Cartan; see [71]).** *Let $\sigma$ be an involutive automorphism of a compact, connected Lie group $\mathfrak{G}$. Denote by $\mathfrak{H}$ its fixed-point set, and assume that on $\mathfrak{G}$ the invariant metric is given. Then the mapping $g\mathfrak{H} \to g\sigma(g^{-1})$ is a diffeomorphism of the manifold $\mathfrak{G}/\mathfrak{H}$ onto the closed, totally geodesic submanifold $V$ in $\mathfrak{G}$, which is a symmetric space in the induced Riemannian metric and called the Cartan model of the space.*

If the group $\mathfrak{G}$ is simply-connected, then the manifold $\mathfrak{G}/\mathfrak{H}$, where $\mathfrak{H}$ is the fixed-point set for the automorphism $\sigma$, is also simply-connected. Since any embedding $i : V \to \tilde{A}(V)$ determines one and only one automorphism $\theta$ up to conjugacy, it follows from Proposition 19.1 that $i$ is Cartan in the sense that the submanifold $V$ admits a representation in the form $\{g\sigma(g^{-1})\}$, where $\sigma = e^{\theta}$ and the element $g$ ranges over the whole of the group $\tilde{A}(V)$. It is easy to show that the action of $\tilde{A}(V)$ on $V$ can be represented as $g(a) = ga\sigma(g^{-1})$, where $a \in V, g \in \tilde{A}(V)$. If $g \in V$, then $g(a) = gag$, and if $g \in \mathfrak{H}$, then $g(a) = gag^{-1}$, i.e., the subgroup $\mathfrak{H}$ acts by means of rotations, whereas $V$ acts on itself via translations. In the following, $\mathfrak{G}$ will denote the universal covering group of the maximal-connected isometry group $I_0(V)$ of the compact, simply-connected, and irreducible space $V$.

Consider an embedding $p : \mathfrak{G} \to V \subset \mathfrak{G}, p(g) = g\sigma(g^{-1})$.

**Lemma 19.1.** *The continuous mapping $p$ determines a principal fibre bundle $\mathfrak{H} \to \mathfrak{G} \xrightarrow{p} V$ with the fibre $\mathfrak{H}$.*

**Proof.** Consider the cosets $g\mathfrak{H}$ relative to $\mathfrak{H}$. If $g_1$ and $g_2$ belong to the same coset, it is obvious that $p(g_1) = p(g_2)$. Conversely, if $p(g_1) = p(g_2)$, then $g_1\sigma(g_1^{-1}) = g_2\sigma(g_2^{-1})$ or $\sigma(k^{-1}) = k^{-1}$, where $k = g_2^{-1}g_1$, i.e., $k \in \mathfrak{H}$ and $g_1 = g_2h$, since the fixed-point set is connected.

Q.E.D.

**Lemma 19.2.** *The intersection of each coset $g_0\mathfrak{H}$ with submanifold $V$ is non-empty.*

**Proof.** Suppose, on the contrary, that there exists such a class $g_0\mathfrak{H}$ that $(g_0\mathfrak{H}) \cap V = \emptyset$. Consider a point $m_0 = p(g_0\mathfrak{H})$ on $V$. Let $m' = \sqrt{m_0}$ be such a point from $V$ that $(m')^2 = m_0$. If there are several such points, then we take any one of them. If $v \in V$, then $v = g\sigma(g^{-1})$ for any $g \in \mathfrak{G}$; therefore, $\sigma(v) = (g\sigma(g^{-1}))^{-1}$, i.e., $\sigma(v) = v^{-1}$, the choice of the element $g$ being immaterial. The mapping $p$ is of the group $\mathfrak{G}$ onto $V$, which permits us to consider the composition map $pi : V \to V$, $pi(v) = v\sigma(v^{-1})$, or $pi(v) = v^2$, and the mapping $pi$ acts on $V$ as "squaring" in the sense of the operation in the group $\mathfrak{G}$. Since $m_0 = (m')^2$, we have $m_0 = pi(m')$. Consider the class $m'\mathfrak{H}$. Then $m' \in m'\mathfrak{H} \cap V$, and $p(m'\mathfrak{H}) = m_0$, i.e., the full inverse image of $m_0$ under the projection $p$ contains points of the two cosets $m'\mathfrak{H}$ and $g_0\mathfrak{H}$. It follows from Lemma 19.1 that $m'\mathfrak{H} = g_0\mathfrak{H}$; therefore, $g_0\mathfrak{H} \cap V \neq \emptyset$, and $\mathfrak{H}$ contains at least $m'$, which is contrary to the original assumption.

Q.E.D.

It follow from Lemma 19.2 that an arbitrary coset $g\mathfrak{H}$ can be represented as $m\mathfrak{H}$, where $m \in V$. If $g \in \mathfrak{G}$, then we denote by $\sqrt{g}$ such an element $g'$ that $(g')^2 = g$, and by $\{\sqrt{g}\}$ the set of all such elements from $\mathfrak{G}$.

**Lemma 19.3.** *Let $m\mathfrak{H}$ be an arbitrary coset, $m \in V$. Then $m\mathfrak{H} \cap V = \{\sqrt{m^2}\} \cap V$.*

**Proof.** We show that $\{\sqrt{m^2}\} \cap V \subset m\mathfrak{H} \cap V$. In fact, let $m_1, m \in V$, and $m_1^2 = m^2$. Then $m_1\sigma(m_1^{-1}) = m\sigma(m^{-1})$ (see above), and $(m^{-1}m_1)\sigma(m_1^{-1}m) = e$, i.e., $k\sigma(k^{-1}) = e$, where $k = m^{-1}m_1$. Since $p(k) = e$, we have $k \in \mathfrak{H}$ and $m_1 = mh$, where $h \in \mathfrak{H}$. Conversely, we prove that $\{\sqrt{m^2}\} \cap V \supset m\mathfrak{H} \cap V$. Let $v \in m\mathfrak{H}, v \in V$. Then $v = mh, h \in \mathfrak{H}$ and $v^2 = v\sigma(v^{-1}) = m^2$, or $v \in \{\sqrt{m^2}\}$.

Q.E.D.

Note that the "zero" coset, subgroup $\mathfrak{H}$, intersects $V$ in the set of all such points $v$ that $v^2 = e$. The "squaring" mapping could have been determined irrespective of the embedding of $V \subset \mathfrak{G}$, by only making use of the representation of an arbitrary point $v \in V$ as of one on the geodesic emanating from a fixed point $e \in V$, and then by doubling the value $t_0$ of parameter $t$, associated with $v$ on the geodesic. Above, we have proved the correctness of this definition for symmetric spaces.

**Proposition 19.1.** *Let pi be the "squaring" mapping of the symmetric space $V = \mathfrak{G}/\mathfrak{H}$. The totally geodesic submanifold $V \subset \mathfrak{G}$ realizes a nontrivial cycle in $H_*(\mathfrak{G}; \mathbf{R})$ if and only if the degree of the mapping $pi : V \to V$ is other than zero.*

**Proof.** The sufficiency is obvious. We now prove the necessity of the condition. Denote by $f$ the following mapping of the group $\mathfrak{G}$ on itself, viz., $f(g) = g^2$. Since $pi(v) = v^2$, where $v \in V$, the commutative diagram

$$\begin{array}{ccc} \mathfrak{G} & \xrightarrow{f} & \mathfrak{G} \\ {\scriptstyle i}\uparrow & & \uparrow{\scriptstyle i} \\ V & \xrightarrow{pi} & V \end{array}$$

arises.

Consider a homomorphism $f_* : H_*(\mathfrak{G}; \mathbf{R}) \to H_*(\mathfrak{G}; \mathbf{R})$. It is generally known that $f_*(x) = 2x$ for any primitive element $x \in H_*(\mathfrak{G}; \mathbf{R})$, therefore, $f_*$ is an isomorphism. If we assume that $\deg(pi) = 0$, then $(pi)_*[V] = 0$ and $(ipi)_*[V] = 0$, i.e., $(fi)_*[V] = 0$, which just makes $i_*[V] = 0$.

Q.E.D.

## §20 Classification Theorem Describing Totally Geodesic Submanifolds Realizing Nontrivial (Co)Cycles in Compact Lie Group (Co)Homology

### 20.1. The Statement of the Classification Theorem

We have shown above that the principal problem of the present chapter is to describe (co)homologically nontrivial, locally minimal, and totally geodesic submanifolds in compact Lie groups and, generally, in symmetric spaces. This is reduced to the solution of the problem when an irreducible, compact, and symmetric space $V$ in its isometry group $\tilde{I}_o(V)$ as a Cartan model realizes a nontrivial (co)cycle in $H_*(\tilde{I}_o(V))$.

**Theorem 20.1.1.** *Let $i : V \to \tilde{I}_0(V) \cong \tilde{A}(V)$ be an embedding of a compact, irreducible, simply-connected, and symmetric Riemannian space $V$ as a totally geodesic submanifold in the simply-connected, maximal, and connected isometry group $\tilde{I}_0(V)$. This embedding is Cartan, i.e., obtained from the standard Cartan embedding by applying a certain automorphism of the group $\tilde{I}_0(V)$. All the cases where $i$ is (co)homologically nontrivial are listed below, viz.,*

(1) of symmetric spaces $V$ of Type I, only the following spaces realize nontrivial (co)cycle in $\tilde{I}_0 V$ :

1a. $V = \mathrm{SU}(2m+1)/\mathrm{SO}(2m+1), m \geqslant 1$, in the group $I_0(V) = \tilde{I}_0(V) = \mathrm{SU}(2m+1), i_*[V] \neq 0$ in $H^*(\tilde{I}_0(V), \mathbf{Z}_p), p \neq 2$, $p$ is prime if $p \neq 0$;

1b. $V = \mathrm{SU}(2m)/\mathrm{Sp}(m), m \geqslant 2$, in the group $I_0(V) = \mathrm{SU}(2m)$, $i_*[V] \neq 0$ in $H^*(\tilde{I}_0(V), \mathbf{Z})$;

1c. $V = S^{2l-1} =$ (sphere) $= \mathrm{SO}(2l)/\mathrm{SO}(2l-1)$, $l \geqslant 4$, in the group $\tilde{I}_0(V) = \mathrm{Spin}(2l)$, $i_*[V] \neq 0$ in $H^*(\tilde{I}_0(V), \mathbf{Z}_p)$, $p \neq 2$, $p$ is prime if $p \neq 0$;

1d. $V = E_6/F_4$ in the group $\tilde{I}_0(V) = E_6$, $i_*[V] \neq 0$ in $H^*(\tilde{I}_0(V); \mathbf{Z}_p)$, $p \geqslant 7$, and is prime or $p = 0$.

All other spaces of Type I realize no nontrivial cocycle in $H^*(\tilde{I}_0(V); \mathbf{R})$, i.e., $i_*[V] = 0$.

(2) Any symmetric space $V$ of Type II always realizes a nontrivial (co)cycle in $H^*(I_0(V); \mathbf{R})$.

**Theorem 20.1.2.** *Let $V$ be an arbitrary compact, simply-connected, and totally geodesic submanifold in a Lie group $\mathfrak{G}$, and let $V = K \times V_{m+1} \times \cdots \times V_s = K \times V'$ be the decomposition into the direct product (see Theorem 18.1). Then $V$ under the embedding $i : V \to \tilde{A}(V)$ realizes a nontrivial cycle in $H^*(\tilde{A}(V); \mathbf{R})$ if and only if the subgroup $K \times \mathfrak{H}$, where $\mathfrak{H}$ is the stability subgroup of $V'$ for real coefficients (recall that $\mathfrak{H} \subset \tilde{A}(V') \subset \tilde{A}(V)$), is totally nonhomologous to zero in the group $\tilde{A}(V)$ for real coefficients.*

## 20.2. The Case of Spaces of Type II

Proposition 19.1 reduces the original problem to a purely geometric one of computation of the degree of the smooth mapping $pi : V \to V; (pi)(v) = v^2$. We start with symmetric spaces of Type II.

Spaces of Type II are precisely compact, connected, and simple Lie groups $V$ with the bi-invariant Riemannian metric, while their one-connectedness may not be assumed. The group $I_0(V)$ is isomorphic to the direct product $V \times V$ (see [198]). The involutive automorphism $\sigma : \mathfrak{G} \to \mathfrak{G}$, $\mathfrak{G} = I_o(V)$ determining the symmetric space $V$ is of the form $\sigma(v_1, v_2) = (v_2, v_1)$, and the fixed point set for $\sigma$ is the subgroup $\mathfrak{H} = \{(v, v)\}$; i.e., $\mathfrak{H}$ is the diagonal in the direct product. Let $G$ and $K$ be the Lie algebras of the groups $\mathfrak{G}$

and $V$, respectively. Then $G = K \oplus K$. On the other hand, $G$ decomposes into the direct sum of two proper subspaces relative to the automorphism $\theta$, $\theta(X,Y) = (X,Y), (X,Y) \in G, G = B + H, \theta(H) = H$.

Since $\sigma(v, v^{-1}) = (v, v^{-1})^{-1}$, $V$ is embedded into the group $V \times V$ as a totally geodesic submanifold of all points of the form $(v, v^{-1})$. The mapping $p : \mathfrak{G} \to V$ is of the form $p(g) = (v_1 v_2^{-1}, v_2 v_1^{-1}) \in i(V)$, where $g = (v_1, v_2)$; hence, $pi(g) = g^2$ and $p$ establishes a diffeomorphism between $V$ and $V \times V/\mathfrak{H}$. Note that since the embedding $\mathfrak{H} \subset \mathfrak{G}$ is the embedding of $V$ as of the diagonal, the subgroup $\mathfrak{H}$ is totally nonhomologous to zero in $\mathfrak{G}$. Let $H_*(V; \mathbf{R}) = \Lambda(x_1, \ldots, x_r)$, $r$ be the rank of $V$, and the elements $x_1, \ldots, x_r$ the primitive generators of the ring $H_*(V; \mathbf{R})$. Then $j_*(x_\alpha) = x_\alpha \otimes 1 + 1 \otimes x_\alpha$, where $j : \mathfrak{H} \to \mathfrak{G}$ is an embedding; therefore, the monomorphic image of $H_*(V; \mathbf{R})$ under the homomorphism $j_*$ is the subring $j_* H_*(V; \mathbf{R}) = \Lambda(x_1 \otimes 1 + 1 \otimes x_1, \ldots, x_r \otimes 1 + 1 \otimes x_r)$.

The degree of the mapping $f_k : V \to V, f_k(v) = v^k$ is known to be equal to $k^r$, where $r$ is the rank of an arbitrary compact Lie group $V$. We have $pi = f_2$; therefore, $\deg(pi) = 2^r$, and due to the properties of the mapping $pi$, the manifold $V$ realizes a nontrivial cycle in $H_*(\mathfrak{G}; \mathbf{R})$. We now compute the element $i_*[V]$. Consider the mapping $\alpha : \mathfrak{G} \to \mathfrak{G}, \alpha(v_1, v_2) = (v_1, v_2^{-1})$. Then the homomorphism $\alpha_* : H_*(V) \otimes H_*(V) \to H_*(V) \otimes H_*(V)$ is arranged as $\alpha_*(a \times 1) = a \otimes 1$ for any $a \in H_*(V)$, and $\alpha_*(1 \times x) = -(1 \otimes x)$ for any element $x \in H_*(V)$. Consider in $H_*(\mathfrak{G}; \mathbf{R})$ the elements $\omega_i = \frac{1}{2}(x_i \otimes 1 - 1 \otimes x_i)$. Then $p_*(\omega_i) = x_i$, and it is obvious that $i_*[V] = 2^r \prod_{i=1}^r \omega_i$.

We have thereby proved the following.

**Theorem 20.2.1.** *Let $i : V \to \tilde{I}_0(V)$ be the Cartan embedding of a compact, irreducible, simply-connected, and symmetric Riemannian space of Type II in a simply-connected, maximal isometry group $\tilde{I}_0(V)$. Then the totally geodesic submanifold $i(V)$ always realizes a nontrivial cycle in $H_*(\tilde{I}_0(V); \mathbf{R}) = \Lambda(x_1, \ldots, x_r) \otimes \Lambda(x_1, \ldots, x_r)$, where $r$ is the rank of the $V$, with $i_*[V] = 2^r \prod_{i=1}^r \frac{1}{2}(x_i \otimes 1 - 1 \otimes x_i)$.*

### 20.3. The Case of Spaces of Type I (Co)Homologically Trivial Cartan Models. Properties of the Squaring Map of a Symmetric Space

We shall now study the spaces of Type I (see Table I), but shall start with the description of the cases where the cycles $i_*[V]$ are trivial.

Since the Poincaré polynomial contains only even powers of $t$ for the spaces $A_l\mathrm{III}_m$, $B_l\mathrm{I}_2$, $B_l\mathrm{I}_{2m}$, $B_l\mathrm{II}$, $C_l\mathrm{I}$, $C_l\mathrm{II}_m$, $D_l\mathrm{I}_2$, $D_l\mathrm{I}_{2m}$, $D_l\mathrm{III}$, $E_6\mathrm{II}$, $E_6\mathrm{III}$, $E_7\mathrm{VI}$, $E_7\mathrm{VII}$, $E_8\mathrm{VIII}$, $E_8\mathrm{IX}$, $F_4\mathrm{I}$, $F_4\mathrm{II}$, and $G_2\mathrm{I}$ (see [404]), we have $i_*[V] = 0$. Consider the spaces $D_l I_{2m+1}, l \geqslant 4,\ 1 \leqslant m \leqslant [\frac{l-1}{2}], P(V,t) = (1 + t^{2l-1})Q(t)$, where $Q(t) = Q(-t)$. It follows that $H^*(V; \mathbf{R}) = \Lambda(u_{2l-1}) \otimes Q$, where the subalgebra $Q$ is generated by $q_1, \ldots, q_\tau$ of even degrees. Since $u_{2l-1}^2 = 0$, any element $x \in H^s(V; \mathbf{R}), x \neq 0$, where $s$ is odd, must be of the form $x = u_{2l-1} q, q \in Q$. The group $\mathrm{Spin}(2l)$ is a simply-connected isometry group $\tilde{I}_0(V)$ for the manifold $V$. Consider the homomorphism $i_*$, where $i : V \to \mathrm{Spin}(2l)$ is the Cartan embedding, and compute the elements $i^*(x_\alpha)$, where $\alpha \in \{3, 7, 11, \dots, 4l-5, 2l-1\}$. If $\alpha < 2l - 1$, then $i^*(x_\alpha) = 0$. The element $i^*(x_{2l-1})$ must not necessarily be trivial. Similarly, the generators $x_\alpha$ for $2l - 1 \leqslant \alpha \leqslant 4l - 5$ may not be mapped into either zero and the "special" generator $\bar{x}_{2l-1}$. Since the degrees of all these generators are odd, we have $i^*(x_\alpha) = u_{2l-1}q'_\alpha, 2l-1 \leqslant \alpha \leqslant 4l-5; i^*(\bar{x}_{2l-1}) = a \cdot u_{2l-1}, a \in \mathbf{R}, q'_\alpha \in Q$.

It is easy to see that all the products of the form $x_\alpha x_\beta$ are carried into zero, so that the unique elements of the ring $H^*(\mathrm{Spin}(2l); \mathbf{R})$, which may not be transformed into zero, are $x_\alpha$ for $2l - 1 \leqslant \alpha \leqslant 4l - 5$ and $\bar{x}_{2l-1}$. Since $1 \leqslant m \leqslant [\frac{l-1}{2}]$, we have $l > 2m \geqslant 2$, and, comparing this with the obvious inequality $2m \geqslant \frac{2(m^2+m-1)}{2m-1}$, we obtain $l > \frac{2(m^2+m-1)}{2m-1}$, which gives $4l - 5 < \dim V$, i.e., $i_*[V] = 0$. For the space $E_6 I$, we also have $i_*[V] = 0$, because $\dim V = 42$.

It remains to consider the space ${}^0A_{2m}I, {}^0A_{2m-1}\mathrm{I}, {}^0A_{2m-1}\mathrm{II}, {}^0D_l\mathrm{II}, {}^0E_6\mathrm{IV}$. We first construct the special bases $B(t_i)$ and $B(t^2)$.

Consider an arbitrary compact, symmetric space $V$. Let $T$ be a certain maximal Abelian submanifold in $V$, $\mathfrak{G} = \tilde{I}_0(V)$, and $\mathfrak{H}$ the stability group of a certain point $e \in T$. We assume that $V$ is embedded into $\mathfrak{G}$ as a Cartan model. It is generally known that, for any point $v \in V$, there always exists an element $h \in \mathfrak{H} \subset I_0(V)$ for which $hvh^{-1} \in T$, where $T$ is a fixed maximal flat subspace in $V$.

The manifold $V$ fibres into orbits under the adjoint action of $\mathfrak{H}$, each of them containing points of $T$. If $t$ is an arbitrary point from $T$, then the orbit $O(t)$ is diffeomorphic to the homogenous space $\mathfrak{H}/\mathfrak{H} \cap C(t)$, where $C(x)$ denotes the centralizer of an element $x$. Since $V \subset \mathfrak{G}$, $T$ can be represented as the intersection $T_0 \cap V$, where $T_0$ is the maximal torus in $\mathfrak{G}$. Each orbit $O(t)$ intersects the torus $T$ transversally in a finite number of points, all of them being conjugate of each other under the action on $T$ of the normalizer $N(T)$. Besides, the tangent spaces $T_t(O(t))$ and $T_t(T)$ are orthogonal to one another. We now consider the mapping $pi : V \to V$, $pi(v) = v^2$, and denote it by $f$. It is clear that $f(T) \subset T$.

**Lemma 20.3.1.** *The mapping $f$ commutes with the adjoint action of the group $\mathfrak{H}$ on the manifold $V$.*

The proof is obvious.

Thus, under the mapping $f$, the orbits of points $t$ are mapped into those of $t^2 = f(t)$.

Let $\Omega(V)$ be the set of singular elements of the space $V = \mathfrak{G}/\mathfrak{H}$ (see [198]). The complement $R(V) = V \backslash \Omega(V)$ is called a *regular set* in $V$. Note for the following that the square root of a regular element is again a regular element.

**Lemma 20.3.2.** *Let $t^2 \in T$ be a regular element, and $t$ an arbitrary square root of the element $t^2$. Then, under the mapping $f$, the orbit $O(t)$ is homeomorphically mapped onto the orbit $O(t^2)$.*

**Proof.** Assume that there exist two points $x_1$ and $x_2$ from $O(t)$, so that $x_1 \neq x_2$ and $f(x_1) = f(x_2)$. Since $x_1, x_2 \in O(t)$, we have $x_1 = a_1 t a_1^{-1}, x_2 = a_2 t a_2^{-1}$ for two elements $a_1$ and $a_2$ from the subgroup $\mathfrak{H}$. Then $a_2^{-1} a_1 t^2 = t^2 a_2^{-1} a_1$, i.e., $a_2^{-1} a_1 \in C(t^2) \cap \mathfrak{H}$. Since $t$ and $t^2$ are regular, we have $C(t) \cap \mathfrak{H} = C(t^2) \cap \mathfrak{H}$; therefore, $x_1 = x_2$, which is contrary to the assumption.

Hence, it suffices to investigate the properties of $f$ restricted to the torus $T$. If $t^2 = f(t)$ and $t^2 \in R(V)$, then there exists a sufficiently small neighbourhood $U$ of the point $t^2$, so that $U \subset R(V)$ (the set $\Omega$ being closed in $V$) and $f^{-1}(U) \subset R(V)$. The mapping $f$ is squaring; therefore, it is smooth and regular in the neighbourhood of $f^{-1}(U)$. It is known (see [198]) that $\dim \Omega \leqslant \dim V - 2$; therefore, the set $R$ is connected. Since $f$ is smooth, to calculate $\deg f$, it suffices to find it on the set $R(V)$.

Let the rank of $V$ be 1. If $t^2 \in R(V)$, all the inverse images $t_i$ of this element under $f$ belong to $T$, being all different and $2^l$ in total. Choose an orientation on $T$ and place at all points $t_i$ the $l$-frames whose orientations are the same as those of $T$. We place another $l$-frame with the same orientation at the point $t^2$. Consider the mapping $df : T_{t_i}(T) \to T_{t^2}(T)$. It is then obvious that, under $f$, the orientation of the torus is unaltered, and the $l$-frames at $t_i$ are transformed into the $l$-frame at $t^2$ with the orientation preserved. Fix a point $t_i$ and distinguish the subspace $T_{t_i}(T)$ from the tangent space $T_{t_i}(V)$ in which we have already chosen a basis $e_1(t_i), \ldots, e_l(t_i)$ denoted by $E(t_i)$. Complete the set of vectors $E(t_i)$ up to the basis $B(t_i)$ for the whole space $T_{t_i}(V)$. Consider the mapping $Q_t : \mathfrak{H} \to O(t)$, $t \in T$; $Q_t(h) = hth^{-1}$, $Q_t(\mathfrak{H} \cap C(t)) = t$. Let the differential of the mapping $Q_t$ be $q_t$, and $q_t : H \to T_t(V)$.

Consider the left translation $L_t : g \to tg$ and transfer the tangent space $T_t(V)$ to the point $e$, using the mapping $l_{t^{-1}} = dL_{t^{-1}}$. Then $l_{t^{-1}}(T_t(V)) \subset G$. Let $X \in H$. Then $q_t(X) = l_t(E - \mathrm{Ad}(t))(X)$. In the subalgebra $H$, we take the plane $A$ orthogonal to the subalgebra $N = T_e C(t)$, $H = N + A$. By the definition of the mapping $q_t, \operatorname{Ker} q_t = N$; therefore, $q_t$ is nonsingular on $A$, which is linearly mapped onto the space $T_t(O(t))$.

Select an arbitrary basis $\{A_1, \ldots, A_p\} = A(e)$ in $A$ and take $q_t(A(e)) = A(t)$ as that for $T_t(O(t))$. We now determine a basis $B(t_i)$ for $T_{t_i}(V)$ at each point $t_i$ by $B(t_i) = \{E(t_i); A(t_i)\}$. Similarly, we define a basis $B(t^2)$ at the point $t^2$. Thus, we assume that the orientation of the manifold is determined by $B(t^2)$ at $t^2$. The bases $E(t^2)$ and $E(t_i)$ can be selected so that $df(e_\alpha(t_i)) = 2e_\alpha(t^2)$. Formally, we will write $df(E(t_i)) = 2E(t^2)$.

**Lemma 20.3.3.** *Let $t^2 \in R(V)$, and $B(t_i)$ and $B(t^2)$ be the above bases. Then $df(B(t_i)) = (2E(t^2); A(t^2))$.*

**Proof.** It suffices to show that $df(A(t_i)) = A(t^2)$. We assume that the group $\mathfrak{G}$ is embedded into the unitary group $U(M)$. Consider a point $t_i$ and another infinitely near one $t_i + \varepsilon X$, where the vector $X \in T_{t_i}(V)$ is of the form $X = \sum x_\alpha q_{t_i}(A_\alpha)$.

Consider the equalities

$$f(t_i + \varepsilon X) = f[\mathrm{Ad}(e + \varepsilon(\sum x_\alpha A_\alpha))(t_i)]$$
$$= \mathrm{Ad}(e + \varepsilon(\sum x_\alpha A_\alpha))(t^4) = t^2 + \varepsilon \sum x_\alpha q_{t^2}(A_\alpha).$$

Hence, $df(q_{t_i}(A_\alpha)) = q_{t^2}(A_\alpha)$.

Q.E.D.

Thus, $\deg f = \sum_{i=1}^{2l} cB(t_i)$, where $cB(t_i) = +1$ if the orientation of $V$, determined by $B(t_i)$, coincides with that determined by $B(t^2)$, and $cB(t_i) = -1$ otherwise.

In the subsequent argument, we follow the same general scheme with minor variations due to concrete particulars of symmetric spaces.

**20.4. The Case of Spaces of Type I. Spaces $\mathrm{SU}(k)/\mathrm{SO}(k)$**

Consider the space ${}^0A_{2m}\mathrm{I}$: $V = \mathrm{SU}(2m+1)/\mathrm{SO}(2m+1)$, $m \geqslant 1$. The rank of $V = 2m$ equals the rank of $\mathfrak{G}$, and $\mathfrak{H} = \mathrm{SO}(2m+1)$ is totally nonhomologous to zero in the group $\mathrm{SU}(2m+1)$ for real coefficients, whereas $P(V,t) = (l + t^5)(1 + t^9)\dots(1 + t^{4m+1})$. The space $A_{2m-1}I$ is of the form $V = \mathrm{SU}(2m)/\mathrm{SO}(2m), m \geqslant 2$, the rank of $V = 2m - 1$ equals the rank of $\mathfrak{G}, \mathfrak{H} = \mathrm{SO}(2m)$, with the subgroup $\mathfrak{H}$ no longer being totally nonhomologous to zero, whereas $P(V,t) = (1+t^5)(1+t^9)\dots(1+t^{4m-3})(1+t^{2m})$. In both cases, the isometry group $I_0(V)$ of the manifold $V = \mathrm{SU}(n)/\mathrm{SO}(n), n \geqslant 3$, is $\mathrm{SU}(n)$. The involutive automorphism $\theta$ on the algebra $A_{n-1}$ is of the form $\theta(X) = \bar{X}$, where the bar means complex conjugation; $\theta$ is extended to the involutive automorphism $\theta(g) = \bar{g}$.

Consider the plane $B$ in the algebra $G$, on which $\theta$ is $-E$, and which is made up of symmetric and purely imaginary matrices of order $n$ with trace 0; $B$ is the tangent space to the Cartan model $V = \{g\sigma(g^{-1})\} = \{gg^T\}$. Since $v^T = v, v \in V$, the model of the symmetric space consists of all unitary, symmetric matrices $s$, and only of them. The maximally Abelian subspace $S$ in the space $B$ consists of all the diagonal matrices

$$\begin{Vmatrix} i\varphi_1 & & 0 \\ & \ddots & \\ 0 & & i\varphi_n \end{Vmatrix},$$

where $\varphi_1 + \dots + \varphi_n = 0$, and the rank of $V$ is equal to that of $\mathfrak{G}$; therefore, $S = S_0, T = T_0$.

Consider the intersection $\mathfrak{H} \cap V$. Due to Lemma 19.3, it suffices to describe all such elements $v \in V$ that $v^2 = e$. It is easy to see that $\mathfrak{H} \cap V = \bigcup_{(p)} \mathrm{SO}(n)/S(O(p) \times O(n-p))$, which means that we cannot make use of the zero coset to find the intersection number for two cycles of additional dimensions $V$ and $\mathfrak{H}$. Since $T = T_0, C(t) = T$ if $t \in R(V)$. Let $t \in R(V)$. Then $O(t) = \mathrm{SO}(n)/\Pi, \Pi = \underbrace{\mathbf{Z}_2 \oplus \cdots \oplus \mathbf{Z}_{2(n-1)}}_{n-1}$. The mapping $q_t : H \to T_t(O(t))$ is an isomorphism of linear spaces, and we can take the basis for the subalgebra $H$ as that for $A(e)$.

Consider the covering space $S$ with respect to the coordinates $\varphi_1, \ldots, \varphi_n$; $\varphi_1 + \cdots + \varphi_n = 2\pi$. As coordinates on the torus $T$, we take the linear forms $\varphi_1, \ldots, \varphi_n$ such that $0 \leqslant \varphi_i \leqslant 2\pi$. The domains on the torus, filled with regular elements, are described by the sets $(\varphi_{i_1}, \ldots, \varphi_{i_n})$, where $0 \leqslant \varphi_{i_1} < \varphi_{i_2} < \cdots < \varphi_{i_n} < 2\pi$. Each of these domains is connected and homeomorphic to an open $(n-1)$-dimensional simplex. Considering all possible sets $(i_1, \ldots, i_n)$, we obtain a covering of the set $R(V)$ with Weyl chambers. Consider the Weyl group $W(\mathfrak{G})$ (see §17.2), which commutes the components of $R(V)$ and is simply transitive on the set of all Weyl chambers.

The group $\mathrm{SU}(n)$ is regarded by us in the standard representation of minimal dimension in a linear space $L$ with the basis $e_1, \ldots, e_n$. Consider an element $n_{ij}, i \neq j$, belonging to the subgroup $\mathrm{SO}(n)$, and represented in $L$ as $n_{ij}(e_i) = e_j, n_{ij}(e_j) = -e_i, n_{ij}(e_k) = e_k$ for $k \neq (i,j)$; if $i = j$, then we assume that $n_{ij}$ is the identity transformation. Then the generators of the Weyl group are transformations $w_{ij} : T \to T$ such that $w_{ij}(t) = n_{ij} t n_{ij}^{-1}, t \in T$. Consider a subgroup $W'$ in $\mathrm{SO}(n)$, generated by the elements $n_{ij} (i \neq j)$ and $n_{ii}$. Then $W'$ is isomorphic to the Weyl group. Each set $\sigma = (i_1, \ldots, i_n)$ determines $\sigma = (\varphi_{i_1}, \ldots, \varphi_{i_n})$, i.e., the torus orientation whose sign coincides with the parity of the permutation transforming $\sigma$ into the standard set $(1, 2, \ldots, n)$. We now select a regular element $t_o^2$ for which we compute the degree of the mapping $f$, viz., (1) $0 < \varphi_1 < \varphi_2 < \cdots < \varphi_n < 2\pi$, (2) $\varphi_1 + \cdots + \varphi_n = 2\pi$, (3) $\varphi_1 + \varphi_2 + \cdots + \varphi_{n-2s} < \varphi_{n-2s+1}, 1 \leqslant s \leqslant [\frac{n}{2}]$. Consider the bases $E(t_0^2)$ and $E(t_{0i})$ determining the same torus orientation. For brevity, we denote the orientation of the basis $B(t)$ by $\mathrm{or}\, B(t)$. If two bases determine the same orientation, we will write $\mathrm{or}\, B(t) = \mathrm{or}\, B'(t)$; in the opposite case, we have $\mathrm{or}\, B(t) = -\mathrm{or}\, B'(t)$.

**Lemma 20.4.1.** *Let $t_0^2 \in \Delta_{1,2,\ldots,n} \subset R(V)$. Assume that $t_{oi} \in \Delta_{1,2,\ldots,n}$. Then* $\mathrm{or}\, B(t_{oi}) = \mathrm{or}\, B(t_0^2)$.

**Proof.** Since the domain $\Delta_{1,2,\ldots n}$ is connected, the points $t_0^2$ and $t_{0i}$ can be joined with a smooth path $\gamma(t)$ wholly contained in $\Delta_{1,2,\ldots,n}$. The frame $E(t_{0i})$ can be continuously carried along $\gamma(\tau)$, so that it always remains tangent to the torus and coincides with the frame $E(t_0^2)$ at the moment $\tau = 1$. Consider the orbits $O(\gamma(\tau))$ along $\gamma(\tau)$. Define a continuous deformation of the frame $A(t_{0i})$ along $\gamma(\tau)$ by putting $F_\tau(A(t_{0i})) = A(\gamma(\tau)); \gamma(0) = t_{0i}, \gamma(1) = t_0^2$. At each moment, the frame $A(\gamma(\tau))$ is orthogonal to the torus.

Q.E.D.

Consider the simplex $\Delta_{i_1 \ldots i_n}$. Denote by $\sigma$ the permutation transforming the set $(1, 2, \ldots, n)$ into $(i_1, \ldots, i_n)$.

Then the simplex $\Delta_{1,2,\ldots n}$ is mapped into $\Delta_{i_1 \ldots i_n}$ by the transformation $w_\sigma \in W(\mathfrak{G}), w_\sigma(t) = n_\sigma t n_\sigma^{-1}, n_\sigma \in W', n_\sigma = n_{i_1 j_1} \ldots n_{i_s j_s}$, where $\sigma = \sigma_{i_1 j_1} \ldots \sigma_{i_s j_s}$ is the decomposition of $\sigma$ in terms of elementary permutations.

Let $c(\sigma)$ be the parity of $\sigma$.

**Lemma 20.4.2.** *Let a root $t_{0i}$ belong to the simplex $\Delta_{i_1 \ldots i_n}$, where $(i_i, \ldots, i_n) = \sigma(2, 3, \ldots, n)$. Then* $\operatorname{or} B(t_{0i}) = \operatorname{or} B(t_0^2)$ *if $c(\sigma) = +1$, and* $\operatorname{or} B(t_{0i}) = -\operatorname{or} B(t_0^2)$ *if $c(\sigma) = -1$.*

**Proof.** Consider the mapping $\varphi : \mathfrak{G} \to \mathfrak{G}, \varphi(g) = n_\sigma g n_\sigma^{-1}$. Since $n_\sigma \in \mathfrak{H}$, we have $\varphi(v) \in V$ for any $v \in V$. Consider the mapping $d\varphi : T_{t_0^2}(V) \to T_{\varphi(t_0^2)}(V)$. Since the group $\mathfrak{H}$ is connected, the element $n_\sigma$ can be joined with a smooth path $\gamma(\tau)$ to the unit element $e$, $\gamma(1) = e$, $\gamma(0) = n_\sigma$. It follows that $\operatorname{or} B(t_0^2) = \operatorname{or}(d\varphi B(t_0^2))$. By definition of $\varphi$, it is clear that $\operatorname{or} E(t_0^2) = c(\sigma) \cdot \operatorname{or} d\varphi(E(t_0^2))$.

It turns out that $\varphi$ naturally acts on the orbit space $O(t)$, viz., $\varphi(O(t)) = O(\varphi(t))$ for any $t \in T$. Indeed, let $x \in O(t)$. Then $x = ata^{-1}$, $\varphi(x) = a_1 \varphi(t) a_1^{-1}$, where $a_1 = n_\sigma a n_\sigma^{-1} \in \mathfrak{H}$, because $n_\sigma \in \mathfrak{H}$. We now calculate the mapping $d\varphi : T_t(O(t)) \to T_{\varphi(t)}(O(\varphi(t)))$. It is easy to compute that $d\varphi(A_\alpha(t_0^2)) = q_{\varphi(t_0^2)}(n_\sigma A_\alpha n_\sigma^{-1})$. Since the points $t_{0i}$ and $\varphi(t_0^2)$ belong to the same simplex $\Delta_{i_1 \ldots i_n}$, they can be joined with a coninuous path $\Psi(\tau)$ such that $\Psi(\tau) \in \Delta_{i_1 \ldots i_n}$ for any $\tau$ and $\Psi(0) = \varphi(t_0^2)$, $\Psi(1) = t_{0i}$. According to Lemma 20.4.1, the basis for $d\varphi B(t_0^2)$ can be continuously translated along $\psi(\tau)$ at $t_{0i}$, so that the basis for $d\varphi B(t_0^2)$ is always tangent to $V$ in the deformation process, the frame $d\varphi E(t_0^2)$ remains tangent to $T$, and the frame $\overline{d\varphi A(t_0^2)}$ always orthogonal to the torus. Thus, we obtain at $t_{0i}$ a basis for $\overline{d\varphi B(t_0^2)}$, so that $\operatorname{or} \overline{d\varphi B(t_0^2)} = \operatorname{or} d\varphi B(t_0^2) = \operatorname{or} B(t_0^2)$. The frame $d\varphi E(t_0^2)$ is deformed into $\overline{d\varphi E(t_0^2)}$ determining the same torus orientation as $d\varphi E(t_0^2)$, i.e., $\operatorname{or} \overline{d\varphi E(t_0^2)} = c(\sigma) \operatorname{or} E(t_0^2)$, whereas $d\varphi A(t_0^2)$ is transformed into $\overline{d\varphi A(t_0^2)}$ at the point $t_{0i}$ : $\overline{d\varphi A(t_0^2)} = \{q_{t_{0i}}(n_\sigma A_\alpha n_\sigma^{-1})\}$.

It remains to compare the orientations determined by the two frames $\{q_{t_{0i}}(n_\sigma A_\alpha n_\sigma^{-1})\}$ and $\{q_{t_{0i}}(A_\alpha)\}$ in the tangent space $T_{t_{0i}}(O(t_{0i}))$. Since $n_\sigma \in \mathfrak{H}$, we can join the point $n_\sigma$ with a smooth path wholly in $\omega(\tau)\mathfrak{H}$ to the unit element of the group $e$. Consider two frames $\{A_\alpha\}$ and $\{n_\sigma A_\alpha n_\sigma^{-1}\}$ in the subalgebra $H$, and define nonsingular linear transformation $F_\tau : H \to H$ by $F_\tau(X) = \operatorname{Ad}(\omega(\tau))(X)$. Then $F_1$ is the identity mapping, and the homotopy $F_\tau$ transforms $n_\sigma A(e) n_\sigma^{-1}$ into the frame $A(e)$, with the frames remaining nondegenerate under the deformation.

Considering the mapping $q_{t_{0i}} : H \to \underline{T_{t_{0i}}(O}(t_{0i}))$, we obtain that the relative orientation of the bases $B(t_{0i})$ and $\overline{d\varphi B(t_{0i})}$ is determined by that of $E(t_{0i})$ and $\overline{d\varphi E(t_0^2)}$, i.e., $\operatorname{or} B(t_{0i}) = c(\sigma) \operatorname{or} \overline{d\varphi B(t_0^2)} = c(\sigma) \operatorname{or} B(t_0^2)$.

**Corollary 20.4.1.** *The required degree of the mapping $f : V \to V$, ($V = A_{n-1}I$) equals the sum $\Sigma(T^{n-1}) = \sum_{i=1}^{2^{n-1}} c(\sigma_i)$, where $\sigma_i$ denotes a permutation of the set $(1, 2, \dots n)$ such that $t_{0i} \in \Delta_{i_1 \dots i_n}$, $(i_1, \dots i_n) = \sigma(1, \dots n)$, with $t_{0i}$ ranging over all the inverse images of the point $t_0^2 \in R(V)$.*

**Lemma 20.4.3.** *Let $m \geqslant 1$. Then $\Sigma(T^{2m}) = 2^m$, $\Sigma(T^{2m-1}) = 0$.*

**Proof.** Suppose that the torus $T^{n-1} \subset \mathrm{SU}(n)$ is realized by matrices of the form

$$\begin{Vmatrix} e^{i\varphi_1} & & 0 \\ & \ddots & \\ 0 & & e^{i\varphi_n} \end{Vmatrix}.$$

Denote an element $t \in T$ by $t = (\varphi_1, \dots, \varphi_n)$. Consider the element $t' \in U(n)$, $t' = (\frac{1}{2}\varphi_1, \dots, \frac{1}{2}\varphi_n)$, $\det(t') = -1$, $0 < \frac{1}{2}\varphi_1 < \dots < \frac{1}{2}\varphi_n < \pi$. Though $(t_0^2)'$ does not belong to $\mathrm{SU}(n)$, we can obtain all the roots $t_{0i}$ of $t_0^2$ from it if we add $\pi$ to the odd number of arguments $\{\frac{1}{2}\varphi_\alpha\}$. Consider a root $t_{\alpha'}$. We agree that if the value of the $p$th argument is greater than $\pi$, then we put "+" in the $p$th place; and if it is less than $\pi$, then we put "−".

Let $n = 2$. Then, to obtain all the roots, we have to replace the odd number of "−" in the line $(-, -)$ by "+", which yields $t_{01} = (-, +)$, $t_{02} = (+, -)$. It is clear that $c(\sigma_1) = 1$, $c(\sigma_2) = -1$, and $\Sigma(T^1) = 0$.

Let $n = 3$. Then there are four roots $t_{0i}$: $t_{01} = (+, -, -)$, $t_{02} = (-, +, -)$, $t_{03} = (-, -, +)$, $t_{04} = (+, +, +)$. We have $c(\sigma_1) = 1$, $c(\sigma_2) = -1$, $c(\sigma_3) = 1$, $c(\sigma_4) = 1$, $\Sigma(T^2) = 2$.

Now let $n$ be even, and $n = 2k + 2$, $k \geqslant 1$. We prove that $\Sigma(T^{2k+2}) = \Sigma(T^2)\Sigma(T^{2k})$; $T^{2k+2} = T^2 \times T^{2k}$. Since $\Sigma(T^2) = 2$, it follows that $\Sigma(T^{2m}) = 2^m$. Note that the value $\Sigma(T^{2k+2})$ is not the degree of the mapping $f|_{T^{2k+2}}$. Consider $t_0^2 \in T^{2k+2}$; $t_0^2 = (\varphi_1, \dots, \varphi_{2k+3})$. Let all the conditions of $t_0^2$ be fulfilled. We define the projections $p_1$ and $p_2$ by $p_1 : T^{2k+2} \to T^2$, $p_2 : T^{2k+2} \to T^{2k}$, putting $p_1(t_0^2) = (\varphi_1 + \dots + \varphi_{2k+1}; \varphi_{2k+2}; \varphi_{2k+3})$, $p_2(t_0^2) = (\varphi_1, \dots, \varphi_{2k}; \varphi_{2k+1} + \varphi_{2k+2} + \varphi_{2k+3})$. The elements $p_1(t_{0i})$, $p_2(t_{0i})$ and $p_1(t_0^2)$, $p_2(t_0^2)$ are regular. Consider $p_2(t_0^2)$ for which all the conditions imposed earlier on $t_0^2$ are fulfilled again, thus making it possible to resort to induction. If we apply the element $w_\sigma \in W(\mathrm{SU}(2k+3))$ to the simplex $\Delta_{12\dots(2k+3)}$, the induced action on the simplexes $p_1(\Delta_{12\dots(2k+3)})$ and $p_2(\Delta_{12\dots(2k+3)})$ arises. Thus, the action of the group $W(\mathrm{SU}(2k+3))$ on the torus $T^{2k+2}$ generates the induced action on $T^2$ and $T^{2k}$, coinciding with the action of $W(\mathrm{SU}(2k+1))$ on $T^{2k}$ and $W(\mathrm{SU}(3))$ on $T^2$. Besides, the projections of the roots $t_{0i}$ onto $T^2$ and $T^{2k}$ supply the complete set of roots from the elements $p_1(t_0^2)$ and $p_2(t_0^2)$.

(a′) Consider all the roots $t_{0i} \in T^{2k+2}$ such that $p_1(t_{0i}) = (+, -, -)$, and assume that $(t_{0i})_{2k+1,2k+1} = \exp\left(\frac{1}{2}\varphi_{2k+1}\right)$. Let a root $t_{0i}$ be of the form $t_{0i} = (*, *, \dots, *, -, -, -)$. Then $p_2(t_{0i}) = (*, *, \dots, *, -)$. We have that $\sigma'$

is a permutation reducing the element $(*,*,\ldots,*,*)$ to "normal" form, i.e., mapping it into the simplex $\Delta_{12\ldots 2k}$. Then, in order to reduce the element $(*,*,\ldots,*,*,-)$ to "normal" form, it is necessary to apply another $2s+1$ transpositions. The final permutation $\sigma''$ is of the form $\sigma'' = \xi\sigma$, where $c(\xi) = -1$. After the permutation $\sigma'$, it is necessary to apply another permutation $\eta$ such that the permutation $\sigma(t_{0i}) = \eta\cdot\sigma'$ maps the element $t_{0i}$ into the simplex $\Delta_{12\ldots(2k+3)}$. Since $\eta$ is odd, we have $c(\sigma(t_{0i})) = c(\sigma(p_2(t_{0i})))$.

($a_1''$) Consider all the roots $t_{0i} \in T^{2k+2}$ such that $p_1(t_{0i}) = (+,-,-)$, and assume that $t_{0i} = (*,*,\ldots,*,+,-,-)$, in which case $c(\sigma(t_{0i})) = c(\sigma(p_2(t_{0i})))$, i.e., $c(\sigma(t_{0i})) = c(\sigma(p_2(t_{0i})))$.

($a_2'$) Let $p_1(t_{0i}) = (-,+,-)$. Assume that $t_{0i} = (*,*,\ldots,*,-,+,-)$. Then $c(\sigma(t_{0i})) = -c(\sigma(p_2(t_{0i})))$.

($a_2''$) Let $p_1(t_{0i}) = (-,+,-)$, $t_{0i} = (*,*,\ldots,*,+,+,-)$. Then $c(\sigma(t_{0i})) = c(\sigma(p_2(t_{0i})))$.

($a_3'$) Let $p_1(t_{0i}) = (-,-,+)$, $t_{0i} = (*,*,\ldots,*,*,-,-,+)$. Then $c(\sigma(t_{0i})) = c(\sigma(p_2(t_{0i})))$.

($a_3''$) Let $p_1(t_{0i}) = (-,-,+)$, $t_{0i} = (*,*,\ldots,*,+,-,+)$. Then $c(\sigma(t_{0i})) = -c(\sigma(p_2(t_{0i})))$.

($a_4'$) Let $p_1(t_{0i}) = (+,+,+)$, $t_{0i} = (*,*,\ldots,*,*,-,+,+)$. Then $c(\sigma(t_{0i})) = c(\sigma(p_2(t_{0i})))$.

($a_4''$) Let $p_1(t_{0i}) = (+,+,+)$, $t_{0i} = (*,*,\ldots,*,*,+,+,+)$. Then $c(\sigma(t_{0i})) = c(\sigma(p_2(t_{0i})))$.

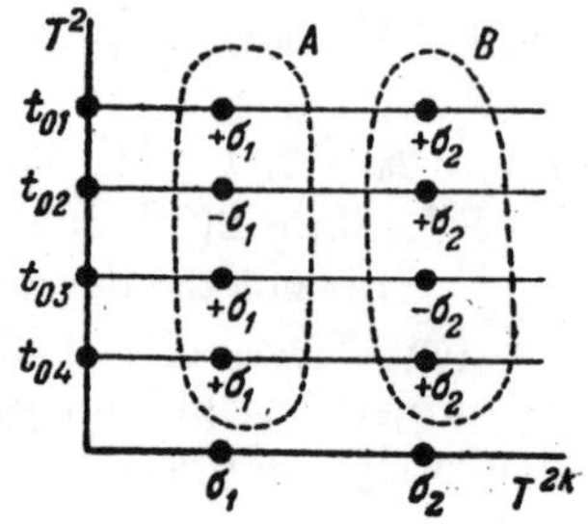


**Fig. 45**

The set of roots $t_{0i} = (*,\ldots,*,*,*,-)$ is denoted by $A$, and that of $t_{0i} = (*,\ldots,*,*,*,+)$ by $B$ (see Fig. 45 representing the direct product $T^2 \times T^{2k}$ and combining all the above cases). Taking the algebraic sum of $t_{0i}$, we obtain the formula $\Sigma(T^{2k+2}) = \Sigma(T^2)\Sigma(T^{2k})$. The above argument can also be repeated for the torus $T^{2k+1} = T^2 \times T^{2k-1}$, which leads to $\Sigma(T^{2k+1}) = \Sigma(T^2)\Sigma(T^{2k-1})$, and since $\Sigma(T^1) = 0$, we have $\Sigma(T^{2m-1}) = 0$. That $\Sigma(T^{2m-1}) = 0$ for an $m \geqslant 1$ could have been proved directly without making use of the formula $\Sigma(T^n) = \Sigma(T^2)\Sigma(T^{n-2})$.

**Corollary 20.4.2.** *The required degree of a mapping $f: V \to V$, where $V = \mathrm{SU}(n)/\mathrm{SO}(n)$, $n \geqslant 3$, is $2^m$ if $n = 2m+1$, and zero if $n = 2m$.*

**20.5. The Case of Spaces of Type I. Spaces $\mathrm{SU}(2m)/\mathrm{Sp}(m)$.**

Consider the symmetric spaces $\mathrm{SU}(2m)/\mathrm{Sp}(m)$, $m \geqslant 2$. Suppose that the rank of $V = m-1$, $\mathfrak{H} = \mathrm{Sp}(m)$ is totally nonhomologous to zero in $\mathrm{SU}(2m)$ for integral coefficients, and $P(V,t) = (1+t^5)(1+t^9)\dots(1+t^{4m-3})$. The group $\mathrm{SU}(2m)$ is the maximal, connected, and simply-connected isometry group. The involutive automorphism in the algebra $G$ is of the form $\Theta(X) = I\bar{X}I^{-1}$, where $I = \begin{Vmatrix} 0 & E \\ -E & 0 \end{Vmatrix}$. The automorphism $\Theta$ is extended to the involution $\sigma(g) = I\bar{g}I^{-1}$. The fixed-point set for the automorphism $\sigma$ is the stability subgroup $\mathfrak{H}$. The totally geodesic submanifold $V$ is formed by all elements of the form $\{g\sigma(g^{-1})\}$, where $\{gIg^T I^{-1}\}$. It is clear that the $V$ is made up of all skew-symmetric matrices $g^T = -g$, and the tangent space $B = T_e(V)$ is of the form $\begin{Vmatrix} Z_1 & Z_2 \\ \bar{Z}_2 & -\bar{Z}_1 \end{Vmatrix}$, where $Z_1 \in \mathrm{su}(m)$, $Z_2 \in \mathrm{so}(m,\mathbf{C})$. The maximal Abelian subspace, torus $T$, is of the form $(\varphi_1,\dots,\varphi_m;\varphi_1,\dots,\varphi_m)$, where $\varphi_1 + \dots + \varphi_m = 2k\pi$, in the notation of Lemma 20.4.3. In our case, the rank of $V$ is less than that of $\mathfrak{G}$, with $T \subset \Omega(\mathfrak{G})$; $t \in T$, $t_{\alpha\alpha} = t_{m+\alpha,m+\alpha}$, $1 \leqslant \alpha \leqslant m$. It is easy to calculate that the intersection $\mathfrak{H} \cap V$ is not discrete, whereas the cycle $\mathfrak{H}$ cannot be used to compute the intersection number for the cycle $V$ and $\mathfrak{H}$. Note that an element $t \in T$ can be completely described by $(\varphi_1,\dots,\varphi_m)$, where $\varphi_1 + \dots + \varphi_m = 2k\pi$. We choose $t \in R(V)$ so that $0 < \varphi_1 < \dots < \varphi_m < 2\pi$. Then $C(t) = S(U(2) \times \dots \times U(2))$ ($m$ factors) and consists of matrices of the form

$$\left\Vert \begin{array}{ccc:ccc} x_1 & & 0 & y_1 & & 0 \\ & \ddots & & & \ddots & \\ 0 & & x_m & 0 & & y_m \\ \hdashline z_1 & & o & w_1 & & 0 \\ & \ddots & & & \ddots & \\ 0 & & z_m & 0 & & w_m \end{array} \right\Vert,$$

where $\begin{Vmatrix} x_i & y_i \\ z_i & w_i \end{Vmatrix} \in U(2)$.

**Lemma 20.5.1.** *If $t \in R(V)$, then $O(t) = \mathrm{Sp}(m)/K$, where $K = \mathrm{Sp}(m) \cap C(t) = \mathrm{SU}(2) \times \dots \times \mathrm{SU}(2)$ ($m$ times).*

**Proof.** Let $c \in C(t) \cap \mathrm{Sp}(m)$, $c = \begin{Vmatrix} X & Y \\ Z & W \end{Vmatrix}$. Then $Ic = \bar{c}I$, which supplies a necessary and sufficient condition for an element $c$ to belong to the subgroup $\mathrm{Sp}(m)$, viz., $Z = -\bar{Y}$, $W = \bar{X}$. Since $c \in C(t)$, we have $z_i = -\bar{y}_i$, $\omega_i = \bar{x}_i$.

Q.E.D.

Consider the mapping $f : V \to V$, $f(v) = v^2$. Let $t_0^2 \in R(V)$. Then there are $2^{m-1}$ solutions of the equation $t^2 = t_0^2$ on the torus $T$. We denote these roots by $t_{0i}$. Construct the bases $B(t_0^2)$ and $B(t_{0i})$ at the points $t_0^2$ and $t_{0i}$ (see above). In our case, the subalgebra $N = \operatorname{Ker} q_t$, $t \in R(V)$ is that of $K = \mathrm{su}(2) \times \cdots \times \mathrm{su}(2)$ ($m$ times). We take the orthogonal complement $A$ to $N$ in the subalgebra $H$, and an arbitrary basis $A_1, \ldots, A_s$, $s = m(2m-1)$ in $A$. Since $df\ B(t_{0i}) = B(t_0^2)$ up to factors involving two vectors of the frame $E(t_{0i})$, it is necessary to compare the orientations of $B(t_{0i})$ and $B(t_0^2)$.

Consider $T \subset T_0$. The general situation is the same as for the torus $(\varphi_1, \ldots, \varphi_m)$. We distinguish from $W(\mathrm{SU}(2m))$ a subgroup $W(V)$ transforming $T$ into intself. Since the elements $t \in T$ are of the form $t = t' \oplus t'$, $t' \in T' \subset \mathrm{SU}(m)$, the generators of $W(V)$ are the transformations $w_{ij} : T \to T$ such that $w_{ij}(t) = n_{ij} t n_{ij}^{-1}$, $n_{ij} = n'_{ij} \oplus n'_{ij}$; $n'_{ij} \in \mathrm{SU}(m)$, and the elements $n'_{ij}$ coincide with $n_{ij}$ considered in the previous item. The elements $w \in W(V)$ associated with an arbitrary permutation $\sigma$ will be denoted by $\omega_\sigma$. Consider in $\mathrm{SU}(2m)$ a subgroup $W''$ generated by $n_{ij}$, $i \neq j$, $n_{ii} = e$. $W''$ is isomorphic to $W(V)$ and acts on the torus $T$ via inner automorphisms. The orientation of the simplex $\Delta_{i_1 \ldots i_m}$ is determined by the parity of $\sigma$ so that $(i_1, \ldots, i_m) = \sigma(1, 2, \ldots, m)$. It is clear that $W'' \subset \mathrm{Sp}(m) \cap N(T)$. Consider $t_0^2 \in R(V)$ and fix some root $t_{0i}$. If $t_{0i} \in \Delta_{12\ldots m}$, then $\operatorname{or} B(t_{0i}) = \operatorname{or} B(t_0^2)$ (see above). Let $t_{0i} \in \Delta_{i_1 \ldots i_m}$, and $(i_1, \ldots, i_m) = \sigma(1, \ldots, m)$. Consider an element $n_\sigma = n'_\sigma \oplus n'_\sigma \in W''$, and define a mapping $\varphi : V \to V$, $\varphi(v) = n_\sigma v n_\sigma^{-1}$, where $\varphi(v) \in V$ for any $v \in V$, since $W'' \subset \mathfrak{H}$.

**Lemma 20.5.2.** *Let* $t_0^2 \in R(V)$. *Then* $\operatorname{or} B(t_0^2) = \operatorname{or} B(t_{0i})$ *for any* $i$, $1 \leqslant i \leqslant 2^{m-1}$.

**Proof.** Similarly to Lemma 20.4.2, we establish that $\operatorname{or} B(t_0^2) = \operatorname{or}(d\varphi B(t_0^2))$. The mapping $\varphi$ acts on the orbit space $O(t)$, viz., $\varphi O(t) = O(\varphi(t))$ for any $t \in T$. We saw in the proof of Lemma 20.4.2 that $d\varphi A_\alpha(t_0^2) = q_{\varphi(t_0^2)}(n_\sigma A_\alpha n_\sigma^{-1})$. Is the mapping $d\varphi : T_{t_0^2}(O(t_0^2)) \to T_{\varphi(t_0^2)}(O(\varphi(t_0^2)))$ nondegenerate? It suffices to show that $n_\sigma A_\alpha n_\sigma^{-1} \in A$ for any $\alpha$. Consider a mapping $\operatorname{Ad}(n_\sigma) : H \to H$. If $n_\sigma = n_{ij}$ is a transformation associated with the elementary permutation, then the subgroup $K = \mathrm{SU}(2) \times \cdots \times \mathrm{SU}(2) = C(t_0^2) \cap \mathfrak{H}$ is transformed into itself, and the action of the transformation $g \to n_\sigma g n_\sigma^{-1}$ on $K$ is reduced to interchanging two sphere $S_i^3$ and $S_j^3$, which means that $\operatorname{Ad}(n_\sigma)(k) = k$; therefore, the orthogonal complement to the subalgegra $k$ is transformed into itself as required. The orbit $\underline{O(t_0^2)}$ is diffeomorphically mapped onto the orbit $\underline{O(\varphi(t_0^2))}$, and we know that $\overline{d\varphi B(t_0^2)} = \overline{\{d\varphi E(t_0^2); \{q_{t_{0i}}(n_\sigma A_\alpha n_\sigma^{-1})\}\}}$ with $\operatorname{or}(\overline{d\varphi E(t_0^2)}) = c(\sigma) \operatorname{or}(E(t_{0i}))$. Since the mapping $q_{t_{0i}}$ is an isomorphism and $A(e)$ and $n_\sigma A(e) n_\sigma^{-1}$ are bases for $A$, it suffices to compare the orientations of the bases $A(e)$ and $n_\sigma A(e) n_\sigma^{-1}$, for which we consider in $H$ a basis $F = \{A(e); \varepsilon_1, \ldots, \varepsilon_{3m}\}$, where $\{\varepsilon_1, \ldots, \varepsilon_{3m}\}$ is a basis for $k$. We apply the transformation $\operatorname{Ad}(n_\sigma)$ to $F$. Since $n_\sigma \in \mathfrak{H}$ and $\mathfrak{H}$ is connected, we then obtain that $\operatorname{or} F = \operatorname{or}(\operatorname{Ad}(n_\sigma)(F))$. The action $\operatorname{Ad}(\sigma)$ on $K$ is known, from which it follows that $\mathrm{or}(\mathrm{Ad}(n_\sigma)(\{\varepsilon_i\})) = c(\sigma)\,\underline{\mathrm{or}(\{\varepsilon_i\})}$, with the consequence that $\mathrm{or}(n_\sigma A(e) n_\sigma^{-1}) = \underline{c(\sigma)\,\mathrm{or}}\,A(e)$. Since $\mathrm{or}(\overline{d\varphi A(t_0^2)}) = c(\sigma)\,\mathrm{or}(A(t_{0i}))$, we obtain $\mathrm{or}\,B(t_{0i}) = \mathrm{or}(\overline{d\varphi B(t_0^2)}) = \mathrm{or}\,B(t_0^2)$. Q.E.D.

**Corollary 20.5.1.** *The required degree of the mapping $f : V \to V$ is $2^{m-1}$, where $V = \mathrm{SU}(2m)/\mathrm{Sp}(m)$, $m \geqslant 2$; therefore, $i_*[V] \neq 0$.*

### 20.6. The Case of Spaces of Type I. Spaces $S^{2l-1} = \mathrm{SO}(2l)/\mathrm{SO}(2l-1)$. Explicit Computation of Cocycles Realizable by Totally Geodesic Submanifolds of Type I

Consider the symmetric spaces $V = \mathrm{SO}(2l)/\mathrm{SO}(2l-1), l \geqslant 4$, where the rank of $V = l$ and $\mathfrak{H} = \mathrm{Spin}(2l-1)$ is totally nonhomologous to zero in the group Spin $(2l)$ for real coefficients, with $P(V,t) = 1 + t^{2l-1}$. The group $\mathfrak{G} = \mathrm{Spin}(2l)$ is the maximal, connected, and simply-connected isometry group of the manifold $V$. The plane $B = T_e(V)$ consists of matrices $b_{ij}$, where $b_{ij} = -b_{ji}$ and $b_{ij} = 0$ if both subscripts are different from 1. Since the rank of $V$ is unity, it is not necessary to investigate the orbit space of $O(t)$. Since $V$ is diffeomorphic to the sphere $S^{2l-1}$, it is evident that $\deg f = 2$ if $f : S^{2l-1} \to S^{2l-1}$, and $\deg f = 0$ if $f : S^{2l} \to S^{2l}$.

**Corollary 20.6.1.** *The required degree of the mapping $f \colon V \to V$, where $V = \mathrm{SO}(2l)/\mathrm{SO}(2l-1), l \geqslant 4$, equals 2; therefore, $i_*[V] \neq 0$.*

It follows from Proposition 19.1.

The elements $(i^*)^{-1}[V]$ can now be calculated in quite a simple manner. We have considered all the symmetric spaces whose isometry groups are not special Lie. It turned out that the unique spaces realizing nontrivial cycles in $H_*(\mathfrak{G}; R)$ are ${}^0A_{2m}\mathrm{I}$, $m \geqslant 1$; ${}^0A_{2m-1}\mathrm{II}$, $m \geqslant 2$; ${}^0D_l\mathrm{II}$, $l \geqslant 4$. The symmetric space ${}^0E_6\mathrm{IV}$ whose isometry group is the special group $E_6$ is studied below.

Consider the principal fibre bundle $\mathfrak{H} \xrightarrow{j} \mathfrak{G} \to V$, where the stability subgroup $\mathfrak{H}$ is connected, and assume that $\mathfrak{H}$ is totally nonhomologous to zero in $\mathfrak{G}$ for a coefficient ring $K$. Then the homomorphism $p^* : H^*(V;K) \to H^*(\mathfrak{G};K)$ is a monomorphism, and we can consider in $H^*(\mathfrak{G};K)$ a subalgebra $p^*(H^*(V;K))$ isomorphic to $H^*(V,K)$. With these assumptions, we have the following.

**Proposition 20.6.1 (see [53]).** *(a) Let $K = \mathbf{Z}; H^*(\mathfrak{G};\mathbf{Z})$ be torsion-free and possess as simple system of primitive generators. (b) Let $K = \mathbf{R}$ or $K = \mathbf{Z}_p$. Assume that $H^*(\mathfrak{G},K)$ possesses a simple system of primitive generators (it being always true for $K = \mathbf{R}$). In both cases, the image of the homomorphism $p^*$ then admits a simple system of primitive generators, which means that we can always select a simple system of primitive generators $x_1,\ldots,x_k$ in the ring $H^*(\mathfrak{G},K)$ (with the above assumptions), so that $x_1,\ldots,x_n$ (where $n \leqslant k$) belong to $p^*(H^*(V;K))$ and are a simple system of generators in this* *subalgebra. For example,* $H^*(\mathrm{SU}(n);\mathbf{Z})$ *and* $H^*(\mathrm{SO}(n);\mathbf{Z}_p)$ *possess simple systems of primitive generators.*

(1) $V = {}^0A_{2m}\mathrm{I}$. Let $K = \mathbf{Z}_p$, where $p \neq 2$ (and can be equal to zero). It is known (see [53]) that the subgroup $\mathfrak{H} = \mathrm{SO}(2m+1), m \geqslant 1$, is totally nonhomologous to zero in the group $\mathfrak{G} = \mathrm{SU}(2m+1)$ for the coefficient ring $\mathbf{Z}_p$. Then we can select a simple system of primitive generators $\{x_3, x_5, \ldots, x_{4m+1}\}$ in $H^*(\mathrm{SU}(2m+1);\mathbf{Z}_p$ such that the subalgebra $p^*(H^*(V;\mathbf{Z}_p)) \cong \Lambda(u_5, u_9, \ldots, u_{4m+1})$ admits the elements $\{x_5, x_g, x_{13}, \ldots, x_{4m+1}\}$ as a simple system of generators, with $p^*(u_\alpha) = x_\alpha; a = (5, 9, 13, \ldots, 4m+1)$.

**Statement 20.6.1.** *Let* $i : V \to \tilde{I}_0(V) \cong \tilde{A}(V)$ *be the above embedding of the space* $V = \mathrm{SU}(2m+1)/\mathrm{SO}(2m+1), m \geqslant 1$, *as of a totally geodesic submanifold into the isometry group* $\tilde{I}_0(V) = \mathrm{SU}(2m+1)$.

*Then the submanifold* $i(V)$ *realizes a nontrivial cocycle in* $H^*(\tilde{I}_0(V);\mathbf{Z}_p)$, *where* $p \neq 2$ *and is prime if* $p \neq 0$. *Consider an element* $\Omega = x_5x_9 \ldots x_{4m+1} \in H^*(\tilde{I}_0(V);\mathbf{Z}_p)$. *Then* $[V] = i^*(N^{-1} \cdot \Omega)$, *where* $N = 2^m (\mathrm{mod}\, p)$.

The proof obviously follows from Corollary 20.4.2 and the choice of generators in $H^*(\tilde{I}_0(V);\mathbf{Z}_p)$.

(2) $V = {}^0A_{2m-1}\mathrm{II}$. Let $K = \mathbf{Z}$. Then the subgroup $\mathfrak{H} = \mathrm{Sp}(m)$ is totally nonhomologous to zero in the group $\mathfrak{G} = \mathrm{SU}(2m)$ for the coefficient ring $\mathbf{Z}$ (see [53]).

**Statement 20.6.2.** *Let* $i : V \to \tilde{I}_0(V) \cong \tilde{A}(V)$ *be the above embedding of the space* $V = \mathrm{SU}(2m)/\mathrm{Sp}(m), m \geqslant 2$ *as a totally geodesic submanifold into the isometry group* $I_0(V) = \mathrm{SU}(2m)$. *Then the submanifold* $I(V)$ *realizes a nontrivial cocycle in* $H^*(I_0(V);\mathbf{Z})$.

*Consider the element* $\Omega = x_5x_9x_{13} \ldots x_{4m-3} \in H^*(I_0(V);\mathbf{Z})$. *Then* $i^*(\Omega) = N[V]$, *where* $N = 2^{m-1}$.

The proof follows from Corollary 20.5.1 and the choice of generators in $H^*(I_0(V);\mathbf{Z})$.

(3) $V = {}^0 D_l\mathrm{II}$. Let $K = \mathbf{Z}_p$, where $p \neq 2$ and prime if $p \neq 0$. Then the subgroup $\mathrm{Spin}(2l-1), l \geqslant 4$, is totally nonhomologous to zero in the group $\mathrm{Spin}(2l)$ for the coefficient ring $\mathbf{Z}_p$ (see [53]).

**Statement 20.6.3.** *Let* $i : V \to I_0(V) \equiv A(V)$ *be the above embedding of the space* $V = \mathrm{SO}(2l)/\mathrm{SO}(2l-1), l \geqslant 4$, *as of a totally geodesic submanifold into the isometry group* $\tilde{I}_0(V) = \mathrm{Spin}(2l)$. *Then the submanifold* $i(V)$ *realizes a nontrivial cocycle in* $H^*(\tilde{I}_0(V);\mathbf{Z}_p)$, *where* $p \neq 2$ *and is prime if* $p \neq 0$. *Consider the element* $\Omega = \bar{x}_{2l-1} \in H^*(\tilde{I}_0(V);\mathbf{Z}_p)$. *Then* $[V] = i^*(\frac{1}{2}\Omega)$.

The proof follows from Corollary 20.6.1.

**Corollary 20.6.2.** *Then the cocycles* $\xi_q = N_a^{-1}\Omega_q = N_q^{-1}x_5x_9x_{13} \ldots x_{4q+1}$, *where* $1 \leqslant q \leqslant m$ *and* $N_q = 2^q (\mathrm{mod}\, p)$, *are realized by totally geodesic submanifolds of type* ${}^0A_{2_q}I$ *in the ring* $H^*(\mathrm{SU}(2m+1);\mathbf{Z}_p), m \geqslant 1, p \neq 2$.

**Corollary 20.6.3.** *The cocycles* $\eta_q = N_q^{-1}\Omega_q = N_q^{-1}x_5x_9\ldots x_{4q-3}$, *where* $2 \leqslant q \leqslant m$ *and* $N_q = 2^{q-1}(\bmod p)$, *are realized by totally geodesic submanifolds of type* ${}^0A_{2_{q-1}}II$ *in the ring* $H^*(\mathrm{SU}(2m);\mathbb{Z}_p), m \geqslant 2, p \neq 2$.

Consider the embedding $j : \mathrm{SU}(2m) \to \mathrm{SU}(2m+1)$. Then the cocycles realizable in $H^*(\mathrm{SU}(2m+1);\mathbb{Z}_p)$ by the submanifolds of type ${}^0A_{2q}\mathrm{I}$, $2 \leqslant q \leqslant m$, and $j({}^0A_{2q-1}\mathrm{II})$, $2 \leqslant q \leqslant m$, are cohomologous, with $\eta_q = j^*(\xi_q)$.

## 20.7. The Case of Spaces of Type I. Space $E_6/F_4$

Consider the space ${}^0E_6\mathrm{IV}$, $V = E_6/F_4$, where the rank of $V$ is 2, $\dim V = 26$, $\mathfrak{H} = F_4$ is totally nonhomologous to zero in the group $\mathfrak{G} = E_6$ for real coefficients, and $P(V,t) = (1+t^9)(1+t^{17})$. The special group $E_6$ with centre $\mathbb{Z}_3$ is the maximal, connected, and simply-connected isometry group of the manifold $V$. Consider the field $F$. Let $K$ be the algebra of Cayley numbers (algebra of octaves) over $F$ (see [217]). We will assume for the present that $F = \mathbb{C}$. Though the algebra $K$ is noncommutative and nonassociative, it satisfies a certain weaker associative condition, or so-called alternative law, viz., $x^2y = x(xy)$ and $yx^2 = (yx)x$ for any elements $x, y \in K$. Consider the linear space over $F$, normally denoted by $M_3^8$, and formed by all $(3 \times 3)$-matrices of the form

$$X = \begin{Vmatrix} \xi_1 & x_3 & \bar{x}_2 \\ \bar{x}_3 & \xi_2 & x_1 \\ x_2 & \bar{x}_1 & \xi_3 \end{Vmatrix},$$

where $x_i \subset K, \xi_i \subset F, 1 \leqslant i \leqslant 3$. The addition and multiplication of matrices by elements of $F$ are defined as always, which turns $M_3^8$ into a 27-dimensional space over $F$. The algebra structure in $M_3^8$ is introduced by the operation $X \cdot Y = \frac{1}{2}(XY + YX)$, where $XY$ and $YX$ are the usual products of $(3 \times 3)$-matrices. The multiplication operator $X \cdot Y$ turns $M_3^8$ into a nonassociative algebra with $X \cdot Y = Y \cdot X$ and $(X^2 \cdot Y) \cdot X = X^2 \cdot (Y \cdot X)$. Let $A \in M_3^8, R_A : M_3^8 \to M_3^8$ be the right translation $R_A(X) = X \cdot A$, and $D_{A,D} = [R_A, R_B] = R_A R_B - R_B R_A$ a linear transformation of $M_3^8$. It follows from the above two indentities that $D_{A,B}$ is the derivation of the algebra $M_3^8$. If $f_4'$ is the Lie algebra over $F$ of all derivations of $M_3^8$, then it is isomorphic to the Lie algebra of the special group $F_4$ (see [216]); therefore, the simply-connected group $F_4$, where $Z(F_4) = 0$, is given a faithful linear representation in the 27-dimensional space over $F$ as the automorphism group of the Jordan algebra $M_3^8$. Consider three orthogonal idempotents

$$e_1 = \begin{Vmatrix} 1 & 0 & 0 \\ 0 & 0 & 0 \\ 0 & 0 & 0 \end{Vmatrix}, \quad e_2 = \begin{Vmatrix} 0 & 0 & 0 \\ 0 & 1 & 0 \\ 0 & 0 & 0 \end{Vmatrix}, \quad e_3 = \begin{Vmatrix} 0 & 0 & 0 \\ 0 & 0 & 0 \\ 0 & 0 & 1 \end{Vmatrix}$$

in $M_3^8$, and distinguish a subalgebra $f_0'$ from $f_4'$, made up of all the derivations

$d$ of $M_3^8$ such that $d(e_1) = d(e_2) = d(e_3) = 0$. It is known that the algebra $f_0'$ is isomorphic to the orthogonal Lie algebra of $D_4$-type (see [216], [217]).

We now extend $f_4'$ by including it into an algebra $e_6'$ isomorphic to the Lie algebra of the special group $E_6$. Denote by $e_6'$ the linear space of all linear transformations of $M_3^8$ of the form $L = R_A + D$, where $A \in M_3^8, \operatorname{Spur} A = 0, D \in f_4'$. The commutator operation in the space $e_6'$ is introduced in the usual manner, viz., $[L_1, L_2] = L_1L_2 - L_2L_1$, where $LN$ denotes the composition of two transformations $L$ and $N$. It is easy to see that if $\operatorname{Spur} B = 0$, then $[R_A, R_B] \in f_4'[D, R_A] = R_{D(A)}$, where $D \in f_4'$. Since $[f_4', f_4'] \subset f_4'$, $e_6'$ turns into the Lie algebra over $F$, with $[B', f_4'] \subset B'$, $[B', B'] \subset f_4'$, where $B'$ denotes the linear space in $e_6'$, formed by the transformations $R_A$. The Lie algebra $e_6'$ irreducibly acts in $M_3^8$ and possesses zero centre; it is, therefore, semisimple. It is known that $e_6'$ is isomorphic to the noncompact Lie algebra of the noncompact group $E_6$ (see [72]). Consider the group $\mathrm{SL}(27; \mathbf{C})$ acting in $M_3^8$. Then $e_6' \subset \mathrm{sl}(27; \mathbf{C})$, and we can consider the subgroup $E_6(\mathbf{C}) = \exp(e_6')$, which is the noncompact form of $E_6$. We now turn to the compact form of $E_6$. Denote the Cartan subalgebra of the Lie algebra $F_4'$ by $H_4'$, and consider in $e_6'$ the subalgebra $H_6' = FR_{e_1-e_2} + FR_{e_2-e_3} + H_4'$, where $R_{e_1-e_2}$ and $R_{e_2-e_3}$ are two complementary vectors to $H_4'$. It then turns out that $H_6'$ is a Cartan subalgebra in the Lie algebra $e_6'$. If $h_j, 1 \leqslant j \leqslant 6$ denotes a basis for $H_6'$, then, in order to find a basis for the Cartan subalgebra of the compact form of $E_6$, it suffices to consider the set of vectors $\{ih_j\}$. We denote by $e_6$ the compact Lie algebra of the group $E_6 \subset \mathrm{SU}(27)$, by $H_6$ its Cartan subalgebra, and by $f_4$ the compact algebra of the group $F_4 \subset E_6$. The Cartan subalgebra $H_6$ in $e_6$ is of the form $H_6 = i\mathbf{R}R_{e_1-e_2} + i\mathbf{R}R_{e_2-e_3} + H_4$; considering $M_3^8$ over $\mathbf{R}$. The plane $B'$ is transformed into a plane $B \subset e_6$ for which the relations $[B, B] \subset f_4, [B, f_4] \subset B$ hold, after the passage to the compact form of $e_6'$, i.e., $B$ is a triple system and coincides with the tangent space $T_e(V)$ to the totally geodesic submanifold $V$, which is a Cartan model of the space $E_6/F_4$. It is clear that the two-dimensional plane $S = i\mathbf{R}R_{e_1-e_2} + i\mathbf{R}R_{e_2-e_3}$ is contained in $B$ and is its maximal Abelian subspace. We now calculate an element $t \in T = \exp S$ as one of the torus $T_0 \subset \mathrm{SU}(27)$. Let $\alpha = e_1 - e_2, \beta = e_2 - e_3$. If we denote coordinates in the plane $S$ by $\varphi_1$ and $\varphi_2$, then each vector $s \in S$ is of the form $s = i\varphi_1 R_\alpha + i\varphi_2 R_\beta$. We represent an element $X \in M_3^8$ in the form

$$X = \xi_1 e_1 + \xi_2 e_2 + \xi_3 e_3 + \begin{Vmatrix} 0 & x_3 & 0 \\ \bar{x}_3 & 0 & 0 \\ 0 & 0 & 0 \end{Vmatrix} + \begin{Vmatrix} 0 & 0 & \bar{x}_2 \\ 0 & 0 & 0 \\ x_2 & 0 & 0 \end{Vmatrix} + \begin{Vmatrix} 0 & 0 & 0 \\ 0 & 0 & x_1 \\ 0 & \bar{x}_1 & 0 \end{Vmatrix} .$$

Accordingly, we choose an orthonormal basis $\{e_1, e_2, e_3; u_1, u_2, \ldots, u_8; v_1, \ldots, v_8; w_1, \ldots, w_8\}$ for $M_3^8$. Then $M_3^8 = L_0 \oplus L_1 \oplus L_2 \oplus L_3$, where $\dim L_0 = 3, \dim L_i = 8$. The operator $t = \exp(s) \in T$ is of the form

$$t = \begin{Vmatrix} e^{i\varphi_1} & 0 & 0 \\ 0 & e^{i\varphi_2} & 0 \\ 0 & 0 & e^{i\varphi_3} \end{Vmatrix} \oplus \begin{Vmatrix} e^{-\frac{i\varphi_1}{2}} E_8 & 0 & 0 \\ 0 & e^{-\frac{i\varphi_2}{2}} E_8 & 0 \\ 0 & 0 & e^{-\frac{i\varphi_3}{2}} E_8 \end{Vmatrix}$$

with respect to this basis, where $\varphi_1 + \varphi_2 + \varphi_3 = 2k\pi, 0 \leqslant \varphi_1 \leqslant 2\pi, 0 \leqslant \varphi_2 \leqslant 2\pi$. The notation of $t$ is identical to that of the elements $t \in T^2 \subset \mathrm{SU}(3)$. Consider $t \in T$ such that $0 < \varphi_1 < \varphi_2 < \varphi_3 < 2\pi, \varphi_1 + \varphi_2 + \varphi_3 = 2\pi$. If $C(t)$ is the centralizer of $t$ in the group $E_6$, then we have $O(t) = F_4/F_4 \cap C(t)$ for the orbit $O(t)$. We now compute the subgroup $F_4 \cap C(t)$. After passing to the compact form of $E_6$, the algebra $f'_0$ is transformed into the real orthogonal algebra so(8).

Consider the subgroup $C = \exp(\mathrm{so}(8))$ in $F_4$. Then the elements $c \in C$ are such that $c(e_i) = e_i, 1 \leqslant i \leqslant 3$. It is known that $C \cong \mathrm{Spin}(8)$ (see [216]).

**Lemma 20.7.1.** *The connected component of the identity element of* $F_4 \cap C(t)$ *coincides with* $C$.

**Proof.** We show that $\mathrm{so}(8) \subset T_e(F_4 \cap C(t))$. Let $d \in \mathrm{so}(8)$, for which it is necessary and sufficient that $d(e_i) = 0, 1 \leqslant i \leqslant 3$. To prove that $d \in T_e(C(t))$, it suffices to show that $[d, s] = 0$ for any operator $s \in S$, which is verified by straightforward calculation (and left to the reader). Conversely, let $g \in F_4 \cap C(t)$, where $0 < \varphi_1 < \varphi_2 < \varphi_3 < 2\pi$. Note that the singularity of an element $t \in T$ is not only in the operator $t$ being scalar on the subspaces $L_1, L_2, L_3$, but also in the fact that its other eigenvalues can coincide; however, this leads to an increase of $C(t)$ in SU(27), but does not increase $F_4 \cap C(t)$.

Since $gt = tg$, the operator $g$ acting in $M_3^8$ is of the form

$$g = \left\Vert \begin{array}{ccc:c} \alpha_1 & 0 & 0 & \\ 0 & \alpha_2 & 0 & B \\ 0 & 0 & \alpha_3 & \\ \hdashline & A & & P \end{array} \right\Vert, \quad \text{where } |\alpha_i| = 1, \ 1 \leqslant i \leqslant 3.$$

Since the element $e_1$ is an idempotent in $M_3^8$, $g(e_i) = g(e_i) \cdot g(e_i)$. It follows from the explicit form of $g$ that $g(e_i) = \alpha_i e_i + X_i$, where $X_i \in L_1 \oplus L_2 \oplus L_3$. Note that, for elements $X \in L_1 \oplus L_2 \oplus L_3$, $\xi_i = 0, 1 \leqslant i \leqslant 3$; therefore, $e_i \cdot X = 0$ for any $X$ and any $i, 1 \leqslant i \leqslant 3$, which entails that $X_i = X_i^2$. We prove that the space $L' = L_1 \oplus L_2 \oplus L_3$ contains no nonzero idempotents.

In fact, if $\| p_{ij} \|$ is the element $X^2$, where $X \in L'$, then $p_{11} = 2|x_3|^2 + 2|x_2|^2, p_{22} = 2|x_3|^2 + 2|x_1|^2, p_{33} = 2|x_2|^2 + 2|x_1|^2$; and since $X = X^2 \in L'$ by assumption, we have $p_{ii} = 0, 1 \leqslant i \leqslant 3$, which yields $x_1 = x_2 = x_3 = 0$, i.e., $X = 0$.

Thus, $g = E_3 \oplus P$.

Q.E.D.

Consider an element $t_0^2 \in R(V)$, and assume that $\varphi_1 + \varphi_2 + \varphi_3 = 2\pi$. We identify the torus $T$ with the torus $T^2 \subset \mathrm{SU}(3)$ and, as earlier, specify an element $t \in T$ by $(\varphi_1, \varphi_2, \varphi_3)$; $t_0^2$ possesses four inverse images under the mapping $f: V \to V: t_{01} = (+,-,-), t_{02} = (-,+,-), t_{03} = (-,-,+), t_{04} = (+,+,+)$. Consider the action of the group $W(V)$ on $T$. Let the Weyl group $W(\mathrm{SU}(27))$ be realized in the group SU(27) as a subgroup $W'$ (see §20.4). Consider the subgroup $W''$ of $W'$, whose elements transform into $T \subset T_0$ itself under the adjoint action. Then $W''$ is generated by elements $n_\sigma$ of the form $n'_\sigma \oplus n''_\sigma$, where the operators $n'_6 \in U(3), n''_\sigma \in U(24)$. We do not describe the explicit form of $n_\sigma$, since it is sufficiently bulky.

**Lemma 20.7.2.** *Each $n_\sigma \in W''$ is an automorphism of the algebra $M_3^8$; therefore, $W''$ is a subgroup in the group $F_4$.*

This is easily established by direct verification with the aid of the explicit form of $n_\sigma$.

Consider $t_0^2 \in R(V)$, and fix some root $t_{0i}$. If $t_{0i} \in \lambda_{123}$, then $\mathrm{or}\, B(t_{0i}) = \mathrm{or}\, B(t_0^2)$ (see Lemma 20.4.1). Let $t_{0i} \in \lambda_{i_1 i_2 i_3}$, where $(i_1, i_2, i_3) = \sigma(1,2,3)$. Consider an element $n_\sigma \in W''$, and define a mapping $\varphi : V \to V, \varphi(v) = n_\sigma v n_\sigma^{-1}$, where $\varphi(v) \in V$ for any $v \in V$, since $W'' \subset \mathfrak{H}$.

**Lemma 20.7.3.** *Let $t_0^2 \in R(V)$. Then $\mathrm{or}\, B(t_0^2) = \mathrm{or}\, B(t_{0i})$ for any $i, 1 \leqslant i \leqslant 4$.*

**Proof.** Since the subgroup $\mathfrak{H} = F_4$ is connected, and the torus $T$ is identified with the torus $T_2$ in the group SU(3), $\mathrm{or}\, B(t_0^2) = \mathrm{or}\, d\varphi B(t_0^2)$. Further, $\mathrm{or}\, E(t_0^2) = c(\sigma) \cdot \mathrm{or}\, d\varphi E(t_0^2)$ and $d\varphi A(t_0^2) = q_{\varphi(t_0^2)}(n_\sigma A(e) n_\sigma^{-1})$. We prove that the mapping $d\varphi : T_{t_0^2} O(t_0^2) \to T_{\varphi(t_0^2)} O(\varphi(t_0^2))$ establishes the isomorphism of these tangent spaces. It suffices to show that $n_\sigma A_\alpha n_\sigma^{-1} \in A$ for any $\alpha$. Consider the mapping $\mathrm{Ad}(n_\sigma) : H \to H$. We derive from Lemma 20.7.1 that the operators $g \in C = (E_4 \cap C(t))_0$ are of the form $E^3 + g''$, where $g''$ acts in the orthogonal complement $L'$ to the plane $L_0$ (the element $X \in L'$ if and only if $\xi_1 = \xi_2 = \xi_3 = 0$). Since $n_\sigma = n'_\sigma \oplus n''_\sigma$, we have $\mathrm{Ad}(n_\sigma)(C) \subset C$ with the consequence that $\mathrm{Ad}(n_\sigma)(A) \subset A$, because $A$ is the orthogonal complement to SO(8) in $H$. Since $n_\sigma \in \mathfrak{H}$, $\mathrm{Ad}(n_\sigma)$ is nondegenerate, and $\mathrm{Im}\, A = A$ as required.

Since $t_{0i}, \varphi(t_0^2) \in \lambda_{i_1 i_2 i_3}$, these points can be joined with a smooth path, and $d\varphi B(t_0^2)$ can be deformed smoothly into $d\varphi B(t_0^2)$ at the point $t_{0i}$, with $\mathrm{or}(d\varphi E(t_0^2)) = c(\sigma)\, \mathrm{or}\, E(t_0^2)$. It remains to compare the orientations of the frames $A(t_{01})$ and $q_{t_{0i}}(n_\sigma A(e) n_\sigma^{-1})$ in the space $T_{t_{0i}}(O(t_{0i}))$, for which it suffices to compare the orientations of the bases $A(e)$ and $n_\sigma A(e) n_\sigma^{-1}$ in the plane $A$. Let $\{s_1, \ldots, s_p\}$ be a basis for the subalgebra SO(8). Then the relative orientation of the bases $A(e)$ and $n_\sigma A(e) n_\sigma^{-1}$ coincides with that of $\{s_1, \ldots, s_p\}$ and $\{n_\sigma s_1 n_\sigma^{-1}, \ldots, n_\sigma s_p, n_\sigma^{-1}\}$. Thus, it is necessary to learn how the group $W''$ acts on the subgroup $C$. Formally, the action is given by $n(g) = \mathrm{Ad}(n)(g)$.

Consider the symmetric group $S_3$ of permutations of three elements $(1,2,3)$. If $\sigma_{ij}$ denote elementary permutations, then $S_3$ consists of the elements

$\{1, \sigma_{12}, \sigma_{23}, \sigma_{12}\sigma_{23}, \sigma_{23}\sigma_{12}, \sigma_{12}\sigma_{23}\sigma_{12}\}$. Denote the elements $\sigma_{12}$ and $\sigma_{23}\sigma_{12}$ by $\varphi_1$ and $\psi_2$, respectively. Then $S_3 = \{1, \varphi_1, \psi_2, \psi_2^2, \varphi_1\psi_2, \varphi_1\psi_2^2\}$. The commutative subgroups $\mathbf{Z}_3(\psi_3), \mathbf{Z}_2(\varphi_1), \mathbf{Z}_2(\varphi_1\psi_2), \mathbf{Z}_2(\varphi_1\psi_2^2)$ are distinguished from $S_3$. As we know, $C$ is isomorphic to the group Spin(8). We now solve the following problem, viz., what is the subgroup $\bar{C} \subset F_4$ of the automorphisms $g \in F_4$ such that $g(L_0) \subset L_0$, where $L_0$ is the three-dimensional subspace of idempotents in $M$? Since $e_i \cdot e_j = 0$ for $i = j$, it follows from $(g(e_i))^2 = g(e_i)$ that $g(e_i) = \sum \alpha_i e_i = \sum \alpha_i^2 e_i$; hence, $\alpha_i = \{0, 1\}$. Finally, $\{g|_{L_0}\} = \{n'_\sigma\}$, which means that the group of all automorphisms of the subalgebra $L_0$ is isomorphic to $S_3$. Since $C \subset \bar{C}$ is a normal subgroup, $\bar{C}/C \cong S_3 = \{n'_\sigma\}$.

Thus, $\bar{C}$ consists of six connected components, the connected component of the unit element being $C$. We now consider in $F_4$ the subgroup $W'' = \{n_\sigma\}$. Since $n_\sigma|_{L_0} = n'_\sigma|_{L_0}$, we have $W'' \subset \bar{C}$, and because the element $n_\sigma$ is completely determined by the permutation $\sigma$, $W''$ is isomorphic to $S_3$, i.e., we realize the quotient group $\bar{C}/C$ as $W'' \subset \bar{C}$, with precisely one element of $W''$ being contained in each connected component of $\bar{C}$. It is clear that the group $\bar{C}$ is the semidirect product of $W''$ and $C$.

Consider the adjoint action of $W''$ on $C$. Since each element $n_\sigma$ acts on $C$ as an automorphism, we obtain a homomorphism $\varepsilon$ of $W''$ into the group Aut($C$) of all automorphisms of $C$. It is easy to show that if $\mathrm{Ad}(n_\sigma) : C \to C$ is an inner automorphism, then $\sigma = 1$ (see [216]); hence, $\varepsilon$ is a monomorphism. It is generally known that the group Aut($C$) consists of six connected components (see [71], [72]); therefore, each connected component of Aut($C$) contains precisely one element of the subgroup $\varepsilon(W'')$. Since $W''$ is isomorphic to $S_3$, we will denote the elements of $W''$ by $\{n(1), n(\varphi_1), \dots, n(\varphi_1\psi_2^2)\}$, and those of $\varepsilon(W'')$ by $\{\bar{I}, \bar{\varphi}_1, \bar{\psi}_2, \dots, \bar{\varphi}_1\bar{\psi}_2^2\}$. It is easy to see that the element $\bar{\psi}_2$ preserves the orientation of $C$. Similarly, the automorphisms $\bar{1}, \bar{\psi}_2^2$ do not alter the orientation of $C$. Since the connected component of the group O(8) not containing the unit element alters the orientation of SO(8) via inner automorphisms $\bar{\varphi}_1, \bar{\varphi}_1\bar{\psi}_2, \bar{\varphi}_1\bar{\psi}_2^2$, the orientation of $C$ also changes by these automorphisms.

Thus, $\bar{1}, \bar{\psi}_2, \bar{\psi}_2^2$ do not alter the orientation of $C$, whereas $\bar{\varphi}_1, \bar{\varphi}_1\bar{\psi}_2, \bar{\varphi}_1\bar{\psi}_2^2$ do alter that of $C$. We now consider the roots $t_{01} = (+,-,-), t_{02} = (-,+,-)$, $t_{03} = (-,-,+), t_{04} = (+,+,+)$; $t_{03}$ and $t_{04}$ are reduced to "normal" form by the identity permutation; therefore, or $B(t_{03}) =$ or $B(t_{04}) =$ or $B(t_0^2)t_{02}$ is reduced to "normal" form by the permutation $\sigma_{23}$; therefore, or $\overline{(d\varphi E(t_0^2))} = -$ or $E(t_0^2) = -$ or $E(t_{02})$. Since $\sigma_{23}$ is associated with the element $n(\varphi_1\psi_2^2)$, or $\overline{(d\varphi B(t_0^2))} =$ or $B(t_{02})$; $t_{01}$ is associated with the permutation $\sigma_{23} \cdot \sigma_{12}$, i.e., or $\overline{(d\varphi B(t_0^2))} =$ or $B(t_{01})$.

Q.E.D.

**Corollary 20.7.1.** *The required degree of the mapping* $f : V \to V$, *where* $V = E_6/F_4$ *equals* 4; *therefore,* $i_*[V] \neq 0$.

We select generators in $H^*(E_6; \mathbf{R})$ precisely in the same way as in §20.6, making use of the total nonhomologousness to zero of the subgroup $F_4$ in the group $E_6$ for $K = \mathbf{R}$.

**Statement 20.7.1.** *Let $i : V \to \tilde{I}_0(V) \cong \tilde{A}(V)$ be the above embedding of the space $V = E_6/F_4$ as of a totally geodesic submanifold into the isometry group $\tilde{I}_0(V) = E_6$. Then the submanifold $i(V)$ realizes a nontrivial cocycle in $H^*(\tilde{I}_0(V); \mathbf{Z}_p)$, where $p \geqslant 7$ and prime, or $p = 0$. Consider the element $\Omega = x_9 \cdot x_{17} \in H^*(\tilde{I}_0(V); \mathbf{Z}_p)$. Then $[V] = i^*(\frac{1}{4}\Omega)$.*

The proof follows from Corollary 20.7.1 and the choice of generators, and also from the fact that the groups $E_6$ and $F_4$ have no $p$-torsion for $p \geqslant 7$ (see above).

**Proof of Theorem 20.1.2.** If $V = V'$, then the statement follows from Theorem 20.1.1. If $V = K \times V'$, then it is necessary to apply Proposition 17.2.1 to the subgroup $K \times \mathfrak{H}$ and to refer to Corollary 19.1.1.

This theorem has the following geometric meaning.

**Proposition 20.7.1.** *Let $V = V'$, and $\mathfrak{H} \subset \tilde{A}(V)$ be the stability subgroup of the manifold $V$. If* ind$(x, y)$ *denotes the intersection number of two cycles $x$ and $y$ of additional dimensions, then* $\operatorname{ind}\{i_*[V], j_*[\mathfrak{H}']\} = \deg f, f : V \to V, f(v) = v^2$, *where the coset $\mathfrak{H}'$ equals $t_0^2 \cdot \mathfrak{H}, t_0^2 \in R(V)$.*

The proof easily follows from the explicit description of the bases $B(t_0^2)$ (see above).

As a rule, we have until now assumed that a totally geodesic submanifold $V$ is simply-connected, which permitted us to make use of the decomposition $V = K \times V'$. We cannot reject it in the general case; however, in certain concrete situations (e.g., in the statement of Theorem 20.2.1.), the one-connectedness assumption can be dropped. Similarly, it can also be dropped in the statement of Theorem 20.1.1.

Here, another corollary arises.

**Proposition 20.7.2.** *Let $V$ be a simply-connected, totally geodesic submanifold in a compact, connected group $\mathfrak{G}$. Assume that $V$ realizes a cohomology generator in $H^*(\mathfrak{G}; \mathbf{R})$. Then $V$ is diffeomorphic to one of the manifolds $S^{2l-1}, l \geqslant 2$; $\mathrm{SU}(2)/\mathrm{SO}(3)$ (of type ${}^0A_2I$), with each of them realizing a nontrivial cocycle in its isometry group.*

The proof follows from Theorems 20.1.1, §18.1, and Statement 19.1.

Another proof of certain results of the present section was subsequently given by O. V. Manturov in [258] on analyzing the $K$-function of homogeneous spaces.

## §21 Classification Theorem Describing Cocycles in the Compact Lie Group Cohomology Realizable by Totally Geodesic Spheres

### 21.1. Classification Theorem Formulation

Here, we solve a problem that is in a certain sense inverse to the one solved above. Until now, we have proceeded from a given totally geodesic submanifold $V$ in a compact Lie group $\mathfrak{G}$, and clarified when it realizes a nontrivial cocycle in the group $H_*(\mathfrak{G})$. The complete solution of the problem was obtained in §20.

We now fix a certain manifold $V_0$. Let $\mathfrak{G}$ be a Lie group. The question arises what nontrivial cycles $x \in H_*(\mathfrak{G})$ may be realized via totally geodesic submanifolds diffeomorphic to $V$. If $\mathfrak{G}$ is the isometry group $I_0(V)$ of the space $V$, provided $V_0 = V_0'$ (see Theorem 18.1), then the complete answer to the question is also contained in §20. However, if $\mathfrak{G}$ is not an isometry group, then an additional analysis is required.

As the "representing" manifold $V_0$, we select a sphere. The choice is justified not only by the sphere being the simplest of spaces $V$, which could realize nontrivial cycles in Lie groups at all (see Theorems 20.1.1 and 20.1.2), but to a considerable extent also by the realization of cycles via spheres being intimately related to that of nontrivial elements of the homotopy groups $\pi_*(\mathfrak{G})$. Moreover, in the case where the ambient manifold is not a group but a symmetric space of type 1, we will immediately solve the important problem of realizing the nontrivial elements of the homotopy groups $\pi_*(M)\otimes\mathbf{R}$ by totally geodesic spheres. Since a totally geodesic sphere can be regarded as an extremal for the multidimensional Dirichlet functional, which is quadratic with respect to derivatives, we therefore solve the problem of finding stationary (critical) points relative to the Dirichlet functional defined on the functional space of mappings of the sphere into a Lie group. This problem has many independent applications whose description is outside the scope of the present monograph.

The above problem of the realization by totally geodesic spheres of nontrivial elements of homotopy groups turns out to be remarkably related to that of finding the maximal number of linearly independent vector fields on spheres. We will describe the relation in detail.

**Theorem 21.1.1.** *Let $\mathfrak{G}$ be a compact, simple Lie group, and $k = k(n) = [1 + \log_2 n]$. Then the unique elements $x \in H_*^{(*)}(\mathfrak{G};\mathbf{R})$ realizable in a Lie group $\mathfrak{G}$ by totally geodesic spheres up to real factors and modulo the kernel of the homomorphism $H^*(\mathfrak{G}) \to H^*(S)$ induced by an embedding of sphere $S$ into $\mathfrak{G}$ are*

*(1)* $(x_3, x_5, x_7, \dots, x_{2k-1})$ *in* $H^*(\mathrm{SU}(n), \mathbf{R})$;

*(2)* $(x_3, x_7, x_{11}, \dots, x_{2k-1})$ *if* $k \equiv 0 (\mathrm{mod}\, 4)$,
$(x_3, x_7, x_{11}, \dots, x_{2k-3})$ *if* $k \equiv 1 (\mathrm{mod}\, 4)$,
$(x_3, x_7, x_{11}, \dots, x_{2k-5})$ *if* $k \equiv 2 (\mathrm{mod}\, 4)$,
*and* $(x_3, x_7, x_{11}, \dots, x_{2k-3})$ *if* $k \equiv 3 (\mathrm{mod}\, 4)$,

*in* $H^*(\mathrm{SO}(n), \mathbf{R}), n \geqslant 8$;
*(3)* $(x_3, x_7, x_{11}, \ldots, x_{2k-1})$ *if* $k \equiv 0 (\bmod 4)$,
$(x_3, x_7, x_{11}, \ldots, x_{2k+1})$ *if* $k \equiv 1 (\bmod 4)$,
$(x_3, x_7, x_{11}, \ldots, x_{2k-1})$ *if* $k \equiv 2 (\bmod 4)$,
*and* $(x_3, x_7, x_{11}, \ldots, x_{2k-3})$ *if* $k \equiv 3 (\bmod 4)$,
*in* $H^*(\mathrm{Sp}(n), \mathbf{R}), n \geqslant 1$;
*(4)* $x_3$ *in* $H^*(G_2; \mathbf{R})$;
*(5)* $x_2$ *in* $H^*(F_4; \mathbf{R})$;
*(6)* $(x_3, x_9)$ *in* $H^*(E_7; \mathbf{R})$;
*(7)* $(x_3, x_{11})$ *in* $H^*(E_7; \mathbf{R})$;
*and* *(8)* $x_3$ *in* $H^*(E_8; \mathbf{R})$.

A similar problem of realizing nontrivial elements of the homotopy groups $\pi_*(M) \otimes \mathbf{R}$ by totally geodesic spheres for all compact, irreducible and symmetric spaces of Type I (i.e., which are not Lie groups) will be solved completely in §22.

## 21.2. Totally Geodesic Spheres Realizing Bott Periodicity

It turns out that the construction of totally geodesic spheres realizing nontrivial cocycles is closely related to Bott periodicity, which happens to be of explicitly "totally geodesic character" and, as follows from the above, is also intimately connected with the spinor representations of orthogonal groups. This relation of Bott periodicity to the extremals for the multidimensional Dirichlet functional permits us, in particular, to interpret Bott periodicity in terms of multidimensional variational problems.

If $U$ is the direct limit of unitary groups $\varinjlim U(n)$, then, by the Bott periodicity theorem, the isomorphisms $\pi_{i-1}(U) = \pi_{i+1}(U)$ hold for $i \geqslant 1$, with $\pi_0(U) = 0, \pi_1(U) = \mathbf{Z}$ (see [276]).

Let $f_{i-1} : S^{i-1} \to \mathrm{SU}(m)$ be the continuous mapping representing the element $1 \in \mathbf{Z} = \pi_{i-1}(\mathrm{SU}(m))$, where $(i-1)$ is odd. We write out the explicit form of the Bott periodicity isomorphism. Consider the group $\mathrm{SU}(2)$, representing its elements as matrices $p = \begin{Vmatrix} \alpha & \beta \\ -\bar{\beta} & \bar{\alpha} \end{Vmatrix}$, where $|\alpha|^2 + |\beta|^2 = 1$, and take a two-dimensional disc $D^2$ on the sphere $S$, determined by the condition $\beta \in \mathbf{R}, \beta \geqslant 0$. We embed it into the group $\mathrm{SU}(2m)$ as follows, viz.,

$$p \to p \otimes E_m = \begin{Vmatrix} \alpha E_m & \beta E_m \\ -\beta E_m & \bar{\alpha} E_m \end{Vmatrix}.$$

Consider the set of points $g(\sigma, \alpha, \beta) = (E_m \oplus f_{i-1}^{-1}(\sigma)) \times (p(\alpha, \beta) \otimes E_m) \cdot (E_m \oplus f_{i-1}(\sigma))$. It is clear that the set $\{g, (\sigma, \alpha, \beta)\}$ is the sphere $S^{i+1} \subset \mathrm{SU}(2m)$, and we can show that the correspondence $S^{i-1} \to S^{i+1}$ describes the periodicity isomorphism. We give the detailed proof in §23. Thus, if $f_{i-1}(\sigma) \in \mathrm{SU}(m)$, then $f_{i+1}(S^{i+1}) = \{g(\sigma, \alpha, \beta)\}$, where $g(\sigma, \alpha, \beta) = \begin{Vmatrix} \alpha E_m & \beta f_{i-1}(\sigma) \\ -\beta f_{i-1}^{-1}(\sigma) & \bar{\alpha} E_m \end{Vmatrix}$.

Instead of the embedding of $D^2$ into the group, we can consider the embedding of the sphere $S^2 \to \mathrm{SU}(2m), p \mapsto p \otimes E_m$, where $p = p(\alpha, \beta)$, $|\alpha|^2 + |\beta|^2 = 1$. The submanifold $\{p \otimes E_m\}$ is totally geodesic is the group $\mathrm{SU}(2m)$, since the sphere $S^{i-1} \subset \mathrm{SU}(m)$ admits an embedding into $\mathrm{SU}(2m)$ as the set of elements $E_m \otimes f_{i-1}(\sigma)$, we obtain the representation $S^{i+1} = \{x \cdot S^2 \cdot x^{-1}\}$ for the sphere $S^{i+1}$, where $x \in S^{i-1} \subset \mathrm{SU}(2m)$, i.e., $S^{i+1}$ is obtained under the adjoint action of the sphere $S^{i-1}$ on the sphere $S$.

Set $m = 2$, and take the standard mapping $f_3(\sigma) = \begin{Vmatrix} x & y \\ -\bar{y} & \bar{x} \end{Vmatrix}$ to be the original $f_3 : S^3 \to \mathrm{SU}(2)$, where $|x|^2 + |y|^2 = 1$. Applying the above procedure, we obtain the mapping $f_{2k+1} : S^{2k+1} \to \mathrm{SU}(2^k)$, which represents the element $1 \in \mathbf{Z} = \pi_{2k+1}(\mathrm{SU}(2^k))$.

**Lemma 21.2.1.** *The sphere $f_{2k+1}(S^{2k+1})$ is a totally geodesic submanifold in the group $\mathrm{SU}(2^k), k \geqslant 1$.*

**Proof.** It suffices to show that the subspace $T_e(S^{2k+1}) \subset \mathrm{su}(2^k)$ is a triple Lie system, and that $S^{2k+1} = \exp T_e(S^{2k+1})$. Denote $\beta f_{2k-1}(\sigma)$ by $g_{2k-1}$. Then the points $g \in S^{2k+1}$ are of the form

$$g = \begin{Vmatrix} \alpha E_{2^{k-1}} & g_{2k-1} \\ -\bar{g}^T_{2k-1} & \bar{\alpha} E_{2^{k-1}} \end{Vmatrix}.$$

If $g_{2k-1} = \|g_{ij}\|$, then $|\alpha|^2 + \sum_i |g_{ij}|^2 = 1$ for any $j, 1 \leqslant j \leqslant 2^{k-1}$.

Represent $\alpha$ as $|\alpha| e^{i\varphi}$. Let $x_1, x_2, \ldots, x_{2k}$ be real parameters describing the sphere $S^{2k-1}$. Introduce the coordinates $\{\varphi, x_1, \ldots, x_{2k}\}$ and select a basis $\{A_1, \ldots, A_{2k+1}\}$ in the subspace $T_e(S^{2k+1})$, where

$$A_1 = \frac{\partial g}{\partial \varphi}\bigg|_e, \quad A_i = \frac{\partial g}{\partial x_{i-1}}\bigg|_e, \quad 2 \leqslant i \leqslant 2k+1.$$

We prove that $A_i^2 = -E_{2^k}$, and $A_i A_j + A_j A_i = 0$ for $i \neq j, 1 \leqslant i, j \leqslant 2k+1$. Let $k = 1$. Then the statement is obvious. Assume that it has been proved for $S^{2k-1}$. Denote the associated basis vectors by $C_1, C_2, \ldots, C_{2k-1}$. Then the vectors $A_i, 1 \leqslant i \leqslant 2k+1$ are of the form

$$A_1 = \begin{Vmatrix} iE_{2^{k-1}} & 0 \\ 0 & -iE_{2^{k-1}} \end{Vmatrix}, \quad A_2 = \begin{Vmatrix} 0 & E_{2^{k-1}} \\ -E_{2^{k-1}} & 0 \end{Vmatrix},$$

$$A_{2+j} = \begin{Vmatrix} 0 & C_j \\ C_j & 0 \end{Vmatrix}, \quad 1 \leqslant j \leqslant 2k-1.$$

We see by direct verification that $A_i A_j + A_j A_i = -2\delta_{ij} E_{2^k}$. Hence, $[[A_i, A_j], A_l] = 0$ if $i \neq l, j \neq l$, and $[[A_i, A_j], A_j] = -4A_i$, i.e., $T_e(S^{2k+1})$ is a Lie triple system. That $S^{2k+1} = \exp T_e(S^{2k+1})$ is verified by straightforward calculation.

## 21.3. Realization of Homotopy Group Elements of the Compact Lie Groups by Totally Geodesic Spheres

Consider the complex space $\mathbf{C}^{n^2}$ of the matrices $(n \times n)$ and turn it into a Hermitian by introducing the metric $\varphi(A,B) = \mathrm{Re}\ \mathrm{Spur}(AB^*), B^* = \bar{B}^T$. Then the unitary group is represented as a smooth submanifold in the sphere $S^{2n^2-1}$ of radius $\sqrt{n}$, with the underlying Hermitian metric inducing on the submanifold $U(n)$ the special Riemannian metric invariant with respect to right and left translations. This simple observation proves to be extremely helpful in the study of the multidimensional Dirichlet functional extremals on submanifolds in compact Lie groups.

Consider the intersection $U(n) \cap u(n)$. Since those and only those $g$ belong to it for which $g^2 = -E_n$, we have $U(n) \cap u(n) = (iE_n) \cup (-iE_n) \cup (\cup_{p=1} G^{\mathbf{C}}_{n,p})$. However, if we consider the intersection $\mathrm{SU}(2m) \cap \mathrm{su}(2m)$, then it consists of one manifold $G^{\mathbf{C}}_{2m,m}$. If $X \in T_e(S^{2k+1})$, then any point of the sphere $S^{2k+1}$ is representable as $\cos|x| \cdot E_{2^k} + \frac{\sin|x|}{|x|} \cdot X$; in particular, $X = \exp(\frac{\pi}{2} \cdot X)$.

It follows that a totally geodesic sphere $S^{2k+1}$ intersects the Lie algebra $\mathrm{su}(2^k)$ in the former's totally geodesic equator $S^{2k} \subset G^{\mathbf{C}}_{2^k,2^{k-1}}$ and is totally geodesic in the submanifold $G^{\mathbf{C}}_{2^k,2^{k-1}}$. The sphere $S^{2k+1}$ is the intersection $\mathrm{SU}(2^k) \cap \Pi_{2k+2}$, where the plane $\Pi_{2k+2}$ in the space $\mathbf{C}^{n^2}$ is spanned by the vectors (matrices) $E_{2^k}, A_1, \ldots, A_{2k+1}$.

**Proposition 21.3.1.** *Let $\mathfrak{G} = \mathrm{SU}(n), n \geqslant 2$, and $k = [1 + \log_2 n]$. Consider the stable homotopy groups $\pi_{2(k-p)-1}(\mathfrak{G})$, where $0 \leqslant p \leqslant k-2$. Then their elements $\{1,2,3,\ldots,2^p\} \in \mathbf{Z} = \pi_{2(k-p)-1}(\mathfrak{G})$ are realized by totally geodesic spheres.*

**Proof.** Consider in the group $\mathrm{SU}(m)$ the subgroup $\mathrm{SU}(2^{k-1})$ embedded standardly, viz., $(E_{n-2^{k-1}}) \oplus \{g\}$. The embedding $\mathrm{SU}(2^{k-1}) \to \mathrm{SU}(n)$ induces an isomorphism of $(2k-1)$-dimensional homotopy groups, which gives in $\mathrm{SU}(n)$ a totally geodesic sphere $S^{2k-1}$ realizing the element $1 (p = 0)$. Consider two embeddings of the group $\mathrm{SU}(2^{k-2})$ into the group $\mathrm{SU}(2^{k-1})$, viz., $i_1(g) = E_{2^{k-2}} \oplus g$ and $l_2(g) = g \oplus E_{2^{kj-2}}$. Then $i_1$ yields a totally geodesic sphere $S^{2k-3}$ realizing the element $1 \in \pi_{2k-3}(\mathrm{SU}(n))$, in turn, realized under $i_2$. Consider the embedding $j(x) = i_1(x) \oplus i_2(x) = i_1(x) \cdot i_2(x), x \in S^{2k-3}$. It is clear that the mapping $j$ determines the element $[i_1] + [i_2]$, so that the embedding $j : S^{2k-3} \to \mathrm{SU}(2^{k-1}) \subset \mathrm{SU}(n)$ yields a totally geodesic sphere $S^{2k-3}$ realizing the element $2 \in \pi_{2k-3}(\mathrm{SU}(n))$. The embeddings $i_1, \ldots, i_{2^p}$ of the group $\mathrm{SU}(2^{k-p-1})$ in $\mathrm{SU}(2^{k-1})$ arise similarly, with $[i_1] = \cdots = [i_{2^p}] = 1 \in \pi_{2(k-p)-1}$ and $[i_1 \ldots i_s] = \{s\} \in \pi_{2(k-p)-1}(\mathrm{SU}(n))$.

Q.E.D.

**Proposition 21.3.2.** *Let $\mathfrak{G} = \mathrm{SO}(n), n \geqslant 8$ and $k = [\log_2 n]$. Consider the stable homotopy groups $\pi_{s_p}(\mathfrak{G})$, where $s_p = 2(k-p)-1, 0 \leqslant p \leqslant k-2$. Then the following elements of these groups are realized by totally geodesic spheres:*

*(1)* $\{1,2,3,4,\ldots,2^p\} \in \mathbf{Z}$ *if* $s_p \equiv 3 \pmod 8$;
*(2)* $\{2,4,6,8,\ldots,2^{p+1}\} \in \mathbf{Z}$ *if* $s_p \equiv 7 \pmod 8$;
*(3)* $\{1\} \in \mathbf{Z}_2$ *if* $s_0 = (2k-1) \equiv 1 \pmod 8$;
*and (4) if* $s_p \equiv 5 \pmod 8$, *then* $\pi_{s_p}(\mathfrak{G}) = 0$.

**Proof.** Let us consider the group $\mathrm{SO}(2^k)$ standardly embedded into the group $\mathrm{SO}(n)$ and the embedding $j : \mathrm{SU}(2^{k-1}) \to \mathrm{SO}(2^k)$, $j(A+iB) = \left\| \begin{smallmatrix} A & B \\ -B & A \end{smallmatrix} \right\|$. Under the embedding $\mathrm{SO}(2^k) \to \mathrm{SO}(n)$, the homotopy group homomorphism is induced, it being an isomorphism for $i \leqslant 2^k - 2, n \geqslant 8$; therefore, the problem is wholly reduced to the study of the homomorphism $j_*$. Let $j : U(r) \to \mathrm{SO}(2r)$ be an embedding. Then the fibration $U(r) \to \mathrm{SO}(2r) \to F_{2r}$ arises, where $f_{2r}$ denotes a manifold diffeomorphic to the submanifold of all complex structures in $\mathrm{SO}(2r)$, i.e., all such elements $g \in \mathrm{SO}(2r)$ that $g^2 = -E$ (see [276]). It is known that $\pi_i(F_{2r}) = \pi_{i+1}(\mathrm{SO}(2r))$ for $i \leqslant 2r-4$. We obtain from the exact homotopy sequence of this fibration that

$$\begin{array}{llll}
\text{a)} & \pi_{8\alpha+1}(U(r)) \to \pi_{8\alpha+1}(\mathrm{SO}(2r)), & j_*(1) = 1, & \mathbb{Z} \to \mathbb{Z}_2; \\
\text{b)} & \pi_{8\alpha+3}(U(r)) \to \pi_{8\alpha+3}(\mathrm{SO}(2r)), & j_*(1) = 1, & \mathbb{Z} \to \mathbb{Z}; \\
\text{c)} & \pi_{8\alpha+5}(U(r)) \to \pi_{8\alpha+5}(\mathrm{SO}(2r)), & j_*(1) = 0, & \mathbb{Z} \to 0; \\
\text{d)} & \pi_{8\alpha+7}(U(r)) \to \pi_{8\alpha+7}(\mathrm{SO}(2r)), & j_*(1) = 2, & \mathbb{Z} \to \mathbb{Z}.
\end{array}$$

Set $r = 2^{k-1}$. Then $2k - 1 \leqslant 2r - 4$ for $k \geqslant 4$, i.e., the statement follows from Proposition 21.3.1. If $k = 3$, then the homomorphism $(j_*)_5$ is not stable, but the group $\pi_5(\mathrm{SO}(8))$ is still stable, and therefore zero.

Q.E.D.

**Proposition 21.3.3.** *Let* $\mathfrak{G} = \mathrm{Sp}(2n), n \geqslant 6$, *and* $k = [1 + \log_2 n]$. *Consider the stable homotopy groups* $\pi_{s_p}(\mathfrak{G})$, *where* $s_p = 2(k-p) - 1, 0 \leqslant p \leqslant k-2$. *Then the following elements are realized by totally geodesic spheres:*
*(1)* $\{1,2,3,4,\ldots,2^p\} \in \mathbf{Z}$ *if* $s_p \equiv 7 \pmod 8$;
*(2)* $\{2,4,6,8,\ldots,2^{p+1}\} \in \mathbf{Z}$ *if* $s_p \equiv 3 \pmod 8$;
*(3)* $\{1\} \in \mathbf{Z}_2$ *if* $s_0 = (2k-1) \equiv 5 \pmod 8$, $n \geqslant 8$;
*and (4) if* $s_p \equiv 1 \pmod 8$, *then* $\pi_{s_p}(\mathfrak{G}) = 0$.

**Proof.** Consider the embedding $j : U(n) \to \mathrm{Sp}(2n)$, where $\mathrm{Sp}(2n)$ is realized as a subgroup in $\mathrm{SU}(2n)$, consisting of the elements $x$ such that $xI = I\bar{x}, I = \left\| \begin{smallmatrix} O & E \\ -E & O \end{smallmatrix} \right\|$. It is clear that $j(U(n)) = \mathrm{SO}(2n) \cap \mathrm{Sp}(2n)$. The problem is reduced to the computation of the homomorphism $j_* : \pi_{s_p}(U(n)) \to \pi_{s_p}(\mathrm{Sp}(2n))$. Since $xI = Ix$ for any $x \in j(U(n))$, the quaternionic automorphism of $x$ transforms into themselves the eigensubspaces of the operator $I$; therefore, the fibration $U(n) \to \mathrm{Sp}(2n) \to K_{2n}$ is given rise (see [276]). It is easy to obtain from the exact homotopy sequence of this fibration that

a) $\pi_{8\alpha+1}(U(n)) \to \pi_{8\alpha+1}(\mathrm{Sp}(2n)), \quad j_*(1) = 0, \quad \mathbf{Z} \to 0;$
b) $\pi_{8\alpha+3}(U(n)) \to \pi_{8\alpha+3}(\mathrm{Sp}(2n)), \quad j_*(1) = 2, \quad \mathbf{Z} \to \mathbf{Z};$
c) $\pi_{8\alpha+5}(U(n)) \to \pi_{8\alpha+5}(\mathrm{Sp}(2n)), \quad j_*(1) = 1, \quad \mathbf{Z} \to \mathbf{Z}_2;$
d) $\pi_{8\alpha+7}(U(n)) \to \pi_{8\alpha+7}(\mathrm{Sp}(2n)), \quad j_*(1) = 1, \quad \mathbf{Z} \to \mathbf{Z}.$

If $n \geqslant 10$, then $2k - 1 \leqslant n - 3$; therefore, the statement follows from Proposition 21.3.1. If $n = 9$, then the homomorphism $j_*$ is unstable in dimension 7. However, for $n \geqslant 6$, $j_*$ is stable in dimension 3, i.e., in the group $\pi_3(\mathrm{Sp}(2n)), 6 \leqslant n < 10$, and we still obtain the required realization. For $n = 8, 9$, we have $5 \leqslant n - 3$ and obtain the desired realizations in the groups Sp(16) and Sp(18).

Consider the commutative diagram

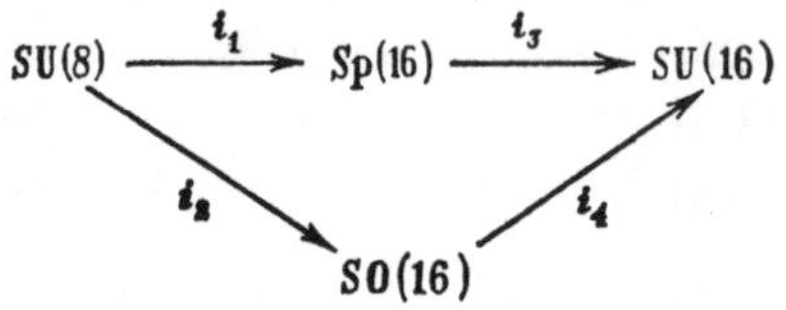


where $i_3 i_1 = i_4 i_2$. Let $i_{1_*} : \pi_7(\mathrm{SU}(8)) \to \pi_7(\mathrm{Sp}(16))$ be the corresponding homomorphism. Since $i_{2_*}$ and $i_{4_*}$ are stable, we have $i_{2_*}(1) = 2$ and $(i_4 i_1)_*(1) = 2$, whence $i_{1_*}(1) = 1$. The case $n = 9$ is considered similarly.

Q.E.D.

## 21.4. Necessary Results Concerning the Spinor and Semispinor Representations of an Orthogonal Group

To solve the cycle realization problem, we need concepts related to orthogonal group spinor and semispinor representations. Furthermore, the general procedure by means of which we will discover totally geodesic spheres realizing nontrivial elements of the homotopy groups $\pi_s(\mathfrak{G})$ will not act in dimension 7, which makes it necessary to investigate the group SO(8).

Consider the complex Euclidean space $R_n^+$, and denote the orthobasis by $\{e_1, \ldots, e_n\}$ and the basis $n$-vectors, which are the skew product of $e_{i_1}, \ldots, e_{i_s}$, by $e_{i_1 \ldots i_s}$. Consider the complex linear space $\mathbf{C}_n^+$ whose elements are the following formal sums, so-called aggregates,

$$A = a + \sum_{i_1} a^{i_1} e_{i_1} + \sum_{i_1 < i_2} a^{i_1 i_2} e_{i_1 i_2} + \ldots + a^{12\ldots n} e_{12\ldots n}.$$

Their multiplication by complex numbers and addition are defined componentwise. We can introduce associative multiplication on $\mathbf{C}_n^+$, with the following properties, viz.,

(1) $(A + B)C = AC + BC$, $C(A + B) = CA + CB$;

(2) $(AB)C = A(BC)$;
(3) $AB = BA = a'B$ if $A = (a', 0, 0, \dots, 0)$, $a' \in \mathbf{C}$;
(4) $aa = |a|^2$, where $a = (0, a, 0, \dots, 0)$;
and (5) $\frac{1}{k!}[a_1 a_2 \dots a_k] = P[a_1, a_2, \dots, a_k]$,
where the alternating product of the vectors $a_1, \dots, a_k$ is on the left, and they are regarded as elements of $\mathbf{C}_n^+$, whereas on the right is the exterior product of $a_1, \dots, a_k$, regarded as an element of $C_n^+$, too. In particular, $e_{i_1} e_{i_2} \dots e_{i_k} = e_{i_1} e_{i_2} \dots e_{i_k}; i_1 < \dots < i_k$. Then $C_n^+$ with this multiplication operator is called a Clifford algebra (see [348]). Consider a smooth submanifold $W'$ in $C_n^+$, made up of aggregates (sometimes called rotation operators) of the form $V = a_k \dots a_2 a_1$, where $a_i$ are any nonisotropic vectors, i.e., $|a_i^2| \neq 0$. It is clear that $V^{-1} = \frac{a_1}{|a_1|} \dots \frac{a_k}{|a_k|}$. $W'$ is disconnected and consists of two connected components, viz., of the submanifold of even rotation operators ($k$ being even) and that of odd ones ($k$ being odd).

Now, let $n = 2\nu$, to which the case of odd $n$ is reduced. Consider the group $\mathrm{O}(2\nu, \mathbf{C})$. Then any orthogonal transformation in the space $R_{2\nu}^+$ can be represented as $y = VxV^{-1}$, where $x, y \in \mathbf{R}_{2\nu}^+$ and $V$ is a certain rotation operator $V \in W'$; $W'$ is a group. If $V = a_k \dots a_2 a_1$, we put $\tilde{V}* = (-1)^k a_1 a_2 \dots a_k$, then and $V\tilde{V}^* = (-1)^k |a_1|^2 \dots |a_k|^2$. By definition, we include those and only those rotation operators into $W$ for which $V\tilde{V}^* = 1$. $W$ is a so-called *spin*or group. We can define a projection $p : W \to \mathrm{O}(2\nu; \mathbf{C})$ by associating each $V \in W$ with a rotation $x \to VxV^{-1}$. Since $p(V) = p(-V)$, $W$ is, obviously, a two-sheeted covering over the group $\mathrm{O}(2\nu; \mathbf{C})$. We will denote the unit element of the component of $W$ by $\mathrm{Spin}(2\nu; \mathbf{C})$; it consists of even rotation operators. If we consider the subalgebra of even aggregates in the algebra $C_n^+$, then it is isomorphic $C_{n-1}^+$ for any $n$ (see [348]). For odd $n$, $C_n^+$ is decomposed into the direct sum of two of its subalgebras isomorphic to $C_{n-1}^+$. Then $C_{2\nu}^+$ admits the so-called spinor representation in $2^\nu$-dimensional complex vector space; more precisely, the Clifford algebra $C_{2\nu}^+$ is isomorphic to the algebra of all linear homogeneous transformations of the complex vector space $C_{2^\nu}$. This isomorphism is unique up to conjugation and is called the spinor representation of $C_{2\nu}^+$. Since the group $\mathrm{Spin}(2\nu; \mathbf{C})$ is realized in $C_{2\nu}^+$ via a submanifold $W_0$, then the spinor representation of $C_{2\nu}^+$ induces a certain representation of $\mathrm{Spin}(2\nu; \mathbf{C})$ called the spinor representation of the orthogonal group $\mathrm{SO}(2\nu; \mathbf{C})$. Consider the tangent space $T_e(W_0)$ to $W_0$ at the point $I$, and translate it parallel to itself to the origin. We obtain a subspace $T$ in $C_{2\nu}^+$, which is the Lie algebra of the group $W_0$ and spanned by all possible simple 2-vectors $e_{ij} = e_i e_j = \frac{1}{2}(e_i e_j - e_j e_i)$; therefore, the spinor representation of $C_{2\nu}^+$ induces that of the algebra $T$. The spinor representation of the group $\mathrm{SO}(2\nu; \mathbf{C})$ is completely reducible, and decomposable into the direct sum of two irreducible representations of dimension $2^{\nu-1}$. If we consider the Clifford algebra $C_{2\nu}^+$ over the field $\mathbf{R}$, then the monomorphism $\mathrm{Spin}(2\nu; \mathbf{C}) \to \mathfrak{G}L(2^\nu; \mathbf{C})$ is replaced by the monomorphism $\mathrm{Spin}(2\nu) \to \mathrm{SU}(2^\nu)$, where $\mathrm{Spin}(2\nu)$ is a compact, simply-connected, and simple Lie group for $2\nu \neq 4$. Thus, the spinor representation of SO($2\nu$) is unitary. The decomposition of a spinor representation into a direct sum is ensured by that of $C_\nu^+$ into two subspaces $X_1 \oplus X_2$, where $X_1$ are even aggregates, and $X_2$ is odd. It is clear that $X_1$ and $X_2$ are the invariant planes of the spinor representation, on each of which the representation of Spin($2\nu$) is, in general, is no more faithful. Since the centre $Z(\mathrm{Spin}(2\nu))$ coincides either with $\mathbf{Z}_4$ or $\mathbf{Z}_2 \oplus \mathbf{Z}_2$, the restrictions of the representation to the planes $X_1$ and $X_2$ are faithful representations of the two semispinor groups $\mathrm{Spin}_1(2\nu)$ and $\mathrm{Spin}_2(2\nu)$. If $\nu$ is odd, then $\mathrm{Spin}_1(2\nu), \mathrm{Spin}_2(2\nu), \mathrm{SO}(2\nu)$ are isomorphic. If $\nu$ is even, then $\mathrm{Spin}_1(2\nu)$ and $\mathrm{Spin}_2(2\nu)$ are homeomorphic, but not isomorphic to the SO($2\nu$).

### 21.5. Spinor Representation of the Orthogonal Group SO(8) and the Cayley Number Automorphism Group

Denote by Aut($G$) and Int($G$), respectively, the groups of all automorphisms and inner automorphisms of a simple Lie algebra $G$. Then Aut(so($2\nu$)) Int(so($2\nu$)) = $\mathbf{Z}_2$ if $\nu = 4$, Aut(so(8))/Int(so(8)) = $S_3$ is the permutation group of the third degree. Let us study Aut(so)(8)). If $\nu = 4$, then the spinor representation of the group SO(8) is decomposed into the sum of two irreducible eight-dimensional representations, the semispinor representation operators on each of the planes $X_1$ and $X_2$, preserving symmetric bilinear forms, which leads to the isomorphism $\mathrm{SO}(8) \cong \mathrm{Spin}_1(8) \cong \mathrm{Spin}_2(8)$ (see [72]).

Consider the spinor representation of the Lie algebra so(8). Let $X \in$ so(8), and let $S$ denote a spinor representation. Then $\bar{S}(X) \in \mathrm{so}(16) \subset \mathrm{su}(16)$ for any $X$, and $\bar{S}(X) = \bar{S}_1(X) \oplus \bar{S}_2(X)$, where the operators $\bar{S}_i(X), i = 1, 2$, act on the subspaces of $X_i$. Define the homomorphism $h : \mathrm{so}(8) \to \mathrm{so}(8)$ by putting $h(X) = \bar{S}_1(X)$. It is clear that $h$ is an automorphism of so(8). It turns out that $h^3 = E$, and $h^2(X) = h(\bar{S}_1(X)) = \bar{S}_2(X), h(\bar{S}_2(X)) = X$ (see [72]).

Consider the algebra of octaves (see §20.7), and select an orthonormal basis $\{1, e_2, e_3, \ldots, e_8\}$, $e_i e_j + e_j e_i = -2\delta_{ij}, 1 \cdot x = x \cdot 1$, for any $x \in K$. Then multiplication in the algebra $K$ is given by the following table, where the arrowheads indicate the sign of the product (Fig. 46). For example, $e_2 e_3 = e_5, e_4, e_2 = -e_6$, etc. The automorphism $h$ leaves the 14-dimensional subalgebra $G_2$ fixed, which, as can be shown by straightforward calculation, is the Lie algebra of the special group $G_2$, the octave algebra automorphism group. Now, consider the elements $A[\alpha a, g]$ in so(8) where $g = (g_2, g_3, \ldots, g_8), a = (a_2, \ldots, a_8), \alpha \in \mathbf{R}, A[\alpha a, g] = \| b_{ij} \|, b_{1p} = \alpha a_p, 2 \leqslant p \leqslant 8, b_{pq} = g_r$, and the values $p, q, r$ are determined by the relation $e_r = e_p e_q, 2 \leqslant p < q$. In other words, the matrix $A[\alpha a, a]$ is determined by the multiplication table in the algebra of octaves. We distinguish three linear subspaces $A[a, a], A[-a, a], A[-2a, 0]$ from so(8), with all these planes lying in the orthogonal complement to $G_2$. It follows from the properties of $h$ that the isomorphisms

$$A[-2a, 0] \xrightarrow{h} -A[-a, a] \xrightarrow{h} A[a, a] \xrightarrow{h} A[-2a, 0]$$

hold.

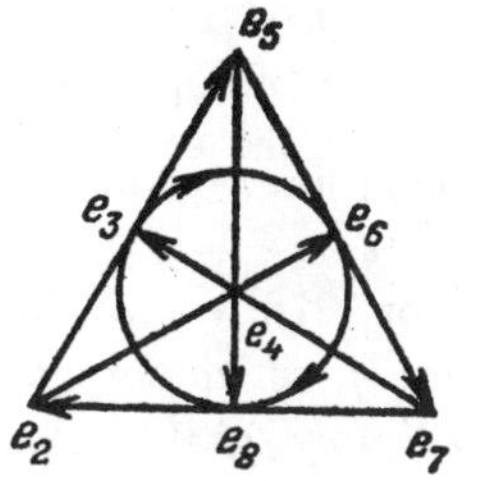


**Fig. 46**

We introduce the new notation $B_0 = A[-2a, 0], B_1 = A[-a, a], B_2 = A[a, a]$. The subspace $B_0$ coincides with the tangent plane to the submanifold $X_7 \subset \mathrm{SO}(8)$ (see §17.3), which is the Cartan model of a symmetric space of type ${}^0D_4\mathrm{II}$ in the group SO(8). It is clear that $B_0$ is a triple system; so are the subspaces $B_1, B_2$, because $B_1 = hB_0,\ B_2 = h^2B_0 = hB_1,\ hB_2 = B_0$. Consider SO(8) and $X_7 = \exp B_0, \exp B_1, \exp B_2$. It is easy to see that $\exp B_1$ and $\exp B_2$ are diffeomorphic to the sphere $S^7$. Since $X_7 = \mathbb{R}P^7$, the automorphism $h$ of the algebra SO(8) cannot be extended to an automorphism of the whole of SO(8).

We have obtained three totally geodesic submanifolds $\exp B_0 = S_0^7$, $\exp B_i = S_i^7, i = 1, 2$, in the group Spin(8), which are diffeomorphic to $S^7$. If $H$ denotes an automorphism of Spin(8) so that $dH = h$, then $S_2^7 = H^2(S_0^7), S_0^7 = H(S_2^7)$. If $\pi$ is the projection of Spin(8) onto SO(8), then $\pi(S_0^7) = \mathbb{R}P^7, \pi(S_i^7) = \exp B_i \subset \mathrm{SO}(8)$.

Denote by $p_1$ and $p_2$, respectively, the projections of Spin(8) onto the groups $\mathrm{Spin}_1(8)$ and $\mathrm{Spin}_2(8)$. Then $h$ generates the three automorphisms

$$H_1\colon \mathrm{SO}(8) \to \mathrm{Spin}_1(8), \quad H_2\colon \mathrm{Spin}_1(8) \to \mathrm{Spin}_2(8),$$
$$H_3\colon \mathrm{Spin}_2(8) \to \mathrm{SO}(8),$$

the automorphism $H_3H_2H_1$ being the identity. It is obvious that $H_1\pi = p_1H, H_2p_1 = p_2H, H_3p_2 = \pi H$ (Fig. 47). Since $Z(\mathrm{Spin}(8)) = \mathbb{Z}_2 \oplus \mathbb{Z}_2$, we have $\exp B_0 = S^7, \exp B_1 = \mathbb{R}P^7, \exp B_2 = S^7$ in Spin(8), and $\exp B_0 = S^7, \exp B_1 = S^7, \exp B_2 = \mathbb{R}P^7$ in Spin(8). It is clear that each of the subspaces $B_i$ is a triple system and generates the whole of the algebra so(8). Regarded as generated by the sphere $S^7 = \exp B_2$, the group SO(8) should be considered as the semispinor group $\mathrm{Spin}_2(8)$, with SO(8) being generated itself by the manifold $\mathbb{R}P^7 = \exp B_0$.

The automorphism $h$ is outer, and generates the subgroup $\{I, h, h^2\}$. It is easy to select such an automorphism $\omega : \mathrm{so}(8) \to \mathrm{so}(8)$ that $\omega^2 = I$, and $h\omega h = \omega$ (its explicit form not being described here). Thus, the group $S_3$

is realized in the group Aut(so(8)) as a subgroup generated by the elements $h$ and $\omega$. It is generally known that $\pi_7(\mathrm{SO}(8)) = \mathbf{Z} \oplus \mathbf{Z}$ (see [278]). To clarify which elements $x \in \pi_7(\mathrm{SO}(8))$ admit a realization by means of totally geodesic submanifolds of type ${}^0D_4\mathrm{II}$, we denote by $e_1$ and $e_2$ the generators of the group $\pi_7(\mathrm{Spin}(8))$, and that of the group $\pi_7(S^7)$ by $e$. We can assume that the relations $j_*(e_1) = e, i_*(e') = e_2$ are fulfilled in the exact sequence

$$\pi_8(S^7) \to \pi_7(\mathrm{Spin}(7)) \xrightarrow{i_*} \pi_7(\mathrm{Spin}(8)) \xrightarrow{j_*} \pi_7(S^7) \to 0,$$

where $e'$ is the generator of the group $\pi_7(\mathrm{Spin}(7)) = \mathbf{Z}$. Since $\pi_8(S^7) = \mathbf{Z}_2$, the exact sequence takes the form $0 \to \mathbf{Z} \to \mathbf{Z} \oplus \mathbf{Z} \to \mathbf{Z} \to 0$ (that $\pi_8(\mathrm{Spin}(7)) = 0$ can be seen either in [53] or [278]). Consider the three totally geodesic spheres $S_0^7, S_1^7$, and $S_2^7$ in Spin(8), which determine the elements $\gamma$, $\alpha$, and $\beta$, respectively, belonging to $\pi_7(\mathrm{Spin}(8))$. It is clear that $\gamma$ is the characteristic class of the standard fibration $\mathrm{Spin}(8) \to \mathrm{Spin}(9) \to S^8$.

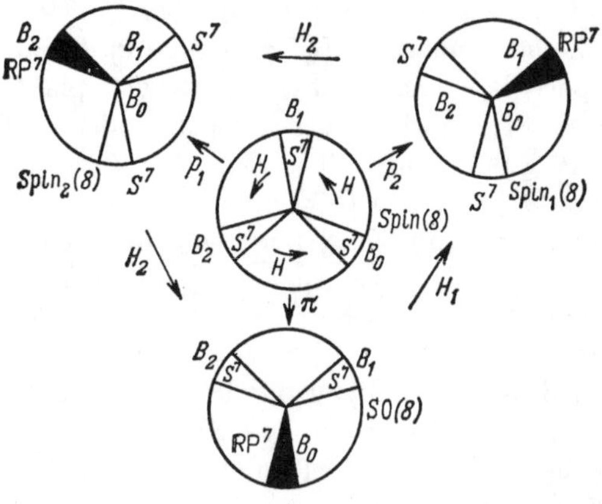


**Fig. 47**

Let $H_*$ be the automorphism of the group $Z \oplus Z$, generated by $H$.

**Proposition 21.5.1.** *Any totally geodesic sphere* $S^7 \subset \mathrm{Spin}(8)$ *can be made coincident via the inner automorphism of the group* Spin(8) *with one of the* $S_0^7, S_1^7$, *and* $S_2^7$. *Besides,* $H_*(\alpha) = -\beta$, *and* $\gamma = \alpha - \beta$, *with no two elements of* $\alpha, \beta$, *and* $\gamma$ *being proportional.*

**Proof.** Let $V$ be an arbitrary totally geodesic submanifold in Spin(8) of type ${}^0D_4\mathrm{II}$. Then the *subspace* $B = T_e(V)$ determines an involutive automorphism $\theta$ which is, therefore, conjugate to the canonical automorphism $\theta_0$ associated with the embedding of the plane $B_0$. Then $V$ is diffeomorphic to $S^7$ and $B_0 = \rho(B)$, where $\rho \in \mathrm{Aut}(\mathrm{so}(8))$. Consider the automorphism $\omega(X) = (-E_1 \oplus E_7) \cdot X \cdot (-E_1 \oplus E_7), X \in \mathrm{so}(8)$. It is clear that $h$ and $\omega$ generate the whole of the group Aut(so(8)) by means of Int(so(8)). Since $\omega$ does not alter the orientation of the sphere $S_0^7$, only $h$ possesses the "essential" action, which proves the first statement.

Further, it is easy to show that $H_*(\alpha) = -\beta, H_*(\beta) = \gamma, H_*(\gamma) = -\alpha$. If $d \in \pi_7(\mathrm{Spin}(8))$, then $d = \varphi e_1 + \psi e_2$ and $j_*(d) = \varphi e$. It follows from the explicit form of the spheres $S_1^7$ and $S_2^7$ that $j_*(\alpha) = e, j_*(\beta) = -e$, and since $\gamma$ is a characteristic class, $j_*(\gamma) = 2e$ (see [399]). Hence, $\alpha = e_1 + xe_2, \beta = -e_1 + ye_2, \gamma = 2e_1 + ze_2; x, y, z \in \mathbf{Z}$. Let the automorphism $H_*$ be represented by an integral matrix $\left\| \begin{smallmatrix} m & p \\ n & q \end{smallmatrix} \right\|$ with respect to the basis $\{e_1, e_2\}$. Then it follows from the condition $H_*^3 = I, H_* \neq I$ that $n \neq 0, q = -(1+m), p = -\frac{m^2+m+1}{n}$. Since $H_*(e_1+xe_2) = e_1 - ye_2, H_*(-e_1+ye_2) = 2e_1+ze_2, H_*(2e_1+ze_2) = -e_1 = xe_2$, it follows that $x = y + z$, i.e., $\gamma = \alpha - \beta$. We now assume that $\gamma = 2\alpha$. Then, applying $H_*$, we obtain $\alpha = 2\beta$, which is impossible. It is proved similarly that no two of $\alpha, \beta$, and $\gamma$ are proportional.

**Corollary 21.5.1.** *Any totally geodesic submanifold of type* ${}^0D_4II$ *in the group* SO(8) *can be made coincident via an inner automorphism of* SO(8) *either with the submanifold* $\mathbf{R}P^7 = \pi(S_0^7)$ *or the spheres* $\pi(S_1^7), \pi(S_2^7)$*, it being* $\gamma = \alpha - \beta$*, and no two of* $\alpha, \beta$*, and* $\gamma$ *being proportional.*

Since $\pi_k(\mathrm{SO}(8)) = \pi_k(\mathrm{Spin}(8)), k > 1$, the corresponding elements of the group $\pi_k(\mathrm{SO}(8))$ have been denoted by the same letters.

## 21.6. Description of Totally Geodesic Spheres Realizing Nontrivial (Co)Cycles in Simple Lie Group Cohomology. The Case of the Group SU($n$)

Consider the group $\mathrm{SU}(n), n \geqslant 2$. Then it follows from Proposition 21.3.1 that the cohomology generators $x_3, x_5, x_7, \ldots, x_{2k-1}$ of the ring $H^*(\mathrm{SU}(n); \mathbf{R})$ are realized by totally geodesic spheres, where $k = [1 + \log_2 n]$. It remains to show that they are unique elements admitting such a realization.

**Proof of Theorem 21.1.1 (item (1)).** Let $S^{2p-1}$ be a totally geodesic sphere in the group $\mathrm{SU}(n)$. Then $\mathrm{SU}(n)$ contains a subgroup $A(V)$ locally isomorphic to the $\mathrm{Spin}(2p)$ if $p \geqslant 2$ (see Theorem 18.1). We prove that the group $A(V)$ cannot be isomorphic either to the group $\mathrm{SO}(2p)$ or the group $\mathrm{PSO}(2p)$ for $p > 4$. Indeed, we assume that $A(V) \cong \mathrm{SO}(2p)$. Then $S^{2p-1}$ generates a triple system $B = T_e(S^{2p-1})$ associated with an involutive automorphism $\theta, \theta(B) = -B$. There exists such an automorphism $\rho$: $\mathrm{so}(2p) \to \mathrm{so}(2p)$ that $\rho\theta\rho^{-1} = \theta_0$, where $\theta_0$ is the standard involutive automorphism determined by the triple system $B_0$. Since $p > 4$, any automorphism of the algebra $\mathrm{so}(2p)$ is uniquely extendible to a certain automorphism of the group $\mathrm{SO}(2p)$. Let $\bar{\rho}$ be such an extension. Then $\bar{\rho}(S^{2p-1}) = \exp B_0 = \mathbf{R}P^{2p-1}$, which is impossible. The case $p = 4$ follows from Corollary 21.5.1. If $p > 4$, then $A(V)$ cannot be isomorphic to $\mathrm{PSO}(2p)$, since otherwise we would obtain a totally geodesic sphere in $\mathrm{SO}(2p)$.

Let $p = 4$. We prove that $A(S^7)$ is not isomorphic to $\mathrm{PSO}(8) = \mathrm{SO}(8)/\mathbf{Z}_2$, where the subgroup $\mathbf{Z}_2$ consists of the elements $\{E_8, -E_8\}$. Assume that

$A(S^7) \cong \mathrm{PSO}(8)$. Then one of the components of the inverse image $\tau^{-1}(S^7) \subset \mathrm{SO}(8), \tau$ being the projection of SO(8) onto PSO(8), is a totally geodesic sphere; therefore, it can be transformed either into the sphere $\pi(S_1^7)$ or the sphere $\pi(S_2^7)$ by means of the inner automorphism $\rho$. Since $\rho$ is inner, it generates an automorphism $\tau\rho$: $\mathrm{PSO}(8) \to \mathrm{PSO}(8)$, whose differential transforms $T_e(S^7)$ either into $B_1$ or $B_2$. But then $\tau\rho(S^7) = \exp B_i$ ($i$ is either 1 or 2), where exp is taken in PSO(8). Consider the spheres $\pi(S_i^7), i = 1, 2$, in SO(8). It is obvious that the element $-E_8$ of the centre belongs to both spheres (see §21.5); therefore, $\tau\pi(S_i^7) \cong \mathbf{R}P^7 = \exp B_i$ in PSO(8). Since $\tau\rho$ must transform $S^7$ into $\exp B_i$, we come to a contradiction. It follows that the group $A(S^{2p-1})$ for $p \geqslant 4$ is isomorphic either to Spin($2p$) or some of the semispinor groups Spin($2p$). Recall that SO(8) can be regarded as a semispinor group (see §21.5).

Consider the group $\mathrm{SU}(n), n \geqslant 4$. Let $S^{2p-1} \subset \mathrm{SU}(n)$, where $p > k$, i.e., $p \geqslant 4$. Then the embedding $A(S^{2p-1}) \subset \mathrm{SU}(n)$ generates the faithful linear representation $C$ of either the group Spin($2p$) or the group $\mathrm{Spin}_i(2p)$, its dimension being $n$. $C$ is decomposed into the direct sum of irreducible representations, viz., $C = \oplus C_r$. We assume first of all that $A(S^{2p-1}) = \mathrm{Spin}_i(2p)$. Then $Z(A(V)) = \mathbf{Z}_2$. At least one of the $C_r$ is a faithful representation of $\mathrm{Spin}_i(2p)$. If it were not so, then each $C_r$ would be a faithful representation of the group PSO($2p$), and then the whole of $C$ would be a faithful representation of PSO($2p$), which is contrary to the original assumption. Meanwhile, it is known that the dimension of an irreducible, faithful linear representation of $\mathrm{Spin}_i(2p)$ is not less than $2^{p-1}$ (see [297], [348]); hence, $2^{p-1} \leqslant \dim C_{r_0} \leqslant n$, i.e., $k < p \leqslant 1 + \log_2 n$, which is contrary to the choice of the value $k$. Thus, $A(S^{2p-1})$ for $p \geqslant 4$ is not isomorphic to $\mathrm{Spin}_i(2p), i = 1, 2$.

Assume now that $A(S^{2p-1}) = \mathrm{Spin}(2p)$. Then a faithful linear representation $C$ of Spin($2p$) of dimension $n$ is given rise. Decomposing $C$ into the direct sum of irreducible representations $C_r$, we obtain that at least one of them should be a faithful representation either of Spin($2p$) or some of $\mathrm{Spin}_i(2p)$, whence we again derive the estimation $2^{p-1} \leqslant n$, which once more contradicts the choice of $k$. Since we have thereby exhausted the stock of locally isomorphic, simple, and compact groups of type $D_p$, the group SU($n$) for $p \geqslant 4$ cannot contain the totally geodesic sphere $S^{2p-1}$, where $p > k$, i.e., statement (1) of Theorem 21.1.1 is proved for $n \geqslant 4$. The proof for the group SU(3) is carried out by elementary means.

## 21.7. The Case of the Groups SO($n$) and Sp($2n$)

We now turn to SO($n$) and Sp($2n$), in which cases there are sharp differences from the group SU($n$). For example, the elements of the ring $H^*(\mathfrak{G}; \mathbf{R})$, whose totally geodesic realization has already been established in Propositions 21.3.2 and 21.3.3, do not exhaust the whole set of elements admitting such realization. We again turn to the analysis of unitary Bott periodicity.

Consider the sphere $f_{2k-1}(S^{2k-1})$ in the group $\mathrm{SU}(2^{k-1})$ and distinguish a subgroup $A(S^{2k-1})$.

**Lemma 21.7.1.** *The group $A(S^{2k-1})$ is isomorphic to one of the semispinor groups* $\mathrm{Spin}_i(2k), k \geqslant 4$.

Note that if $k$ is odd, then we assume $\mathrm{Spin}_i(2k)$ to be isomorphic to $\mathrm{Spin}(2k)$ (see §21.4).

**Proof.** As in the proof of Theorem 21.1.1 (1), we obtain that $A(S^{2k-1})$ is isomorphic either to $\mathrm{Spin}(2k)$ or $\mathrm{Spin}_i(2k)$, the embedding $A(S^{2k-1}) \to \mathrm{SU}(2^{k-1})$ generating a faithful irreducible representation. If $k$ is even, then $\mathrm{Spin}(2k)$ has no faithful irreducible representations of dimension $2^{k-1}$ (see [297]); therefore, $A(S^{2k-1}) \cong \mathrm{Spin}_i(2k)$. If $k$ is odd, then we immediately obtain that $A(S^{2k-1}) \supset \mathrm{Spin}(2k) = \mathrm{Spin}_i(2k)$.

Q.E.D.

The appearance of the subalgebra $\mathrm{so}(2k) \subset \mathrm{su}(2^{k-1})$ generated by the triple system $B = T_e(S^{2k-1})$ can also be seen directly. In §21, we indicated a basis $A_1, \ldots, A_{2k-1}$ such that $A_iA_j + A_jA_i = -2\delta_{ij}E$. The commutator of the plane $B$ is spanned by the basis elements $\{\frac{1}{2}A_iA_j\}, i \neq j$. It is obvious that the subalgebra $[B, B]$ is isomorphic to the algebra $\mathrm{so}(2k-1)$.

Consider a Clifford algebra $C^+_{2k}$ with a basis $e_1, \ldots, e_{2k}$ in the space $R^+_{2k}$. Suppose that it is represented by its spinor image in the space $S_{2^k}$. Since $T_e(W_0)$ is spanned by simple 2-vectors $\{e_{ij}\}$, it suffices to give a spinor representation of the algebra $\mathrm{so}(2k)$ only on the 2-vectors $\{e_i, 2k\}$, since $e_{ij} = -e_{i,2k}e_{j,2k}$. Meanwhile, $e^2_{i,2k} = -1$; therefore, it is clear that the elements $A_1, \ldots, A_{2k-1}$ are the images of $e_{i,2k} (i = 1, \ldots, 2k-1)$ for a semispinor representation of the algebra $C^+_{2k}$.

**Corollary 21.7.1.** *The totally geodesic sphere $f'_{2k-1}(S^{2k-1}) \subset \mathrm{Spin}_i(2k)$ realizes the generator $x'_{2k-1} \in H^*(\mathrm{Spin}_i(2k); \mathbf{R})$. The proof follows from the fact that the mapping $f_{2k-1} : S^{2k-1} \to \mathrm{Spin}_i(2k) \subset \mathrm{SU}(2^{k-1})$ represents the generator of the group $\mathbf{Z} = \pi_{2k-1}(\mathrm{SU}(2^{k-1}))$.*

**Proof of Theorem 21.1.1. (item (2)).** (a) Consider $\mathfrak{G} = \mathrm{SO}(n)$ and assume that $k = [1 + \log_2 n]$ is odd. It follows from Proposition 21.3.2 that the elements $x_3, x_7, \ldots, x_{2k-3}$ are realized by totally geodesic spheres. Assume that $\mathrm{SO}(n)$ contains a totally geodesic sphere $S^{2p-1}$, where $p > k-1$. Since $n \geqslant 8$, we have $p \geqslant 4$. Then the subgroup $A(S^{2p-1})$ is isomorphic either to $\mathrm{Spin}(2p)$ or $\mathrm{Spin}_i(2p)$, and the embedding $A(S^{2p-1}) \to \mathrm{SO}(n)$ generates a faithful representation, which entails that $2^{p-1} \leqslant n$, i.e., $2p-1 \leqslant 1+2\log_2 n$. Since the group $\pi_{2p-1}$ is stable, and if the sphere $S^{2p-1}$ realizes the nontrivial element $H^*(\mathrm{SO}(n); \mathbf{R})$, it means that $2p-1 \geqslant (4\omega - 1) + 4 = 4\omega + 3$. We obtain the inequality $4\omega + 3 \leqslant 1 + 2\log_2 n, k+1 \leqslant 1 + \log_2 n$, which is contrary to the selection of $k$, and item (2) of the theorem is proved for odd $k$.

(b) Let $\mathfrak{G} = \mathrm{SO}(n)$. Suppose that $k$ is even, and $k = [1+\log_2 n]$, $k = 2\omega$, in which case $p \geqslant 4$ and $\mathrm{SO}(n)$ contains a subgroup $A(S^{2p-1})$ isomorphic either to $\mathrm{Spin}(2p)$ or $\mathrm{Spin}_i(2p)$, which again yields the estimation $P \leqslant 1 + \log_2 n$. It follows from Proposition 21.3.2 that the elements $x_3, x_7, \ldots, x_{2k-5}$ are already realized by totally geodesic spheres. Suppose that $2p-1 \geqslant 2k+1$. Then $k+1 \leqslant 1+\log_2 n$, which is contrary to the choice of $k$. It remains to make it clear whether the generator $x_{2k-1}$, where $k = 2\omega$, can be realized by a totally geodesic sphere.

**Lemma 21.7.2.** *Let $k$ be even, and $k \geqslant 4$. Then if $\frac{k(k-1)}{2}$ is even, there exists an embedding $\xi$: $\mathrm{Spin}_i(2k) \to \mathrm{SO}(2^{k-1})$ such that the totally geodesic sphere $\xi f'_{2k-1}(S^{2k-1})$ realizes the generator $x_{2k-1} \in H^*(\mathrm{SO}(2^{k-1}); \mathbf{R})$, and, if $\frac{k(k-1)}{2}$ is odd, then there exists an embedding $\eta$: $\mathrm{Spin}_i(2k) \to \mathrm{Sp}(2^{2-1})$ such that the totally geodesic sphere $\eta f'_{2k-1}(S^{2k-1})$ realizes the generator $x_{2k-1} \in H^*(\mathrm{Sp}(2^{k-1}); \mathbf{R})$.*

**Proof.** Consider the spinor representation $\bar{S}$: $\mathrm{Spin}(2k) \to \mathrm{SU}(2^k)$. Since $k$ is even, the projection $p_i$: $\mathrm{Spin}(2k) \to \mathrm{Spin}_i(2k)$ is defined. Consider transposing aggregates in the algebra $C^+_{2k}$, under which the basis $s$-vector $e_{i_1 i_2 \ldots i_s}$ is replaced by $e_{i_s \ldots i_2 i_1}$. Denote a transposed aggregate by $\tilde{A}$, and consider an automorphism $(\bar{S}(A))^T \to \bar{S}(\tilde{A})$ of the complete matrix algebra on the space $S_{2^k}$, which is isomorphic to $C^+_{2k}$, where $(\bar{S}(A))^T$ designates the usual transposing in $S_{2^k}$. It is known that any automorphism of the $C^+_{2k}$ is inner (see [297], [348]); therefore, $\bar{S}(\tilde{A}) = C(\bar{S}(A))^T C^{-1}$, where $C$ is a certain nonsingular constant matrix. Transposing an aggregate $V \in W_0$ means passing to the inverse element $\tilde{V} = V^{-1}$, and we obtain that $C = (\bar{S}(V))C(\bar{S}(V))^T, V \in W_0$. It is easy to show that the tensor $C$ is determined uniquely up to a constant numerical factor $c \neq 0$. Thus, the subgroup $\mathrm{Spin}(2k) \subset \mathrm{SU}(2^k)$ preserves the bilinear form $C$, which is always either symmetric or skew-symmetric, viz., $c^{\lambda\mu} = (-1)^{\frac{k(k-1)}{2}} c^{\mu\lambda}$ (see [348]). We distinguish from $\mathrm{SU}(2^k)$ the subgroups $\mathrm{SO}(2^k)$ and $\mathrm{Sp}(2^k)$, where $\mathrm{SO}(2^k) = \mathrm{SU}(2^k) \cap \mathrm{SO}(2^k; \mathbf{C}), \mathrm{Sp}(2^k) = \mathrm{SU}(2^k) \cap \mathrm{Sp}(2^k; \mathbf{C})$. Then, if $\rho = \frac{k(k-1)}{2}$ is even, the representation $\bar{S}$ is an embedding of the group $\mathrm{Spin}(2k)$ into the group $\mathrm{SO}(2^k)$; and if $\rho$ is odd, then it is an embedding into the group $\mathrm{Sp}(2^k)$. The situation gets more complicated for the semispinor representation $\bar{S}_i$. The representation space $S_{2^k}$ is decomposed into the sum of two invariant subspaces $S_{2^k} = X_1 \oplus X_2$ (see §21.4); therefore, it is necessary to clarify under which conditions the tensor $C$ admits a decomposition into the direct product $C = C_1 \oplus C_2$, where the tensors $C_i$ are restrictions of $C$ to the planes $X_i$ and, therefore, have the same symmetry type as $C$.

Consider an orthobasis in the space $R^+_{2k}$. Let $\bar{S}$ be a spinor representation. Denote the operators $\bar{S}(e_j)$ by $E'_j$. Then the tensor $C$ is of the form $C = E'_{k+1} \cdot E'_{k+2} \cdots E'_{2k}$ for even $k$ (see [348]), and $C = E'_1 E'_2 \ldots E'_k$ for odd $k$. Since $E'_j$ commute the subspaces $X_1$ and $X_2$, $C$ admits a restriction to $X_1$ and $X_2$ if and only if $k$ is even. We can now sum up in the form of the

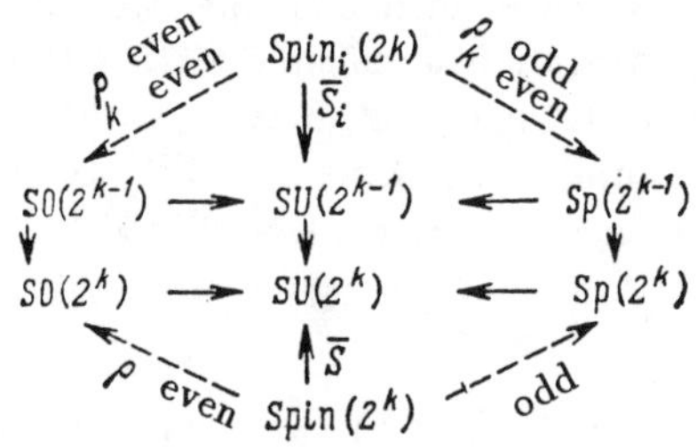


**Fig. 48**

diagram in Fig. 48. Since the mapping $f_{2k-1}$: $S^{2k-1} \to \mathrm{SU}(2^{k-1})$ generates a semispinor representation $\bar{S}_i$ (see Lemma 21.7.1), the mapping $\xi f'_{2k-1}$ realizes the generator of the group $\pi_{2k-1}(\mathrm{SO}(2^{k-1}))$, and the mapping $\eta f'_{2k-1}$ that of the group $\pi_{2k-1}(\mathrm{Sp}(2^{k-1}))$.

Q.E.D.

Consider the group $\mathfrak{G} = \mathrm{SO}(n)$. Let $k = 4\omega$. Then $\rho$ is odd. We derive from Lemma 21.7.2 that there exists a totally geodesic sphere $S^{2k-1}$ in the group $\mathrm{SO}(2^{k-1})$ for $k \geqslant 4$, realizing the generator $x_{2k-1}$. Since $k - 1 \leqslant \log_2 n$, there exists an embedding $j$ : $\mathrm{SO}(2^{k-1}) \to \mathrm{SO}(n)$ inducing the homotopy group monomorphism $\pi_{2k-1}(\mathrm{SO}(2^{k-1})) \to \pi_{2k-1}(\mathrm{SO}(n))$ for $k \geqslant 5$. It follows that the totally geodesic sphere $j\xi f'_{2k-1}(S^{2k-1})$ realizes the generator $x_{2k-1}$.

Now, let $k = 4$. Then the homomorphism $j_*$: $\pi_7(\mathrm{SO}(8)) \to \pi_7(\mathrm{SO}(9))$ is not a monomorphism. It is generally known that the kernel of the homomorphism $j_*{:}\mathbb{Z}\oplus\mathbb{Z} \to \mathbb{Z}$ (see §21.5) is the subgroup generated by the characteristic class $\gamma$ (see [399]). Hence, we derive from Corollary 21.5.1 that $j_*(\gamma) = j_*(\alpha) - j_*(\beta)$, i.e., $j_*(\alpha) = j_*(\beta)$. But, since the elements $\alpha$ and $\beta$ are not proportional to $\gamma$, $j_*(\alpha) \neq 0$ and the totally geodesic sphere $f_\pi(S^7_1)$, as well as the sphere $j\pi(S^7_2)$, realizes the generator $x_7 \in H^*(\mathrm{SO}(9); \mathbb{R})$. Since the group $\pi_7(\mathrm{SO}(9))$ is already stable, the above sphere still realizes the generator $x_7$ under the embedding $\mathrm{SO}(9) \to \mathrm{SO}(n)$.

We have thereby proved the following.

**Lemma 21.7.3.** *Let* $k = [1 + \log_2 n]$, $k \equiv 0 (\mathrm{mod}\, 4)$, $k \geqslant 4$. *Then the generator* $x_{2k-1} \in H^*(\mathrm{SO}(n); \mathbb{R})$ *is realized by a totally geodesic sphere.*

We now prove another auxiliary proposition.

**Lemma 21.7.4.** *Let* $k = [1 + \log_2 n]$ *for even* $k \geqslant 6$. *Assume that* $\mathrm{SO}(n)$ *contains a totally geodesic sphere* $S^{2k-1}$. *Then there exists an embedding* $\mathrm{Spin}_i(2k) \to \mathrm{SO}(2^{k-1})$.

**Proof.** Consider the group $A(S^{2k-1})$. As has already been proved above, it is isomorphic either to $\mathrm{Spin}(2k)$ or some of the groups $\mathrm{Spin}_i(2k)$ for $k > 4$. Decompose a faithful representation $C$ of $A(S^{2k-1})$ into the direct sum of irreducible representations $C = \oplus C_r$. Then there is at least one $C_{r_0}$ among the representations $C_r$, which is an irreducible two-valued representation of $\mathrm{SO}(2k)$; therefore, its dimension is a multiple of $2^{k-1}$ (see [297]), and because the value $k$ was chosen to be maximal, $\dim C_{r_0} = 2^{k-1}$. Since unique two-valued irreducible representations of dimension $2^{k-1}$ of $\mathrm{SO}(2k)$ are semispinor representations $\bar{S}_i$, we have $C_{r_0} = \bar{S}_i$ for a certain $i$.

Q.E.D.

Now, if $k \equiv 2 \pmod 4$, and $\rho$ is odd, we prove that the generator $x_{2k-1}$ cannot be realized by a totally geodesic sphere.

**Lemma 21.7.5.** *Let $k \equiv 2 \pmod 4$, $k = [1 + \log_2 n]$, $k \geqslant 6$. Then there exists no totally geodesic sphere $S^{2k-1}$ in the group $\mathrm{SO}(n)$.*

**Proof.** Assume the contrary, viz., let $\mathrm{SO}(n)$ contain a totally geodesic sphere $S^{2k-1}$. By Lemma 21.7.4, there exists an embedding $\zeta$: $\mathrm{Spin}_i(2k) \to \mathrm{SO}(2^{k-1})$ the arisen representation being irreducible. Consider the composition mapping $\kappa\zeta$: $\mathrm{Spin}_i(2k) \to \mathrm{SU}(2^{k-1})$, where $\kappa$: $\mathrm{SO}(2^{k-1}) \to \mathrm{SU}(2^{k-1})$ is the standard embedding. We then obtain a faithful representation of the group $\mathrm{Spin}_i(2k)$ of dimension $2^{k-1}$, which is obviously irreducible and, therefore, equivalent to one of the semispinor representations $\bar{S}_i$. Since $\rho$ is odd, the representation $\kappa\zeta$ possesses a skew-symmetric bilinear invariant $C$ (see Lemma 21.7.2): therefore, $\kappa\zeta(\mathrm{Spin}_i(2k)) \subset \mathrm{Sp}(2^{k-1})$, i.e., $\kappa\zeta(\mathrm{Spin}_i(2k)) \subset \mathrm{SO}(2^{k-1}) \cap \mathrm{Sp}(2^{k-1})$, and we have obtained a faithful irreducible representation of the group $\mathrm{Spin}_i(2k)$ in the group $U(2^{k-2})$, which is impossible, since the least dimension of such a representation is $2^{k-1}$.

Q.E.D.

Lemmas 21.7.3 and 21.7.5 complete the proof of Theorem 21.1.1(2). We do not dwell on the proof of Theorem 21.1.1(3), since it repeats the above argument in its basic features. The proof of items (4)–(8) of Theorem 21.1.1 is carried out in accordance with the same scheme (see [88]). Theorem 21.1.1 has thus been proved completely.

## §22 Classification Theorem Describing Elements of Homotopy Groups of Symmetric Spaces of Type I, Realizable by Totally Geodesic Spheres

### 22.1. Classification Theorem Statement

The results of §22 describe totally geodesic spheres in compact, irreducible, and symmetric spaces of Type II. Since the study of totally geodesic spheres in a compact, simply-connected, and symmetric space reduced to that of irreducible spaces of Types I and II (see above) to clarify the general situation, it remains to investigate irreducible spaces of Type I, which will be done in the present section. All of them are listed in Table 1. Combining certain series, we obtain the following

$$\mathrm{SU}(p+q)/S(U(p)\times U(q)),\quad \mathrm{SO}(p+q)/S(O(p)\times O(q)),$$
$$\mathrm{Sp}(2(p+q))/\mathrm{Sp}(2p)\times \mathrm{Sp}(2q),\quad \mathrm{SU}(n)/\mathrm{SO}(n),$$
$$\mathrm{SU}(2n)/\mathrm{Sp}(2n),\quad \mathrm{SO}(2n)/U(n),\quad \mathrm{Sp}(2n)/U(n),$$

the special ones being excluded.

Denote the generators of the group $\pi_s(V)\otimes\mathbf{R}$ by $\{x_s, y_s, \dots\}$. Then $\pi_*(V)\otimes\mathbf{R} = \oplus_s\mathbf{R}(x_s, y_s, \dots)$, where $\mathbf{R}(x_s, y_s)$ denotes the group $\pi_s(V)\otimes\mathbf{R}$. Let $Q_N(V)$ be the subgroup in $\pi_*(V)\otimes\mathbf{R}$, made up of all such elements $x$ that $\dim x < N$. Suppose that $k = k(n) = [1 + \log_2 n]$. We define the function

$$f_i(n) = f_i(2^{k-1}) = \begin{cases} 2k - i - 1, & \text{if} \quad k \equiv 0 (\mathrm{mod}\, 4), \\ 2k - i - 3, & \text{if} \quad k \equiv 1 (\mathrm{mod}\, 4), \\ 2k - i - 5, & \text{if} \quad k \equiv 2 (\mathrm{mod}\, 4), \\ 2k - i - 3, & \text{if} \quad k \equiv 3 (\mathrm{mod}\, 4). \end{cases}$$

**Theorem 22.1.1.** *(A) Let $V$ be a compact, irreducible, and symmetric space of Type I, whose motion group is not special Lie. Then unique elements of the group $\pi_*(V)\otimes\mathbf{R}$ realizable by totally geodesic spheres are the following (here $k = k(n)$), viz.,*

*(1) if $V = \mathrm{SU}(2n)/S(U(n)\times U(n)), k \geqslant 3, Q_{2k}(V) = \oplus_{1\leqslant\alpha\leqslant k}\mathbf{R}(x_{2\alpha})$, then $\{x_{2\alpha}\}$, where $1 \leqslant \alpha \leqslant k$,*

*(2) if $V = \mathrm{SO}(2n/S(\mathrm{O}(n)\times \mathrm{O}(n)), k \geqslant 6, Q_{2k}(V) = \oplus_{1\leqslant\alpha\leqslant k/2}\mathbf{R}(x_{4\alpha})$, then $\{x_{4\alpha}\}$, where $4 \leqslant 4\alpha \leqslant f_7(16n)$,*

*(3) if $V = \mathrm{Sp}(2n)/(\mathrm{Sp}(n)\times \mathrm{Sp}(n)), n = 2s, k \geqslant 8, Q_{2k}(V) = \oplus_{1\leqslant\alpha\leqslant k/2}\mathbf{R}(x_{4\alpha})$, then $\{x_{4\alpha}\}$, where $4 \leqslant 4\alpha \leqslant f_3(4n)$,*

*(4) if $V = \mathrm{SU}(n)/\mathrm{SO}(n), k \geqslant 5, Q_{2k-1}(V) = \oplus_{1\leqslant\alpha\leqslant(k-1)/2}\mathbf{R}(x_{4\alpha+1})$, then $\{x_{4\alpha+1}\}$, where $5 \leqslant 4\alpha+1 \leqslant f_6(8n)$,*

*(5) if $V = \mathrm{SU}(2n)/\mathrm{Sp}(2n), k \geqslant 5, Q_{2k+1}(V) = \oplus_{1\leqslant\alpha\leqslant k/2}\mathbf{R}(x_{4\alpha+1})$, then $\{x_{4\alpha+1}\}$, where $5 \leqslant 4\alpha+1 \leqslant f_2(4n)$,*

*(6) if $V = \mathrm{SO}(2n)/U(n), k \geqslant 5, Q_{2k}(V) = \oplus_{\alpha\leqslant(k-1)/2}\mathbf{R}(x_{4\alpha+2})$, then $\{x_{4\alpha+2}\}$, where $2 \leqslant 4\alpha+2 \leqslant f_1(2n)$,*

*(7) if $V = \mathrm{Sp}(2n)/U(n), k \geqslant 5, Q_{2k}(V) = \oplus_{1\leqslant\alpha\leqslant(k-1)/2}\mathbf{R}(x_{4\alpha+2})$, then $\{x_{4\alpha+2}\}$, where $2 \leqslant 4\alpha+2 \leqslant f_5(8n)$.*

*(B) Let $V$ be a compact, irreducible, and symmetric space of Type I, whose motion group is special Lie (see Table 1). Then unique elements $x \in H^*(V;\mathbf{R})$ realizable by totally geodesic spheres are*

*1)* $x_2$ in $H^*(\mathrm{Ad}\, E_6/T^1\cdot \mathrm{Spin}(10))$;
*2)* $x_9$ in $H^*(E_6/F_4)$;
*3)* $x_2$ in $H^*(\mathrm{Ad}\, E_7/T^1\cdot E_6)$.

**22.2. Proof of the Classification Theorem. Relation between the Number of Linearly Independent Fields on Spheres and that of the Elements of Homotopy Groups Realizable by Totally Geodesic Spheres**

In §21.3, we have represented the group SU($2p$) as a smooth submanifold in the sphere $S^{8p^2-1}$ of radius $\sqrt{2p}$. Similarly, we embed the group SO($2p$) into the sphere $S^{4p^2-1}$ of radius $\sqrt{2p}$. Central plane sections of SU($2p$) and SO($2p$) will be defined as the intersection $\Pi_s \cap \mathfrak{G}$, where $\Pi_s$ is a subspace of dimension $s$, which passes through the origin. We prove that the maximum possible dimension of the sphere $S^{s-1} = \Pi_s \cap \mathfrak{G}$, which is a central plane section and, therefore, a totally geodesic sphere in the group $\mathfrak{G}$, cannot exceed the values $s(4p)$ for SU($2p$) and $s(2p)$ for SO($2p$), where $s(r)$ is the maximal number of linearly independent vector fields on the sphere $S^{r-1}$. Moreover, these values are attained. A similar estimation $s - 1 \leqslant s(4p)$ also holds for the group Sp($2p$).

Let $S^{s-1} = \Pi_s \cap \mathrm{SU}(2p)$. (In the following, we will speak, for brevity, of $C$-sections.) It is clear that there exists such an element $g_0 \in \mathrm{SU}(2p)$ that the translation $g \to g_0 g$ of SU($2p$) on itself carries $S^{8p^2-1}$ into another $C$-section $g_0 S^{s-1}$ which contains both elements $E_{2p}$ and $-E_{2p} \in \mathrm{SU}(2p)$. Then $g_0 S^{s-1}$ is covered with geodesics which are minimal in $S^{8p^2-1}$ and join the two poles $E_{2p}$ and $-E_{2p}$. These trajectories are also geodesics in SU($2p$), and their length is $\pi\sqrt{2p}$. It follows (from Lemma 21.2.1 and Lemma 21.1 in [276]) that each of them does meet a totally geodesic submanifold $G^{\mathbf{C}}_{2p,p}$; hence, the sphere $g_0 S^{s-1}$ intersects the Lie algebra su($2p$) in the equator $'S^{s-2}$. We can select an orthogonal basis $A_1, \ldots, A_{s-1}$, all of whose vectors $A_i$ belong to $'S^{s-2}$, in the plane $'\Pi_{s-1} = \Pi_s \cap \mathrm{su}(2p)$ determining the sphere $'S^{s-2}$. Since $A_i^2 = -E_{2p}$, it easily follows from the condition of orthogonality that $A_i A_j + A_j A_i = 2\delta_{ij} E_{2p}$; hence, the required estimate.

The proof for the groups SO($2p$) and Sp($2p$) is carried out exactly in the same manner.

In §21.2, we have constructed the spheres $S^{2k-1} \subset \mathrm{SU}(2^k)$ and thereby obtained $C$-sections in the group SO($2^{k+1}$).

Denote by $q$ the largest dimension of the spheres $S^s$ that are $C$-sections in SO($2^{k+1}$), and realize generators in $H^*(\mathrm{SO}(2^{k+1}); \mathbf{R})$.

Then

(1) $s(2^{k+1}) = 2k + 1 > q = 2k - 1, \quad k \equiv 0 (\mathrm{mod}\, 4)$,

(2) $s(2^{k+1}) = 2k + 1 = q = 2k + 1, \quad k \equiv 1 (\mathrm{mod}\, 4)$,

(3) $s(2^{k+1}) = 2k + 3 = q = 2k + 3, \quad k \equiv 2 (\mathrm{mod}\, 4)$,

and

(4) $s(2^{k+1}) = 2k + 2 > q = 2k + 1, \quad k \equiv 3 (\mathrm{mod}\, 4)$.

Thus, the approximate remarkable equality holds, viz., $f_i(n) \cong s(n) - i$ for $n = 2^{k+1}$.

It can be seen from the proof of Proposition 21.3.1 that only the greatest element $2^p$ is realized as $C$-sections of the elements $\{1, 2, 3, \ldots, 2^p\}$ in the group $\pi_{2(k-p)+1}(\mathrm{SU}(2^k)), 0 \leqslant p \leqslant k-1$. Considering the embedding $j\colon \mathrm{SU}(2^k) \to \mathrm{SO}(2^{k+1})$, we obtain that the following elements are realized by $C$-sections in the group $\mathrm{SO}(2^{k+1})$ for $k \geqslant 3$, $s = 2(k-p)+1$, viz.,

(1) $2^p \in \mathbf{Z} = \pi_{s_p}(\mathrm{SO}(2^{k+1}))$, $s_p \equiv 3 (\mathrm{mod}\, 8)$, $0 \leqslant p \leqslant k-2$,

(2) $2^{p+1} \in \mathbf{Z} = \pi_{s_p}(\mathrm{SO}(2^{k+1}))$, $s_p \equiv 7 (\mathrm{mod}\, 8)$, $0 \leqslant p \leqslant k-2$,

(3) $1 \in \mathbf{Z} = \pi_{2k+1}(\mathrm{SO}(2^{k+1}))$ if $k \equiv 0 (\mathrm{mod}\, 4)$; then $s_0 = (2k+1) \equiv 1 (\mathrm{mod}\, 8)$,

and

(4) $\pi_{s_p}(\mathrm{SO}(2^{k+1})) = 0$, $s_p \equiv 5 (\mathrm{mod}\, 8)$, $0 \leqslant p \leqslant k-2$.

However, the above spheres do not totally exhaust the geodesic ones in $\mathrm{SO}(2^{k+1})$, which realize nontrivial cycles. Let $k \equiv 2 (\mathrm{mod}\, 4)$. Then it follows from Lemma 21.7.2 that there exists an embedding $S^{2k+3} \to \mathrm{SO}(2^{k+1})$ realizing the generator $x_{2k+3}$ in $H^*(\mathrm{SO}(2^{k+1}); \mathbf{R})$, and which is a $C$-section.

**Lemma 22.2.1.** *Consider the manifold* $V = \mathrm{SU}(2n)/S(U(n) \times U(n))$, *where* $2^k \leqslant 2n < 2^{k+1}, k \geqslant 3, s_p = 2(k-p), 0 \leqslant p \leqslant k-1$. *Then the elements* $2^p$ *are realized by totally geodesic spheres in the groups* $\pi_{s_p}(V)$.

**Proof.** Consider the manifold $V' = G^{\mathbf{C}}_{2^k, 2^{k-1}}$ realized in the group $\mathrm{SU}(2^k)$ as the intersection $\mathrm{SU}(2^k) \cap \mathrm{su}(2^k)$. Let $S^3, \ldots, S^{2k+1}$ be totally geodesic spheres realizing $2^p$ in the groups $\pi_{2(k-p)+1}(\mathrm{SU}(2^k)), 0 \leqslant p \leqslant k-1$, and which are $C$-sections. All the spheres $S^{2\alpha+1} (1 \leqslant \alpha \leqslant k)$ are contained in the space of minimal geodesics $\Omega(\mathrm{SU}(2^k); E_{2^k}; -E_{2^k})$ (see [276]). Each $S^{2\alpha+1}$ cuts a totally geodesic sphere $S^{2\alpha}$ out of $G^{\mathbf{C}}_{2^k, 2^{k-1}}$. Due to Theorem 23.3 from [276], $S^{2\alpha}$ determines $2^p$ in the groups $\mathbf{Z} = \pi_{2\alpha}(G^{\mathbf{C}}_{2^k, 2^{k-1}})$ for $\alpha = k-p, k \geqslant 1$.

We now consider the manifold $V = G^{\mathbf{C}}_{2n,n}$, where $2^k \leqslant 2n < 2^{k+1}$, and determine the mapping $j\colon G^{\mathbf{C}}_{2^k, 2^{k-1}} \to G^{\mathbf{C}}_{2n,n}$ by setting $j(A) = (iE_{n-2^{k-1}}) \oplus (-iE_{n_2^{k-1}}) \oplus A$. Since the submanifold $j(G^{\mathbf{C}}_{2^k, 2^{k-1}})$ is totally geodesic, all $S^2, \ldots, S^{2k}$ are transformed into totally geodesic spheres realizing nontrivial elements of the groups $\pi_{2\alpha}(G^{\mathbf{C}}_{2n,n}), 1 \leqslant \alpha \leqslant k$.

Q.E.D.

To study symmetric spaces of Type I, which are different from $G^{\mathbf{C}}_{2n,n}$, we shall resort to orthogonal Bott periodicity. Consider the group $\mathrm{SO}(n)$, and put $n = 2^{k+1}, k \geqslant 3$, distinguishing from $\mathrm{SO}(n)$ a smooth submanifold $\Omega_1(n)$ consisting of all complex structures $I$ in the space $\mathbf{R}^n$. Take the space of minimal geodesics $\gamma$ in $\mathrm{SO}(n)$, joining $E$ to $-E$. Then it is homeomorphic to $\Omega_1(n)$ (see [276]). Fix the elements $\{I_1, I_2, \ldots, I_{p-1}\} \in \Omega_1(n)$ such that $I_\alpha I_\beta + I_\beta I_\alpha = -2\delta_{\alpha\beta} E$, and determine the space $\Omega_p(n)$ as the set of all such complex structures $I$ that $II_\alpha + I_\alpha I = 0$ for $1 \leqslant \alpha \leqslant p-1$. We obtain the embeddings $\Omega_p(n) \to \Omega_{p-1}(n) \to \cdots \to \Omega_1(n) \to \mathrm{SO}(n)$. It was proved in [276] that each space $\Omega_p(n)$ is a totally geodesic submanifold in the group $\mathrm{SO}(n)$, the space of minimal geodesics which join $I_q$ to $-I_q$

in $\Omega_q(n)$ being homeomorphic to $\Omega_{q+1}(n)$ for $0 \leqslant q < p$. There arises an embedding $\Omega_{q+1} \to \Omega\Omega_q$, $\Omega\Omega_q$ denoting the loop space. The manifolds $\Omega_p(n)$ are diffeomorphic to the following symmetric spaces, viz.,

$$\Omega_0(n) = \mathrm{SO}(2^{k+1}), \quad \Omega_1(n) = \mathrm{SO}(2^{k+1})/U(2^k),$$

$$\Omega_2(n) = U(2^k)/\mathrm{Sp}(2^k), \quad \Omega_3(n) = \bigcup_s \mathrm{Sp}(2^k)/(\mathrm{Sp}(2s) \times \mathrm{Sp}(2^k - 2s)),$$

$$\Omega_4(n) = \mathrm{Sp}(2^{k-1}), \quad \Omega_5(n) = \mathrm{Sp}(2^{k-1})/U(2^{k-2}),$$

$$\Omega_6(n) = U(2^{k-2})/O(2^{k-2}), \quad \Omega_7(n) = \bigcup_s \mathrm{SO}(2^{k-2})/S(O(s) \times O(2^{k-2} - s)),$$

$$\Omega_8(n) = O(2^{k-3}).$$

We see that the manifolds $\Omega_p$ contain symmetric spaces of the types of interest with nonspecial motion groups. Consider the following values $m_p$, viz., $m_0 = 1, m_1 = 2, m_2 = m_3 = 4, m_4 = m_5 = m_6 = m_7 = 8, m_8 = 16$. Let $p \neq 0, p \not\equiv 2 (\operatorname{mod} 4)$. Then one important statement is valid, viz., the embedding $\Omega_{p+1}(n) \to \Omega\Omega_p(n)$ induces a homotopy group isomorphism in dimensions not exceeding $\frac{n}{m_{p+1}} - 3$, whereas the embedding $\Omega_1(n) \to \Omega\Omega_0(n)$ induces the corresponding isomorphism in dimensions not less or equal to $n-4$. In the case where $p \equiv 2 (\operatorname{mod} 4)$, the estimate is somewhat more complicated, and will be given where necessary.

**Lemma 22.2.2.** *Consider the manifold* $\Omega_1(2^{k+1}) = \mathrm{SO}(2^{k+1})/U(2^k)$. *Let* $k \geqslant 3, s_p = 2(k-p), -1 \leqslant p \leqslant k-1$. *Then the following elements of the groups* $\pi_{s_p}(\Omega_1(2^{k+1}))$ *are realized by totally geodesic spheres, viz.,*

*(1)* $2^p \in \mathbf{Z}$ *if* $s_p \equiv 2 (\operatorname{mod} 8)$, $p \geqslant 0$,

*(2)* $2^{p+1} \in \mathbf{Z}$ *if* $s_p \equiv 6 (\operatorname{mod} 8)$, $p \geqslant 0$,

*(3)* $1 \in \mathbf{Z}_2$ *if* $s_0 = (2k) \equiv 0 (\operatorname{mod} 8)$; *then* $k \equiv 0 (\operatorname{mod} 4)$,

*and*

*(4)* $1 \in \mathbf{Z}$ *if* $s_{-1} = (2k+2) \equiv 6 (\operatorname{mod} 8)$; *then* $k \equiv 2 (\operatorname{mod} 4)$.

**Proof.** Consider the submanifold $\Omega_1$ and totally geodesic spheres in the group $\mathrm{SO}(2^{k+1})$, which are $C$-sections. The equators of each of these are contained in $\Omega_1$ and associated with the original spheres under the periodicity isomorphism. The description of the elements in cases (1)–(4) is obtained from the corresponding case of the $C$-sections in $\mathrm{SO}(2^{k+1})$ (see above).

Q.E.D.

**Lemma 22.2.3.** *Consider the manifold* $\Omega_2(2^{k+1}) = U(2^k)/\mathrm{Sp}(2^k)$.

*Let* $k \geqslant 5, s_p = 2(k-p) - 1, -1 \leqslant p \leqslant k-1$.

*Then the following elements are realized in the groups* $\pi_{s_p}(\Omega_2(2^{k+1}))$ *by totally geodesic spheres, viz.,*

*(1)* $2^p \in \mathbf{Z}$ *if* $s_p \equiv 1 (\operatorname{mod} 8)$, $p \geqslant 0$,

*(2)* $2^{p+1} \in \mathbf{Z}$ *if* $s_p \equiv 5 (\operatorname{mod} 8)$, $p \geqslant 0$,

*(3)* $1 \in \mathbf{Z}_2$ *if* $s_0 = (2k-1) \equiv 7 (\mathrm{mod}\, 8)$; *then* $k \equiv 0 (\mathrm{mod}\, 4)$,

*and*

*(4)* $1 \in \mathbf{Z}$ *if* $s_{-1} = (2k+1) \equiv 5 (\mathrm{mod}\, 8)$; *then* $k \equiv 2 (\mathrm{mod}\, 4)$.

**Proof.** Since the equator, i.e., the sphere of the sphere $S^{2(k-p)}$ of the sphere $S^{2(k-p)+1}$ consists of such elements $g$ that $g^2 = -E$, we can choose $2(k-p)+1$ anticommuting structures $A_1^{(p)}, A_2^{(p)}, \ldots, A_{2(k-p)+1}^{(p)}$ on $S^{2(k-p)}$. To determine the space $\Omega_2$, we need to select some complex structure. Fix the element $I_1 = A_{2(k-p)+1}^{(p)}$, and consider on $S^{2(k-p)}$ the equator $S^{2(k-p)-1}$ generated by $A_1^{(p)}, \ldots, A_{2(k-p)}^{(p)}$. Then $S^{2(k-p)-1} \subset \Omega_2(2^{k+1})$. The geodesics $\gamma$ joining $I_1$ to $-I_1$ along $S^{2(k-p)}$ are minimal in $\mathrm{SO}(2^{k+1})$; therefore, the sphere $S^{2(k-p)-1}$ determines the same element in the group $\pi_{s_p}(\Omega_2(2^{k+1}))$ as the $S^{2(k-p)}$ in the group $\pi_{2(k-p)}(\Omega_1(2^{k+1}))$ for $s_p \leqslant 2^{k-1} - 3$. Repeating the same construction for each $p$, $-1 \leqslant p \leqslant k-1$ ($p = -1$ only provided that $k \equiv 2 (\mathrm{mod}\, 4)$), we obtain the required set of totally geodesic spheres in $\Omega_2(2^{k+1})$.

**Lemma 22.2.4.** *Consider the manifold* $\Omega_3(2^{k+1})$, *and distinguish from it a component of greatest dimension* $\Omega_3^0(2^{k+1}) = \mathrm{Sp}(2^k)/\mathrm{Sp}(2^{k-1}) \times \mathrm{Sp}(2^{k-1})$. *Let* $k \geqslant 8, s_p = 2(k-p)-2, -1 \leqslant p \leqslant k-2$. *Then the following elements are realized by totally geodesic spheres in the groups* $\pi_{s_p}(\Omega_3^0(2^{k+1}))$, *viz.,*

*(1)* $2^p \in \mathbf{Z}$ *if* $s_p \equiv 0 (\mathrm{mod}\, 8)$, $p \geqslant 0$,

*(2)* $2^{p+1} \in \mathbf{Z}$ *if* $s_p \equiv 4 (\mathrm{mod}\, 8)$, $p \geqslant 0$,

*(3)* $1 \in \mathbf{Z}_2$ *if* $s_0 = (2k-2) \equiv 6 (\mathrm{mod}\, 8)$; *then* $k \equiv 0 (\mathrm{mod}\, 4)$,

*and*

*(4)* $1 \in \mathbf{Z}$ *if* $s_{-1} = (2k) \equiv 4 (\mathrm{mod}\, 8)$; *then* $k \equiv 2 (\mathrm{mod}\, 4)$.

**Proof.** The space $\Omega\Omega_2$ consists of an infinite number of components and is still more and more accurately approximated with the increase of $k$ by the disconnected space $\Omega_3(2^{k+1})$ containing a finite number of components. It suffices to confine ourselves to $\Omega_3^0(2^{k+1})$ and the corresponding component $\Omega^0\Omega_2(2^{k+1})$. Since it is basically similar to the proof of Lemma 22.2.3, we will not dwell on the further argument.

We omit the proofs of Lemma 22.2.5–22.2.8, because they are different from the above reasoning only in the necessity to closely watch the indices of the geodesics filling the fixed spheres, and also to reject duly superfluous nonsymmetric components of loop spaces.

**Lemma 22.2.5.** *Consider the manifold* $\Omega_4(2^{k+1}) = \mathrm{Sp}(2^{k-1})$. *Let* $k \geqslant 7, s_p = 2(k-p)-3, -1 \leqslant p \leqslant k-2$. *Then the following elements of the groups* $\pi_{s_p}(\Omega_4(2^{k+1}))$ *are realized by totally geodesic spheres, viz.,*

*(1)* $2^p \in \mathbf{Z}$ *if* $s_p \equiv 7 (\mathrm{mod}\, 8)$, $p \geqslant 0$,

*(2)* $2^{p+1} \in \mathbf{Z}$ *if* $s_p \equiv 3 (\mathrm{mod}\, 8)$, $p \geqslant 0$,

*(3)* $1 \in \mathbf{Z}_2$ *if* $s_0 = (2k-3) \equiv 5 (\mathrm{mod}\, 8)$; *then* $k \equiv 0 (\mathrm{mod}\, 4)$,

*and*

(4) $1 \in \mathbf{Z}$ *if* $s_{-1} = (2k-1) \equiv 3 \bmod 8)$; *then* $k \equiv 2 (\bmod 4)$.

**Lemma 22.2.6.** *Consider the manifold* $\Omega_5(2^{k+1}) = \mathrm{Sp}(2^{k-1})/U(2^{k-2})$. *Let* $k \geqslant 6, s_p = 2(k-p)-4, -1 \leqslant p \leqslant k-3$. *Then the following elements in the groups* $\pi_{s_p}(\Omega_5(2^{k+1}))$ *are realized by totally geodesic spheres, viz.,*

(1) $2^p \in \mathbf{Z}$ *if* $s_p \equiv 6 (\bmod 8)$, $p \geqslant 0$,

(2) $2^{p+1} \in \mathbf{Z}$ *if* $s_p \equiv 2 (\bmod 8)$, $p \geqslant 0$,

(3) $1 \in \mathbf{Z}_2$ *if* $s_0 = (2k-4) \equiv 4 (\bmod 8)$; *then* $k \equiv 0 (\bmod 4)$,

*and*

(4) $1 \in \mathbf{Z}$ *if* $s_{-1} = (2k-2) \equiv 2 (\bmod 8)$; *then* $k \equiv 2 (\bmod 4)$.

**Lemma 22.2.7.** *Consider the manifold* $\Omega_6(2^{k+1}) = U(2^{k-2})/O(2^{k-2})$. *Let* $k \geqslant 6, s_p = 2(k-p)-5, -1 \leqslant p \leqslant k-3$. *Then the following elements in the groups* $\pi_{s_p}(\Omega_6(2^{k+1}))$ *are realized by totally geodesic spheres, viz.,*

(1) $2^p \in \mathbf{Z}$ *if* $s_p \equiv 5 (\bmod 8)$, $p \geqslant 0$,

(2) $2^{p+1} \in \mathbf{Z}$ *if* $s_p \equiv 1 (\bmod 8)$, $p \geqslant 0$,

(3) $1 \in \mathbf{Z}_2$ *if* $s_0 = (2k-5) \equiv 3 (\bmod 8)$; *then* $k \equiv 0 (\bmod 4)$,

*and*

(4) $1 \in \mathbf{Z}$ *if* $s_{-1} = (2k-3) \equiv 1 (\bmod 8)$; *then* $k \equiv 2 (\bmod 4)$.

**Lemma 22.2.8.** *Consider the manifold* $\Omega_7(2^{k+1})$, *and distinguish from it the component of greatest dimension* $\Omega_7^0(2^{k+1}) = \mathrm{SO}(2^{k-2})/S(O(2^{k-3}) \times \mathrm{O}(2^{k-3}))$ *in it. Let* $k \geqslant 8, s_p = 2(k-p)-6, -1 \leqslant p \leqslant k-4$. *Then the following elements in the groups* $\pi_{s_p}(\Omega_7^0(2^{k+1}))$ *are realized by totally geodesic spheres, viz.,*

(1) $2^p \in \mathbf{Z}$ *if* $s_p \equiv 4 (\bmod 8)$, $p \geqslant 0$,

(2) $2^{p+1} \in \mathbf{Z}$ *if* $s_p \equiv 0 (\bmod 8)$, $p \geqslant 0$,

(3) $1 \in \mathbf{Z}_2$ *if* $s_0 = (2k-6) \equiv 2 (\bmod 8)$; *then* $k \equiv 0 (\bmod 4)$,

*and*

(4) $1 \in \mathbf{Z}$ *if* $s_{-1} = (2k-4) \equiv 0 (\bmod 8)$; *then* $k \equiv 2 (\bmod 4)$.

We now turn to the proof of Theorem 22.1.1. It remains to show only the uniqueness up to the scalar multiplier of the above realizations for the groups $\pi_*(V) \otimes \mathbf{R}$.

**Lemma 22.2.9.** *Let* $S^{2p}$ *be a totally geodesic sphere in an arbitrary compact group* $\mathfrak{G}$. *Consider the subgroup* $A(S^{2p})$ *(see Theorem 18.1), with* $p \leqslant 4$. *Then* $A(S^{2p})$ *is isomorphic to the group* $\mathrm{Spin}(2p+1)$.

**Proof.** This fact is analogous to the statement already used above that $A(S^{2p-1})$ is isomorphic either to $\mathrm{Spin}(2p)$ or $\mathrm{Spin}_i(2p)$.

**Lemma 22.2.10.** *Let* $p \geqslant 4$, *and* $p = [\log_2 n]$. *Assume that the totally geodesic sphere* $S^{2p}$ *is contained in the group* $\mathrm{SO}(n)$. *Then there exists an embedding* $\mathrm{Spin}(2p+1) \to \mathrm{SO}(2^p)$.

This is proved similarly to Lemma 21.7.4.

**Lemma 22.2.11.** *Let $p \geqslant 4, p = [1 + \log_2 n]$. Assume that a totally geodesic sphere $S^{2p}$ exists in the group* $\mathrm{Sp}(2n)$. *Then there exists an embedding* $\mathrm{Spin}(2p+1) \to \mathrm{Sp}(2^p)$.

**Lemma 22.2.12.** *Let $p = [\log_2 n]$, and $p \geqslant 4$. Assume that $\frac{p(p-1)}{2}$ is odd if $p$ is even, and $\frac{p(p+1)}{2}$ is odd if $p$ is odd. Then there exists no totally geodesic sphere $S^{2p}$ in the group* $\mathrm{SO}(n)$.

**Proof.** Assume the contrary, viz., let $S^{2p}$ be contained in $\mathrm{SO}(n)$. Then, according to Lemma 22.2.10, there exists an embedding $\mathrm{Spin}(2p+1) \to \mathrm{SO}(2^p) \subset \mathrm{SU}(2^p)$. The composition embedding $\mathrm{Spin}(2p+1) \to \mathrm{SU}(2^p)$ generates an irreducible representation $C'$ of the group $\mathrm{Spin}(2p+1)$ of dimension $2^p$. It is known that there exists only one faithful representation of $\mathrm{Spin}(2p+1)$ of dimension $2^p$ (see [297]), viz., the spinor representations $\bar{S}$. Thus, $C'$ is equivalent to $\bar{S}$. It was established in [348] that $\bar{S}$ possesses two bilinear invariants $C$ and $E'$ depending on whether $p$ is even or not. More precisely, if $p$ is even, then

$$C = (\bar{S}(V))\, C\, (\bar{S}(V))^T, \quad V \in \mathrm{Spin}(2p+1), \; C^T = (-1)^{\frac{p(p-1)}{2}} C;$$

and if $p$ is odd, then

$$E' = (\bar{S}(V))\, E'\, (\bar{S}(V))^T, \quad E'^T = (-1)^{\frac{p(p+1)}{2}} \cdot E'.$$

Due to the assumptions of the lemma, the representation $C' \sim \bar{S}$ either preserves the skew-symmetric invariant $C$ if $p$ is even, or the skew-symmetric invariant $E'$ if $p$ is odd. Hence, the subgroup $\mathrm{Spin}(2p+1)$ is contained in both cases in the subgroup $\mathrm{Sp}(2^p) = \mathrm{SU}(2^p) \cap \mathrm{Sp}(2^p; \mathbf{C})$, whence $\mathrm{Spin}(2p+1) \subset \mathrm{SO}(2^p) \cap \mathrm{Sp}(2^p) = U(2^{p-1})$, and we have obtained a faithful representation of $\mathrm{Spin}(2p+1)$ in $U(2^{p-1})$, which is impossible.

Q.E.D.

**Lemma 22.2.13.** *Let $p = [1 + \log_2 n], p \geqslant 4$. Assume that $\frac{p(p-1)}{2}$ is even if $p$ is even, and that $\frac{p(p+1)}{2}$ is even if $p$ is odd. Then there exists no totally geodesic sphere $S^{2p}$ in the group* $\mathrm{Sp}(2n)$.

The proof is similar to the previous one.

**Proof of Theorem 22.1.1 (A), (1).** It remains to show that a nontrivial element $x \in \pi_*(V) \otimes \mathbf{R}$ not proportional to any $x_{2\alpha}, 1 \leqslant \alpha \leqslant k$, cannot be realized by any totally geodesic sphere. Assume the contrary, viz., let a totally geodesic sphere $S^{2p}$, where $p \geqslant k+1$, be contained in $V$. Embedding $V$ as a totally geodesic submanifold in the group $\mathrm{SU}(2n)$ (meaning the Cartan model), we obtain a totally geodesic sphere $S^{2p}$ in $\mathrm{SU}(2n)$. By Lemma 22.2.9, there arises a faithful representation of the group $\mathrm{Spin}(2p+1)$. On the other hand, $2n \geqslant 2^p$ (see [297]), i.e., $2^{k+1} > 2n \geqslant 2^p \geqslant 2^{k+1}$, which is impossible.

**Proof of Theorem 22.1.1 (A), (6).** Consider the manifold $\Omega_1(2^{k+1})$, and construct an embedding $i : \Omega_1(2^{k+1}) \to \Omega_1(2^{k+1}+n')$, where $n' = 2s$, for which we have to fix in the group $\mathrm{SO}(n')$ a complex structure $I'_1$, and put $i(g) = g \oplus I'_1, g \in \Omega_1(2^{k+1})$. Since $\Omega_1(2^{k+1}+n') = \mathrm{SO}(2n)/U(n), n = 2^{k+s}$, we have constructed an embedding $\mathrm{SO}(2^{k+1})/U(2^k) \to \mathrm{SO}(2n)/U(n)$, where $2^k \leqslant n < 2^{k+1}$. The embedding $\Omega_1(2^{k+1}) \to \Omega_1(2n)$ induces a homotopy group isomorphism for $i \leqslant 2^{k+1} - 4$. Thus, the following elements of the group $\pi_*(V) \otimes \mathbf{R}$ are realized by totally geodesic spheres in the space $V = \mathrm{SO}(2n)/U(n)$, where $2^{k-1} \leqslant n < 2^k$ and we have replaced $k$ by $k-1, k \geqslant 4$ (see Lemma 22.2.2), viz.,

(a) $\{x_2, x_6, x_{10}, \ldots, x_{2k-2}\}$, $k \equiv 0 (\mathrm{mod}\, 4)$,

(b) $\{x_2, x_6, x_{10}, \ldots, x_{2k-4}\}$, $k \equiv 1 (\mathrm{mod}\, 4)$,

(c) $\{x_2, x_6, x_{10}, \ldots, x_{2k-2}\}$, $k \equiv 2 (\mathrm{mod}\, 4)$,

and

(d) $\{x_2, x_6, x_{10}, \ldots, x_{2k}\}$, $k \equiv 3 (\mathrm{mod}\, 4)$.

Consider a Cartan model of the space $V$ in the group $\mathrm{SO}(2n)$. Hence, if we admit the existence in $V$ of a totally geodesic sphere $S^{2p}, p \geqslant k+1$, then it is transformed into a totally geodesic sphere in $\mathrm{SO}(2n)$. Since $p \geqslant k+1 \geqslant 5$, due to Lemma 22.2.9, there exists an embedding $\mathrm{Spin}(2p+1) \to \mathrm{SO}(2n)$, and then $2^{k+1} \leqslant 2^p \leqslant 2n$, which is impossible. In cases (a), (c), and (d), the theorem is proved. It remains to consider case (b), but here the problem of the generator $x_{2k}$ is not clear. Assume that $V$ contains a totally geodesic sphere $S^{2k}$. Then we obtain a totally geodesic sphere $S^{2k}$ in $\mathrm{SO}(2n)$, which is impossible according to Lemma 22.2.12.

**Proof of Theorem 22.1.1 (A), (5).** Let us consider the manifold $\Omega_2(2^{k+1}) = U(2^k)/\mathrm{Sp}(2^k), k \geqslant 5$, which contains a totally geodesic submanifold $\Omega'_2(2^{k+1}) = \mathrm{SU}(2^k)/\mathrm{Sp}(2^k)$ containing, in turn, all totally geodesic spheres from Lemma 22.2.3 for $s_p > 1$. A totally geodesic embedding $i : \Omega'_2(2^{k+1}) \to \Omega'_2(4n) = \mathrm{SU}(2n)/\mathrm{Sp}(2n)$, where $4n = 2^{k+1} + n', n' = 4s$, is constructed easily. The actual realization is carried out by means of Lemma 22.2.3. Further, it follows from Lemma 22.2.9 that there exists in $V$ no totally geodesic sphere $S^{2p}$, where $p \geqslant k+1$. The theorem is thereby proved in the case where $k \not\equiv 0 (\mathrm{mod}\, 4)$, but then the problem of the generator $X_{2k+1}$ is not clear.

Assume that the space $V = \mathrm{SU}(2n)/\mathrm{Sp}(2n)$ contains a totally geodesic sphere $S^{2k+1}$. Consider a Cartan model $V \subset \mathrm{SU}(2n)$, and translate $S^{2k+1}$ so that it contains the unit element $e$. Denote by $S^{2k} \subset S^{2k+1}$ the totally geodesic equator in $S^{2k+1}$, i.e., the collection of points $\{\gamma\left(\frac{1}{2}\right)\}$, where $\gamma(0) = e, \gamma(1) = g_0$, $g_0$ being the point "diametrically opposite" to the point $e$. Then, for any point $g \in S^{2k}$, we have $g^2 = g_0$, and due to Lemma 19.3, all such points belong to the same coset relative to the subgroup $\mathrm{Sp}(2n)$, i.e., we obtain a totally geodesic sphere $S^{2k}$ in $\mathrm{Sp}(2n)$, which is impossible according to Lemma 22.2.13.

The proof of items (2), (3), (4), and (7) of Theorem 22.1.1.(A) is in many respects similar to above cases (1), (5), and (6), and we do not give it here with all the particulars.

The second part (B) of Theorem 22.1.1 has been proved in [88] by applying the same technique and methods. Theorem 22.1.1 is thus proved.

It is interesting to note that the function $f_i(n)$ constructed above has been shown to be directly related to the maximal number of linearly independent vector fields on spheres, and it is "universal" for any compact, irreducible, and symmetric space in spite of great algebraic differences between spaces on Types I and II.

# CHAPTER 5

# VARIATIONAL METHODS FOR CERTAIN TOPOLOGICAL PROBLEMS

## §23 Bott Periodicity from the Dirichlet Multidimensional Functional Standpoint

### 23.1. Explicit Description of the Bott Periodicity Isomorphism for the Unitary Group

Particularly interesting is the study of multidimensional functionals (e.g., Dirichlet and volume functionals) on the functional spaces of smooth manifold mappings to compact Lie groups and symmetric spaces. Thus, it is shown in the present section that known Bott periodicity for the groups $U(n)$ (resp. $\mathrm{O}(n), \mathrm{Sp}(n)$) most naturally arises in investigating stationary (critical) points of the Dirichlet functional on the space $\Pi(D^2, U(n))$ of mappings of the two-dimensional disc $D^2$ with fixed and sufficiently symmetric boundary (resp. $\Pi(D^8, \mathrm{O}(n))$, $\Pi(D^8, \mathrm{Sp}(n))$) into a group, which permits us to obtain the corresponding periodicity isomorphism not in several, as is usually done, but in one "step." It turns out, meanwhile, that Bott periodicity is carried to the manifold of the Dirichlet multidimensional functional absolute minima.

First, we describe Bott periodicity in convenient form. Consider the unitary group $\mathrm{SU}(2m)$. Denote by $\Omega(\mathrm{SU}(2m); E_{2m} - E_{2m})$ (with $E_{2m} \in \mathrm{SU}(2m)$ being the identity transformation) the space of piecewise smooth paths emanating from the point $E_{2m}$ and ending at the point $-E_{2m}$, and the complete space of all continuous paths from $E_{2m}$ to $-E_{2m}$, by $\Omega^*(\mathrm{SU}(2m); E_{2m}, -E_{2m})$. Then the embedding $\Omega \to \Omega^*$ is a homotopy equivalence (see [276]).

Consider a subspace $\widetilde{\Omega}$ of the space $\Omega^*(\mathrm{SU}(2m); E_{2m}, -E_{2m})$, which is formed by all minimal geodesics from $E_{2m}$ to $-E_{2m}$. It is known that $\widetilde{\Omega}$ is homeomorphic to the complex Grassmannian manifold $G^{\mathbf{C}}_{2m,m}$ (see §§21,

22). Besides, the natural embedding

$$G^{\mathbf{C}}_{2m,m} \to \Omega(\mathrm{SU}(2m); E_{2m}, -E_{2m}) \to \Omega^*(\mathrm{SU}(2m); E_{2m}, -E_{2m})$$

induces a homotopy group isomorphism in dimensions not exceeding $2m$ (see [276]). On the other hand, it follows from the standard fibration $U(m) \to V_{2m,m} \to G^{\mathbf{C}}_{2m,m}$ that the boundary operator $\partial : \pi_i(G^{\mathbf{C}}_{2m,m}) \to \pi_{i-1}(U(m))$ is an isomorphism for all $i \leqslant 2m$. Finally, combining the two isomorphisms, we just obtain the periodicity isomorphism

$$\pi_{i-1}(U(m)) \overset{\partial}{\underset{\approx}{\leftarrow}} \pi_i(G^{\mathbf{C}}_{2m,m}) \underset{\approx}{\rightarrow} \pi_i[\Omega^*(\mathrm{SU}(2m);\ E_{2m},\ -E_{2m})] \approx \pi_{i+1}(\mathrm{SU}(2m)).$$

To write the isomorphism chain explicitly, suppose that $f_{i-1} : S^{i-1} \to U(m)$ is a continuous mapping representing the class $[f] \in \pi_{i-1}(U(m))$. Construct $f_{i+1} : S^{i+1} \to \mathrm{SU}(2m)$ from $f_{i-1}$, represent the group SU(2) as that of the matrices $\{p\}$, where $p = \begin{Vmatrix} \alpha & \beta \\ -\bar{\beta} & \bar{\alpha} \end{Vmatrix}$, $|\alpha|^2 + |\beta|^2 = 1$, and distinguish a two-dimensional disc $'D^2$ from SU(2) by the condition $p \in {}'D^2$, $\beta \in \mathbf{R}$, $\beta \geqslant 0$. Embed $'D^2$ into $\mathrm{SU}(2m)$ as follows

$$'i\colon\ p \to p \otimes E_m = \begin{Vmatrix} \alpha E_m & \beta E_m \\ -\beta E_m & \bar{\alpha} E_m \end{Vmatrix},$$

and then obtain a curve $'\gamma(\beta) = \{p(\alpha, \beta) | \alpha = i\tau, \tau \in \mathbf{R}, \tau \geqslant 0\}$ on $'D^2$. We put $\gamma(\beta) = {}'i[{}'\gamma(\beta)]$ and represent points of the space $G^{\mathbf{C}}_{2m,m}$ as invariant planes associated with the eigenvalue $\lambda = i$ associated with the operators $g : \mathbf{C}^{2m} \to \mathbf{C}^{2m}$, $g \in \mathrm{SU}(2m)$, $g^2 = -E_{2m}$. Then we have $\gamma^2 = -E_{2m}$ for the point $\gamma = \gamma(\beta)$, i.e., $\gamma(\beta) \in G^{\mathbf{C}}_{2m,m} \subset \mathrm{SU}(2m)$ for $0 \leqslant \beta \leqslant 1$. Consider the set of elements $g$ such that

$$g = g(\sigma,\ i\tau,\ \beta) = [E_m \oplus f_{i-1}^{-1}(\sigma)] \cdot [p(i\tau,\ \beta) \otimes E_m] \cdot [E_m \oplus f_{i-1}(\sigma)]$$

in the manifold $G^{\mathbf{C}}_{2m,m}$, where $\sigma \in S^{i-1}$, $f_{i-1}(\sigma) \in U(m)$. For $\beta = 1$, we obtain a mapping of the sphere $S^{i-1}$ given by the formula

$$\sigma \longmapsto \begin{Vmatrix} 0 & f_{i-1}(\sigma) \\ -f_{i-1}^{-1}(\sigma) & 0 \end{Vmatrix},$$

meanwhile, the set $\{g(\sigma, i\tau, \beta)\}$ for $0 \leqslant \beta \leqslant 1$ is representable as the image of the sphere $S^i$, where $\{g(\sigma, i\tau, \beta)\} \subset G^{\mathbf{C}}_{2m,m}$, $\partial S^i = S^{i-1}$ and $\partial : \pi_i(G^{\mathbf{C}}_{2m,m}) \to \pi_{i-1}(U(m))$. We now consider the set of points $\{g(\sigma, \alpha, \beta)\}$, where $g(\sigma, \alpha, \beta) = [E_m \oplus f_{i-1}^{-1}(\sigma)][p(\alpha, \beta) \otimes E_m] \times [E_m \oplus f_{i-1}(\sigma)]$.

Then $\{g(\sigma, \alpha, \beta)\}$ can be represented as the image of the sphere $S^{i+1}$ under a certain mapping $f_{i+1} : S^{i+1} \to \{g(\sigma, \alpha, \beta)\} \subset \mathrm{SU}(2m)$. Thus, if $f_{i+1}(\sigma) \in U(m)$, then $f_{i+1}(S^{i+1}) \subset \mathrm{SU}(2m)$, and it obviously follows from the above construction that the correspondence $f_{i-1} \to f_{i+1}$ is just what generates the required periodicity isomorphism.

Note that

$$g(\sigma, \alpha, \beta) = \left\| \begin{matrix} \alpha E_m & \beta f_{i-1}(\sigma) \\ -\beta f_{i-1}^{-1}(\sigma) & \bar{\alpha} E_m \end{matrix} \right\|.$$

If $m = 2$, then the identity mapping $f_3(\sigma) = \left\| \begin{smallmatrix} x & y \\ -\bar{y} & \bar{x} \end{smallmatrix} \right\|, |x|^2 + |y|^2 = 1$ can be taken as the original mapping $f_3 : S^3 \to \mathrm{SU}(2)$. We now have $[f_3] = 1 \in \pi_3(\mathrm{SU}(2))$. Passing to $m = 2^2, 2^3, 2^4, \ldots$, we obtain the mappings $f_{2k+1} : S^{2k+1} \to \mathrm{SU}(2^k)$, where $[f_{2k+1}] = 1 \in \pi_{2k+1}(\mathrm{SU}(2^k)), k \geqslant 1$.

## 23.2. Unitary Periodicity and One-Dimensional Functionals

Note that in the above (usual) approach, the unitary periodicity isomorphism splits up into the decomposition of two, each of which increases by unity the dimension of the sphere and, therefore, that of the homotopy group. That the required increase of the dimension of the sphere by 2 is obtained on carrying out the two steps agrees with the method of the usual proof of the periodicity theorem, in which the one-dimensional action and length functionals are employed, defined on the functional space of mappings of the interval $D^1$ (i.e., one-dimensional disc). Consider this process in more detail. Let $D^1$ be fixed, where $\partial D^1 = S^0$ (zero-dimensional sphere, pair of points). Then $\Omega^*(\mathrm{SU}(2m); E_{2m}, -E_{2m}) = \Pi_1$ is the space of continuous mappings $f$ of $D^1$ to the group $\mathrm{SU}(2m)$, under which $f|_{S^0} \equiv i_0|_{S^0}$, where $i_0(S^0) = \{E_{2m}, -E_{2m}\}$. The action functional on the space $\Pi_1' = \Omega(\mathrm{SU}(2m); E_{2m}, -E_{2m})$ is defined by $E_0^1(\omega) = \int_0^1 \left\| \frac{d\omega}{dt} \right\|^2 dt$, where $= \omega(0) = E_{2m}$, $\omega(1) = -E_{2m}$. The length functional $l_0^1(\omega) = \int_0^1 \left\| \frac{d\omega}{dt} \right\| dt$ is naturally related to the latter. It is known that the critical (stationary) points of the length functional 1 can be discovered by means of the functional $E$. The set of points on which the action functional $E$ and, therefore, 1 attain an absolute minimum is a subspace in $\Pi_1'$, homeomorphic to the Grassmannian manifold $G_{2m,m}^{\mathbf{C}}$; hence, as it follows from Morse theory, the 2m-dimensional skeleton of the space $\Pi_1$ is the homotopy equivalent to that of the space $(G)_{2m,m}^{\mathbf{C}}$. In other words, we can say that the "analytic" part of the periodicity isomorphism is in the isomorphism $\pi_i(G_{2m,m}^{\mathbf{C}}) \approx \pi_i(\Pi_1') \approx \pi_i(\Pi_1) \approx \pi_{i+1}(\mathrm{SU}(2m))$, since the second step $\pi_i(G_{2m,m}^{\mathbf{C}}) \approx \pi_{i-1}(U(m))$ is a corollary of a purely homotopy statement having no relation to $E$ and its critical points.

It turns out that the periodicity isomorphism can be obtained in "one step," but not in two, if we make use of a convenient two-dimensional and not a one-dimensional functional.

## 23.3. The Periodicity Theorem for a Unitary Group is Based on the Dirichlet Functional Two-Dimensional Extremals

We obtain the periodicity isomorphism, considering two-dimensional functionals on a particular mapping space. We take the circle $S_0^1 = \left\| \begin{matrix} \alpha E_m & 0 \\ 0 & \bar{\alpha} E_m \end{matrix} \right\|$, $|\alpha| = 1$, which is a one-parameter subgroup in the unitary group $\mathrm{SU}(2m)$, and fix it on the analogy of the one-dimensional case where the zero-dimensional sphere $S^0 = \{E_{2m}, -E_{2m}\}$ is fixed in $\mathrm{SU}(2m)$. Further, let $D^2$ be the two-dimensional disc with boundary $S^1$ in its standard Euclidean metric. Fix the mapping $j_0 : S^1 \to \mathrm{SU}(2m)$ transforming $S^1$ isometrically into the circle $S_0^1$ in the group. Let $i'' : D^2 \to' D^2$ be a smooth mapping preserving a system of polar coordinates on the disc, so that $i'i''(s) = j_0(s)$ for $s \in S^1$.

Denote the space of all continuous mappings $f : D^2 \to \mathrm{SU}(2m)$ by $\Pi_2$, so that $f|_{S^1} \equiv j_0$. It has the homotopy type of a (cell) CW-complex. Consider the subspace $\Pi_2' \subset \Pi_2$ formed by all mappings $f$ belonging to the Sobolev space $H_1^2(D^2)$. Recall the definition of this functional space.

Let $G$ be a domain in $\mathbf{R}^\nu(x^1, \ldots, x^\nu)$. We will say that a function $u : G \to \mathbf{R}$ is of the class $H_m^p(G)$ if and only if (1) $u \in L_p(G)$ and (2) there exist functions $r_\alpha \in L_p(G)$, $\alpha = (\alpha_1, \ldots, \alpha_\nu), 0 \leqslant |\alpha| \leqslant M$, such that $\int_G g(x) r_a(x)dx = (-1)^{|\alpha|} \int_G [D^\alpha g(x)] u(x) dx$ for any $g \in C_0^\infty(G)$, where $|\alpha| = \alpha_1 + \cdots + \alpha_\nu$,

$$D^\alpha g = \frac{\partial^{|\alpha|}(g)}{(\partial x^1)^{\alpha_1} \ldots (\partial x^\nu)^{\alpha_\nu}}.$$

If $m = 1$, then $\alpha = \alpha_1$, and if $f : D^2 \to \mathrm{SU}(m)$, then $f \in H_1^2(D^2)$ in that and only that case where all the coordinate functions specifying the mapping belong to the functional space $H_1^2(D^2)$. We replace the requirement for the piecewise smoothness of a mapping $f$ in the one-dimensional case by requiring that $f$ belong to the class $H_1^2(D^2)$.

We now define the Dirichlet functional $D : \Pi_2' \to \mathbf{R}$ on the space $\Pi_2'$, associating each mapping $f \in \Pi_2'$ with the value of the Dirichlet integral $D[f]$ on $f$. Recall the definition of the Dirichlet functional. Let $u \in H_m^p(G)$. Then the functions $r_\alpha(x)$, $\alpha = (\alpha_1, \ldots, \alpha_\nu)$, in general, defined nonuniquely (but whose nonuniqueness is immaterial) are called the *derivatives* of the function $u$ and are denoted by $D^\alpha(u)$ or $u_{,\alpha}$. If $\alpha = 0$, then $u_{,\alpha} = u$. Now, suppose that $M$ and $V$ are two Riemannian manifolds with metric tensors $g_{ij}(x)$, $x \in M$, $\hat{g}_{\alpha\beta}(v)$, $v \in V$. Each mapping $f : V \to M$ is associated with mixed-type tensors, where $f \in H_1^2[V, M]$. For example, $x_\alpha^i = x^i_{,\alpha}$, where $x^i$ are the local curvilinear coordinates of points $x = f(v) \in M$, and the differentiation is understood in the above sense. Denote the complete covariant derivative of the mixed tensor by $\nabla_\alpha$. Putting $(x_\alpha^i, y_\beta^i) = \hat{g}^{\alpha\beta} g_{ij} x_\alpha^i y_\beta^j$, we define the scalar product of two tensors $x_\alpha^i$and $y_\beta^j$. Now, let $f \in H_1^n[V, M]$. We set $D[f] = \int_V \left[\frac{(x_\alpha^i x_\beta^j)}{n}\right]^{n/2} dv$, where $dv$ is the volume element on $V$, and $n = \dim V$.

A mapping $f \in H_1^n[V, M]$ is said to be *harmonic* if $\delta D[f; \eta] = 0$ for any vector field $\eta(f)$ of the class $H_1^n$, defined on the image $f(V)$. The corresponding Euler equation for the Dirichlet functional $D[f]$ is of the form $\nabla^\alpha \nabla_\alpha x^i = 0$. Here we take the two-dimensional disc $D^2$ as $V$. Then $\hat{g}^{\alpha\beta}(v) = \delta^{\alpha\beta}$, and $D[f]$ takes the form

$$D[f] = \frac{1}{2} \int\limits_V [(x_1^i, x_1^j) + (x_2^i, x_2^j)]\, dv = \frac{1}{2} \int\limits_V g_{ij} (x_1^i x_1^j + x_2^i x_2^j)\, dv.$$

The first variation $\delta D$ is of the form $\delta D[f; \eta] = \int_V (x_\alpha^i, \nabla_\beta \eta^i) dv$. If $D^2$ is parametrized by means of Euclidean coordinates $u, v$, then

$$D[f] = \frac{1}{2} \int\limits_V [(x_u, x_u) + (x_v, x_v)]\, du\, dv,$$

$$x = (x^1, \ldots, x^p), \quad p = \dim M,$$

$$\delta D[f; \eta] = \int\limits_V \left[\left(\frac{D\eta}{du}, x_u\right) + \left(\frac{D\eta}{dv}, x_v\right)\right] du\, dv, \quad \eta \in H_1^2.$$

We also consider the area functional $\mathrm{vol}_2 f$ on the space $\Pi_2'$, associating each mapping $f \in \Pi_2'$ with the value of the integral

$$\int\limits_V \sqrt{\det \Omega}\, du\, dv, \text{ where } \Omega = \begin{Vmatrix} (x_u, x_u) & (x_u, x_v) \\ (x_u, x_v) & (x_v, x_v) \end{Vmatrix}.$$

As we have seen earlier, the inequality $\mathrm{vol}_2 F \leqslant D[f]$ holds, with the equality holding if, and only if the mapping $f$ is generalized conformal (see [289]). Note that, here, too, an analogy is valid with the one-dimensional case of the space of paths with fixed ends, viz., the action $E_0^1$ and length $l$ functionals are related by a similar inequality $[l_0^1(\omega)]^2 \leqslant E_0^1(\omega)$, with the equality holding if and only if $\omega$ is the minimal geodesic from the point $\omega(0)$ to $\omega(1)$.

As well as the action functional, the Dirichlet functional permits us to reject the "superfluous" fixed volume functional points, or all those mappings that are different from a harmonic one by continuous change of parameters in the two-dimensional disc, which certainly leaves the value of the area functional unaltered, but does alter the Dirichlet functional.

Note for the following that the isomorphism $\beta_2 : \pi_s(\Pi_2) \approx \pi_{s+2}(\mathrm{SU}(2m))$ holds, and that the space $\Pi_2$ is homotopy equivalent to the space $\widetilde{\Pi}_2$ of all continuous mappings $S^2 \mapsto \mathrm{SU}(2m)$ with a fixed point.

**Theorem 23.3.1.** *Consider the group* $\mathrm{SU}(2m)$ *and the space* $\Pi_2$ *of all continuous mappings of the two-dimensional disc* $D^2$ *to the group* $\mathrm{SU}(2m)$ *with fixed boundary* $S^1$*, and also to the subspace* $\Pi_2' \subset \Pi_2$ *of Sobolev mappings. Let* $W$ *be the set of all points* $f$ *in the space* $\Pi_2'$ *at which the Dirichlet functional* $D[f]$ *attains an absolute minimum. Then (a)* $W$ *is homeomorphic to the group* $U(m)$*, and (b) the embedding* $i : W \to \Pi_2' \to \Pi_2$ *induces the* *homotopy group isomorphism $(i_*)_s : \pi_s(U(m)) \to \pi_s(\Pi_2)$ for $s \leqslant 2m$; therefore, the 2m-dimensional skeleton of $\Pi_2$ is homotopy-equivalent to that of $U(m)$, and the composition $\beta_2 \circ (i_*)_s : \pi_s(U(m)) \overset{\approx}{\to} \pi_{s+2}(\mathrm{SU}(2m))$ is a Bott periodicity isomorphism for $s \leqslant 2m$.*

**Proof.** Consider the two-dimensional sphere

$$S_0^2 = \begin{Vmatrix} \alpha E_m & \beta E_m \\ -\beta E_m & \bar{\alpha} E_m \end{Vmatrix}, \quad \beta \in \mathbb{R}, \quad |\alpha|^2 + |\beta|^2 = 1,$$

in the group $\mathrm{SU}(2m)$. One of the half-spheres, viz., $\beta \geqslant 0$, coincides with the disc $D_0^2$ whose embedding into $\mathrm{SU}(2m)$ was carried out above. The circle $S_0^1$ is the equator $\beta = 0$ of the sphere $S_0^2$. Since the embedding of the sphere $S_0^2 \to \mathrm{SU}(2m)$ is extended to the monomorphism $\mathrm{SU}(2) \to \mathrm{SU}(2m)$, it is obvious that $S_0^2$ is a geodesic submanifold in $\mathrm{SU}(2m)$ and locally minimal *a fortiori*. Therefore, $D_0^2$, too, is a totally geodesic submanifold in $\mathrm{SU}(2m)$. Consider the set $W'$ of totally geodesic discs $D^2(x) \subset \mathrm{SU}(2m)$ of the form $D^2(x) = xD_0^2x^{-1}$, where $x \in \mathrm{SU}(2m)$ and $xsx^{-1} \equiv s$ for any $s \in S_0^1$.

**Lemma 23.3.1.** *The set $W'$ is homeomorphic to the space $U(m)$.*

**Proof.** Let $D^2(x) \in W'$. Then $xs = sx$ for any $s \in S_0^1$. Since $S_0^1 = \{\alpha E_m \oplus \bar{\alpha} E_m\}$, it follows that $x = A \oplus D$, where $A, D \in U(m)$, i.e., $x = (E_m \oplus DA^{-1}) \cdot (A \oplus A) = x_1 \cdot (A \oplus A)$, $x_1 = E_m \oplus DA^{-1}$. Since $(A \oplus A) \cdot d = d \cdot (A \oplus A)$ for any $d \in D_0^2$ and $A \in U(m)$, we have

$$D^2(x) = D^2(x_1) = \begin{Vmatrix} \alpha E_m & \beta C \\ -\beta C^{-1} & \bar{\alpha} E_m \end{Vmatrix}, \quad C = DA^{-1}.$$

Since $\beta \geqslant 0$, the matrix $C$ is determined uniquely by this condition. Thus, we have associated each disc $D^2(x)$ with an element $C \in U(m)$, $C = C[D^2(x)]$. Let $C[D^2(x)] = C[D^2(x')]$. Then it is obvious that $x' \cdot x^{-1} \in \{A \oplus A\}$; therefore, $D^2(x)$ and $D^2(x')$ coincide. Conversely, if $C \in U(m)$, then $C = C[D^2(x)]$, where $x = E_m \oplus C$, i.e., the correspondence $D^2(x) \to C[D^2(x)]$ constructed by us is just the required homeomorphism between $W'$ and $U(m)$.

Q.E.D.

Construct an embedding $i : U(m) \to \Pi_2'$. Let $g \in U(m)$. Then the disc

$$D^2(E_m \oplus g) = \begin{Vmatrix} \alpha E_m & \beta g \\ -\beta g^{-1} & \bar{\alpha} E_m \end{Vmatrix}$$

is constructed from this element uniquely, and, if $g_1 \neq g_2$, then $D^2(E_m \oplus g_1) \cap D^2(E_m \oplus g_2) = S_0^1$. Let $i_0 : D^2 \to D_0^2$ be the above fixed mapping $i_0 = i' \circ i''$, $i_0|_{S^1} \equiv j_0 : S^1 \to S_0^1$. Put $i(g)(\xi) = (E_m \oplus g) \cdot i_0(\xi) \cdot (E_m \oplus g^{-1})$, where $\xi \in D^2$. It is clear that $i : g \longmapsto i(g)$ is the required embedding $U(m) \to \Pi_2'$.

It follows from Lemma 23.3.1 that the set of mappings $i(U(m)) \subset \Pi'_2$ coincides with that of the form $\{\mathrm{Ad}_x \circ i_0\}$, where the elements $x$ range over the whole group $G = \{A \oplus A\} \subset U(2m), G \approx U(m)$, i.e., $i(U(m))$ is the orbit of the point $i_0 \in \Pi'_2$ under the adjoint action of the group $G$ on the mapping set $\Pi_2$.

**Lemma 23.3.2.** *The homomorphism $\beta^2 \circ (i_*)_s : \pi_s(U(m)) \to \pi_{s+2}(\mathrm{SU}(2m))$ coincides with the periodicity homorphism.*

**Proof.** Let $f : S^s \to U(m)$, $f \in [f] \in \pi_s(U(m))$, $\sigma \in S^s$. Then

$$[(\beta_2)_\# \circ (i_\#)_s](f)(\sigma) = \bigcup_{\sigma \in S^s} (D^2[E_m \oplus f(\sigma)]) = \begin{Vmatrix} \alpha E_m & \beta f(\sigma) \\ -\beta f^{-1}(\sigma) & \bar{\alpha} E_m \end{Vmatrix}.$$

It immediately follows from §23.1 and Morse theory that the homomorphism $\beta_2 \circ (i_*)_s$ is precisely the periodicity isomorphism if $s \leqslant 2m$. Since $\beta_2$ is an isomorphism in any dimension, it follows that homomorphism $(i_*)_s : \pi_s(U(m)) \to \pi_s(\Pi_2)$ is also an isomorphism for $s \leqslant 2m$; therefore, the $2m$-dimensional skeleton of the space $\Pi_2$ is homotopy equivalent to that of the space $i(U(m))$.

Q.E.D.

Thus, the embedding $i : U(m) \to \Pi_2$ satisfies all the necessary requirements. It remains to show that $i(U(m)) = W$.

Consider the Euclidean space $\mathbf{R}^{8m^2}$ associated with the complex space $\mathbf{C}^{4m^2}$ of all $2m \times 2m$-matrices, which is endowed with the form $\varphi(A, B) = \mathrm{Re\,Spur}(AB^*)$. Then, as has already been noted, the group $\mathrm{SU}(2m)$ is isometrically embeddable into the sphere $S^{8m^2-1}$ of radius $\sqrt{2m}$ as a smooth submanifold on which the Euclidean metric induces the special bi-invariant Riemannian metric on $\mathrm{SU}(2m)$. Therefore, many metric relations on $\mathrm{SU}(2m)$ can be conveniently treated from the standpoint of the ambient sphere $S^{8m^2-1}$. The first corollary of this remark is, e.g., that there exist no infinitesimal variations of the disc $D_0^2$ in $\mathrm{SU}(2m)$, leaving the boundary $S_0^1 = \partial D_0^2$ fixed, and such that the perturbed disc $\widetilde{D}_0^2$ is minimal in $\mathrm{SU}(2m)$, but not totally geodesic. In fact, let such a variation exist. Note that the circumference $S_0^1 \subset \mathrm{SU}(2m) \subset S^{8m^2-1}$ is that of a great circle in $S^{8m^2-1}$, whereas $D_0^2$ is a central plane section of the ambient sphere by a three-dimensional plane through the origin in $R^{8m^2}$. Since $\widetilde{D}_0^2$ is not totally geodesic in $\mathrm{SU}(2m)$, it is not totally geodesic in $S^{8m^2-1}$ either, and not obtained from $D_0^2$ by rotation about $S_0^1$. It is then obvious that its area is strictly greater than that of $D_0^2$ to linear approximation, or $\delta(\mathrm{vol}_2) > 0$, and $\widetilde{D}_0^2$ is not minimal, which is contrary to the assumption. Thus, the variation of any disc $D^2(x) \in W'$, which is fixed on its boundary circle, either leaves $D^2(x)$ totally geodesic in the group, and then the variation is reduced to a rotation of the disc about $S_0^1$, or violates its minimality and *a fortiori* total geodesicity.

**Lemma 23.3.3.** *The relation $i(U(m)) \subset W$ is valid.*

**Proof.** Since its mapping $f \in i(U(m))$ is of the form $f = \mathrm{Ad}_x \circ i_0$, $x \in G$, it suffices to verify that the point $i_0$ is an absolute minimium for the Dirichlet functional. Since $\mathrm{SU}(2m) \subset S^{8m^2-1}$, and $D_0^2$ is a central plane section of the sphere $S^{8m^2-1}$, $i_0$ is an absolute minimum for the functional $\mathrm{vol}_2$. Any minimal vector being also harmonic, $i_0$ is a critical point for the Dirichlet functional, too. (Note that the generalized harmonicity of $i_0$ is obvious from the latter's construction.) Because $\mathrm{vol}_2[f] \leqslant D[f]$ always, it is clear that $i_0$ is the absolute minimum for the Dirichlet functional.

Q.E.D.

**Lemma 23.3.4.** *The relation $i(U(m)) \supset W$ holds; therefore, $i(U(m)) = W$, where $W$ is the set of absolute minima for the Dirichlet functional.*

**Proof.** Let $f : D^2 \to \mathrm{SU}(2m)$, $f|_{S^1} \equiv j_0$ be an absolute minimum of the functional $D$. It follows from Lemma 23.3.3 that the value of $D$ at absolute minimum points is $D[i_0]$, and that it equals $\mathrm{vol}_2[i_0]$. Since $\mathrm{vol}_2[f] \leqslant D[f] = D[i_0] = \mathrm{vol}_2[i_0]$, we have $\mathrm{vol}_2[f] \leqslant \mathrm{vol}_2[i_0]$; but since this relation can be considered in the metric on the sphere $S^{8m^2-1}$, it is obvious that $\mathrm{vol}_2[f] = \mathrm{vol}_2[i_0]$, and then $f(D^2) \subset S^{8m^2-1}$ is a plane central section. Besides, the mapping $f$ is harmonic. Extend the totally geodesic disc $f(D^2)$ to the sphere $S^2$ which is totally geodesic in $S^{8m^2-1}$ and *a fortiori* in the group $\mathrm{SU}(2m)$. We obtain two totally geodesic spheres $S_0^2$ and $\widetilde{S}^2$, $S_0^2 \cap \widetilde{S}^2 \supset S_0^1 \ni E_{2m}$, in $\mathrm{SU}(2m)$. The subgroups $G_1$ and $G_2$ which are isomorphic to the group $\mathrm{SU}(2)$ are minimal subgroups containing $S_0^2$ and$\widetilde{S}_2$. The embeddings $\alpha_1 : G_1 \to \mathrm{SU}(2m)$, $\alpha_2 : G_2 \to \mathrm{SU}(2m)$ determine two faithful representations of $\mathrm{SU}(2)$ in $\mathrm{SU}(2m)$. Since the rank of $\mathrm{SU}(2)$ is unity, we can assume that the circumference $S_0^1$ is the image of a one-dimensional torus $T^1 \approx S^1 \subset \mathrm{SU}(2)$, $S_0^1 \subset T^{2m-1}$, where $T^{2m-1}$ is the maximal torus in $\mathrm{SU}(2m)$. Because two representations $j_1$ and $j_2$ coincide on $T^1$, they are equivalent, i.e., there exists an element $x \in \mathrm{SU}(2m)$ such that $j_1 = \mathrm{Ad}_x \circ j_2$. Two spheres $S_0^2$ and $x\widetilde{S}^2x^{-1}$ embedded into the group $G_1$ can be made coincident by another inner automorphism $\mathrm{Ad}_{x_1}$. We then obtain in $S_0^2$ two geodesics $S_0^1$ and $x_1 x S_0^1 x^{-1} x_1^{-1}$. Therefore, there exists an element $x_2 \in G_1$ such that $S_0^1 \equiv x_2 x_1 x S_0^1 x^{-1} x_1^{-1} x_2^{-1}$, and accordingly, the automorphism $\mathrm{Ad}_y$, where $y = x_2 x_1 x$ reduces the mapping $f$ into the mapping $i_0$, leaving the circumference $S_0^1$ fixed, i.e., $f \in i(U(m))$.

Q.E.D.

The proof of Theorem 23.3.1 is thereby completed.

Note that all the points of the set $W$ are not simply minimal for the area and the Dirichlet functional, but even "totally geodesic." This fact was also valid in the one-dimensional case, but there the minimality of any trajectory mechanically implied its geodesicity. However, in the two-dimensional case, the total geodesicity does not at all follow from the minimality of a disc.

Moreover, unique totally geodesic discs $D^2$ with boundary $S_0^1$ are those of the set $W'$. In other words, if $f \in \Pi_2^1$ is a critical point for the Dirichlet functional and, moreover, if $f(D^2)$ is a totally geodesic disc, then $f \in W$.

### 23.4. The Periodicity Theorem for an Orthogonal Group is Based on the 8-Dimensional Dirichlet Functional Extremals

Consider the Euclidean space $\mathbf{R}^{p^2}$, real $p \times p$-matrices, and $\varphi(A, B) = \mathrm{Spur}(A \cdot B^T)$. Then the group $\mathrm{SO}(p)$ is isometrically embeddable into the sphere $S^{p^2-1}$ of radius $\sqrt{p}$ as a smooth submanifold on which the underlying Euclidean metric $\varphi(A, B)$ induces the bi-invariant Riemannian metric. The Lie algebra $\mathrm{so}(p)$ of the group $\mathrm{SO}(p)$ can be embedded into $\mathbf{R}^{p^2}$ as a linear subspace of matrices $X$, $X^T = -X$, and the intersection $\mathrm{so}(p) \cap \mathrm{SO}(p)$ is a compact, symmetric space $\mathrm{SO}(p)/U(p/2)$ if $p$ is even. Denote $\mathrm{so}(p) \cap \mathrm{SO}(p)$ by $\Omega_1(p)$. It is then obvious that $\Omega_1(p)$ consists of only those elements $g \in \mathrm{SO}(p)$ for which $g^2 = -E$, and $\Omega_1(p)$ is the set of complex structures in $\mathrm{SO}(p)$ (see [276]).

Now, we put $p = 16r$. Then there exist eight anticommuting complex structures in the group, which we denote by $I_1, I_2, \ldots, I_8$; $I_s^2 = -E$; $I_s I_k + I_k I_s = 0$, $k \neq s$. All the vectors $I_s (1 \leqslant s \leqslant 8)$ lie in the plane $\mathrm{so}(16r)$, and, due to the anticommutative condition, they are pairwise orthogonal. Besides, each $I_s$ is orthogonal to the vector $E \in \mathrm{SO}(16r)$; therefore, the sphere $S_0^8 = \{x \in \mathrm{SO}(16r) | x = a^0 E + a^1 I_1 + \cdots + a^8 I_8;\ (a^0)^2 + (a^1)^2 + \cdots + (a^8)^2 = 1\}$ is a plane section of the sphere $S^q$, where $q = 256r^2 - 1$, passing through the origin, and hence, totally geodesic both in $S^q$ and $\mathrm{SO}(16r) \subset S^q$. It is clear that $S_0^8 \cap \mathrm{so}(16r) = S_0^8 \cap \Omega_1(16r) = \bar{S}_0^7$, where $\bar{S}_0^7$ is the totally geodesic equator $a^0 = 0$. Fix the totally geodesic sphere $S_0^7 = \{x = a^0 E + a^1 I_1 + \cdots + a^7 I_7;\ (a^0)^2 + \cdots + (a^7)^2 = 1\}$ in the group $\mathrm{SO}(16r)$, with $S_0^7$ being the boundary of the totally geodesic disc $D_0^8 \subset S_0^8$, $D_0^8 = \{x \in S_0^8; a^8 \geqslant 0\}$. Let $D^8$ be the standard disc in the Euclidean metric, $S^7 = \partial D^8$, $i''$ the standard mapping of $D^8$ onto the hemisphere (which is the identity on the boundary $\partial D^8$), $i'$ the unique isometric embedding of the hemisphere $i''(D^8)$ into $\mathrm{SO}(16r)$, coinciding on $i''(S^7)$ with the fixed isometric embedding $j_0 : S^7 \to S_0^7$. Put $i_0 = i' \circ i''$, $i_0 : D^8 \to \mathrm{SO}(16r)$. Consider the space $\Pi_8$ of all continuous mappings $f : D^8 \to \mathrm{SO}(16r)$ such that $f|_{S^7} \equiv j_0$. Let $\Pi_8' \subset \Pi_8$ be the subspace of all mappings $f$ of class $H_1^8(D^8)$. Consider the volume $\mathrm{vol}_8 f = \int_{D^8} \sqrt{\det \Omega}\, dv$ and the Dirichlet $D[f] = \int_{D^8} \left[\frac{1}{8} \sum_{\alpha=1}^8 g^{ij} (x_\alpha^i \cdot x_\alpha^j)\right]^4 dv$ functionals on $\Pi_8'$. Denote the standard isomorphism $\pi_s(\Pi_8) \approx \pi_{s+8}(\mathrm{SO}(16r))$ by $\beta_8$.

**Theorem 23.4.1.** *Consider the group* $\mathrm{SO}(16r)$, *spaces* $\Pi_8$ *and* $\Pi_8'$, *and the set* $W$ *of all those points* $f$ *in* $\Pi_8$ *at which the Dirichlet functional* $D[f]$ *attains an absolute minimum. Then (a)* $W$ *is homeomorphic to the group* $\mathrm{O}(r)$, *and (b) the embedding* $i : W \to \Pi_8' \to \Pi_8$ *induces the homotopy group isomorphism* $(i_*)_s : \Pi_s(\mathrm{O}(r)) \to \pi_s(\Pi_8)$ *for* $s \leqslant r - 2$; *therefore, the* $(r-2)$*-dimensional* *skeleton of* $\Pi_8$ *is homotopy equivalent to that of* $O(r)$, *and the composition* $\beta_8 \circ (i_*)_s : \pi_s(O(r)) \xrightarrow{\approx} \pi_{s+8}(SO(16r))$ *is the Bott periodicity isomorphism for* $s \leqslant r-2$.

**Remark.** Since $\pi_2(U(2m)) = 0$, the space $\Pi_2$ is connected. Because $\pi_8(SO(16r)) = \mathbf{Z}_2$, $\Pi_8$ is disconnected and consists of two connected components. As will be seen from the proof, $W$ also consists of two components, each component of $\Pi_8$ containing one component of $W$, and is contractible just onto the latter as $r \to \infty$.

**Proof of the theorem.** Consider the set $\Omega_8$ of all complex structures $I$ in the group $SO(16r)$, anticommuting with the structures $I_1, I_2, \dots, I_7$, i.e., with each point of the six-dimensional sphere $S_0^6 \subset S_0^7\{a^0 = 0\}$. E.g., $I_8 \in \Omega_8$. It is generally known that $\Omega_8$ consists of two connected components, and is homeomorphic to the group $O(r)$ (see [276]); therefore, $\Omega_8$ is contained in the plane orthogonal to the vectors $E, I_1, \dots, I_7$. It is clear that $S_0^8 \cap \Omega_8 = \{I_8, -I_8\}$; therefore, $D_0^8 \cap \Omega_8 = I_8$ (i.e., one point).

Associate each point $x \in \Omega_8$ with a totally geodesic sphere $S_0^8(x)$ with $S_0^7$ as the equator. If $x \in \Omega_8$, then $x$ is orthogonal to the vectors $E, I_1, \dots, I_7$ ($xI_s = -I_s x, 1 \leqslant s \leqslant 7$), and $E$ to all complex structures. Therefore, the sphere spanned by the basis vectors $E, I_1, \dots, I_7, x$ is a central plane section of the sphere $S^q$, and totally geodesic in $SO(16r)$. Consider the disc $D^8(x) = \{y \in S^8(x);\, y = y^0 E + \dots + y^7 I_7 + y^8 x;\, y^8 \geqslant 0\}$ in $S^8(x)$. Then each $x \in \Omega_8$ is associated with only one totally geodesic disc $D^8(x)$ such that $\partial D^8(x) = S_0^7$, and, if $x_1 \neq x_2$, then $D^8(x_1) \cap D^8(x_2) = S_0^7$. As well as in the case of unitary periodicity, we can define an embedding $i : O(r) \cong \Omega_8 \to \Pi_8' \to \Pi_8$, since, for any disc $D^8(x)$, $x \in \Omega_8$, there exists a unique isometry $\omega(x) : i''(D^8) \to D^8(x)$, $\omega(x) \circ i''|_{S^7} \equiv j_0$; then $i(x) = \omega(x) \cdot i''$.

**Lemma 23.4.1.** *The embedding* $i : O(r) \to \Pi_8$ *induces the homotopy group isomorphism up to dimension* $r-2$.

**Proof.** Let $f : S^s \to O(r)$. Then we obtain the set $\{D^8(x)\}$, $x \in f(S^s)$, $\Pi_8' \ni i(x)$, in $SO(16r)$. Since the sphere $S_0^7$ is fixed, the set $\bar{S} = \bigcup_{x \in f(S^s)} D^8(x)$ arises in $SO(16r)$, which determines a mapping $F : S^{s+8} \to SO(16r)$ such that $F|S^s \equiv f$, where $S^s$ is the equator in the sphere $S^{s+8}$. Now, consider the sequence of zero-dimensional spheres $S_k^0 = \{I_k, -I_k\}$, $1 \leqslant k \leqslant 7$. Fixing $S_0^7$, we can construct a correspondence $\gamma_7 : x \to D^1(x)$, where $x \in \Omega_8$ and $D^1(x)$ is a minimal geodesic with mid-point $x$ from the point $I_7$ to the point $-I_7$. Then $D^1(x) \in \Omega_7$ (see [276]), and there exists a mapping $F_7 : S^{s+1} \to \Omega_7$ such that $F_7(S^{s+1}) = \bigcup_{x \in f(S^s)} D^1(x)$, $F_7|_{S^7} \equiv f$, whereas it follows from Morse theory that the mapping $f \to F_7$ determines the isomorphism $\pi_s(\Omega_8) \xrightarrow{\approx} \pi_{s+1}(\Omega_7)$. Having fixed the sphere $S_0^6$, we obtain the correspondence $\gamma_6 : y \to D^1(y)$, $y \in \Omega_7$. Meanwhile, there exists a mapping $F_6 : S^{s+2} \to \Omega_6$ such that $F_6(S^{s+2}) = \bigcup_{y \in F_7(S^{s+1})} D^1(y)$, $F_6|_{S^{s+1}} \equiv F_7$. Continuing the process, we obtain the correspondences $\gamma_7, \gamma_6, \dots, \gamma_1, \gamma_0$, where $E = I_0$. The mapping

$F_0 : S^{s+8} \to \Omega_0 = \mathrm{SO}(16)$ corresponds to $f$ under the periodicity isomorphism $F_0(S^{s+8}) = F(S^{s+8})$, since $\bigcup_{x\in f(S^7)}[\gamma_0 \circ \gamma_1 \circ \cdots \circ \gamma_7(x)] = \bar{S}$. Therefore, we can assume that $F_0 = F$, which completes the proof of the lemma, since $\pi_s(\Pi_8) \overset{\beta_8}{\approx} \pi_{s+8}(\mathrm{SO}(16r))$.

All the statements of item (b) of Theorem 23.4.1 are thereby fulfilled for the subspace $i(\mathrm{O}(r)) \subset \Pi_8$. It remains to prove that $W = i(\mathrm{O}(r))$.

**Lemma 23.4.2.** *The relation $i(\mathrm{O}(r)) \subset W$ holds.*

**Proof.** Since $i(x)D^8$ is a central plane section, the statement of the lemma is completely analogous to Lemma 23.3.3.

**Lemma 23.4.3.** *The relation $i(O(r)) = W$ holds.*

**Proof.** Let $f \in W$, i.e., let the Dirichlet functional assume its minimal value on the mapping $f$. Let $i_0 : D^8 \to D_0^8$ (see above); it is then obvious that $\mathrm{vol}_8\, i_0 = D[i_0]$. Since $\mathrm{vol}_8 f \leqslant D[f] = D[i_0] = \mathrm{vol}_8\, i_0$, it is established precisely as in the proof of Lemma 23.3.4 that the image $f(D^8)$ is a central plane section containing the sphere $S_0^7$. Let $x \in f(D^8)$, and let the vector $x$ be orthogonal to the vectors $E, I_1, \ldots, I_7$. Then $x = \gamma(\frac{1}{2})$, where $\gamma$ is a geodesic on the disc $f(D^8)$, $\gamma(0) = E$, $\gamma(1) = -E$. Since the length $l(\gamma)$ equals the length $l(\gamma')$, where the geodesic $\gamma' \subset f(D^8)$ is such that $\gamma'(0) = E$, $\gamma'(1) = -E$, $\gamma'(\frac{1}{2}) = I_1$, $\gamma$ is a minimal geodesic from the point $E$ to the point $-E$ in the group $\mathrm{SO}(16r)$; therefore, $x = \gamma(\frac{1}{2}) \in \Omega_1$ (see [276]), i.e., $x^2 = -E$. Because $x$ is orthogonal to the vectors $I_s (1 \leqslant s \leqslant 7)$, we have $\frac{1}{\sqrt{2}}(x + I_s) \in \Omega_1$, i.e., $\frac{1}{2}(x + I_s)^2 = -E$, whence $xI_s + I_s x = 0$, or $x \in \Omega_8$, but then $f \in i(O(r))$ due to $f(D^8) = D^8(x)$.

Q.E.D.

The proof of Theorem 23.4.1 is thus complete.

It is clear that a perfectly similar theorem holds also for the case of a symplectic group $\mathrm{Sp}(n)$.

In the case of the unitary periodicity, we stated that the set $i(U(m)) \subset \Pi_2$ was the orbit of a point $i_0 \in \Pi_2$ under the adjoint action of the group $G \subset U(2m)$, $G \cong U(m)$ on the mapping space $\Pi_2$. For orthogonal periodicity, there also exists such a representation for $i(\mathrm{O}(r))$, though it is not used in the proof.

**Lemma 23.4.4.** *The set $W = i(\mathrm{O}(r))$ is the orbit of a point $i_0 \in \Pi_8$ under the adjoint action of the group $G \subset \mathrm{SO}(16r)$ on the mapping space $\Pi_8$, where $G = I_8\Omega_8 \approx \mathrm{O}(r)$.*

**Proof.** It suffices to establish that, for any totally geodesic disc $D^8(x)$, $x \in \Omega_8$, there exists an element $g \in \mathrm{SO}(16r)$ such that $gI_s = I_s g$ $(1 \leqslant s \leqslant 7)$ and $gxg^{-1} = I_8$. We have for it that $g\Omega_8 g^{-1} \subset \Omega_8$ and $(gD_0^8 g^{-1}) \cap \Omega_8 = gI_8 g^{-1}$, i.e., $gD^8(x)g^{-1} = D^8(gxg^{-1})$. Let $R$ be the subgroup of all elements

$g \in SO(16r)$ such that $gI_s = I_s g (1 \leqslant s \leqslant 7)$. Suppose that $p(g) = gI_8g^{-1}$ is the natural projection $p : R \to \Omega_8$. Consider a translation $g \to I_8 g$ in the group $SO(16r)$. Let $g \in R$, $g = \exp A$, $A \in T_E(R)$. Since $gI_s = I_s g$, we have $AI_s = I_s A$.

But it is then easy to see that $I_8 g$ anticommutes with the elements $I_s (1 \leqslant s \leqslant 7)$ (see [34]), i.e., $I_8 g \in \Omega_8$, $I_8 R \subset \Omega_8$. Conversely, let $I_8 \exp A \in \Omega_8$. Then $AI_s = I_s A$ $(1 \leqslant s \leqslant 7)$, i.e., $gI_s = I_s g$, where $g = \exp A$, $g \in R$, $I_8 R \supset \Omega_8$, $\Omega_8 = I_8 R$; therefore, $p$ is a diffeomorphism, and, for any $x \in \Omega_8$, there exists an element $g \in R$ such that $x = gI_8 g^{-1}$.

Q.E.D.

It follows from the two basic theorems of the section that the mechanism of the appearance both of unitary and orthogonal periodicity is the same, and the final result depends only on the dimensions (two or eight) of the discs whose mapping spaces we are investigating.

It would be interesting to obtain direct proofs, not making use of the information related to the one-dimensional length and action functionals. Such a direct proof would immediately follow from the contractability of the $2m$-dimensional skeleton of the space $\Pi_2$ (resp. $(r-2)$-dimensional skeleton of the space $\Pi_8$) on the subspace $i(U(m))$ (resp. $i(O(r))$), which is the set of absolute minimum points for the Dirichlet functional. It is precisely the corresponding contractability theorem for the one-dimensional action functional that has made it possible to realize the isomorphism in Bott periodicity. A similar statement for multidimensional functionals is absent, and this is the main difficulty hampering a direct proof of Theorems 23.3.1 and 23.4.1.

## §24 Three Geometric Problems of Variational Calculus

### 24.1. Minimal Cones and Singular Points of Minimal Surfaces

Consider the circle $S^1$ and the direct product of two zero-dimensional spheres $S^0 \times S^0 \subset S^1$ (Fig. 49). In Fig. 49a, a one-dimensional minimal surface (two line segments) is represented whose boundary coincides with $S^0 \times S^0$. On the other hand, the one-dimensional "cone" (two diameters represented in Fig. 49b) is not obviously a minimal one-dimensional surface, since there exists a variation decreasing the length of the curve at the vertex of the "cone," the origin. The quadruple point is seen to be composed into two triple points (Fig. 49c).

Consider now the two-dimensional sphere $S^2$ and a contour in it, the direct product $S^0 \times S^1$. The two-dimensional minimal surface catenoid, the surface of revolution with boundary $S^0 \times S^1$ formed by a catenary curve, is represented in Fig. 50a. On the other hand, just as in the one-dimensional case, it is clear that the cone is not minimal from the standpoint of the two-dimensional area functional, since there exists a variation at its vertex, decreasing the area

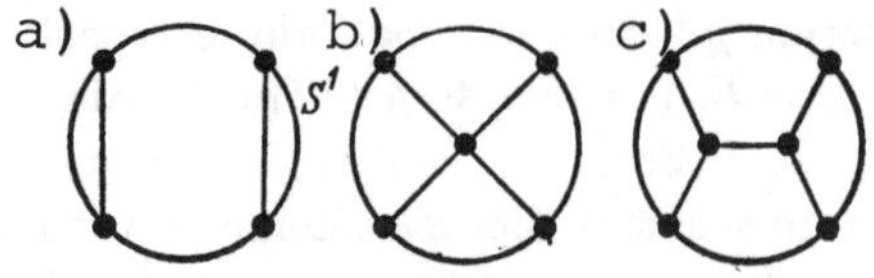


Fig. 49

(Fig. 50b). (A variation can be understood in terms of homology variations.) By comparing the two-dimensional minimal surface (the catenoid in Fig. 50a) with the one-dimensional one (two line segments in Fig. 49a), we see that the two-dimensional film sags and forms a neck whose diameter is less than that of the boundary circles. However, in the one-dimensional case, the "neck" had the same "diameter" as the boundary zero-dimensional spheres.

Consider the three-dimensional sphere $S^3$ and the "contour" in it, the two-dimensional submanifold $S^1 \times S^1 = T^2$ (torus), standardly embedded with the Euclidean-induced Riemannian metric. It can be computed (we omit the actual procedure) that the three-dimensional minimal film schematically represented in Fig. 51 has a still narrower neck than that of the two-dimensional catenoid. In other words, an interesting effect is manifest, viz., as the dimension of the boundary contour grows, the minimal film sags still more and more (Fig. 52). The exact statement of this phenomenon and corresponding formulas and computations will be given below. Intuitively it is clear that, with the growth of dimension in this monotone process, a moment will come when the minimal film with boundary $S^p \times S^q$ embedded into the sphere $S^{p+q+1}$ sags so much that it collapses and turns into a cone with the vertex at the origin. It is obvious that the vertex is a singular point of the minimal surface. In the inverse process when the dimension decreases, the resolution of singularity occurs, and the singular surface, the cone, turns into a nonsingular one, the "catenoid," on which a vanishing cycle appears.

Thus, we can expect the existence of globally minimal surfaces in sufficiently large dimensions, i.e., cones with an essential singular point in the origin, whose neighbourhood is not homeomorphic to a disc.

The question arises: In what dimensions do minimal cones in Euclidean space exist?

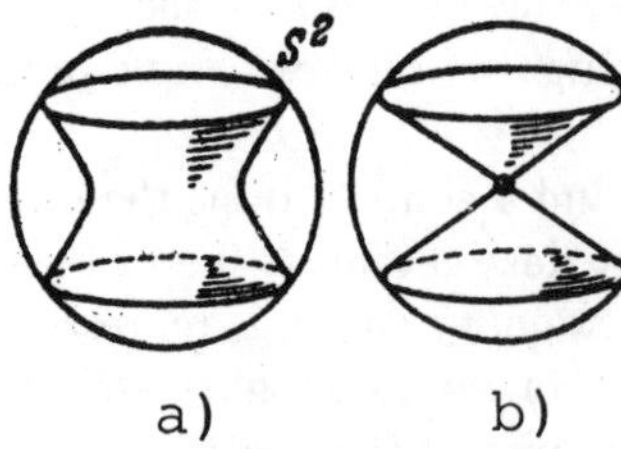


Fig. 50

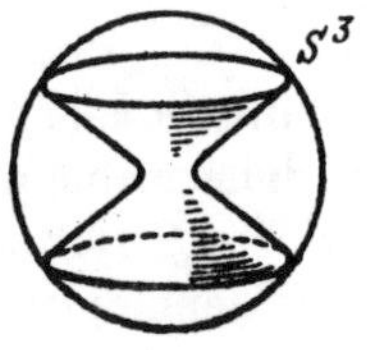


Fig. 51

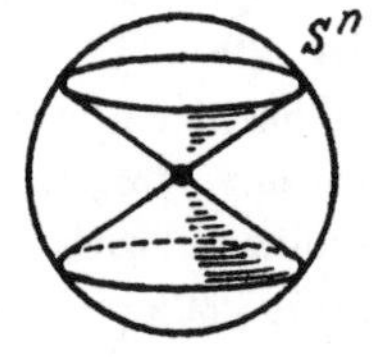


Fig. 52

It turns out that certain other important geometric and analytic problems in the theory of differential equations and the geometry of Lie groups and algebras depend on the solution of this problem. Simple examples given above show that, in small dimensions, we can expect no minimal cones different from the standard disc which is, certainly, a minimal cone. This assumption is justified.

Here, we give the known "minimal cone theorem" (see [391]).

**Theorem 24.1.1 (F. J. Almgren, J. Simons).** *Let $A^{n-2}$ be a closed, locally minimal submanifold in the standard sphere $S^{n-1}$ standardly embedded into Euclidean space $\mathbb{R}^n$, and not be standard, totally geodesic sphere (equator) $S^{n-2} \subset S^{n-1}$. Then, if $n \leqslant 7$, the cone $CA$, i.e., the $(n-1)$-dimensional surface formed by all radii from the point $O$ to the points of $A^{n-2}$, is not a globally minimal surface, or, in other words, it is not the least extremal for the $(n-1)$-dimensional volume functional considered on the class of surfaces $X \in \mathcal{O}(H_{n-2}(A))$ with fixed boundary, $A^{n-2}$.*

We briefly describe the proof (see the particulars in [391]). It suffices to construct a variation of the cone, which would be zero on the boundary (submanifold) $A^{n-2} \subset S^{n-1}$, and to decrease the volume of $CA$, for which we should study the formula for the second variation of the functional $\mathrm{vol}_{n-1}$, carried out in [391]. It is easy to verify that if $A^{n-2}$ is a locally minimal submanifold in the sphere $S^{n-1}$, then $CA \backslash O$ is also a locally minimal submanifold in $\mathbb{R}^n$ relative to variations with small supports; therefore, to discover a variation decreasing the volume of the cone, we need to consider the perturbations possessing a suffficiently large support, which can, in general, shift the cone vertex, e.g., carry out the resolution of the singularity described above. Here, we can understand by a "cone variation" the one introduced in §6, i.e., the cone and perturbed surface must, e.g., be homologous in the sense of the usual homology. In [391], cone variations are understood in the sense of the "currents."

Let $x \in A$, $t \in [0,1]$. Then we can introduce on $CA$ the coordinates $(x,t)$, where $(x,0)$ is the vertex $O$ of $CA$ for any $x \in A$, and all the points of the form $(x,1)$ belong to the boundary $A$, $A \subset S^{n-1}$. Denote by $CA_\varepsilon$ the submanifolds $CA \backslash (D^n(O,\varepsilon) \cap CA) \subset \mathbb{R}^n$, where $D^n(O,\varepsilon)$ is a ball of radius $\varepsilon$ with centre at $O$ (Fig. 53). If we fix a unit vector field $N(x,t)$ normal to the submanifold $CA$, then any smooth variation $V(x,t)$ of $CA_\varepsilon$ with fixed boundaries $A$ and $A_\varepsilon$ is uniquely determined by a smooth function $F(x,t)$ such that $F(x,1) = F(x,\varepsilon) = 0$ for all $x \in A$, i.e., $V(x,t) = F(x,t) \cdot N(x,t)$.

The problem is reduced to the study of the properties of F(x,t). Here, we only give the final result. The reasoning neccessary for its proof is of purely analytic character (see [391]).

Let $A^{n-2} \subset S^{n-1}$ be a closed, locally minimal submanifold in $S^{n-1}$, which is not a totally geodesic sphere (equator) $S^{n-2}$. Then we can choose $F(x,t)$ so that $I(V,V)$, the second variation of the volume functional, is strictly less

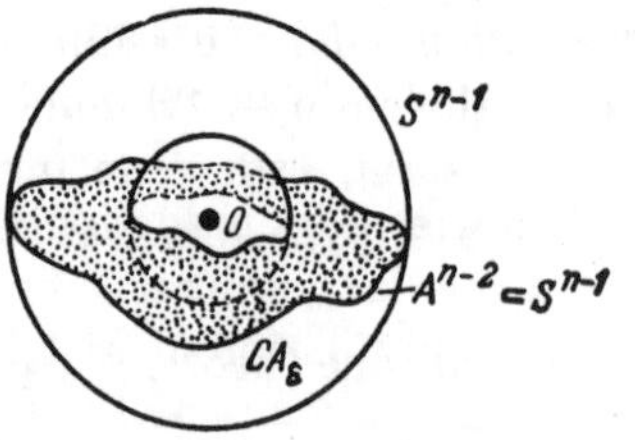


**Fig. 53**

than zero if and only if $-(n-2)+(\frac{n-3}{2})^2+(\frac{\pi}{\log\varepsilon})^2<0$.

Theorem 24.1.1 immediately follows from this auxiliary assertion, since $-5+4+(\frac{\pi}{\log\varepsilon})^2$ can be made negative by having chosen sufficiently small $\varepsilon$.

Thus, for $n \leqslant 7$, $CA$ is not minimal among all surfaces with boundary $A^{n-2}$. As is seen from the scheme of the proof and the estimate for $I(V,V)$, the proof is not valid for $n=8$. In fact, it turns out that there already exist cones in eight-dimensional Euclidean space, which below are shown to be not only locally, but also globally minimal surfaces. An example of such a cone is the direct product of two spheres $S^3(\frac{\sqrt{2}}{2}) \times S^3(\frac{\sqrt{2}}{2})$, which can be naturally embedded as a manifold into the sphere $S^7$ as a locally minimal submanifold $A$ (see the proof in [391] and in the subsequent sections), the explicit embedding being

$$\sum_{i=1}^{4}(x^i)^2=\sum_{i=5}^{8}(x^i)^2 \cap \left(\sum_{i=1}^{8}(x^i)^2=1\right)=A.$$

Then each variation of $CA$, preserving its boundary $A$, originally increases the cone volume, i.e., $CA$ is a minimal surface with a singular point in the origin.

The above theorem about cones makes it possible to prove the interior regularity of globally minimal surfaces of codimension one in Euclidean space. We give one such result. Though dealing with currents, varifolds, and chains, we will still make use of the globally minimal surface language worked out, since it is equivalent to the above.

**Proposition 24.1.1 (see [391]).** *Fix an $(n-2)$-dimensional compact, oriented submanifold $A$ in Euclidean space $\mathbf{R}^n$, where $n \leqslant 7$. Denote the class of surfaces spanning $A$ in the sense of the current theory by $\mathcal{O}(A)=\mathcal{O}(H_{n-2}(A))$. Then there exists a globally minimal surface $X_0 \in \mathcal{O}(A)$ with the least $(n-1)$-dimensional volume, so that it is a real analytic minimal submanifold in $\mathbf{R}^n$ at all of its interior points.*

**Sketch of the proof.** We first show, in terms of current theory, that there exists a globally minimal surface $X_0$ with prescribed boundary $A$ (see [135]).

The surface turns out to be a smooth submanifold in $\mathbf{R}^n$ almost everywhere in the sense of $\mathrm{vol}_{n-1}$, meaning the minimal current support. We then show that the set of tangent cones which should be minimal with respect to their boundary (if it is fixed) is correctly defined at each singular point of the surface, not lying on $A$. It is then shown by induction that each of the cones can be regarded as that over a regular minimal submanifold $A^{n-2}$ in the standard sphere $S^{n-1}$. We then apply Theorem 24.1.1, according to which such a cone can be only the standard disc cutting the totally geodesic sphere $S^{n-1}$ out of $S^{n-2}$. Finally, it is proved that if a tangent cone at a singular point is a disc, then the surface is regular at the point (see the other corollaries of the "cone theorem" in [136]).

Note that the self-intersection of a minimal immersed submanifold is certainly not regarded as a singular point of the surface in the above results, since then the tangent cone turns into a set of standard discs, each of which is associated with its surface sheet passing through the self-intersection point.

## 24.2. The Equivariant Plateau Problem

Consider a Riemannian manifold $M^n$ on which its isometry group acts differentiably. (Recall that an isometry group is Lie.) We will consider subgroups $G$ in the isometry group $I_0(M)$, the connected component of the unit element in $I(M)$. Let $V^p \subset M^n$ be a certain surface invariant with respect to the action of a group $G$. We call such surfaces *$G$-invariant.* The natural question arises: Can we find a globally minimal one among such surfaces, in addition realizing a nontrivial (co)cycle? In other words, can we guarantee the existence of a minimal surface which would possess a prescribed symmetry group with boundary invariant with respect to the group?

**Proposition 24.2.1 (W. Y. Hsiang, H. B. Lawson; see [214]).** *A $G$-invariant manifold $V \subset M$ is locally minimal relative to all small variations $\eta$ if and only if it is locally minimal with respect to small equivariant variations $\eta_G$, i.e., invariant with respect to the group action.*

This important observation permits us to reduce the problem of the discovery of minimal submanifolds in $M$ to the one on the orbit space $M/G$.

Recall that if $x \in M$ and $G(x)$ is the orbit of a point $x$ under the action of the group $G$, the $G(x) = G/G_x$, where $G_x$ is the stability subgroup of $x$ in $G$. Two orbits $G(x)$ and $G(y)$ are said to be *of the same type* if the subgroups $G_x$ and $G_y$ are conjugate to each other in $G$, i.e., $G_x = gG_yg^{-1}$ for a certain $g \in G$. The conjugacy classes of the subgroups $\{G_x\}$ are called *orbit types.* A class $(G_y)$ is said to *follow* a class $(G_x)$ if there exists $(G_y)$ such that $G_x \supseteq gG_yg^{-1}$. We will write $(G_y) \geqslant (G_x)$. It is generally known that if $M$ is connected, then there exists a unique class $(H)$ such that $(H) \geqslant (G_x)$ for any $x \in M$. Further, the union of all orbits from this class is an open, dense submanifold $M^* = \{x \in M | G_x \in (H)\}$.

(1) The class $(H)$ is called the *principal orbit type.*

(2) If $(H') \neq (H)$, but $\dim H' = \dim H$, then the class $(H')$ is called an exceptional orbit type.

(3) If $(H'') \neq (H)$, and $\dim H'' > \dim H$, then the class $(H'')$ is called a singular orbit type.

It is proved quite simply that if $\bigcup_{\alpha \in D}(H_\alpha)$ is an arbitrary set of classes $(H_\alpha)$, each of which is not the principal orbit type, and if the set $K = \{x \in M | G_x \in (H_\alpha)$ for a certain $\alpha \in D$ is a manifold, then $K$ is a locally minimal submanifold. It follows that a singular set in $M$ is a locally minimal surface stratified by locally minimal surfaces of lesser dimensions. In particular, if a certain orbit $G(x_0)$ is such that there are no more of the same type in the neighbourhood of the point $x_0$, then $G(x_0)$ automatically is a locally minimal submanifold.

Let $\pi : M \to M/G$ be the canonical projection onto the orbit space $M/G$. Then an open dense subset $M/G$ is contained in $M/G$ and is a submanifold. If $\varphi(X, Y)$ is a scalar product on $T_x(M)$, then we can construct a new scalar product (Riemannian metric) on $M^*/G$, for which we fix the distribution $R$ of planes normal to the orbits $G(x)$. Let $X'$, $Y' \in T_{\pi(x)}[M^*/G]$. Then there exist unique vectors $X, Y$, with the inverse images of $X', Y'$ belonging to $R$, and we can put $\varphi'(X', Y') = \varphi(X, Y)$. Thereby, the distance between the points from $M^*/G$ equals the length of the orthogonal trajectory joining the corresponding orbits. Furthermore, we define the volume function $v(a)$ on $M/G$ by putting

$$v(a) = \begin{cases} \text{volume of } \pi^{-1}(a) \text{ if } \pi^{-1}(a) \text{ is the principal orbit;} \\ m\times \text{ volume of } \pi^{-1}(a) \text{ if } \pi^{-1}(a) \text{ is an exeptional orbit,} \\ m \text{ being the number of points in the homogeneous space} \\ H'/H \text{ and } H' \text{ corresponding to the point } x \in \pi^{-1}(a); \\ 0 \text{ if } \pi^{-1}(a) \text{ is a singlar orbit.} \end{cases}$$

Let $V^p \subset M$ be a $G$-invariant submanifold. Put $k = p - \nu = \dim V - \nu$, where $\nu$ is the dimension of the principal orbit. Then the submanifold $V \cap M^*$ is sent into the $k$-dimensional submanifold in $M^*/G$ by the projection $\pi : V \to M/G$. If $ds$ is a metric on $M^*$, and $\tilde{ds}$ is the one on $M/G$, then we construct a new Riemannian metric on the quotient by putting $dl_k = v^{1/k} \cdot \tilde{ds}$. Since $v(a) \to 0$ as the point $a$ tends to the singular orbit image, we can assume that the metric $dl_k$ is given on the whole space $M/G$ (Fig. 54).

Note that the volume of a $G$-invariant submanifold $V \subset M$ equals that of the manifold $V^*/G$, where $V^* = V \cap M^*$, calculated on the space $M^*/G$ in the metric $dl_k$, which entails the following.

**Proposition 24.2.2 (see [214]).** *Let $V^p \subset M$ be a certain $G$-invariant submanifold. A submanifold $V$ is locally minimal in $M$ if and only if the submanifold*

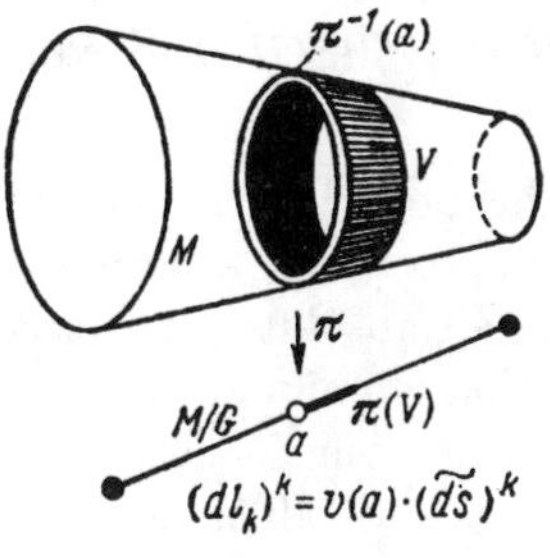


**Fig. 54**

*$V^*/G \subset M^*/G$ is locally minimal in the quotient $M/G$ with respect to the metric $dl_k$, where $k = p - \nu$.*

**Remark.** Let $k = 0$. Then $V \subset M$ coincides with a certain orbit $G(x)$. Therefore, the locally minimal orbit $G(x)$ in $M$ should be found by considering a smooth volume function $v(a)$ on the manifold $M^*/G$ and by finding all its critical points that are just extremal, or locally minimal, orbits, which, in turn, entails the following.

**Proposition 24.2.3 (see [214]).** *Each compact homogeneous space $G/H$ can be immersed into a certain standard sphere $S^{n-1}$ as a locally minimal submanifold.*

**Remark.** If, for $V \subset M$, we have $k = 1$, then the submanifold $V$ is locally minimal if and only if the trajectory $\pi(V^*) \subset M^*/G$ is a geodesic with respect to the conformal metric $dl_1 = v(a)\widetilde{ds}$. This case will be given special attention below.

Let $G$ be a compact Lie group acting orthogonally on $R^n$, and preserve the orientation (i.e., $G \subset \mathrm{SO}(n)$). Suppose that $A^{n-2} \subset \mathbf{R}^n$ is a compact, oriented, closed and $G$-invariant submanifold in $\mathbf{R}^n$ (then $A^{n-2} \subset S^{n-1}$). Consider $A^{n-2} \subset S^{n-1}$ as a "contour," the boundary in the multidimensional Plateau problem. Fix the class of surfaces $X \in \mathcal{O}(A) = \mathcal{O}(H_{n-2}(A))$ which span the boundary $A$ in the sense of the usual homology, with the homomorphism $H_{n-2}(A) \to H_{n-2}(X)$ being trivial. Then there always exists a globally minimal surface $X_0^{n-1} \in \mathcal{O}(A)$ with the least volume. It is natural to expect that a particular symmetry type of the boundary $A$ (e.g., the $G$-invariance of the boundary) implies the appropriate symmetry of the minimal surface spanning the boundary. This assumption is justified for the case of manifolds $A$ embedded into the sphere.

**Theorem 24.2.1 (Lawson; see [238]).** *Let $G$ be a closed subgroup in the group $\mathrm{SO}(n)$, and let a manifold $A^{n-2}$ be embedded into the sphere $S^{n-1} \subset \mathbf{R}^n$ and $G$-invariant. Then there exists a globally minimal surface $X_0^{n-1}$ with boundary $A^{n-2}$ understood in the sense of the theory of currents, which is*

***invariant relative to the group $G$. If it is unique, then the solution of the Plateau problem is also unique in the class of all surfaces $X^{n-1} \subset O(A)$, now not necessarily $G$-invariant.***

Thus, in the case where the boundary $A$ admits a nontrivial symmetry group, it is sometimes sufficient to find a minimal surface in the class of $G$-invariant films in order to find an absolutely minimal surface spanning $A$. If the obtained $G$-invariant solution is unique, then this implies the uniqueness also in the class of all surfaces, i.e., our $G$-invariant surface now also realizes an absolute minimum in the class of all nonsymmetric films (see [238] for various details in the language of currents). The requirement that the group $G$ should be contained in the group $\mathrm{SO}(n)$ is essential. It is impossible to reject it, which is shown by the following simple example. Take as $A$ a two-dimensional torus $T^2$ embedded into the three-dimensional sphere $S^3$ as $T^2 = \{(z, w) \in \mathbf{C}^2, |z| = |w|\}$, and as $G$ the group $\mathbf{Z}_2 \subset \mathrm{O}(4)$ whose generator is realized by the orthogonal mapping $(z, w) \to (w, \bar{z})$. Then, as can be computed, the minimal surfaces spanning $T^2$ in thefour-dimensional space $\mathbf{R}^4$ are not invariant relative to the action of $G$, though the torus is invariant (see [135],[238]).

Thus, in the case of a $G$-invariant boundary $A$, where $G \subset \mathrm{SO}(n)$, it suffices to find a $G$-invariant minimal surface if we want to find a globally minimal one. If the latter is unique, then it automatically realizes an absolute minimum in the whole class $O(A)$. We will make use of this circumstance in the cone problem.

Assume that $A^{n-2} \subset S^{n-1}$ is the principal orbit of the action of $G$. Then $\pi(A) \subset \mathbf{R}^n/G$ is a certain point $q$, assuming that $A$ is connected. The problem of finding a minimal film spanning $A^{n-2}$ and invariant under the action of $G$ is thereby equivalent to that of finding the shortest geodesic from $q$ to the boundary of the two-dimensional manifold $\mathbf{R}^{n*}/G$ considered with the metric $v^2(a)\widetilde{ds^2}$. The two problems can be distinguished, viz., (A) of giving the least of all possible orthogonal actions of connected, compact Lie groups $G$ on $\mathbf{R}^n$ with principal orbits of co-dimension two ($k = 1$), and (B) of describing within the framework of this list all the geodesics $\gamma$ of interest from $q$ to the boundary $\partial\mathbf{R}^{n*}/G$. It turns out that the metric $\widetilde{ds}$ on the quotient space $\mathbf{R}^{n*}/G$ is always Euclidean, and therefore, $dl_1^2 = v^2(a)(dx^2 + dy^2)$, where $(x, y)$ are Cartesian coordinates on $\mathbf{R}^{n*}/G \subset \mathbf{R}^2(x, y)$.

We may make use of a simple mechanical analogue. Since $v(a) = n(x, y)$ is a smooth function vanishing on the boundary $\partial\mathbf{R}^{n*}/G$, the geodesic metrics $dl_1$ are precisely the trajectories of light rays passing in accordance with the Fermat principle in two-dimensional continuous transparent medium filling the cone $\mathbf{R}^{n*}/G$ with refractive index $n(x, y) = c/v(x, y)$, where $c$ is the light velocity and $v(x, y)$ that of a ray at a point $(x, y)$. The transparent medium is supposed to be isotropic at each point, but nonhomogeneous on the cone. The Fermat principle states that a light ray passing from a point $A$ to a point $B$ chooses a path with the least time to reach $A$ with fixed energy.

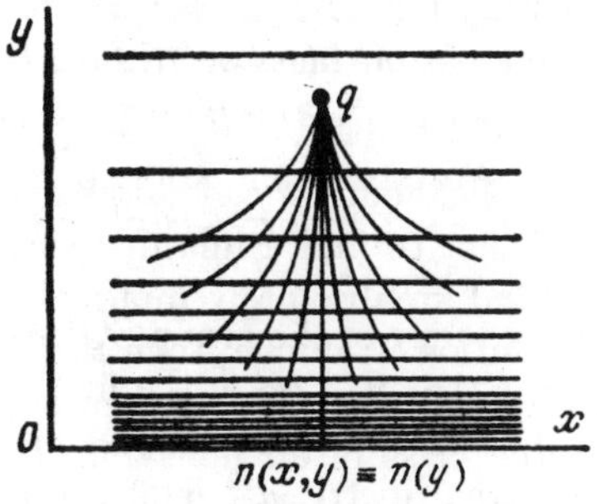


**Fig. 55**

One general remark: If

$$D=\{(x,\ y)\in\mathbb{R}^2;\ y\geqslant 0\},\quad n(x,\ y)\equiv n(y),\ n(y_1)>n(y_2)$$

for $y_1 > y_2$ and $n(0) = 0$, then the light rays emanating from a fixed point $q \in D$ to the boundary $\partial D$ pass as shown in Fig. 55. We now list the compact Lie groups $G$ orthogonally acting on the space $R^n$ with co-dimension two. We also indicate the stability subgroups $H$ associated with the principal orbits $G/H$ of the action, viz.,

1) $(SO(r)\times SO(s))/(SO(r-1)\times SO(s-1))$; 2) $(SO(2)\times SO(k))/(\mathbb{Z}_2\times SO(k-2))$; 3) $(SU(2)\times SU(k))/(T^1\times SU(k-2))$; 4) $(Sp(2)\times Sp(k))/(Sp^2(1)\times Sp(k-2))$; 5) $U(5)/(SU(2)\times SU(2)\times T^1)$; 6) $SO(3)/\mathbb{Z}_2^2$; 7) $SU(3)/T^2$; 8) $Sp(3)/Sp^3(1)$; 9) $Sp(2)/T^2$; 10) $G_2/T^2$; 11) $F_4/Spin(8)$; 12) $(Spin(10)\times U(1))/(SU(4)\times T^1)$.

How to embed $H \to G$ is described in the works of W. Hsiang and H. Lawson in the language of representation theory (see [214], [238]). The above spaces (1) and (2) are different by the representation $H \to G$.

We now come back to the cones. Put $n = 2m$, and consider $C_{2m-1} = \left\{\sum_{i=1}^m x_i^2 = \sum_{j=m+1}^{2m} x_j^2\right\}$ in $\mathbb{R}^{2m}$. Then $C_{2m-1} \cap S^{2m-1} = S^{m-1} \times S^{m-1}$, and $\partial C_{2m-1} = S^{m-1} \times S^{m-1}$. It is easy to show that $A^{n-2} = S^{m-1} \times S^{m-1}$ is embedded into the sphere $S^{n-1}$ as a locally minimal submanifold; therefore, $C_{2m-1}$ is locally minimal, i.e., it annihilates the Euler operator for the volume functional. As we already know, it is not minimal for $m < 4$, since a contracting deformation decreasing the volume exists at its vertex. However, for $m \geqslant 4$, the situation changes apruptly.

Let $A^{2m-2} = S^{m-1} \times S^{m-1}$ be the standard minimal submanifold in the sphere $S^{2m-1}$ (see above), and $m \geqslant 4$. Then any small variation of $C_{2m-1} = CA^{2m-2}$ increases its volume, i.e., $C_{2m-1}$ is minimal with respect to any small variation, and therefore, is a local minimum (see [391]). The $C_{2m-1}$ for $m \geqslant 4$ are thereby first candidates for rejecting the theorem on the interior regularity of minimal surfaces of co-dimension one if $n \geqslant 8$. The problem of their global minimality for fixed boundary $S^{m-1} \times S^{m-1}$ is still open and can be completely solved on the basis of the equivariant Plateau problem.

As an example, consider one series $(G, H)$. For example, put $r = s$ in series (1) of the above list. We study the corresponding minimal surfaces $T_0$ with boundary $\partial T_0 = S^{m-1} \times S^{m-1} \subset S^{2m-1} \subset \mathbf{R}^{2m}$. Here, we assume that $\mathbf{R}^{2m} = \mathbf{R}^m(x) \oplus \mathbf{R}^m(y)$, where $x = (x^1, \ldots, x^m)$, $y = (y^1, \ldots, y^m)$ and $A^{2m-2} = \{|x| = |y|\}$, with one copy of the group $\mathrm{SO}(m)$ acting on $\mathbf{R}^m(x)$ and another on $\mathbf{R}^m(y)$. It is clear that $\mathbf{R}^{2m}/G$ is the first quadrant on the plane $\mathbf{R}^2(x \geqslant 0, y \geqslant 0)$. We will denote the coordinates on the plane $\mathbf{R}^2 \subset \mathbf{R}^{2m}$ also by $x$ and $y$. The conformal metric $dl_1^2$ is of the form $dl_1^2 = (xy)^{4m-4} \cdot (dx^2 + dy^2)$. The manifold $A^{2m-2}$ is represented on $K = \mathbf{R}^{2m}/G$ by a point $q$ with coordinates $(1,1)$. The absolutely minimal surface $T_0$ with boundary $A^{2m-2}$ is represented in $K$ by the minimal geodesic from $q$ to the boundary of the domain $K$. It is easily proved that $A^{2m-2} \subset S^{2m-1}$ is a locally minimal submanifold, this orbit being a critical point of the orbit volume function. It follows that the bisector $x = y$ is the geodesic joining of $q$ to $O$ (which is also a consequence of the metric $dl_1$ being symmetric with respect to the variables $x$ and $y$); therefore, the cone $CA^{2m-2}$ is locally extremal for any $m$ and annihilates the Euler operator. However, the problem of finding a minimal geodesic and, therefore, an absolutely minimal surface $T_0$ requires a special and sufficiently nontrivial investigation.

It can be shown by straightforward computation (verify!) that if the minimal geodesic $\gamma_0$ from $q$ to $\partial K$ passes onto the boundary $\partial K \backslash \{O\}$, then it meets the corresponding coordinate axis at right angles; therefore, the corresponding minimal surface $T_0 = \pi_*^{-1}(\gamma_0)$ is an analytic submanifold without singularities, with boundary $A^{2m-2}$.

We will assume that dimension $m$ is a continuous parameter, $1 \leqslant m < \infty$. For $m = 1$, we have $dl_1^2 = dx^2 + dy^2$, and the light rays emanating from the point source $q$ are represented in Fig. 56; $Oq$ is not a minimal geodesic: such are the lines $Qq$ and $Q'q$. Let $m = \varepsilon + 1$, $\varepsilon > 0$. Then we show that the points $Q$ and $Q'$ start moving in the direction of the vertex $O$, and the distribution of light rays emanating from the point source looks qualitatively as in Fig. 57. Supplying a qualitative explanation, we give the analytic proof below. In fact, consider the level lines $n(x, y) = \text{const}$. Then, since the refractive index decreases if we move closer to the boundary $\partial K$, the light beam is deformed in accordance with the Fermat principle (see Fig. 55), and looks as in the qualitative diagram in Fig. 57. The minimal geodesics $Qq$ and $Q'q$ sag in the direction of $O$. A light ray different from $Qq$ and $Q'q$ cannot reach the boundary of the domain, for, in its interaction with the level lines of the refractive index $n(x, y)$, it is obvious that a moment when the ray bends always comes, the sign of the second derivative changes, and the ray alters its direction, beginning to move back inside the domain towards the bisector of the quadrant and meeting it in a point conjugate to $q$ (see the exact computation below).

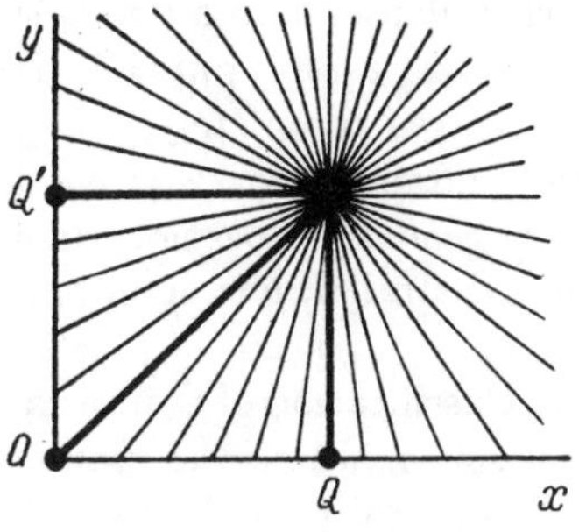


**Fig. 56**

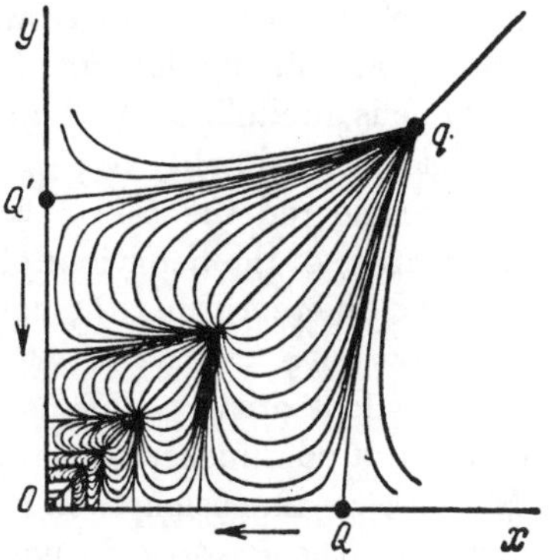


**Fig. 57**

For $m > 0$, a quadrilateral $OQ'qQ$ occurs, and the behaviour of the light rays, or geodesics, inside $OQ'qQ$ is different from that outside. It can be easily seen that, with the growth of $m$, the points $Q$ and $Q'$ shift towards the cone vertex $O$, and $OQ'qQ$ starts collapsing onto the bisector $Oq$. Direct computation shows that the final flattening of the quadrilateral occurs for $m = m_0 = 2.5 \times \sqrt{2} \simeq 3.9$ (Fig. 58), and the qualitative picture of the light ray distribution remains invariable for $m > m_0$.

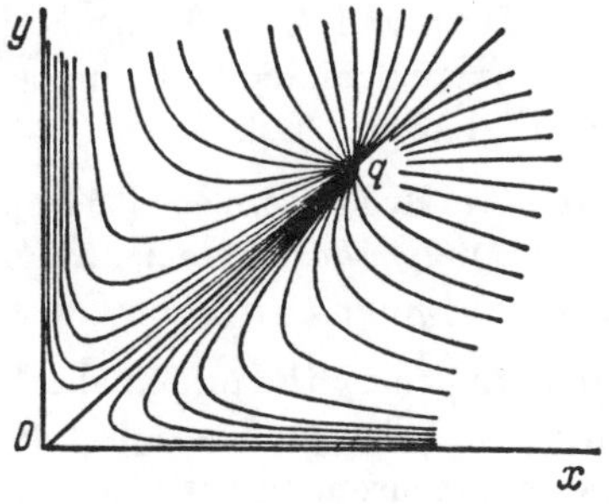


**Fig. 58**

Therefore, for $m \geqslant 4$, there exists a unique geodesic joining of the point $q$ to the boundary $\partial K$, viz., the interval $Oq$. Hence, it is minimal; that is why the cones $C_{2m-1} = C(S^{m-1} \times S^{m-1}) = \pi^{-1}(Oq)$ are $G$-invariant minimal surfaces. Due to the uniqueness of the geodesic $Oq$, i.e., that of a $G$-invariant solution, where $G = \mathrm{SO}(m) \times \mathrm{SO}(m)$, it follows from Theorem 24.2.1 that, for $m \geqslant 4$, the cone $C_{2m-1}$ is a globally minimal surface and a minimal integral current in the terminology of [135], with the boundary $S^{m-1} \times S^{m-1}$ possessing one singular point $O$, the cone vertex.

The geodesic $Qq$ represents the minimal film $\pi^{-1}(Qq)$ with the boundary $S^{m-1} \times S^{m-1}$. For $m = 1$, it consists of two line segments, starts sagging in the direction of the origin as $m$ grows, and its neck gradually approaches $O$ (cf. catenoid), and finally, for $m \simeq 3.9$, the neck collapses into a point, the vanishing cycle is annihilated, and minimal cones $C_{2m-1}$ arise for $m \geqslant 4$. The problem of global minimality of $C_{2m-1}$ is thereby solved completely.

We have considered the case $r = s$ in detail; however, for $r \neq s$, the qualitative picture of the behaviour of the pencil of geodesics does not change much, with the difference that the point $q$ lies on the straight line determined by the equation $\tan\alpha = r/s$.

We shall now look at the problem of a complete classification of $G$-invariant minimal cones of co-dimension one in Euclidean space, where the group $G$ acts in $\mathbf{R}^n$ with co-dimension two. It was solved practically totally by [214], [238]. However, these works do not supply a complete cone classification, since certain important particular cases remain unsolved, among which are, as it turns out, globally minimal and earlier unknown cones. The complete solution of the problem and final classification theorem will be obtained on the basis of the above scheme, i.e., the study of conjugate points and qualitative behaviour of the pencil of geodesics emanating from a point which was critical for the orbit volume function. It then turns out that the qualitative picture of the distribution of the pencil of geodesics, and the related surgery mechanism for minimal surfaces as dimension grows, are universal and "serve" all the other minimal cones of co-dimension one from the list of [214] (see also [162]). The corresponding computation and complete analysis of the Jacobi equation were carried out by A. V. Tyrin and A. T. Fomenko [162]; the stability and nonstability of several cones was calculated by P. Simoes [461], Bruce Solomon, and Leon Simon, then by G. Lawlor [462].

**Proposition 24.2.4 (W. Hsiang, H. Lawson; see [214], [238]).** *Let $G \subset \mathrm{SO}(n)$ be a connected, compact subgroup whose principal orbits of action on $\mathbf{R}^n$ are of co-dimension two. Then the group $G$ is one of those listed in Table 2.*

The principal orbits determined by the principal orbit type (see above) are generic of maximal dimension. The quotient space $\mathbf{R}^n/G$ in all the cases of the action of a group of co-dimension two is a cone on the two-dimensional plane, i.e., of the form $C(\alpha) = \{(x,y) \in \mathbf{R}^2 : 0 \leqslant \tan(\frac{y}{x}) \leqslant \alpha, x \geqslant 0\}$. Let $\pi : \mathbf{R}^n \to \mathbf{R}^n/G \approx C(\alpha)$ be the standard projection onto the orbit space (see above). Introduce the orbit volume function $v : \mathbf{R}^n/G \approx C(\alpha) \to \mathbf{R}^1$ by the formula $v(q) = \mathrm{vol}_{n-2}\,\pi^{-1}(q)$, where $q \in \mathbf{R}^n/G$ (see [214], [238]). Denote by $dl^2 = v(x,y)(dx^2 + dy^2)$ the conformal metric on the cone $C(\alpha) = \mathbf{R}^n/G$ (see above). It is obvious that it degenerates on the boundary of $C(\alpha)$, since orbits of lesser dimensions are generated by the boundary of a two-dimensional cone, compared with generic orbits. If $\gamma$ is a curve on $C(\alpha)$, then its length in the metric $dl$ determines the volume of an orbit, i.e., $\mathrm{vol}_{n-1}\,\pi^{-1}(\gamma) = l(\gamma)$. If $\gamma$ is a geodesic, then its inverse image $\pi^{-1}(\gamma)$ is a locally minimal surface $\mathbf{R}^n$. Recall that if a manifold $M^{n-2}$ is realized as a certain principal orbit in $\mathbf{R}^n$, then the projection $\pi$ establishes a one-to-one correspondence between $G$-invariant surfaces with boundary $M^{n-2}$ and geodesics with respect to the metric $dl$, emanating from the point $q = \pi(M^{n-2})$ to the boundary of $C(\alpha)$ (see [238]). The uniqueness of such a geodesic for a fixed starting point in the domain $C(\alpha)$ entails that of the solution to the absolute minimum problem in the class of all films not necessarily invariant under the group action, i.e., the given invariant solution is also globally minimal.

**Table 2**

| № | $G$ | $\dim_{\mathbb{R}} \varphi$ | $H$ | $\alpha$ | $\lambda = (\text{vol})^2$ |
|---|---|---|---|---|---|
| 1 | $SO(r) \times SO(s)$ | $r+s$ | $SO(r-1) \times$ $\times SO(s-1)$ | $\frac{\pi}{2}$ | $x^{2r-2}\, y^{2s-2}$ |
| 2 | $SO(2) \times SO(k)$ | $2k$ | $\mathbb{Z}_2 \times SO(k-2)$ | $\frac{\pi}{4}$ | $(xy)^{2k-4}\,(x^2-y^2)^2$ |
| 3 | $SU(2) \times SU(k)$ | $4k$ | $T^1 \times SU(k-2)$ | $\frac{\pi}{4}$ | $(xy)^{4k-6}\,(x^2-y^2)^4$ |
| 4 | $Sp(2) \times Sp(k)$ | $8k$ | $(Sp(1))^2 \times Sp(k-2)$ | $\frac{\pi}{4}$ | $(xy)^{8k-10}\,(x^2-y^2)^8$ |
| 5 | $U(5)$ | 20 | $SU(2) \times SU(2) \times T^1$ | $\frac{\pi}{4}$ | $(xy)^2\, \text{Im}\,\{(x+iy)^4\}^8$ |
| 6 | $SO(3)$ | 5 | $\mathbb{Z}_2 \times \mathbb{Z}_2$ | $\frac{\pi}{3}$ | $\text{Im}\,\{(x+iy\ ^3\}^2$ |
| 7 | $SU(3)$ | 8 | $T^2$ | $\frac{\pi}{3}$ | $\text{Im}\,\{(x+iy)^3\}^4$ |
| 8 | $Sp(3)$ | 14 | $(Sp(1))^3$ | $\frac{\pi}{3}$ | $\text{Im}\,\{(x+iy)^3\}^8$ |
| 9 | $Sp(2)$ | 10 | $T^2$ | $\frac{\pi}{4}$ | $\text{Im}\,\{(x+iy)^4\}^4$ |
| 10 | $G_2$ | 14 | $T^2$ | $\frac{\pi}{6}$ | $\text{Im}\,\{(x+iy)^6\}^4$ |
| 11 | $F_4$ | 26 | $Spin(8)$ | $\frac{\pi}{3}$ | $\text{Im}\,\{(x+iy)^3\}^8$ |
| 12 | $Spin(10)$ $\times U(1)$ | 32 | $SU(4) \times T^1$ | $\frac{\pi}{3}$ | $\text{Im}\,\{(x+iy)^3\}^{20}$ |

It suffices to test only locally minimal cones for global minimality. It is not hard to see that here such cones are those over orbits of maximal volume in the sphere $S^{n-1}$, which determine maximum points for the volume function on the orbit space. Under the projection $\pi$, the cone over an orbit is transformed into a line segment on the quotient space $C(\alpha)$, emanating from the point $q = \pi(M^{n-2})$ to the vertex $O$ of $C(\alpha)$.

In Table 2, $\dim_{\mathbb{R}} \varphi$ denotes the dimension of the linear representation $\varphi : G \to \mathrm{SO}(n) \subset GL(n)$, $\alpha$ the vertex angle of the two-dimensional quotient space $C(\alpha) \approx \mathbb{R}^n/G$.

Further, $V^2$ is the square of the volume function $v$, and $H$ is the principal stability subgroup of the action, i.e., associated with the principal orbit type. Thus, the manifold $M^{n-2}$ is diffeomorphic to the homogeneous space $G/H$ (see [214], [238] for more detailed information about the representation of $\varphi$).

**Theorem 24.2.2 Classification Theorem for Minimal Cones of Co-Dimension (W. Hsiang, H. Lawson, A. V. Tyrin, A. T. Fomenko, P. Simoes, B. Solomon, L. Simon, and G.Lawlor).** *Unique globally minimal surfaces with boundary $M$, where $M$ are the orbits represented in Table 2, are the cones over the following manifolds $G/H = M$, viz.,*

a) $S^{r-1} \times S^{s-1} = \dfrac{\mathrm{SO}(r) \times \mathrm{SO}(s)}{\mathrm{SO}(r-1) \times \mathrm{SO}(s-1)}$ in $\mathbb{R}^{r+s}$ for $r+s \geqslant 8$;

b) $\dfrac{\mathrm{SO}(2) \times \mathrm{SO}(k)}{\mathbb{Z}_2 \times \mathrm{SO}(k-2)}$ in $\mathbb{R}^{2k}$ for $k \geqslant 8$;

c) $\dfrac{\mathrm{SU}(2) \times \mathrm{SU}(k)}{T^1 \times \mathrm{SU}(k-2)}$ in $\mathbb{R}^{4k}$ for $k \geqslant 4$;

d) $\dfrac{\mathrm{Sp}(2) \times \mathrm{Sp}(k)}{(\mathrm{Sp}(1))^2 \times \mathrm{Sp}(k-2)}$ in $\mathbb{R}^{8k}$ for $k \geqslant 2$;

e) $\dfrac{U(5)}{\mathrm{SU}(2) \times \mathrm{SU}(2) \times T^1}$ in $\mathbb{R}^{20}$;

f) $\mathrm{Sp}(3)/(\mathrm{Sp}(1))^3$ in $\mathbb{R}^{14}$;

g) $F_4/\mathrm{Spin}(8)$ in $\mathbb{R}^{26}$;

h) $\dfrac{\mathrm{Spin}(10) \times U(1)}{\mathrm{SU}(4) \times T^1}$ in $\mathbb{R}^{32}$.

*For all the other manifolds $G/H$ from Table 2, the corresponding cones over them are not minimal, and there exists a volume-decreasing variation. The above globally minimal cones are $G$-invariant with respect to the corresponding groups from Table 2.*

**Remark.** Compared with H. Lawson's work [238], the news is the global minimality of the cones over the manifolds $S^4 \times S^2$ in $\mathbb{R}^8$, $S^6 \times S^1$ in $\mathbb{R}^9$, $(\mathrm{SO}(2) \times \mathrm{SO}(8))/(\mathbb{Z}_2 \times \mathrm{SO}(6))$ in $\mathbb{R}^{16}$, and $(\mathrm{SO}(2) \times \mathrm{SO}(9))/\mathbb{Z}_2 \times \mathrm{SO}(7))$ in $\mathbb{R}^{18}$, $(\mathrm{SU}(2) \times \mathrm{SU}(4))/(T^1 \times \mathrm{SU}(2))$ in $\mathbb{R}^{16}$, and also the nonminimality of the cones for the groups in the second line of the table for $4 \leqslant k \leqslant 7$, from the third line for $k = 2, 3$ and sixth, seventh, ninth, and tenth lines. Theorem 24.2.2 thereby extends the results of [238] and supplies a complete classification of invariant cones of co-dimension two.

Let $G$ be one of the classification list groups (see Table 2). Consider a locally minimal cone $CM$ over the orbit of maximal volume $M^{n-2}$ in the sphere $S^{n-1}$. As was shown above, the cone is associated with the geodesic $Oq$ on the quotient space $C(\alpha)$; more precisely, with that of its line segments, which reaches the boundary. To clarify the circumstances in which this segment is a geodesic of least length from the point $q$ to the boundary of $C(\alpha)$, we describe the qualitative behaviour of the pencil of all geodesics emanating from $q$. We find all the points conjugate to $q$ along the segment of $Oq$.

**Lemma 24.2.1.** *Let a domain $U$ with the conformal metric $ds^2 = \lambda(x,y)$ $(dx^2+dy^2)$, $q \in U$, with $\gamma(t)$ being a geodesic, $\gamma(0)=q$, be given in the plane $\mathbf{R}^2 = \mathbf{R}^2(x,y)$. The points conjugate to $q$ along $\gamma$ and are associated just with those values of the parameter $t$, for which the function $\omega(t)$ defined as the solution of the equation*

$$\ddot{\omega} - K\omega = 0, \qquad \omega(0) = 0 \tag{1}$$

*vanishes, where $K$ is the scalar curvature. With respect to conformal coordinates, the identity*

$$-K = \frac{1}{\lambda} \cdot R^1_{2,12} \tag{2}$$

*holds, where $R^1_{2,12}$ is the Riemann tensor component.*

The proof is reduced to staightforward computation.

For simplicity, we will assume that the parameter $t$ along the geodesic is natural, i.e., $|\dot{\gamma}(t)| \equiv 1$. Simple computations show that

$$R^1_{2,12} = -\frac{1}{2\lambda}\left(\frac{\partial^2\lambda}{\partial x^2} + \frac{\partial^2\lambda}{\partial y^2}\right) + \frac{1}{2\lambda^2}\left(\left(\frac{\partial\lambda}{\partial x}\right)^2 + \left(\frac{\partial\lambda}{\partial y}\right)^2\right). \tag{3}$$

In our case, $\lambda = v^2$. Note that the function $v^2$ is always a homogeneous polynomial of degree $2n-4$ (see Table 2). Let $y = \theta(x)$ be a function determining a geodesic on the plane with respect to Cartesian coordinates. The coordinate $x$ on the trajectory $\gamma$ depends on the parameter $t$, and the functions $R^1_{2,12}$ and $\lambda = v^2$ depend on $x$. It can be seen from (3) that the equality $R^1_{2,12} = \frac{c'}{x^2}$ holds on $\gamma$, where $c'$ is a numerical coefficient; therefore, $-K = \frac{R^1_{2,12}}{\lambda} = \beta x^{-2n-4+2}$, where $\beta$ is a certain numerical factor.

To study the dependence of $x$ on $\gamma$, we have $1 = |V|^2 = \lambda(x^2+y^2) = x^n\dot{x}^2\lambda(1,\theta)(1+\theta^2)$. Hence, if $p$ means the dimension $2n-4$, we obtain $\left(x^{\frac{p+2}{2}}\right)^{\cdot} = \frac{p+2}{2}\cdot\left(\frac{1}{1+\theta^2}\cdot\frac{1}{\lambda(1,\theta)}\right)^{\frac{1}{2}}$. Thus, $x^{\frac{p+2}{2}} = At + B$, where

$$A=\frac{p+2}{2}\cdot\left(\frac{1}{1+\theta^2}\cdot\frac{1}{\lambda(1,\theta)}\right)^{1/2}, \tag{4}$$

and $B$ is determined from the condition $\gamma(0) = q$, whence $-K = \frac{C}{(At+B)^2}$. Thus, equation (1) takes the form

$$\ddot{\omega}+\frac{C}{(At+B)^2}\,\omega=0, \tag{5}$$

where $A$ and $C$ are determined from (4). Equation (5) is reduced to $\ddot{\omega} = \frac{D}{t^2}\omega$ by the obvious substitution $u = At + B$, and we obtain that $D = \frac{-C}{A^2}$. We write explicitly the solutions of the last equation, viz.,

$$\omega = \begin{cases} C_1 t^{\frac{1}{2}+s} + C_2 t^{\frac{1}{2}-s} & \text{for } 4D+1>0 \\ C_1\sqrt{t} + C_2\sqrt{t}\ln t & \text{for } 4D+1=0 \\ C_1\sqrt{t}\cos(s\ln t) + C_2\sqrt{t}\sin(s\ln t) & \text{for } s=\sqrt{|4D+1|} \end{cases}$$

(e.g., see [222]). It follows immediately that either $\gamma$ has no points conjugate to $q$ at all, or there are infinitely many of them, and they condensate in the vicinity of the limit point $O$. If $\frac{4C}{A^2} > 1$, then the conjugate points are there; otherwise, they are absent.

The study of the coefficients $A$ and $C$ permits us to decide in each concrete case which of these two probabilities is valid. Consider one by one all the cases in Table 2.

**Series 1.** We have

$$G=SO(r)\times SO(s)\subset SO(r+s),\quad r\geqslant 2,\ s\geqslant 2,$$
$$\mathbb{R}^n/G=\{(x,y)\in\mathbb{R}^2:\ x\geqslant 0,\ y\geqslant 0\},$$
$$v^2=x^{2r-2}\cdot y^{2s-2}=x^p y^m,\quad y=\sqrt{\frac{m}{p}}\,x$$

being a geodesic. It is easy to compute that $R^1_{2,12} = \frac{1}{2}(\frac{p}{x^2} + \frac{m}{y^2})$. Therefore, $-D = \frac{C}{A^2} = \frac{4(p+m)}{(p+m+2)^2}$. Since conjugate points exist if and only if $\frac{4C}{A^2} > 1$, we obtain as a necessary and sufficient condition for conjugate points to exist that $16(p+m) > (p+m)^2 + 4(p+m) + 4$, i.e., $5 - 2\sqrt{2} < r + s < 5 + 2\sqrt{2}$ ($p = 2r - 2, m = 2s - 2$). We consider only the case of $r > 1$, $s > 1$, and obtain accordingly that, for $r + s < 5 + 2\sqrt{2} \simeq 7.82$, conjugate points are there. Meanwhile, we assume that $r$ and $s$ are continuous parameters. As $r + s \to 5 + 2\sqrt{2}$ to the left of this critical value, conjugate points start approaching $O$ again, and vanish for $r + s \geqslant 5 + 2\sqrt{2}$.

**Series 5–12.** All the corresponding conformal metrics are reduced to the form $ds^2 = y^m(x^2+y^2)^{\frac{1-p}{p}} \cdot (dx^2+dy^2)$, the metric being given on the upper half-plane $y > 0$. In the fifth case, we have $p = 4$, $m = 8 + \frac{1}{2}$, and $p$ and $m$ coincide with the corresponding values in the other cases if we write the metric in the form $ds^2 = \{\mathrm{Im}(x+iy)^p\}^m(dx^2+dy^2)$. Consider the point $q = \pi(M^{n-2}) = e^p$ in the quotient space, and replace $W = z^p$. Express the metric of interest with respect to polar coordinates $W = re^{i\varphi}$. We then obtain $ds^2 = r^{m+\frac{2}{p}-2}\sin^m\theta \cdot (r^2d\theta^2 + dr^2)$. To compute $C$, we notice that the component $R^1_{2,12}$ and the function $\lambda$ depend on $y$ on the $Oy$-axis, with $y$, in turn, being a function of $t$. If $f(y) = Cy^n$, then we put $K(f) = C$. Hence,

$$K\left(R^1_{2,12}|_{Oy}\right) = \frac{1}{2}K\left(\left[\left(\frac{\partial\lambda}{\partial x}\right)^2 + \left(\frac{\partial\lambda}{\partial y}\right)^2\right]_{Oy}\right) - \frac{1}{2}K\left(\left[\frac{\partial^2\lambda}{\partial x^2} + \frac{\partial^2\lambda}{\partial y^2}\right]_{Oy}\right) = \frac{1}{2}m.$$

To compute $A$ on the $Oy$-axis, we obtain

$$\lambda(\dot{x}^2+\dot{y}^2) = 1, \quad \lambda\dot{y}^2 = 1, \quad A = \left(y^{\frac{m}{2}+\frac{1}{p}}\right)^{\bullet} = \frac{m}{2} + \frac{1}{p}.$$

Therefore, $\frac{C}{A^2} = \frac{1}{A^2}K(R^1_{2,12}|_{Oy}) = \frac{2p^2m}{(mp+2)^2}$, and we immediately see that $\frac{4C}{A^2} > 1$ only in the sixth, seventh, ninth, and tenth cases, where there are conjugate points. In the fifth, eighth, eleventh, and twelfth series, there are no conjugate points.

We now consider the second, third, and fourth series. For convenience, we will rewrite the metrics in the form $ds^2 = \frac{x^py^m}{\sqrt{x^2+y^2}} \times (dx^2+dy^2)$ on the quadrant $x \geqslant 0$, $y \geqslant 0$. The straight line $y = \sqrt{\frac{m}{p}}x$ is the geodesic. Note that, in the prior metric, the slope of the geodesic was $\sqrt{\frac{p+m}{m}} - \sqrt{\frac{p}{m}}$. In the second series, we have $p = 2$, $m = 2k-4$; in the third series, $p = 4$, $m = 4k-6$; and $p = 8$, $m = 8k-10$ in the fourth. As well as in the first series, it is easy to see that $R^1_{2,12} = \frac{1}{2}(\frac{p}{x^2} + \frac{m}{y^2})$ by applying the explicit formula for $R^1_{2,12}$. We obtain $x^{p+m-1}\dot{x}^2\lambda(1,\theta) \times (1+\theta^2) = |V| = 1$ for $A$; hence, $A = \left(x^{\frac{p+m+1}{2}}\right)^{\bullet} = \frac{p+m+1}{2\sqrt{\lambda(1,\theta)(1+\theta^2)}}$. Since $C = \frac{1}{2\lambda(1,\theta)}(p + \frac{m}{\theta^2})$, we have $\frac{4C}{A^2} = \frac{16(p+m)}{(p+m+1)^2}$. Therefore, conjugate points exist if and only if $\frac{4C}{A^2} > 1$, or if and only if

$$\begin{array}{ccccc} 7-4\sqrt{3} & \leqslant & p+m & \leqslant & 7+4\sqrt{3} \\ \wr & & & & \wr \\ 2{,}072 & & & & 13{,}928 \end{array}$$

Thus, in the second series, for $2k < 9+4\sqrt{3} \simeq 15.928$, there are conjugate points. Conjugate points are absent for $2k \geqslant 9+4\sqrt{3}$. In particular, if we consider only whole $k$, i.e., the geometric case, then there are no conjugate points, beginning with $k = 8$. In the third series, there are conjugate points for $4k < 9+4\sqrt{3} \simeq 15.928$, and they are absent for $4k \geqslant 9+4\sqrt{3}$. For integral $k$, there are no conjugate points, starting with $k = 4$. In the fourth series, we have $k \geqslant 2$. Therefore, there are no conjugate points for all whole $k$.

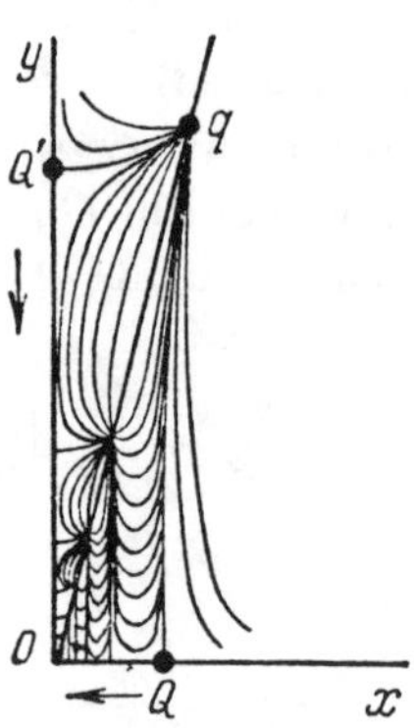


**Fig. 59**

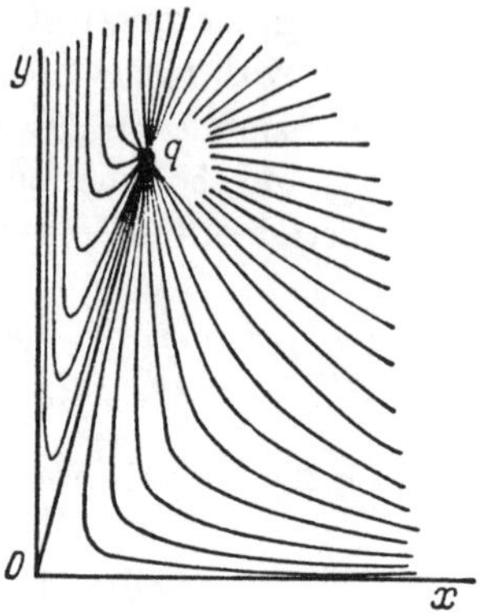


**Fig. 60**

Now, making use of the obtained information about the conjugate-point distribution, we clarify the qualitative behaviour of geodesics for all metrics from Table 2.

**Series 1.** We assume that $p$ and $m$ ($r$ and $s$) are continuous parameters, $m/p = \text{const}$. For $r+s < 5+2\sqrt{2}$, the picture of the distribution of geodesics emanating from the points $q$ is as shown in Fig. 59. The geodesics $qQ$ and $qQ'$ meet the boundary of the quotient $C(\alpha)$ at an angle of $\pi/2$, and those different from $qQ$ and $qQ'$ cannot immediately reach the boundary, since any geodesic different from $Oq$ should reach it at $\pi/2$; therefore, they return to $Oq$, intersecting it at a point conjugate to $q$. As $r+s$ grow, the points $Q$ and $Q'$ approach $O$, and, for $r+s = 5+2\sqrt{2} \simeq 7.82$, the conjugate points vanish (see Fig. 60). The picture then remains qualitatively invariable, the only geodesic reaching the quotient space boundary is the segment $Oq$. (Series 6 is shown in Fig. 61.)

Show that the conjugate points imply the existence of variations decreasing the volume of a locally minimal cone for variations with small supports, not affecting the origin. It is obvious that there exist variations on the quotient, which decrease the length of geodesics passing through the origin. Recall that the length of a curve in $\mathbf{R}^n/G$ equals the volume of its inverse image in $\mathbf{R}^n$. Hence, having taken a variation on the quotient, we shall obtain that of a frustum of the cone with stationary boundary, decreasing its volume. Extending it identically to the domain at the vertex, we obtain the required volume-decreasing variation even without shifting the cone vertex!

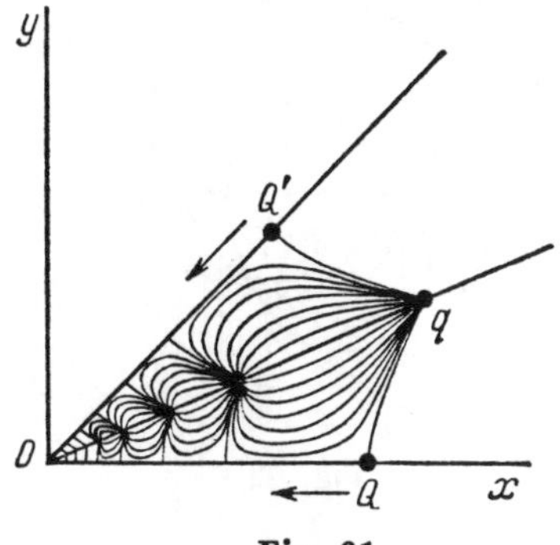


Fig. 61

### 24.3. Representation of Equivariant Singularities as Singular Points of Closed Minimal Surfaces Embedded into Symmetric Spaces

It turns out that the minimal cones which we studied in §24.2 are tangent, approximating the behaviour of certain closed minimal surfaces at their singular points. We now solve the following problem, namely: What are those globally minimal closed surfaces into which the above local singularities can be glued naturally? In other words, how can we extend (or integrate) these to closed minimal surfaces in a Riemannian manifold? We illustrate this by completely describing all stationary (locally minimal) closed, smooth, and $\mathrm{SO}(n-2)$-invariant submanifolds in the standard sphere $S^{n-1}$.

Let $M = S^n \subset \mathbf{R}^{n+1}$, $G = \mathrm{SO}(n-1)$ act via the representation $\rho = t_{n-1} \oplus 2\Theta_1$, where $t_{n-1}$ is the standard action of $\mathbf{R}^{n-1}$ and $\Theta_1$ the identity representation. It is clear that $S^n/G$ is a two-dimensional disc $D^2$ with the standard spherical metric $\widetilde{ds}^2 = d\Theta^2 + \cos^2\Theta d\varphi^2$, where $\theta$ is the radius and $\varphi$ is the angle of rotation in the polar coordinate system. When $\Theta = 0$, we obtain the boundary of $D^2$. Straightforward computation yields $dl_1^2 = \sin^{2n}\Theta(d\Theta^2 + \cos^2\Theta d\varphi^2)$. To describe all $G$-invariant minimal hypersurfaces in $S^n$ completely, we have to find all closed geodesics on the disc $D^2(\Theta,\varphi)$ with the metric $dl_1^2$, which was done by T. Otsuki [318], and W. Hsiang and H. Lawson in [214]. Put $a_0 = \cos^{-1}(1/\sqrt{n+1})$. The closed geodesic can be of two types, viz., diameters when $\varphi = \text{const}$ and the trajectories of the form

$$\varphi = \pm \int_a^{\Theta(\varphi)} \frac{d\Theta}{\cos\Theta \sqrt{\dfrac{\sin^{2n}(\Theta)\cdot\cos^2(\Theta)}{\sin^{2n}(\Theta_0)\cdot\cos^2(\Theta_0)} - 1}}, \qquad \Theta_0 = a_0,$$

where $a \leqslant a_0$. As $a \to a_0$, the geodesic tends to the dotted circle in Fig. 62. Therefore, the corresponding $G$-invariant minimal surfaces are the equators $S^{n-1} \subset S^n$ and the submanifolds $S^1 \times S^{n-2} \subset S^n$ with self-intersections.

In this problem, we can completely describe the behaviour of geodesics on

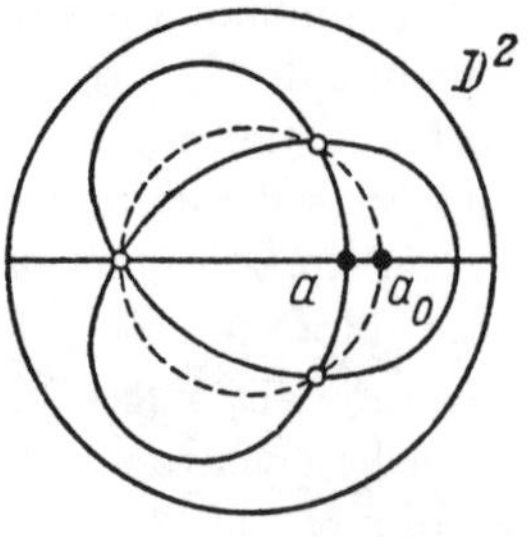


**Fig. 62**

the two-dimensional disc on whose boundary the metric $dl_1^2$ is annihilated, i.e., we actually consider the two-dimensional sphere $S^2$ with a fixed metric with one singularity. In the general case of symmetric spaces, the sphere turns out to be replaced by a two-dimensional torus on which a Riemannian metric with singularity is also given (see below).

Let $M^n = G/H$ be a compact, symmetric space. Consider the corresponding Cartan decomposition for the Lie motion group algebra (see the description in §17.4). The subgroup $H$ acts on $M \cong V$ via the rotations $v = hvh^{-1}$, fibering the manifold $V$ into orbits (as before, $M$ means the Cartan model of the symmetric space, embedded into the isometry group).

Let $B$ be the tangent plane to the Cartan model in the Lie algebra, and $K \subset B$ the Weyl chamber of the symmetric space (see §17.4). Consider the adjoint action of $H$ on the plane $B : b \to hbh^{-1}$. An orbit $O(k)$ passes through each point $k \in K$, orthogonal to a plane $P \subset B$, where $P$ is a Cartan subalgebra, or the maximal commutative plane. Let $C = \bigcup_{b \in K} O(b)$, and $\widetilde{C} = \exp(C) \subset M^n$. Then the set $\widetilde{C}$ is homeomorphic to $C$ in the case of a simply-connected $M$ (see §17.4), and the complement $M \backslash \widetilde{C}$ is precisely the set of all points in $M$, first conjugate of $e$, where $e = \exp(0)$, the centre of the ball. Thus, any compact, simply-connected and symmetric space $M$ can be obtained from the $n$-dimensional ball $C$ by factorizing its boundary sphere $\partial C$ with respect to a certain convenient action of the compact group. Since the whole of the manifold $M$ is swept out by the maximal torus $\widetilde{P} = \exp(P)$ in its rotation about $e$ by the adjoint action of $H$, almost all the events of interest will be studied just on $P$, its dimension being called the *rank* of the symmetric space.

We now clarify the minimal equivariant surface structure in a symmetric space.

Above, we have given the complete list of all equivariant minimal cones of co-dimension one in Euclidean space, which turn out to be "linearizations" of global minimal films placed in symmetric spaces of rank two. More precisely, we naturally identify the Euclidean space $\mathbf{R}^n$ with the tangent space $B$ to the symmetric space $M = G/H$. The minimal cone $CA^{n-2}$ is then the tangent minimal cone associated with a singular point of the minimal hypersurface, contained in $M^n$, with the manifold $A^{n-2}$ coinciding with the homogeneous space $H/N$, the principal orbit of the adjoint action of the stability group $H$ on $M$ (see Fig. 63). These closed minimal surfaces admit a very simple description in terms of a certain smooth function given on the manifold $M^n$, viz., they are its level surfaces. More precisely, let $f(t)$ be a smooth function (which we exhibit below) given on the maximal torus $\widetilde{P} \subset M$ of dimension two. A new function $\tilde{f}(x)$ can be constructed from $\tilde{f}(t)$ on the whole of $M$ by putting $\tilde{f}(x) = f(t)$, where $t = hxh^{-1}$ for a certain $h \in H$. Meanwhile, the original function $f(t)$ on the torus should be invariant with respect to the Weyl group which acts on it. In other words, $\tilde{f}(x)$ is constant on the orbits of the adjoint action of $H$ on $M$.

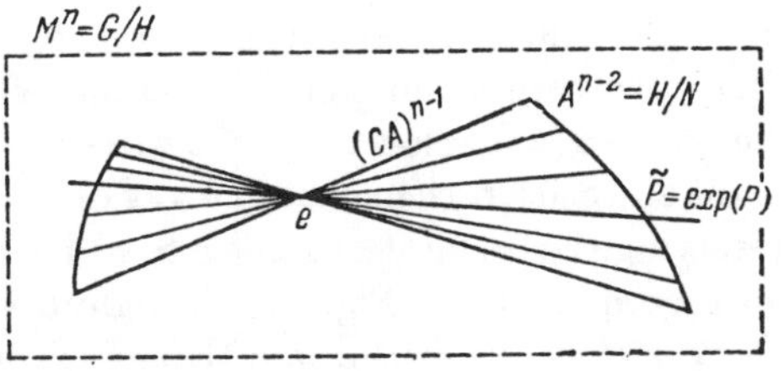


**Fig. 63**

Consider the level surfaces $\tilde{f}(x) = \text{const}$. The foliation of $M$ is defined by closed hypersurfaces which are manifolds without singularities except a unique minimal hypersurface. The latter contains singular points of the above types, viz., cones. Since $\tilde{f}(x)$ is $H$-invariant, all its level surfaces are also $H$-invariant. What $f(t)$ suffices to be given on the torus $\widetilde{P}$ to obtain a complete description of all minimal cones of co-dimension one?

As it turns out, it is sufficient to consider $\widetilde{P}'$ functions of the form $\text{Re}(z^p)$, $p = 2, 3, 4, \ldots$. In particular, if $p = 2$, then we obtain minimal cones over the sphere product, which were studied in detail in §24.2.

Thus, all equivariant singularities of minimal surfaces of co-dimension one are described by integral degrees of the complex variable $z$ on a two-dimensional torus with the standard complex structure.

Consider in detail only the case of cones over the direct sphere product, i.e., where $p = 2$. For $p > 2$, the reasoning is similar. Take the basic series $G/H = (\text{SO}(r+1) \times \text{SO}(s+1))/(\text{SO}(r) \times \text{SO}(s))$, and put $M = G/H$. Then $M$ is a compact, symmetric space of rank two, $M = S^r \times S^s$, $N = \text{SO}(r-1) \times \text{SO}(s-1)$, and, therefore, the principal orbit of the action of $H$ on $M$ is $H/N = S^{r-1} \times S^{s-1}$.

If we consider the tangent space $B = T_e(M)$, then we see that the cone over $A^{n-2} = C(S^{r-1} \times S^{s-1})$ coincides with the one considered in §24.2.

Since we search for $N$-invariant minimal hypersurfaces in $M$, it suffices to study geodesics on a two-dimensional torus $T^2 \subset M$ in the metric $dl_1^2 = v^2(a)\widetilde{ds}^2$. It is known that $\widetilde{ds}^2$ is the usual Euclidean metric on the torus (see §17.4); therefore, it remains to calculate the volume function. Straightforward computation shows that $v(x,y) = \sin^{r-1}(x)\sin^{s-1}(y)$, where $x, y$ are Cartesian coordinates on the torus, in which case it can be easily seen that $M/N$ is a plane square with sides $(0,\pi)$, and an open cell $K$, the Weyl chamber. Here, there are only two distinct orthogonal roots of the same length, though with different multiplicities $r-1$ and $s-1$. We have to study the behaviour of the geodesics on $(0,\pi)$ with the metric $dl_1^2 = \sin^{2r-2}(x) \cdot \sin^{2r-2}(y)$. Regarding $r$ and $s$ as continuous parameters, we put $r + s = n = \text{const}$. If $r = s$, then the diagonal $x = y$ is, obviously, a geodesic (the level lines of the function $f$ on the torus are represented in Fig. 64). The torus is represented as the plane to be factorized. One of the level lines is the union of two geodesic diagonals in the metric $dl_1$. Since the level lines are orthogonal to the boundary of the square, the constructed function is invariant with respect to the Weyl group; because the $Ox$- and $Oy$-axes are associated with the roots $a_1$ and $a_2$, the torus is obtained from the square $(-\pi,\pi)$ by identifying the opposite sides, and preserving the orientation. Let $r > s$. Then the level lines of $f$ start deforming in accordance with Fig. 65. Straightforward computation shows that the geodesic $x = y$ is deformed into the one given by the equation $y\tan y = \frac{s}{r}x\tan x$. The geodesics $\gamma$ and sides of the inside square are given as thick lines, passing through the vertex at an angle $\alpha$, $\tan\alpha = \sqrt{s/r}$. However, if $r < s$, then the picture is obtained by a reflection in the diagonal. Let $s$ vary continuously from 0 to $\infty$. The corresponding evolution of the geodesic representing a closed minimal surface in $S^r \times S^s$ is shown in Fig. 66. The function $f(x)$ obtained on the torus possesses one minimum, with one maximum, and two saddles, with all the critical points being nondegenerate, and the saddles (i.e., zeroes of order two) are described by the vector field $\operatorname{grad}\operatorname{Re}(z^2)$. The inverse image of a geodesic $\gamma$ under the projection $\pi : M \to M/N$ is the $N$-invariant minimal surface of co-dimension one in $S^r \times S^s$.

Extend $f(x)$ from the torus to the whole of the manifold $S^r \times S^s$ and to an $N$-invariant function $\tilde{f}(x)$ whose values fill the interval $[0,1]$. Then the level hypersurfaces fibre $M$, and we obtain for $\tilde{f}(x) = \frac{1}{2}$ an $N$-invariant minimal closed surface, the inverse image of the trajectory $\gamma$, containing precisely two singular points. The submanifold $\{\tilde{f} = 0\}$ is the sphere $S^s$, the submanifold $\{\tilde{f} = 1\}$ is the sphere $S^r$, and both are minimal extremal isolated orbits of the action of the group $N$. If $\tilde{f}(x) = c$, $c \neq (0, 1/2, 1)$, then the hypersurface $\tilde{f}(x) = c$ has no singular points. The $N$-invariant surface $\{\tilde{f}(x) = \frac{1}{2}\}$ is not globally minimal; however, it is minimal with respect to any variations whose supports embrace no more than a "half" of the whole surface (Fig. 66).

It is proved similarly that all the other singularities described above can be included into closed minimal surfaces in the corresponding symmetric space. All these hypersurfaces are level surfaces of smooth functions invariant with

Fig. 64

Fig. 65

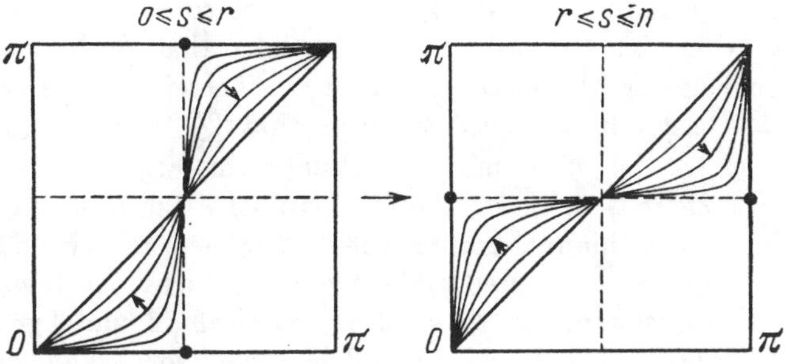


Fig. 66

respect to the action of $N$ and the Weyl group. Their level lines can be represented on the two-dimensional torus as the integral curves of vector fields of the form $\operatorname{grad}\operatorname{Re}(z^p)$. Then the minimal hypersurface containing conic singularities will be given by the separatrix diagram of the degenerate saddle, or zero of the order $p$, of the flow $\operatorname{grad}\operatorname{Re}(z^p)$.

## 24.4. On the Existence of Nonlinear Functions Whose Graphs in Euclidean Space Are Minimal Surfaces

Let $x^n = f(x^1, \ldots, x^{n-1})$ be a real smooth real-value function defined on the Euclidean space $\mathbf{R}^{n-1}(x^1, \ldots, x^{n-1})$. Then its graph in the space $\mathbf{R}^n(x^1, \ldots, x^n)$ is a submanifold of co-dimension one. Suppose that it is a locally minimal surface, i.e., the function $f$ satisfies the differential equation

$$\sum_{i=1}^{n-1} \frac{\partial}{\partial x^i}\left(\frac{f_{x^i}}{\sqrt{1+\sum_{j=1}^{n-1}(f_{x^j})^2}}\right)=0.$$

The question arises: Must the function $f$ be linear if it is defined on the whole of the hyperplane $\mathbf{R}^{n-1}$? The problem bears the name of S. N. Bernstein who gave a positive answer for $n = 3$, and proved that, if $z = f(x,y)$ is the graph of a smooth function given on the whole plane $\mathbf{R}^2$, if the Gaussian curvature $K$ of this graph is nonpositive, and if there is a point at which $K$ is strictly negative, then $\sup_{(x,y)\in\mathbf{R}^2} |f| = +\infty$. The solution of the above problem follows easily. In fact, if $z = f(x,y)$ is a locally minimal surface, then we can consider the function $u = \tan^{-1}(\frac{\partial f}{\partial x})$; the function $f$ is a solution to the equation $(1 + f_x^2)f_{yy} - 2f_x f_y f_{xy} + (1 + f_y^2)f_{xx} = 0$. (Recall that the minimal surface equation in three-dimensional Euclidean space acquires just this form.) A simple calculation demonstrates that the graph of $u(x,y)$ has nonpositive Gaussian curvature $K$. If $u$ is not constant, then there exists a point at which $K < 0$, with the consequence that $\sup|u| = +\infty$, which is impossible due to the boundedness of $u$. Therefore, $u$ is constant, but then $\frac{\partial f}{\partial x} = \text{const}$. It is proved similarly that $\frac{\partial f}{\partial y} = \text{const}$, in which case $f = ax + by + c$, i.e., $f$ is linear. Certainly, this argument is not valid in dimensions greater than three; therefore, to solve the problem of existence of nonlinear locally minimal graphs, the development of new and sufficiently complex machinery is required. Meanwhile, it turns out that the answer depends on dimension $n$, viz., for small $n$, any locally minimal graph is linear, whereas, for large $n$, there exist essentially nonlinear locally, and even globally, minimal graphs.

More precisely, the following is valid.

**Theorem 24.4.1 (E. Bombieri, E. De Georgi, and E. Giusti; see [51]).** *Let $x^n = f(x^1,\ldots,x^{n-1})$ be a smooth function defined everywhere on the hyperplane $\mathbf{R}^{n-1}$ in $\mathbf{R}^n$, and let its graph be a locally minimal surface in $\mathbf{R}^n$ with zero mean curvature. Then the function $f$ is linear for $n \leqslant 8$. If, however, $n \geqslant 9$, then there exists a nonlinear function $f$ whose graphs are locally, and globally, minimal surfaces.*

It is remarkable that the solution to this problem is closely related to that of minimal cone existence (see the previous sections). We give the outline of the proof for $n \leqslant 8$.

We can assume that the graph $X^{n-1}$ of $x^n = f(x^1,\ldots x^{n-1})$ passes through the origin in $\mathbf{R}^n$. It turns out that its part inside any compact boundary on the graph realizes an absolute minimum of $\mathrm{vol}_{n-1}$. In other words, $X^{n-1}$ is a globally minimal surface in $\mathbf{R}^n$ relative to perturbations with compact supports in the homology class. We take the intersection of $X^{n-1}$ with the ball of radius $r$, centre at the origin $O$. Carry out the similarity transformation with the coefficent $l/r$, which yields a surface $X_r^{n-1}$ enclosed in the ball of unit radius with boundary on the sphere of unit radius, the boundary being a locally minimal submanifold in the sphere (see Fig. 67). If $r$ varies, then the surface $X_r$ in the ball $D^n(O,1)$ will, in general, be altered, too. Consider the "limit" of these minimal surfaces as $r \to \infty$, which turns out to exist and to be a cone $CA$ over a certain $(n-2)$-dimensional set $A$ in the sphere of unit radius. Since the original graph is globally minimal, the cone is also globally minimal with respect to its boundary. But, by Theorem 24.1.1 and the boundedness from above of dimension $(n \leqslant 7)$, the cone should be a disc, from which it follows quite easily that the graph is a hyperplane, thus completing the proof. In dimension 8, a certain additional argument is required, which we omit and refer the reader to [391], for example.

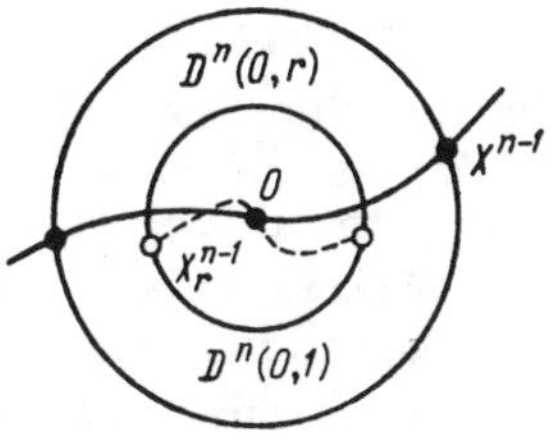


**Fig. 67**

The existence of nonlinear minimal graphs in dimensions greater than 8 has been proved in [51], and is of more analytic character. We will not dwell on it here.

### 24.5. Harmonic Mappings of Spheres in Nontrivial Homotopy Classes

In the present section, we consider harmonic mappings of Riemannian manifolds and solve the problem of finding a harmonic mapping in a certain prescribed homotopy class of mappings. Since harmonic mappings are extremal functions for the multidimensional Dirichlet functional, we first describe its basic properties.

Denote two smooth Riemannian manifolds by $M$ and $N$. Suppose that $M$ is compact and orientable. We shall be interested in the properties of mappings $f: M \to N$ from the standpoint of the Dirichlet functional. Let $TM$ and $TN$ be the tangent bundles over $M$ and $N$, respectively, and $C^\infty(\xi)$ the space of $C^\infty$ cross sections of a certain differentiable vector bundle $\xi$ over $M$. Furthermore, if $C^\infty(M,N)$ is the space of smooth mappings $f: M \to N$, then, for each $f \in C^\infty(M,N)$, a Riemannian vector bundle over $M$ is defined, being induced by $TN$ via $f$, which means that its fibre over a point $x \in M$ is the linear space $T_{f(x)}N$. Meanwhile, the Riemannian structure is carried from $TN$.

Let $\mathrm{Hom}(TM, f^*TN)$ be the vector bundle over $M$ of linear homomorphisms of the fibrations $TM$ and $f^*TN$. In other words, its fibre over a point $x \in M$ is the space of continuous linear mappings of $T_xM$ to $T_{f(x)}N$. We now describe the Riemannian structure in $\mathrm{Hom}(TM, f^*TN)$. If $a, b \in C^\infty(\mathrm{Hom}(TM, f^*TN))$, and $a, b$ are $C^\infty$ cross sections, then they can be written in the matrix form $a^i_j, b^i_j$ with respect to local coordinate systems on $M$ and $N$. Now, if $g_{ij}, \hat{g}_{ij}$ are the metric tensors on $M$ and $N$, written with respect to the same coordinates, respectively, then the Riemannian structure is introduced by the formula

$$(a,b) = g^{ij}\, \hat{g}_{\alpha\beta}\, a^\alpha_i\, b^\beta_j. \tag{1}$$

Thus, both metric tensors of the Riemannian manifolds in question are involved. Since, for each smooth mapping $f \in C^\infty(M, N)$, its differential $df$ is defined, it is evident that $df \in C^\infty(\mathrm{Hom}(TM, f^*TN))$, which permits us to introduce a nonnegative smooth function $\|df\|^2 = (df, df)$ by making use of Riemannian structure (1). Because $M$ is compact and orientable, we can integrate the function with respect to the whole of $M$. By definition, we take the obtained value as that of the Dirichlet functional on the mapping $f$, viz.,

$$D[f] = \int_M \|df\|^2 * 1, \quad D\colon C^\infty(M, N) \to \mathbb{R}^1_+, \ \mathbb{R}^1_+ = \{x \geqslant 0\}, \tag{2}$$

where $*1$ means the standard Riemannian volume form on $M$, which we have written by resorting to the operator $*$ transforming exterior forms of degree $k$ into those of degree $m-k$, with $m = \dim M$. Then the $m$-dimensional Riemannian volume form is the image of the scalar function 1 on applying the operator $*$. This way of writing it is convenient for further computation. Though the given construction of the Dirichlet functional (2) is different from the definition of §23, and since we do not raise the integrand to a power, the two approaches are equivalent in many cases from the extremals standpoint. Raising the integrand to a power is convenient if we compare the Dirichlet integral with the volume functional (see §23), but it is immaterial in the present section.

**Definition 24.5.1.** The Dirichlet functional extremals $D[f]$ are called *harmonic mappings.*

Note that they coincide with those defined in §23. Moreover, it follows that if $N = \mathbf{R}^1$, the real line with the standard Euclidean metric, and $\varphi \in C^\infty(M, \mathbf{R}^1) = C^\infty(M)$, then $\|d\varphi\| = |\operatorname{grad}\varphi|$, and we obtain the usual Dirichlet functional defined on the scalar functions

$$D\colon C^\infty(M)\to\mathbb{R}^1_+,\quad D[\varphi]=\int_M |\operatorname{grad}\varphi|^2 *1. \tag{3}$$

Since the extremals of (3) are harmonic functions on a manifold, or the elements of the kernel of the Laplace operator $\Delta$ on $C^\infty(M)$, Definition 24.5.1 yields the natural extension of the concept of harmonic functions to the case of smooth mappings to a certain Riemannian manifold.

We shall derive the Euler–Lagrange equations explicitly for the Dirichlet functional (2), indicate certain important particular solutions, and then consider the problem of discovery of harmonic mappings in Euclidean sphere mapping classes (for more detailed familiarity with harmonic mapping theory, we recommend the survey in [118]).

Assume now that symmetric connections corresponding to the Riemannian metric are introduced on the tangent bundles $TM$ and $TN$. Define the Riemannian connection $\overset{*}{\nabla}$ on the bundle $f^*TN$ induced by a mapping $f: M\to N$. If $v\in C^\infty(f^*TN)$ and $E_1,\dots E_n$ are local-basis smooth vector fields in $TN$, then the decomposition $v=v^iE_i$ can be written locally, where $v^i\in C^\infty(U)$, i.e., $v^i$ are functions defined locally on $U\subset M$.

Let $a\in T_xM$. We then put

$$\overset{*}{\nabla}_a v=(av^i)\,E_i+v^i\hat{\nabla}_{df(a)}E_i, \tag{4}$$

where $av^i$ denotes the action of the vector field $a$ as of the differential operator on the function $v^i$, and $\widehat{\nabla}$ is a connection on $TN$. That equality (4), in fact, determines a Riemannian connection on the fibration $f^*TN$ follows from the properties of the operation, given below. In many respects their proof is similar to that for analogous connections introduced in §2 (the reader can verify it by himself or refer to [181]).

Thus, let $a,b\in C^\infty(TM)$, $\varphi$, $\psi\in C^\infty(M)$, $v$, $w\in C^\infty(f^*TN)$. Then

$$\overset{*}{\nabla}_{\varphi a+\psi b}v=\varphi\overset{*}{\nabla}_a v+\psi\overset{*}{\nabla}_b v,\quad \overset{*}{\nabla}_a(v+w)=\overset{*}{\nabla}_a v+\overset{*}{\nabla}_a w,$$

$$\overset{*}{\nabla}_a(\varphi v)=(a\varphi)\,v+\varphi\overset{*}{\nabla}_a v,\quad a\,(v,\ w)=(\overset{*}{\nabla}_a v,\ w)+(v,\ \overset{*}{\nabla}_a w).$$

The latter equality means that the introduced connection is Riemannian, i.e., corresponds to the metric.

We now turn to the fibration $\mathrm{Hom}(TM,f^*TN)$ and define the Riemannian connection $\widetilde{\nabla}$ on it as follows. Let $a,b\in C^\infty(TM)$, $h\in C^\infty(\mathrm{Hom}(TM,f^*TN))$. We then set

$$(\tilde{\nabla}_a h)\,b=\overset{*}{\nabla}_a h\,(b)-h\,(\nabla_a b), \tag{5}$$

where $\nabla$ is a connection on the fibration $TM$.

We list the main properties of this connection, the proofs being similar to the argument in §2 or referred to [181] (as well as in the previous case). The form $\tilde{\nabla} h$ is a bilinear mapping from the direct product $C^\infty(TM) \times C^\infty(TM)$ to the space $C^\infty(f^*TN)$ relative to the addition to and multiplication by scalar functions. Besides, if $h' \in C^\infty(\mathrm{Hom}(TM, f^*TN))$, $\varphi \in C^\infty(M)$, then we have

$$\tilde{\nabla}_a(h+h') = \tilde{\nabla}_a h + \tilde{\nabla}_a h', \quad \tilde{\nabla}_a(\varphi h) = (a \cdot \varphi) \cdot h + \varphi \tilde{\nabla}_a h,$$
$$a(h, h') = (\tilde{\nabla}_a h, h') + (h, \tilde{\nabla}_a h').$$

The latter equality means that the introduced connection is Riemannian.

**Definition 24.5.2.** The gradient of the differential of a mapping $f: M \to N$, i.e., an expression of the form $\tilde{\nabla} df$, is called its *second fundamental form* $A(f)$. The gradient is a smooth bilinear form on the direct product $TM \times TM$ with values in the induced fibration $f^*TN$.

**Definition 24.5.3.** The trace of the second fundamental form of a mapping $f: M \to N$, i.e., an expression of the form $H(f) = \mathrm{tr}\, A(f) \in C^\infty(f^*TN)$ is called its *mean curvature* $H(f)$.

It is shown below that the two definitions naturally generalize those of the second fundamental form and mean curvature for the case of immersing the manifold $M$ into the manifold $N$ to the case of an arbitrary smooth mapping. If $f: M \to N$ is an immersion, then Definitions 24.5.2 and 24.5.3 turn into those given in §2. We now state without proof the properties of the second fundamental form, which are necessary for the following.

**Lemma 24.5.1.** *The second form $A(f)$ is symmetric on $TM \times TM$.*

**Lemma 24.5.2.** *If $f$ is an isometric immersion of a manifold $M$ into a manifold $N$, then the form $A(f)$ and the mean curvature $H(f)$ coincide with the second fundamental form $A$ and the mean curvature $H$ of the submanifold $f(M)$ in $N$ (see the definitions in §2). Namely, if $a, b \in T_x M$, then $A(f)_x(a, b) = A_{f(x)}(df(a), df(b))$.*

We now turn to the derivation of the Euler–Lagrange equations for the Dirichlet functional and study of totally geodesic mappings along with locally minimal immersions.

**Lemma 24.5.3.** *The equation $H(f) = 0$, where $H(f)$ is the mean curvature of a mapping $f$, is said to be Euler–Lagrange for the Dirichlet functional (2). In other words, $f$ is harmonic if and only if its mean curvature is zero.*

**Definition 24.5.4.** A mapping $f \in C^\infty(M, N)$ is said to be *totally geodesic* if it transforms geodesic manifolds $M$ into geodesic manifolds $N$.

**Lemma 24.5.4.** *A mapping $f \in C^\infty(M, N)$ is totally geodesic if and only if $A(f) = 0$, i.e., its the second fundamental form is identically zero.*

**Proof.** Let $\gamma(t)$ be a geodesic in $M$, i.e., $\nabla_{\dot\gamma}\dot\gamma = 0$. Consider the curve $\Gamma(t) = f(\gamma(t))$ in $N$. Then $\dot\Gamma = df(\dot\gamma)$. Therefore, by definition of the connections $\overset{*}{\nabla}, \tilde\nabla$, we have

$$\hat\nabla_{\dot\Gamma}\dot\Gamma = \hat\nabla_{df(\dot\gamma)} df(\dot\gamma) = \overset{*}{\nabla}_{\dot\gamma} df(\dot\gamma) = (\tilde\nabla_{\dot\gamma} df)\dot\gamma = A(f)(\dot\gamma, \dot\gamma).$$

Hence, if $A(f) = 0$, then the curve $\Gamma(t)$ is a geodesic in $N$; vice versa, if $\Gamma(t)$ is a geodesic in $N$, the $A(f)(a,a) = 0$ for any $a \in C^\infty(TM)$. It follows that $A(f) = 0$, because $A(f)$ is a symmetric form.

Q.E.D.

**Corollary 24.5.1.** *Totally geodesic mappings and locally minimal isometric immersions are harmonic mappings.*

**Proof.** If $f$ is a totally geodesic mapping, then, by Lemma 24.5.4, we have $A(f) = 0$. Therefore, $H(f) = \operatorname{tr} A(f) = 0$. If $f$ is a locally minimal immersion, then it isometrically maps $M$ onto $f(M)$, and the mean curvature of the manifold $f(M)$ in $N$ is zero. According to Lemma 25. 4. 2, this statement is equivalent to the vanishing of the mean curvature $H(f)$ of the mapping $f$.

Q.E.D.

One of the principal geometric problems of Riemannian manifold harmonic mapping theory is that of finding a harmonic representative in the homotopy class; namely, can a given mapping $f \in C^\infty(M,N)$ be continuously deformed into a harmonic mapping? In terms of the previous notation, the question can be asked differently: Does there exist a mapping $F \in C^\infty(M \times [0,1], N)$ such that $F|_{M\times\{0\}} = f$, $F|_{M\times\{1\}}$ is harmonic? Below, we list certain important cases where the answer is positive, and dwell at length on the discovery of harmonic mappings in the sphere mapping homotopy classes, i.e., the harmonic realization of elements of the homotopy groups $\pi_i(S^n)$. It is easy to give examples of such homotopy classes of mappings of $M$ to $N$ in which there is no harmonic mapping, or representative.

Let $M, N$ be two smooth, compact, closed, and orientable Riemannian manifolds. The positive answer to the above question can be obtained in the cases (1) $\dim M = 1$, (2) $\dim N = 1$, (3) $\dim M = 2$, $\pi_2(N) = 0$, and (4) $N$ has a nonpositive sectional curvature (e.g., see [118]).

In (4), we suppose that $R(\cdot,\cdot)$ is the curvature tensor of the symmetric Riemannian connection in the tangent bundle $TN$ of $N$. The value $\rho_x(X,Y) = -\langle R(X,Y)X,Y\rangle_x$ is called sectional curvature along a normed element of area spanned by two vectors $X$, $Y \in T_xN$, where the angle bracket means the scalar product on $T_xN$. The manifold $N$ is said to be of nonpositive sectional curvature in $\rho_x(X,Y) \leqslant 0$ at each point $x \in N$; $X,Y \in T_xN$. Note that the Riemannian metric of nonpositive curvature can be introduced on two-dimensional, closed, and orientable surfaces of genus $g \geqslant 1$. There are no metrics with nonpositive sectional curvature on the spheres $S^n$ for $n \geqslant 2$. It is rather simple to give an example of a mapping having no harmonic one homotopic to it for which we have to consider a smooth mapping of a two-dimensional torus to a two-dimensional sphere (e.g., see [118]) with arbitrary Riemannian metrics.

**Theorem 24.5.1 (R. Smith; see [397]).** *The list below contains those homotopy groups of spheres, for which the harmonic realizations have already been found, and each element of which, or homotopy class, contains a harmonic mapping, viz.,*

1) $\pi_n(S^n) = \mathbb{Z}$, $n = 1, 2, \ldots, 7$; 2) $\pi_{n+1}(S^n) = \mathbb{Z}_2$, $n = 3, 4, \ldots, 8$;
3) $\pi_7(S^5) = \mathbb{Z}_2$; 4) $\pi_9(S^6) = \mathbb{Z}_{24}$; 5) $\pi_7(S^3) = \mathbb{Z}_2$; 6) $\pi_{15}(S^9) = \mathbb{Z}_2$.

*Furthermore, there exist groups in which only part of the elements possess harmonic realization. For example, (7) in the group $\pi_3(S^2) = \mathbf{Z}$, elements of the form $\pm k^2$, where $k$ is an integer, are realized harmonically; (8) in the group $\pi_7(S^4) = \mathbf{Z} \oplus \mathbf{Z}_{12}$, elements of the form $(\pm k^2, 0)$, where $k$ is an integer, are realized harmonically; (9) in the groups $\pi_{n+3}(S^n) = \mathbf{Z}_{24}$, for $n = 5, 6, \ldots, 10$, only unit elements are realized harmonically, except the group $\pi_9(S^6)$ which is completely harmonically realized (see above).*

We call a mapping $f : S^n \to S^p$ a *harmonic polynomial* if $f = F|_{S^n}$, where $F : \mathbf{R}^{n+1} \to \mathbf{R}^{p+1}$, is such that the functions $F^i (i = 1, \ldots, p+1)$ are harmonic homogeneous polynomials of the same homogeneity degree, in which case $f$ also turns out to be harmonic (see [397]).

We call a mapping $f_1 * f_2 : S^{n+m+1} \to S^{p+q+1}$ of the form

$$f_1 * f_2(x, y) = \left(|x| f_1\left(\frac{x}{|x|}\right), |y| f_2\left(\frac{y}{|y|}\right)\right)$$

with respect to Cartesian coordinates $(x, y) \in \mathbf{R}^{n+1} \times \mathbf{R}^{m+1}$ and $(x', y') \in \mathbf{R}^{p+1} \times \mathbf{R}^{q+1}$ the *join* of two mappings of Euclidean spheres $f_1 : S^n \to S^p$ and $f_2 : S^m \to S^q$.

**Theorem 24.5.2 (Join Theorem; see [397]).** *Let $f_1 : S^n \to S^p$ and $f_2 : S^m \to S^q$ be harmonic polynomials of homogeneity degrees $k_1$ and $k_2$, respectively.*

*(1) If $k_1 > \theta(n-1)$, $k_2 > \theta(m-1)$, where $\theta = \frac{\sqrt{2}-1}{2}$, then there exists a harmonic mapping $\Phi : S^{n+m+1} \to S^{p+q+1}$ homotopic to the join $f_1 * f_2$, and*

*(2) if $k_1 = k_2$, $n = m$, then there also exists a harmonic mapping homotopic to the join $f_1 * f_2$.*

## 24.6. A Rough Outline of Certain Recent Results on the Link of Harmonic Mapping Properties to the Topology of Manifolds

Here we briefly survey the new results obtained by Pluzhnikov, Tyrin, and White.

Recall that a harmonic mapping $f : M \to N$ of a Riemannian manifold is said to be *stable* if the second variation form of the Dirichlet functional $D$ is positive semidefinite, calculated at a "point" $f$. The results of M. I. Vyshik [423] and S. Smale [396], obtained within the framework of general elliptic theory show that any harmonic mapping realizes a local minimum of the Dirichlet functional in the class of variations with sufficiently small supports. Hence, the necessity to consider "global" variations scattered across the whole manifold $M$. It can be immediately seen from the second variation formula of the Dirichlet functional that all $f : M \to N$ are stable if $M$ is closed, and the sectional curvature of $N$ is nonpositive. On the other hand, E. Mazet [263] and R. Smith [398] have shown that the identity mappings of Riemannian manifolds, which are the second simplest harmonic mappings after constants, are not always stable. P. Leung [249] and A. V. Tyrin [415]–[416] have obtained whole series of Riemannian manifolds $N$ such that any nonconstant harmonic mapping $f : M \to N$ is unstable, with $M$ being connected and compact. A. I. Pluzhnikov [331]–[334], [435] investigated the "dual" situation and obtained a number of series of homogeneous Riemannian manifolds $M$ such that all nonconstant harmonic mappings $f : M \to N$ are *a priori* unstable for any Riemannian manifold $N$.

Pluzhnikov has considered the problem of *a priori* instability of harmonic mappings $f : M \to N$, where $M$ is a homogeneous Riemannian manifold of positive dimension with an isometry group that acts transitively. As the following theorem shows, the instability of the identity mapping of $M$ plays the determining role.

**Theorem 24.6.1 ([333], [334]).** *Let $M$ be a connected, irreducible, compact, and homogeneous Riemannian manifold, and let the identity mapping of $M$ be unstable. Then all the nonconstant harmonic mappings $f : M \to N$ from $M$ into an arbitrary smooth Riemannian manifold $N$ are unstable.*

Variations $f_t$ of a mapping $f$, which decrease the Dirichlet integral for small $t$, are of the form $f \circ \varphi_t$ in Theorem 24.6.1, where $\varphi_t$ is a one-parameter diffeomorphism group of $M$ so that $D(\varphi_t) < D(\varphi_0)$ for sufficiently small $|t| \neq 0$. Note that irreducible, homogeneous Riemannian manifolds admit a complete classification (see O.V. Manturov's paper [257]).

The theorem makes it possible to indicate a number of spaces $C^\infty(M, N)$ in which all the critical "points" of the Dirichlet functional are "saddles," except constants, or mappings into a point.

**Theorem 24.6.2 ([333], [334]).** *Let $M$ be a connected, compact, irreducible, nonsingular, and symmetric space endowed with the standard Riemannian metric. Then, for any smooth Riemannian manifold $N$, the space $C^\infty(M, N)$ has no nonconstant stable harmonic mappings if and only if $M$ belongs to one of the following series (a)–(e), viz.,*

*(a) standard sphere $S^n$ of dimension $n \geqslant 4$,*

*(b) Grassmannian manifolds over the division ring of quaternions* $\mathrm{Sp}(p+q)/(\mathrm{Sp}(p) \times \mathrm{Sp}(q))$,

*(c) spaces* $\mathrm{SU}(2m)/\mathrm{Sp}(m)$ *associated with the standard embedding of the subgroup* $\mathrm{Sp}(m)$ *into* $\mathrm{SU}(2m)$, $m \geqslant 2$,

*(d) special unitary groups* $\mathrm{SU}(n)$, $n \geqslant 2$, *and*

*(e) symplectic groups* $\mathrm{Sp}(n)$.

**Remark.** The sphere $S^3 = \mathrm{SU}(2)$ is entered in series (d) as a Lie group.

The instability of all nonconstant harmonic mappings $f : S^n \to N (n \geqslant 3)$ was proved earlier by Y. Xin [429].

The proof is based on Theorem 24.6.1 and the results obtained by R. Smith [398] and A.V. Tyrin [415] and [416], who solved the problem of the instability of identity mappings of simply-connected, compact, irreducible nonsingular, and symmetric spaces of Types I and II, respectively. The following result is also due to A.V. Tyrin [415], [416]. Let $M = G/H$ be a compact, homogenous, and irreducible Riemannian manifold with the $G$-invariant metric, so that its identity mapping is unstable. Then all nonconstant harmonic smooth mappings $f : N \to M$ from an arbitrary connected, compact Riemannian manifold $N$ are unstable.

We now give examples of harmonic mappings of symmetric spaces, which turn out to be unstable due to Theorem 24.6.2, viz., embedding mappings of the Cartan models of symmetric spaces (a), (b), and (c); nontrivial homomorphisms of Lie groups (d); (e) those into an arbitrary Lie group with the bi-invariant Riemannian metric, totally geodesic mappings of spheres of dimensions greater than two to symmetric spaces, constructed by the author in [153], [154], and [157]; projection maps of the fibre bundles $G \to G/H$, associated with homogeneous Riemannian manifolds $G/H$ with motion groups (d), (e). The Hopf fibrations $S^{2n-1} \to S^n$ $(n = 2, 4, 8)$ and the harmonic realizations of nontrivial elements of the homotopy groups $\pi_n(S^n)(3 \leqslant n \leqslant 7$, $\pi_{m+1}(S^m)(3 \leqslant m \leqslant 8)$, obtained in R. Smith's work [397].

We now give another result by A.I. Pluzhnikov demonstrating that it is sometimes possible to distinguish in topological terms the classes of those manifolds all of whose nontrivial harmonic mappings are unstable. Recall that a connected manifold is said to be *2-connected* if its first and second homotopy groups are trivial.

**Theorem 24.6.3.** *All the manifolds in Theorem 24.6.2 are symmetric space* $M$ *of Type I, i.e., which is not a group, is 2-connected if and only if all nonconstant harmonic mappings* $f : M \to N$ *to an arbitrary smooth Riemannian manifold* $N$ *are unstable.*

Recall that the dimension of the maximal subspace in the space of vector fields along $f$, on which the Dirichlet functional $D$ second variation form is negative-definite at the "point" $f$ is called the *index* $\mathrm{ind}(f)$ of the harmonic mapping $f : M \to N$. If $M$ is closed, i.e., compact and without boundary, then the index of any harmonic mapping $f : M \to N$ is finite. The positiveness of the index of a mapping is always equivalent to the instability of the mapping.

**Theorem 24.6.4 ([331], [332]).** *Let $f : S^n \to N$ be a nonconstant harmonic mapping of a sphere of positive radius and dimension $n \geqslant 3$, standardly embedded into $R^{n-1}$, to an arbitrary smooth Riemannian manifold $N$. Then* $\operatorname{ind}(f) \geqslant n+1$. *The corresponding space of vector fields along $f$, which generate variations of $f$ decreasing locally the value of $D$, is generated by fields of the form $df(\operatorname{grad}\theta)$, where $\theta$ is the restriction of an arbitrary nonzero linear function in $\mathbf{R}^{n-1}$ to $S^n$.*

A weaker version of the above theorem was obtained by J. Eells and L. Lemaire, viz., $\operatorname{ind}(f) \geqslant k+1$, where $k$ is the maximal rank of the mapping $f$.

Note that the gradient of the restrictions to $S^n$ of a nonzero linear function in $\mathbf{R}^{n-1}$ generates a one-parameter group $\varphi_t$ of conformal diffeomorphisms of the sphere. The variation $f_t$ of $f$, decreasing the Dirichlet integral for small $t$, is of the form $f_t = f \circ \varphi_t$.

The estimate obtained by A. I. Pluzhnikov for the index of a harmonic mapping of the sphere cannot by generally improved due to the equality $\operatorname{ind}(\mathrm{id}_s) = n+1$ holding for the identity mapping $\mathrm{id}_s$ of the sphere of dimension $n \geqslant 3$ (see E. Mazet [263], R. Smith [398]).

Further, A. I. Pluzhnikov considered a number of problems related to the Dirichlet functional minimization, viz.,

$$D(f) \to \inf, \quad f \in F \subset C^\infty(M,N).$$

Its solutions are smooth mappings $f \in F$ of Riemannian manifolds realizing a minimum of $D$ on the subset $F$ of the mapping space. Assume that $M$ is compact, fix a certain mapping $\psi \in C^\infty(M,N)$, and distinguish two principal concretizations of the above problem, viz.,

1. (minimal) realization: Assuming that $\partial M = \emptyset$, it is required to minimize $D$ in the homotopy class $F = [\psi]$ identified with the corresponding connected component of the space $C^\infty(M,N)$,

2. (minimal) "spanning," or Dirichlet, problem: Assuming $\partial M \neq \varnothing$, a minimum of $D$ is desired on the set

$$F = C^\infty_\psi(M,N) = \{f \in C^\infty(M,N) : f|\partial M = \psi|\partial M\}$$

(absolute version) or on the connected component of the space $C^\infty_\psi(M,N)$, containing the mapping $\psi$ (relative version).

Both problems are deeply nontrivial, and have an old history. Let $N$ be closed, i.e., compact and without boundary. The classical results of variational calculus arising from the works of J. Hadamard, E. Cartan and D. Hilbert are statements regarding the solvability of the problems in the case $\dim M = 1$. (The proof can be based on the Hopf and the Rinow theorem "on the shortest," which provides a solution to the absolute version of the Dirichlet problem; see [208]. A modern solution of the realization problem by the method of steepest descent is contained, e.g., in S. I. Alber [4] and W. Klingenberg [226].) The realization problem and both versions of the "spanning" problems also possess solutions if the sectional curvature $N$ is nonpositive (see S. I. Alber[3–4], J. Eells and J. Sampson, [119], R. Hamilton [190]). Solving the two-dimensional Plateau problem, C. Morrey showed in [289] that the absolute version of the Dirichlet problem is solvable in dimension $\dim M = 2$. The author constructed a family homeomorphic to $U(m)$ of totally geodesic mappings of the two-dimensional functional disk $D^2$ to group $\mathrm{SU}(2m)(m \geqslant 1)$ (the metric being Euclidean), which is intimately related to unitary Bott periodicity, and coincides with the set of minima of $D$ in the space $C^\infty_\psi(D^2, \mathrm{SU}(2m))$ generated by the geodesic embedding of the circle $\psi|\partial D^2$, extended to the monomorphism $\mathrm{SU}(2) \to \mathrm{SU}(2m)$ [154]. The solvability of the two dimensional realization problem with $\dim M = 2$ has been proved quite recently, assuming additionally that $\pi_2(N) = 0$ (see Sacks and Uhlenbeck [364], L. Lemaire [248]). J. Sacks and K. Uhlenbeck proved that the nontriviality of $\pi_2(N)$ entails the existence of at least one nonconstant harmonic mapping $f : S^2 \to N$ which is globaly minimal in its homotopy class. The following important result was obtained by S. Hildebrandt, H. Kaul and K. Widman [202]. Denote by $B(r)$ a geodesic ball in $N$ of radius $r > 0$, lying in a the normal neighbourhood of any of its points (normal neighbourhood of a point is any domain of a normal coordinate system with the centre at this point). Let $\psi(\partial M) \subset B(r)$, and $r < \pi/2\sqrt{k}$, where the constant $k > 0$ majorizes the sectional curvature function on $N$. Then there is in $C^\infty_\psi(M, N)$ an absolutely minimal harmonic mapping $f$, with $f(M) \subset B(r)$. Examples of the solutions of the realization problem are holomorphic mappings of closed Kähler manifolds (see A. Lichnerowicz [254]).

It was proved by A. I. Pluzhnikov [333]–[335], [435] (and later by B. White [434]) that the minimal realization problem has no nontrivial solutions in $C^\infty(M, N)$ if at least one of the closed manifolds $M$ and $N$ is *2-connected*, i.e., whose first and second homotopy groups are trivial. These results show that nonconstant harmonic mappings that are minimal in their homotopy class exist only among those connected, closed Riemannian manifolds of whose first two homotopy groups at least one is nontrivial.

It is necessary to note the absence in the literature of examples of spaces $C^\infty_\psi(M, N)$ containing no absolutely minimal harmonic mappings. There are situations (see L. Lemaire [248], H. Karcher and J. Wood [224]), where the condition $\psi/\partial M = \text{const}$ entails the constancy of any harmonic mapping $f : M \to N$. The relative version of the minimal "spanning" problem then has nontrivial solutions.

Within the framework of the minimal realization problem, the first negative result was obtained by J. Eells and J. Sampson in [119], who proved that $\inf D = 0$ in any homotopy class of smooth mappings of the standard sphere $S^n$ in itself for $n \geqslant 3$, where $\inf D$ denotes the *infimum of the set of values of the Dirichlet functional* on the subset of the mapping space, indicated in the text. We will call a homotopy class, or the corresponding connected component of the space $C^\infty(M, N)$, *nontrivial* if it contains no locally constant mappings, since $D(f) = 0$ only for locally constant mappings, the minimal realization problem is unsolvable in those nontrivial homotopy classes on which $\inf D = 0$. Denote by $[\mathrm{id}_M]$ the homotopy class of the identity mapping of a Riemannian manifold $M$, and put

$$D(M) = \inf\{D(\varphi) | \varphi \in [\mathrm{id}_M]\}.$$

Let $M$ be compact, and $D(M) = 0$. It was shown by J. Eells and L. Lemaire in [211] that, first, the equality $D(M) = 0$ remains valid if we replace the Riemannian metric on $M$, and second, for any Riemannian manifold $N$ and Riemannian compact one $N'$, $\inf D = 0$ in all homotopy classes of smooth mappings $M \to N$ and those homotopy classes of smooth mappings $N' \to M$, which contain Riemannian submersions. Koiso [307] posed the problem of describing the set of closed manifolds $M$ such that $D(M) = 0$ for an arbitrary choice of the Riemannian metric. Min–Oo has recently shown (with the result being announced by J. Eells and L. Lemaire) that the set contains all simply-connected compact Lie groups in addition to the spheres $S^n (n \geqslant 3)$.

The following theorem solves the problem of describing the class of closed, i.e., compact and without boundary, Riemannian manifolds $M$ satisfying the equality $D(M) = 0$.

**Theorem 24.6.5 ([333], [435]).** *Let $M$ be a closed, smooth Riemannian manifold. Then $D(M) = 0$ if and only if each connected component of $M$ is 2-connected, i.e., its first and second homotopy groups are trivial.*

Theorem 24.6.5 was proved by A. I. Pluzhnikov in 1984 (see the announcement in [169]) and reveals the relation of the Dirichlet functional to the topological properties of manifolds. The result was later sharpened by B. White [434] (see below).

The class of manifolds possessing, for any Riemannian metric on them, smooth mappings $\varphi \in [\mathrm{id}_M]$, with arbitrarily small Dirichlet integral $D(\varphi)$, turned out to be quite large. By Theorem 24.6.5, in addition to spheres of dimensions greater than three and simply-connected, compact Lie groups, e.g., Stiefel and Grassmannian manifolds over the quaternions, should belong here as well as symmetric spaces from series (c) indicated in Theorem 24.6.2, their direct products and disjoint unions.

It follows from Theorem 24.6.5 that the minimal realization problem cannot be solved in $[\mathrm{id}_M]$ if the connected, closed Riemannian manifold $M$ of positive dimension is 2-connected. On the other hand, the following proposition arises.

**Corollary 24.6.1.** *If at least one of the homotopy groups $\pi_1(M)$ and $\pi_2(M)$ of a connected, closed, and smooth manifold $M$ is nontrivial, then $D(M) > 0$ for any Riemannian metric on $M$.*

The identity mapping itself is known to be a solution of the minimal realization problem in $[\mathrm{id}_M]$ if the closed Riemmanian manifold $M$ has nonpositive curvature (see S. I. Alber[3]–[4], P. Hartman [191]) or is Kähler (see A. Lichnerowicz [254]). In both cases, the "non-2-connectedness" condition stated in Corollary 24.6.1 is valid ($\dim M \geqslant 1$).

A close relation of the problems regarding the existence of globally minimal and stable harmonic mappings becomes clear in the class of compact, symmetric spaces of Type I, i.e., which are not Lie groups. The following is a direct consequence of Theorems 24.6.1, 24.6.3, and 24.6.5.

**Corollary 24.6.2.** *Let $M$ be an arbitrary connected, compact, irreducible, nonsingular, and symmetric space of Type I, endowed with the standard Riemannian metric. Then*

*(1) all nonconstant harmonic mappings from $M$ to an arbitrary smooth Riemannian manifold are unstable if and only if $D(M) = 0$, and (2) the identity mapping $[\mathrm{id}_M]$ is stable if and only if $D(M) > 0$.*

On the other hand, spinor groups Spin($n$) for $n \geqslant 7$ with the standard metric are examples of Riemannian manifolds with the *stable identity mapping,* so that $\inf D = 0$ *in the homotopy class of this mapping* in accordance with Theorem 24.6.5.

The following theorem proved by A. I. Pluzhnikov [333], [335], and [435] demonstrates the importance of the identity mapping homotopy class in the Dirichlet functional minimization problem and is an improvement of the above result of J. Eells and L. Lemaire, since it contains no extra requirement of the Riemannian submersion in the homotopy class.

**Theorem 24.6.6.** *Let $M$ be a compact, smooth Riemannian manifold. Then the following are equivalent, viz.,*

*(a)* $D(M) = 0$,

*(b)* $\inf D = 0$ *in all homotopy classes from $[M, N]$ for an arbitrary smooth Riemannian manifold $N$, and*

*(c)* $\inf D = 0$ *in all homotopy classes from $[N', M]$ for any arbitrary compact, smooth Riemannian manifold $N'$.*

Taking into account Theorem 24.6.5, we derive the following.

**Corollary 24.6.3.** *Let $M$ be a closed, 2-connected, and smooth manifold. Then statements (b) and (c) of Theorem 24.6.6 hold in any Riemannian metric on $M$.*

Note that the equality $\inf D = 0$, valid in a certain homotopy class of mappings of compact Riemannian manifolds, also holds if the Riemannian metrics are replaced.

The proof of Theorem 24.6.6 is based on considering the compositions $f \circ \varphi_i$ and $\varphi_i \circ f'$, where $f$ and $f'$ are representatives of the homotopy classes of mappings from $M$ to $N$ and from $N'$ to $M$, respectively, $\varphi_i \in [\mathrm{id}_M] (i = 1, 2, \dots)$, and $D(\varphi_i) \to 0$ as $i \to \infty$.

Consider a connected, closed Riemannian manifold $M$ and an arbitrary nontrivial homotopy class $\alpha$ belonging to $[M, N]$ or $[N', M]$ in the notation of Theorem 24.6.6. Construct a sequence $f_i \in \alpha (i = 1, 2, \ldots)$ such that $D(f_i) \to \inf_\alpha D$ as $i \to \infty$. If $M$ is 2-connected, then it cannot possess a limit in the space of smooth mappings due to Corollary 24.6.3. (As we shall see below, sequences of this kind converge in the corresponding Sobolev spaces to locally constant mappings.) Thus, the nontriviality of at least one of the homotopy groups $\pi_1(M)$ and $\pi_2(M)$ is a necessary topological condition for the existence of nonconstant globally minimal harmonic mappings in the spaces $C^\infty(M, N)$ and $C^\infty(N', M)$.

**Corollary 24.6.4.** *Let $M$ and $N$ be arbitrary connected, compact, and smooth Riemannian manifolds, and $f \in C^\infty(M, N)$ a harmonic mapping globally minimal in its homotopy class. Then $f = \text{const}$ (mapping into a point) if at least one of $M$ and $N$ is 2-connected and without boundary.*

This result testifies to a limited area of applications of "minimizing" methods of steepest descent type to the construction of nonconstant harmonic mappings due to nontrivial homotopy classes in which an energy functional minimum is sure not to be attained. The study and development of methods supplying nonminimal harmonic mappings acquires special importance. The problem is outside the framework of the present investigation, and we confine ourselves to indicating, e.g., constructive methods for harmonic mappings of spheres (see A. T. Fomenko [157] and R. Smith [397]) and the theory of harmonic mappings of Riemannian surfaces developing rapidly in recent years (see, e.g., J. Eells and J. Wood [121], S. Erdem and J. Wood [130]).

By means of Theorems 24.6.5 and 24.6.6, we can point out a number of situations where no globally minimal harmonic mappings in a given homotopy class are available along with the separation from zero of energy functional values. D. V. Anosov noted that the unattainability of $\inf D$ in the class of the identity mapping of the manifold $S^1 \times S^n (n \geqslant 3)$, though $D(S^1 \times S^n) > 0$, can be derived from Corollary 24.6.3. A more thorough investigation of this observation led us to the following (see A. I. Pluzhnikov [435]).

**Theorem 24.6.7.** *Consider the compact, connected, and smooth Riemannian manifolds $M, N$, and $Q$, and denote by $\pi_N$ and $\pi_Q$ the canonical projections of the Riemannian direct product $N \times Q$ onto the first and second factors, respectively. Let $Q$ be 2-connected, and $\partial Q = \emptyset$. Then*

$$\inf_{[F]} D = \inf_{[\pi_N \circ F]} D$$

*for any $F \in C^\infty(M, N \times Q)$. Meanwhile, $\inf D$ is not attained on $[F]$ if the homotopy class $[\pi_Q \circ F]$ is nontrivial.*

**Corollary 24.6.5.** *Let the conditions of Theorem 24.6.7 be fulfilled, $\partial M = \partial N = \emptyset$, and the sectional curvature $N$ be nonpositive. Assume that both mappings $\pi_N \circ F$ and $\pi_Q \circ F$ are homotopically nontrivial. Then $\inf D$ is* *positive in the class* $[F]$ *but not attained in any of the smooth mappings from* $[F]$.

**Theorem 24.6.8.** *Consider compact, connected, and smooth Riemannian manifolds* $M, N, P, Q$, *and two arbitrary smooth mappings* $f : M \to N$ *and* $\varphi : P \to Q$. *Denote the corresponding induced mapping from* $M \times P$ *into* $N \times Q$ *by* $f \times \varphi$. *Let at least one of the* $P$ *and* $Q$ *be 2-connected and without boundary. Hence,*

$$\inf_{[f\times\varphi]} D = \mathrm{vol}(P) \cdot \inf_{[f]} D,$$

*where* $\mathrm{vol}(P)$ *is the volume of* $P$ *(if* $\dim P = 0$, *then we put* $\mathrm{vol}(P) = 1$*). Meanwhile,* $\inf D$ *is not attained on* $[f \times \varphi]$ *if the homotopy class* $[\varphi]$ *is nontrivial.*

**Corollary 24.6.6.** *Let the conditions of Theorem 24.6.8 be fulfilled, and a globally minimal harmonic mapping be contained in each connected component of the space* $C^\infty(M, N)$. *Assume that both mappings* $f$ *and* $\varphi$ *are homotopy nontrivial. Then* $\inf D > 0$ *on* $[f \times \varphi]$ *but is not attained on any smooth mapping from this homotopy class.*

A number of harmonic, nonminimal mappings of the form $f \times \varphi$ was constructed by L. Lemaire in [247]. If we compare Corollary 24.6.6 with the above results on minimal realization, then we see that the homotopy classes of mappings from $M \times P$ into $N \times Q$ with attainable $\inf D > 0$ arise ($M$ and $N$ being closed, and $C^\infty(P, Q)$ disconnected) if (1) $M = S^1$, $\pi_1(N) \neq 0$, (2) $\dim M = 2$, $\pi_1(N) \neq 0$, $\pi_2(N) = 0$, and (3) $\pi_1(M) \neq 0$, where the curvature of $N$ is nonpositive. For example, we can take this combination of closed manifolds, viz., $M = S^1, P = S^m (m \geqslant 3), \pi_1(N) \neq 0, \pi_m(Q) \neq 0$.

Returning to the problem of possible minimal realization of the homotopy class of the identity mapping of a "2-connected" manifold (see Corollary 24.6.1), note the following.

**Corollary 24.6.7.** *Let* $M$ *and* $P$ *be two connected, closed, and smooth Riemannian manifolds,* $\dim P > 0$, *the groups* $\pi_1(P)$ *and* $\pi_2(P)$ *be trivial, and at least one of the groups* $\pi_1(M)$ *and* $\pi_2(M)$ *be nontrivial. Then* $\inf D$ *is positive in the identity mapping homotopy class for the manifold* $M \times P$ *but is not realized on any smooth mapping from this homotopy class.*

In investigating the behaviour of the Dirichlet functional in various homotopy classes, the decomposition $f = f_2 \circ f_1$ is used of the considered mapping into the composition of a certain fixed mapping from a given homotopy class and a mapping homotopic to the identity.

The following theorem is devoted to the construction of a special diffeotopy whose existence is the central moment of the proof of Theorem 24.6.5.

**Theorem 24.6.9 (A. I. Pluzhnikov [333], [435]).** *Let* $M$ *be an arbitrary connected, closed, 2-connected, and smooth manifold of dimension* $\geqslant 5$ *or one of* *the standard spheres $S^3, S^4$. Then there exists a diffeotopy $\varphi_t (0 \leqslant t < 1)$ of $M$, so that $\varphi_0 = \mathrm{id}_M$, and*

$$\lim_{t \to 1} D(\varphi_t) = 0$$

*for any Riemannian metric on $M$.*

Hence, $\inf D = 0$ on the connected component of the identity element of the diffeomorphism group of $M$, satisfying the conditions of Theorem 24.6.9. It is obvious then that $D(M) = 0$. Combining the statements of Theorems 24.6.9 and 24.6.5, we obtain that $D(M) = 0$ is equivalent in the class of connected, closed manifolds of dimensions greater than four to the existence of the diffeotopy described in Theorem 24.6.9. As for small dimensions, note the following well-known facts. Any connected, closed, 2-connected, and smooth manifold in dimension four is homeomorphic to $S^4$ and homotopy-equivalent to $S^3$ in dimension three. Such manifolds are absent in dimensions one and two.

The main idea of the proof of Theorem 24.6.9 is outlined as follows. It is obtained by building up the diffeotopy $\Phi_t$ constructively and makes use of the existence of the Morse function $\mu$ without the critical points of index $n-1$ and $n-2$, $n = \dim M$, on a manifold $M$ satisfying the conditions of the theorem. $\varphi_t$ models a deformation of $M$, determined by the gradient flow of $\mu$, and is constructed as the composition of "local" deformations, each of which is concentrated on one of the Smale handles of the partition of $M$, induced by $\mu$. If we do not take into account the "gluing" errors, the deformation of an individual handle in coordinates $(x, y) \in \mathbf{R}^p \times \mathbf{R}^q (p + q = n)$ is of the form $(x, y) \to (x, f_t(y))$, where $f_t$ is a mapping of "radial" type, and $q \geqslant 3$. Meanwhile, we use the following observation due to the quadraticity of the Dirichlet functional integrand with respect to derivatives. If $f_t(y) = (1-t)^{-1} \cdot y$, then the Dirichlet integral of the restriction $f_t$ to the inverse image $f_t^{-1} D^q$ of the unit disk in Euclidean space is equal to $(q/2) \cdot \operatorname{vol} D^q \cdot (1-t)^{q-2}$, and tends to zero as $t \to 1$ if $q \geqslant 3$. In the simplest example with the spheres $S^n, n \geqslant 3$, we can take any nonzero real linear functional on $\mathbf{R}^{n-1}$ as $\mu$, in which case the diffeotopy under construction is of the form $\varphi_t = \beta_t \circ \alpha_t$, where $\alpha_t$ "inflates" the neighbourhood of radius $1-t$ of the minimum point $\mu^{-1}(-1)$ onto the lower-hemisphere $S_- = \{\mu \leqslant 0\}$, whereas $\beta_t$ "compresses" the upper-hemisphere $S_+ = S^n \backslash S_-$ into the neighbourhood of radius $1-t$ of the maximum point $\mu^{-1}(1)$. The proof ends in the standard procedure for estimating the integral by means of the decomposition

$$D(\varphi_t) = D(\varphi_t | \alpha_t^{-1} S_-) + D(\varphi_t | \alpha_t^{-1} S_+),$$

where the domain of intergration is small in the first addend, and in the second, the integrand is.

The methods developed by A. I. Pluzhnikov for analyzing the Dirichlet functional are then applied to investigating the Sobolev spaces $W_2^1(M, N)$ of generalized mappings of smooth Riemannian manifolds. Each of the spaces is the "natural domain" of the Dirichlet functional and plays in the case $\dim N > 1$ the same role that the linear Sobolev space $W_2^1(M)$ of functions on M does in the theory of harmonic functions ($N = R^1$) square-integrable along with their differential.

Consider two arbitrary compact, smooth Riemannian manifolds $M$ and $N$, and fix an isometric embedding of $N$ into Euclidean space $\mathbf{R}^k$ of sufficiently large dimension. We define $W_p^m(M,N)$ as a topological subspace in the linear topological space $W_p^m(M,\mathbf{R}^k)$, consisting of such classes of generalized mappings $f : M \to \mathbf{R}^k$ that $f(x) \in N$ almost for all $x \in M$. It can be shown that the topology induced on $W_p^m(M,N)$ does not depend on the choice of an embedding of $N$ into $\mathbf{R}^k$. The norms $\|\cdot\|_m$ can be introduced in the Hilbert spaces $W_2^m(M,\mathbf{R}^k)$, so that the Dirichlet functional will have the simple form $D^R = \|\cdot\|_1^2 - \|\cdot\|_0^2$ for $m \geqslant 1$, where $\|\cdot\|_0$ coincides with the $L_2$-norm. Meanwhile, the Dirichlet functional on the Sobolev space of mappings from $M$ to $N$ is defined as the restriction of $D^R$ to $W_2^m(M,N)$. The definition yields the standard Dirichlet functional on smooth mappings.

A close relation of $D$ to the norm on the underlying linear Sobolev space permits us to apply the above results to analyzing the topological structure of the "nonlinear" space $W_2^1(M,N)$, which, as is shown by the following statements, depends on the topological properties of the manifolds $M$ and $N$.

**Theorem 24.6.10.** *Let $M$ be a connected, closed, and smooth manifold. Then the 2-connectedness of $M$ is equivalent to any of the two statements below, holding for all connected, compact, and smooth Riemannian manifolds $N$ and an arbitrary choice of the Riemannian metric on $M$, viz.,*

*(a) the closure of any connected component of the space $C^\infty(M,N)$ in $W_2^1(M,N)$ contains the set of constant mappings from $M$ to $N$, and*

*(b) the same statement, with $M$ and $N$ interchanged.*

**Corollary 24.6.8.** *Let $M$ and $N$ be two arbitrary connected, closed, and smooth Riemannian manifolds, with at least one of them being 2-connected. Then the intersection of the closures in $W_2^1(M,N)$ of all connected components of the space $C^\infty(M,N)$ is nonempty and contains the constant mapping set. In particular, the closure of $C^\infty(M,N)$ in $W_2^1(M,N)$ is connected.*

The following theorem strengthens a number of the above results and has been independently proved by A. I. Pluzhnikov and B. White [434]. It applies a topological criterion for $\inf D$ to be zero in a given homotopy class.

**Theorem 24.6.11.** *Let $f : M \to N$ be an arbitrary smooth mapping of a closed Riemannian manifold $M$ to a Riemannian manifold $N$. In the homotopy class* $[f]$, $\inf D = 0$ *if and only if the restriction $f|_{[M]^2}$ of $f$ to the two-dimensional skeleton of $M$ is contractible to a point.*

Hence, it follows, in particular, that Theorem 24.6.5 is valid, since the restriction of the identity mapping to a two-dimensional skeleton is contractible if and only if the manifold is 2-connected.

Further research has shown the fruitfulness of the use of the above properties of the Sobolev space $W_p^m(M,N)$ in Dirichlet functional theory. The following strong theorem describes the topology of these spaces and generalizes certain statements above.

We call two mappings $f, h \in C^\infty(M,N)$ *k-homotopic* if their restrictions to the k-dimensional skeleton $M^k$ of the manifld $M$ are homotopic.

**Theorem 24.6.12 ([434]).** *Consider two closed Riemannian manifolds $M$ and $N$. The mappings $f, h \in C^\infty(M,N)$ are in the same connected component of the closure of the space $C^\infty(M,N)$ if and only if they are 2- homotopic, the connected component coinciding with the closure of the set $[f] \subset C^\infty(M,N)$. In particular, $\inf_{[f]} D = \inf_{[f_2]} D$, where $[f_2]$ is the set of smooth mappings 2-homotopic to $f$.*

## 24.7. Properties of the Density of Smooth Mappings of Manifolds

**Definition 24.7.1.** A smooth nonnegative function $e(f) = \frac{1}{2} \parallel df \parallel^2$ on $M$ is called the *integral density* of a smooth mapping $f : M \to N$ of smooth finite-dimensional Riemannian manifolds. The integral of $e(f)$ with respect to a measurable subset $B \subseteq M$

$$D(f|B) = \int_B e(f)dv$$

is called the *B-integral* of $f$.

In local coordinates, the integral density function is computed according to

$$e(f) = \tfrac{1}{2} g^{ij} f_i^\alpha f_j^\beta \sigma_{\alpha\beta}, \tag{1}$$

where $g$ and $\sigma$ are the metric tensors of $M$ and $N$, respectively, whereas the $M$-integral of the mapping $f$ coincides with the value $D(f)$ of the Dirichlet functional.

**Lemma 24.7.1.** *Let $f : M \to N$ and $h : N \to P$ be two smooth mappings of Riemannian manifolds. Then*

$$e(h \circ f) \leqslant 2e(f) \cdot f^* e(h).$$

**Proof.** Fix an arbitrary point $x \in M$ and orthonormal frames in tangent spaces $T_xM$, $T_yN$, and $T_{h(y)}P$, where $y = f(x)$. It follows from the Schwartz inequality that

$$||dh(y) \circ df(x)|| \leqslant ||dh(y)|| \cdot ||df(x)||,$$

where $|| \cdot ||$ is the Hilbert–Schmidt norm of a linear operator. It remains to make use of the inequality $d(h \circ f) = dh \circ df$ and Definition 24.7.1. Lemma 24.7.1 is thus proved.

**Lemma 24.7.2.** *Let $f : M \to N$ and $h : N \to P$ be two smooth mappings of Riemannian manifolds. Then*

$$D(h \circ f|B) \leqslant 2D(f|B) \cdot \sup_{f(B)} e(h)$$

*for any measurable subset $B \subset M$.*

**Proof.** In accordance with Definition 24.7.1 and Lemma 24.7.1,

$$E(h \circ f|B) \leqslant 2 \int_B e(f) \cdot f^* e(h) dv.$$

Both functions in the integrand are nonnegative. It remains to remove the maximum of $f^* e(h)$ from under the integral sign, and Lemma 24.7.2 is thus proved.

**Definition 24.7.2.** Consider a smooth submersion $f : M \to N$ of Riemannian manifolds, and assume that $\dim N > 0$. Denote by $TM^h$ the subbundle of the tangent bundle $TM$ of the manifold $M$ consisting of horizontal subspaces which are the orthogonal complements to the spaces $(\operatorname{Ker} df)_x$ $x \in M$. Denote by $g^h_M$ and $f^* g^h_N$ the restrictions of the original and $f$-induced metrics on $M$ to the fibres of $TM^h$. We call a smooth positive function $\rho(f)$ on $M$, defined by the formula

$$\rho(f) = \sqrt{\frac{\det(g^h_M)}{\det(f^* g^h_N)}},$$

the *volume density* of the submersion $f$. If $\dim N = 0$, then we put $\rho(f) = 1$.

The above formula involves metric tensor matrices that depend on the chosen local coordinate system. However, the volume density function itself does not depend on the choice of a coordinate system and is a global, smooth function defined correctly on the manifold $M$, which can be easily seen by direct verification. If $f$ is a diffeomorphism, then the function $1/\rho(f)$ coincides with the Radon–Nikodým derivative, induced by the mapping $f$, of smooth positive measure on $M$ to the original measure, viz.,

$$1/\rho(f) = f^* dv_N / dv_M. \tag{2}$$

**Lemma 24.7.3.** *Let $f : M \to N$ and $h : N \to P$ be two smooth mappings of Riemannian manifolds, and $f$ a submersion. Then, for any measurable subset $B \subset M$,*

$$D(h \circ f|B) \leqslant 2D(h|f(B)) \cdot \sup_{f(B)} \Psi_f,$$

*where the function $\Psi_f$ on $f(B)$ is positive, and defined by the equality*

$$\Psi_f(y) = \int_{f^{-1}(y) \cap B} e(f) \rho(f) dv \quad (y \in f(B)).$$

*Moreover, if $B$ is compact, then $\Psi_f$ is bounded on $f(B)$.*

The proof is based on the well-known Fubini theorem. We define a function $\theta$ on the set $f(B)$ by the formula

$$\theta(y) = \int_{f^{-1}(y)\cap B} e(h \circ f)\rho(f)dv.$$

The function $\rho(f)$ was chosen so that the application of the Fubini theorem for $\dim N > 0$ leads to

$$D(h \circ f|B) = \int_{f(B)} \theta dv. \tag{3}$$

If $\dim N = 0$, both sides of the latter formula vanish. It follows from Lemma 24.7.1 that $\theta \leqslant 2e(h) \cdot \psi_f$, and the required inequality is obtained on substituting this estimate in (3).

Let a set $B \subset M$ be compact. With the functions $e(f)$ and $\rho(f)$ being bounded on $B$, it suffices to show, if we want to prove the boundedness of $\psi_f$ on $f(B)$, that all subsets $f^{-1}(y) \cap B$ are bounded in volume by the same constant for $y \in f(B)$. Cover $B$ with a system of such open subsets $U_\alpha$ in which the submersion $f$ is of the form of the projection $\mathbf{R}^{n-k} \times \mathbf{R}^k \to \mathbf{R}^k$, and choose a finite subcovering $U_1, \ldots, U_e$. It is obvious that, for each $U_\alpha$, there is a value $c_\alpha$ for which $\operatorname{vol}(f^{-1}(y) \cap U_\alpha) \leqslant c_\alpha$ for all $y \in f(U_\alpha)$. Put $c = l \cdot \max\{c_\alpha | \alpha = 1, \ldots, l\}$. Then $\operatorname{vol}(f^{-1}(y) \cap B) \leqslant c$ for all $y \in f(B)$.

Q.E.D.

We call a smooth nonnegative function $e\rho(f)$ on $M$, defined by

$$e\rho(f) = e(f) \cdot \rho(f), \tag{4}$$

the *mixed density* of a diffeomorphism $f : M \to N$ of Riemannian manifolds, where $e(f)$ is the integral density, and $\rho(f)$ is the volume density of $f$ (see Definitions 24.7.1 and 24.7.2).

**Lemma 24.7.4.** *Let $h, f : M \to M$ be two diffeomorphisms of a Riemannian manifold $M$. Then*

*(a)* $\rho(h \circ f) = \rho(f) \cdot f^*\rho(h)$,

*(b)* $e\rho(h \circ f) \leqslant 2e\rho(f) \cdot f^*e\rho(h)$,

*and*

*(c)* $D(h \circ f|B) \leqslant 2D(h|f(B)) \cdot \sup_B e\rho(f)$

*for any measurable subset $B \subset M$.*

**Proof.** (a) follows from Definition 24.7.2 for the function $\rho(f)$ and the following obvious chain of equalities

$$\rho(h \circ f) = \left(\frac{\det(g)}{\det(f^*g)}\right)^{1/2} \cdot \left(\frac{\det(f^*g)}{\det(f^*h^*g)}\right)^{1/2} = \rho(f) \cdot f^* \left(\frac{\det(g)}{\det(h^*g)}\right)^{1/2},$$

where $g$ is the metric on the manifold $M$.

Inequality (b) follows from (a) and Lemma 24.7.1, whereas (c) follows from Lemma 24.7.3, since $\psi_f(f(x)) = e\rho(f)_{(x)}$ in the present situation.

Q.E.D.

The latter lemma (to be proved in this section) touches upon the dependence of the characteristics of the mapping $f : M \to N$, introduced above, on the choice of Riemannian metrics on the manifolds $M$ and $N$. We shall need it in considering diffeomorphisms.

**Lemma 24.7.5.** *Let $f : M \to N$ be a diffeomorphism of compact Riemannian manifolds. Denote by $\tilde{g}_M$ and $\tilde{g}_N$ certain Riemannian metrics on $M$ and $N$, respectively, which are other than the original. Then there exist positive constants $C_1, C_2$, and $C_3$, not depending on $f$, and such that*

*(a)* $\tilde{e}(f) \leqslant c_1 \cdot e(f)$,

*(b)* $\widetilde{D}(f|B) \leqslant c_2 \cdot D(f|B)$,

*(c)* $\tilde{\rho}(f) \leqslant c_3 \rho(f)$,

*where $B$ is an arbitrary measurable subset in $M$, the tilde indicates integral density, and the $B$-integral and volume density in terms of the metrics $\tilde{g}_M$ and $\tilde{g}_N$.*

**Proof.** Denote by $\widetilde{M}$ and $\widetilde{N}$ two replicas of the smooth manifolds $M$ and $N$, endowed with Riemannian metrics $\tilde{g}_M$ and $\tilde{g}_N$, respectively. Let $\alpha : \widetilde{M} \to M$ and $\beta : N \to \widetilde{N}$ be two canonical diffeomorphisms induced by identifying the replicas. It is clear that the inequalities (a)–(c) to be proved are equivalent to the following, viz.,

$$e(\beta \circ f \circ \alpha) \leqslant c_1 \cdot \alpha^* e(f), \tag{5}$$

$$D(\beta \circ f \circ \alpha | \alpha^{-1} B) \leqslant c_2 \cdot D(f|B), \tag{6}$$

and

$$\rho(\beta \circ f \circ \alpha) \leqslant c_3 \cdot \alpha^* \rho(f). \tag{7}$$

It follows from Lemma 24.7.1 that

$$e(\beta \circ f \circ \alpha) \leqslant 2e(\alpha) \cdot \alpha^* e(\beta \circ f) \leqslant 4e(\alpha) \cdot \alpha^* f^* e(\beta) \cdot \alpha^* e(f).$$

Hence, (5) holds if we put

$$C_1 = 4 \max_{\widetilde{M}} e(\alpha) \cdot \max_N e(\beta).$$

We derive from Lemma 24.7.2 and (b) of Lemma 24.7.4 that

$$D(\beta \circ f \circ \alpha | \alpha^{-1} B) \leqslant C_2' D(\beta \circ f | B) \leqslant C_2' C_2'' D(f|B),$$

where

$$C_2' = 2 \max_{\widetilde{M}} e\rho(\alpha), \qquad C_2'' = 2 \max_N e(\beta),$$

which proves inequality (6) with $C_2 = C_2' \cdot C_2''$. Estimate (7) follows from equality (a) of Lemma 24.7.4. In fact,

$$\rho(\beta \circ f \circ \alpha) = \rho(\alpha) \cdot \alpha^* f^* \rho(\beta) \cdot \alpha^* \rho(f);$$

therefore, it suffices to set

$$C_3 = \max_{\widetilde{M}} \rho(\alpha) \cdot \max_N \rho(\beta).$$

Q.E.D.

### 24.8. The Behaviour of the Dirichlet Functional on the 2-Connected Manifold Diffeomorphism Group. Proof of Theorem 24.6.9

The present section is devoted to the computation of the infimum of the set of the Dirichlet functional $D$ values on the diffeomorphism group of a connected, closed (i.e., compact and without any boundary), and 2-connected Riemannian manifold. (A connected manifold is said to be *2-connected* if its first and second homotopy groups are trivial.) Here, the basic result is Theorem 24.6.9 due to A. I. Pluzhnikov [333], and [435]. In particular, the theorem statement means that $\inf D = 0$, i.e., the values of $D$ are not separated from zero on the component of the diffeomorphism group unit element of a manifold $M$ satisfying the conditions of Theorem 24.6.9. Therefore, $\inf D = 0$ in the homotopy class of the identity mapping $\mathrm{id}_M$ of $M$, irrespective of the choice of a Riemannian metric on $M$. A number of corollaries of this fact will be derived in the subsequent section. J. Eells and L. Lemaire have proved the statement of the theorem earlier for the spheres $S^n, n \geqslant 3$, [119].

The proof is obtained by constructing the required diffeotopy explicitly and includes a series of lemmas below.

We start with the construction of the required special diffeotopy of Euclidean space. Denote by $D^q(r)$ a compact ball of radius $r$ in the space $\mathbf{R}^q$ with centre in the origin.

**Lemma 24.8.1.** *For $q \geqslant 3$, there exists a diffeotopy $\varphi_t (0 \leqslant t \leqslant 1)$ of the space $\mathbf{R}^q$, so that $\varphi_0 = \mathrm{id}$ and the conditions hold:*

*(a) The diffeomorphisms $\varphi_t$ are the identity outside $D^q(1)$,*

*(b) $D(\varphi_t | A_t) \to 0$ as $t \to 1$,*

*(c) $\mathrm{vol}(A_t) \to 0$ as $t \to 1$,*

*(d) $\rho(\varphi_t) \leqslant C_1$ on the whole of $\mathbf{R}^q$ and for all $t$,*

*and*

*(e) $e\rho(\varphi_t) \leqslant C_2$ on $\mathbf{R}^q \backslash B_t$ for all $t$,*

*where $A_t = \varphi_t^{-1} D^q(1/2)$, $B_t = \varphi_t^{-1} D^q(1/4)$,*

*and the positive constants $C_1$ and $C_2$ do not depend on $t$.*

**Proof.** Let $(s, r)$ be spherical coordinates in the space $\mathbf{R}^q \backslash \{0\}$ ($s$ being an element of the unit sphere). Consider an arbitrary diffeomorphism $\varphi$ of the space $\mathbf{R}^q$ of the form

$$\rho(s, r) = (s, R(r)), \tag{1}$$

where $R$ is a certain smooth monotonically increasing function on the half-line $\{r \geqslant 0\}$, vanishing if $r = 0$. It is generally known that the matrices of the metric tensors $g_R$ and $g_S$ on $\mathbf{R}^q$ and $S^{q-1}$, repectively, are related by

$$g_R(s,r) = \begin{pmatrix} r^2 \cdot g_S(s) & 0\vdots 0 \\ 0\ldots 0 & 1 \end{pmatrix}. \tag{2}$$

It is obvious from (1) that the differential of the diffeomorphism $\varphi$ with respect to $(s, r)$ is expressed by

$$d\varphi(S,r) = \mathrm{diag}(1,\ldots,1,R'(r)).$$

Therefore,

$$\varphi^* g_R(s,r) = \begin{pmatrix} R^2 \cdot g_S(s) & 0\vdots 0 \\ 0\ldots 0 & (R')^2 \end{pmatrix}. \tag{3}$$

We now compute the volume density of $\varphi$. Taking the square root of the quotient of the determinants of (2) and (3), we obtain

$$\rho(\varphi) = \left(\frac{r}{R}\right)^{q-1} \cdot \frac{1}{R'}. \tag{4}$$

The integral density of $\varphi$ equals half the trace of the product of matrix (3) and the one inverse to (2), viz.,

$$e(\varphi) = \frac{1}{2}[(q-1)(\frac{R}{r})^2 + (R')^2]. \tag{5}$$

Now, let $R_t(r) (0 \leqslant t < 1)$ be a smooth family of smooth monotonically increasing funtions on the half-line $\{r \geqslant 0\}$ such that $R_0(r) \equiv r$, and the function $R - r$ vanishes for all $t$ if $r = 0$ and $r \geqslant 1$ along with all its derivatives. Then the family of mappings

$$\varphi_t(s,r) = (s, R_t(r)) \quad (0 \leqslant t < 1) \tag{6}$$

forms a diffeotopy of $\mathbf{R}^q$, starting with the identity mapping $\varphi_0$ and satisfying the requirements of (a) of the lemma. If we compare equalities (6) and (1), then we see that formulas (4) and (5) can be used for the diffeotopy $\varphi_t$. Denote by $\varepsilon_t$ the number determined by the equality $R_t(\varepsilon_t) = 1/2$, which exists for each $t$, and show that to satisfy the requirements of (b)–(e), it suffices that the family $R_t(r)$ should satisfy the following extra conditions, viz.,

$$R_t \geqslant r, \; R_t' \geqslant A, \tag{7}$$

$$R_t' \leqslant B \text{ for } r \geqslant \varepsilon_t, \tag{8}$$

$$R_t' \leqslant \frac{C}{r} \text{ on } [0, \varepsilon_t], \tag{9}$$

$$R_t' \leqslant \frac{D}{r} R_t \text{ on } [\delta_t, \varepsilon_t], \tag{10}$$

and

$$\varepsilon_t \to 0 \text{ as } t \to 1, \tag{11}$$

where $A, B, C$, and $D$ are positive constants not depending on $t$, and the values $\delta_t$ are determined by the equality $R_t(\delta_t) = 1/4$. Indeed, consider condition (b) for the $A_t$-integral of the mapping $\varphi_t$. It follows from formula (6) that

$$D(\varphi_t|A_t) = \kappa_q \int_0^{\varepsilon_t} e(\varphi_t) r^{q-1} dr,$$

where $\kappa_q$ is the volume of the $(q-1)$-dimensional sphere of unit radius. Since $R_t \leqslant 1$ on $[0,1]$, the substitution of (9) in (5) leads to the inequality $e(\varphi_t) \leqslant \frac{\text{const}}{r^2}$ holding on $[0, \varepsilon_t]$. Thus,

$$D(\varphi_t|A_t) \leqslant \text{const}\, \frac{\varepsilon_t^{q-2}}{q-2},$$

implying (b) for $q \geqslant 3$ if we take into account (11). (c) also follows from (11), since

$$\text{vol}(A_b) = \kappa_q \int_0^{\varepsilon_t} r^{q-1} dr = \kappa_q \frac{\varepsilon_t^q}{q}.$$

The substitution of inequalities (7) in formula (4) leads to estimation (d). It follows from (4) and (5) that

$$e\rho(\varphi_t) = \frac{q-1}{2} \left(\frac{r}{R_t}\right)^{q-3} \cdot \frac{1}{R_t'} + \frac{1}{2} \left(\frac{r}{R_t}\right)^{q-1} \cdot R_t'.$$

The boundedness of the first addend on the right-hand side of the latter equality for $q \geqslant 3$ follows from (7), whereas that of the second one from (7) and (8) for $r \geqslant \varepsilon_t$, and from (7) and (10) for $r \in [\delta_t, \varepsilon_t]$, thus proving (e).

To complete the proof of Lemma 24.8.1, it remains to see that the above family of functions $R_t(r)$ does exist. We introduce the notation

$$a_t = \frac{1-t}{4}, \; b_t = \frac{3+t}{4}, \; t \in [0,1).$$

Consider the corresponding family of piecewise linear, continuous functions $\bar{R}_t(r)$ on the half-line $\{r \geqslant 0\}$ equal to $r$ outside the intervals $(a_t, b_t)$, linear on $[a_t, 3a_t/2]$ and $[3a_2/2, b_t]$, and equal to $1/2$ for $r = \varepsilon_t$, where $\varepsilon_t = 2a_t \to 0$ as $t \to 1$. It is obvious that they are monotonically increasing, and that $\bar{R}_0(r) \equiv r$. Trivial computation demonstrates that (7)–(10) hold on the linear portions. Thus, we can take the result of smoothing the functions $\bar{R}_t(r)$, changing them in sufficiently small neighbourhoods of the break points $a_t, 3a_t/2$ and $b_t$, so that the derivative $R_t'$ is monotonic in the neighbourhoods.

Q.E.D.

For convenience, we introduce the following.

**Definition 24.8.1.** Let $M$ be a closed (i.e., compact and without boundary), smooth manifold. We call a compact subset $K \subseteq M$ *regular* if $\overline{\operatorname{int} K} = K$, and there exists a diffeotopy $\varphi_t (0 \leqslant t < 1)$ of $M$, so that $\varphi_0 = \mathrm{id}_M$ and

(a) $D(\varphi_t | K_t) \to 0$ as $t \to 1$,

(b) $e\rho(\varphi_t) \leqslant C$ on $M \backslash K_t$,

in any Riemannian metric on $M$, where $K_t = \varphi_t^{-1} K$ and $C$ is a constant not depending on $t$. Meanwhile, we will call $\varphi_t$ itself *regular* with respect to $K$.

Note that the statement of Theorem 24.6.9 means the regularity of the whole of $M$ satisfying the conditions of the theorem.

**Lemma 24.8.2.** *Let the conditions of Definition 24.8.1 be fulfilled for a compact subset $K$ of a closed manifold $M$ and for a diffeotopy $\varphi_t$ at least in one Riemannian metric on $M$. Then $\varphi_t$ is a diffeotopy regular with respect to $K$. Moreover, if $f : M \to N$ is a closed manifold diffeomorphism, and $K \subseteq M$ is a regular subset, then $f(K) \subseteq N$ is also a regular subset.*

**Proof.** That conditions (a) and (b) in Definition 24.8.1 are fulfilled in any Riemannian metric on $M$ follows from Lemma 24.7.5, which proves the first statement. As to the second one, if $\varphi_t$ is a diffeotopy of $M$, which is regular with respect to a compact subset $K \subseteq M$, we fix a Riemannian metric on $M$ and introduce on $N$ a metric such that $f$ is an isometry. All the requirements of Defintion 24.8.1 are fulfilled for the diffeotopy $f \circ \varphi_t \circ f^{-1}$ and subset $f(K) \subseteq N$, since it is obvious that the composition with the isometry preserves the density functions, and $f(K)_t = f(K_t)$. Therefore, they are also fulfilled in any other Riemannian metric on $N$.

Q.E.D.

The concept of the regularity of a subset or diffeotopy introduced is useful due to the following.

**Lemma 24.8.3.** *Let a smooth, closed manifold $M$ possess such an open subset $U$, which is diffeomorphic to the disk and has a regular complement $M \backslash U$. Then $M$ is regular, i.e., there exists a diffeotopy $\Phi_t (0 \leqslant t < 1)$ of $M$ such that $\Phi_0 = \mathrm{id}_M$, and $D(\Phi_t) \to 0$ as $t \to 1$ for any Riemannian metric on $M$.*

**Proof.** Fix a Riemannian metric on $M$. Since the closure of the subset $M \backslash \bar{U} \neq \emptyset$ coincides with $M \backslash U$ (see Definition 24.8.1), there exists an open subset $V \subset M$ diffeomorphic to the disk, containing the closure of the set $U$. Denote by $\psi_t (0 \leqslant t < 1)$ a smooth homotopy of $M$, which is the identity outside $V$, and contracts $U$ to a point, viz., $\psi_0 = \mathrm{id}_M$, $\psi_1(U)$ being a point belonging to $U$. It is not difficult to select the homotopy so that $\psi_t$ is a diffeotopy of $M$ for $t < 1$. Therefore,

$$D(\psi_t | U) \to 0 \text{ as } t \to 1, \tag{12}$$

since the $U$-integral of the mapping $\psi_1$ is zero.

Denote by $\varphi_t (0 \leqslant t < 1)$ a diffeotopy of $M$, which is regular with respect to $M \backslash U$. We determine a new diffeotopy $\Phi_t$ of $M$ by the equality $\Phi_t = \psi_t \circ \varphi_t$. It is obvious that $\Phi_0 = \mathrm{id}_M$. Put $K_t = \varphi_t^{-1}(M \backslash U)$ and represent the Dirichlet integral $D(\Phi_t)$ as the sum of the $K_t$- and $(M \backslash K_t)$-integrals. By making use of Lemma 24.7.2, inequality (c) of Lemma 24.7.4, and the obvious equality $\varphi_t(M \backslash K_t) = U$, we obtain

$$D(\Phi_t) \leqslant 2D(\varphi_t | K_t) \max_M e(\psi_t) + 2D(\psi_t | U) \sup_{M \backslash K_t} e\rho(\varphi_t).$$

The first addend on the right-hand side tends to zero as $t \to 1$ due to (a) of Definition 24.8.1 and the boundedness of the function $e(\psi_t)$ on the compact set $M \times [0,1]$, whereas the second due to (12) and (b) of Definition 24.8.1. That $D(\Phi_t) \to 0$ as $t \to 1$ in any Riemannian metric on $M$ follows from Lemma 24.8.2.

Q.E.D.

The following result is the principal step in proving Theorem 24.6.9.

**Lemma 24.8.4.** *Let $M$ be a connected, closed, and smooth manifold of dimension $n \geqslant 3$, and $\mu$ a smooth Morse function on $M$ with one local minimum, one local maximum, and one critical point on each critical level. Let $\mu$ have no critical points with indices $n-1$ and $n-2$. Then, for each noncritical value $a \in R$ of $\mu$, the set $M^a = \mu^{-1}(-\infty, a) \subset M$ is regular (see Definition 24.8.1).*

The proof is obtained by the induction on the number of critical points of $\mu$, contained in $M^a$. Let there be only one such point. Therefore, it is a minimum, whereas $M^a$ is diffeomorphic to the $n$-dimensional disk. Fix a chart $U \subset M$ containing $M^a$, so that $M^a = \{|x| \leqslant \frac{1}{2}\}$ are local coordinates on $U$. Carry the Euclidean metric from $\mathbf{R}^n$ to the subset $D = \{|x| \leqslant 1\} \subset U$, and extend it arbitrarily to the Riemannian metric on the whole of $M$. Denote by $\theta_t$ the diffeotopy of $\mathbf{R}^n$, carried to the domain $U$, and constructed in Lemma 24.8.1. Construct a diffeotopy of $M$ by the equality

$$\varphi_t = \begin{cases} \theta_t & \text{on } U, \\ \mathrm{id}_M & \text{on } M \backslash U \quad (0 \leqslant t < 1). \end{cases}$$

Therefore, $\varphi_0 = \mathrm{id}_M$, and all the diffeomorphisms $\varphi_t$ are the identity on the subset $(M \backslash U) \cup (U \backslash D)$ (see (a) of Lemma 28.4.1). Therefore, that (a) and (b) of Definition 24.8.1 are fulfilled follows from (b) and (e) of Lemma 24.8.1 and the obvious equality $e\rho(\mathrm{id}_M) \equiv n/2$. It follows from Lemma 24.8.2 that the diffeotopy $\varphi_t$ is regular with respect to $M^a$, i.e., $M^a$ is a regular subset.

The induction step is carried out via considering the noncritical values $a < b$ of $\mu$ such that the subset $M^b \backslash M^a$ contains precisely one critical point of $\mu$ of index lg. It is generally well known that the manifold $M^b$ is diffeomorphic to the manifold $M^a \cup H_\lambda^h$ obtained by "gluing" a Smale handle of index $\lambda$

to $M^a$. To describe the structure of the set $H^n_\lambda$, denote by $\omega(t)$ a function on the interval $[0,1]$, continuous for $t \leqslant 1$ and smooth for $t < 1$, satisfying the conditions (a) $\omega = 1/2$ for $t \leqslant 1/2$, (b) $\omega(t)$ increases for $t \geqslant 1/2$, and $\omega(1) = 1$, (c) the inverse function $\omega^{-1(\tau)}$ for $\tau > 1/2$ is infinitely differentiable and vanishes along with all its derivatives for $\tau = 1$. The handle $H^n_\lambda$ is homeomorphic to the disk $D^n$, and determined by the formula

$$H^n_\lambda = \{(x,y) \in \mathbf{R}^\lambda \times \mathbf{R}^{n-\lambda} \,|\, |x| \leqslant 1,\ |y| \leqslant \tfrac{1}{2}\omega(|x|)\}. \tag{13}$$

The gluing of $H^n_\lambda$ to $M^a$ is done by means of a smooth mapping of the subset $\{|x| = 1, |y| \leqslant 1\}$ of $H^n_\lambda$ into the boundary $\partial M^a$. The above properties of the function $\omega$ provide for the smoothness of the manifold $M^a \cup H^n_\lambda$. We may assume that $H^n_\lambda$ is embedded into the closure of the subset $M \backslash M^a$ and that there exists a diffeomorphism of $M$, sending $M^b$ into $M^a \cup H^n_\lambda$. It follows from Lemma 24.8.2 that to prove Lemma 24.8.4, it suffices to see that the regularity of the subset $M^a$ implies that of $M^a \cup H^n_\lambda$, whereas the conditions listed in Definition 24.8.1 must be fulfilled at least in one Riemannian metric on $M$.

Extend the coordinates $(x,y)$ of $H^n_\lambda$ to a local coordinate system in a certain domain $U \subset M$, so that $U \cap M^a = \{|x| \geqslant 1\}$. The inequality $n - \lambda \geqslant 3$ permits us to consider the diffeotopy $\theta_\tau$ of the space $\mathbf{R}^{n-\lambda}$, constructed in Lemma 24.8.1. Denote by $\Theta_\tau : \mathbf{R}^{n-\lambda} \times \mathbf{R}^1 \to \mathbf{R}^{n-\lambda}$ a two-parameter diffeotopy $\mathbf{R}^{n-\lambda}$ such that $\theta_\tau(y,u) = \Theta_\tau(y)$ for $u \leqslant 1, \Theta_\tau(y,u) = y$ for $u \geqslant 2$, and $\Theta_\tau(y,u) = y$ for $|y| > 1$. Finally, we define a diffeotopy of $M$ by the equality

$$\psi_\tau = \begin{cases} (x,y) \to (x, \Theta_\tau(y,|x|)) & \text{on } U \\ \mathrm{id}_M & \text{on } M\backslash U \end{cases}, \tag{14}$$

where $0 \leqslant \tau < 1$. We introduce the notation

$$Q = \{(x,y) \in U \,||x| \leqslant 1, |y| \leqslant 1\} \subset M,$$

carry the Euclidean metric from $\mathbf{R}^\lambda \times \mathbf{R}^{n-\lambda}$ to $Q$, and extend it to the Riemannian metric on the whole of $M$. Put $H_\tau = \psi_\tau^{-1} H^n_\lambda$, and investigate the behaviour of the characteristics $D(\psi_\tau | H_\tau)$ and $ep(\psi_\tau)$ of $\psi_\tau$ for $\tau \to 1$. Meanwhile, it is necessary to take into account the following inclusions due to (13) and the properties of the function $\omega$, viz.,

$$\{|x| \leqslant 1, |y| \leqslant 1/4\} \subset H^n_\lambda \subset Q, \tag{15}$$

$$H^n_\lambda \subset \{|x| \leqslant 1, |y| \leqslant 1/2\} \subset Q. \tag{16}$$

It follows from (14) and the properties of $\Theta_\tau$ that

$$e(\psi_\tau) = \lambda/2 + e(\theta_\tau)$$

and

$$\rho(\psi_\tau) = \rho(\theta_\tau)$$

on the set $Q$.

Taking into account (16), we obtain

$$D(\theta_\tau|H_\tau) \leqslant \frac{\lambda}{2}\bar{\kappa}_\lambda \cdot \mathrm{vol}(A_\tau) + \bar{\kappa}_\lambda \cdot D(\theta_\tau|A_\tau),$$

where $A_\tau = \theta_\tau^{-1}D^{n-\lambda}(1/2)$, and $\bar{\kappa}_\lambda$ is the volume of the disk $D^\lambda(1)$. Therefore,

$$D(\psi_\tau|H_\tau) \to 0 \quad (\tau \to 1), \tag{17}$$

due to (b) and (c) of Lemma 24.8.1. Taking into account (15), we obtain that the function $e\rho(\theta_\tau)$ is bounded on $Q\backslash H_\tau$ from above by the maximum of the function $\frac{\lambda}{2}\rho(\Theta_\tau) + e\rho(\theta_\tau)$ on the set $D^{n-\lambda}(1)\backslash B_\tau$, where $B_\tau = \theta_\tau^{-1}D^{n-\lambda}(\frac{1}{4})$. Hence, there exists a constant $C$ independent of $\tau$, so that $e\rho(\psi_\tau) \leqslant c$ on $Q\backslash H_\tau$ due to (d) and (e) of Lemma 24.8.1. Moreover, since all the diffeomorphisms $\psi_\tau$ are the identity outside the set $M^a \cup Q$,

$$e\rho(\psi_\tau) \leqslant c \text{ on } M\backslash(M^a \cup H_\tau^n). \tag{18}$$

Denote by $\varphi_t (0 \leqslant t < 1)$ a diffeotopy of the manifold $M$, which exists due to the induction hypothesis, and regular with respect to $M^a$. Condition (a) in Definition 24.8.1 permits us to choose a smooth monotonic change of the parameter $t(\tau)$ such that $t(0) = 0$, $t(1) = 1$ and

$$D(\varphi_{t(\tau)}|M_\tau^a) \max_M e(\psi_\tau) \to 0 \quad (\tau \to 1), \tag{19}$$

where $M_\tau^a = \varphi_{t(\tau)}^{-1} M^a$. We now define a new diffeotopy of $M$ by the equality

$$\theta_\tau = \psi_\tau \circ \varphi_{t(\tau)} \quad (0 \leqslant \tau < 1),$$

and prove that $\theta_\tau$ satisfies requirements (a) and (b) of Definition 24.8.1 for the compact submanifold $M^a \cup H_\lambda^n$.

(a) Introduce the notation $P_\tau = \theta_\tau^{-1}(M^a \cup H_\lambda^n)$. It was indicated above that $M^a \cap U = \{|x| \geqslant 1\}$; therefore, $\psi_\tau(M^a) = M^a$, and $P_\tau = M_\tau^a \cup \varphi_{t(\tau)}^{-1} H_\tau$. Represent the $P_\tau$-integral of the diffeomorphism $\theta_\tau$ as the sum of the $M_\tau^a$- and the $\varphi_{t(\tau)}^{-1}H_\tau$-integrals. Taking into account the inclusion $\varphi_{t(\tau)}^{-1}H_\tau \subset M\backslash M_\tau^a$, we obtain on the basis of Lemma 24.7.2 and inequality (c) of Lemma 24.7.4 that

$$D(\theta_\tau|P_\tau) \leqslant 2D(\varphi_{t(\tau)}|M_\tau^a) \max\nolimits_M e(\psi_\tau) + 2D(\psi_\tau|H_\tau) \sup\nolimits_{M\backslash M_\tau^a} e\rho(\varphi_{t(\tau)}).$$

The first addend on the right-hand side tends to zero as $\tau \to 1$ due to (19), whereas the second due to (17) and (b) of Definition 24.8.1. Consequently,

$$D(\Phi_\tau|P_\tau) \to 0 \text{ as } \tau \to 1. \tag{20}$$

(b) The estimate

$$e\rho(\Phi_\tau) \leqslant 2e\rho(\varphi_{t(\tau)}) \cdot \varphi^*_{t(\tau)} e\rho(\psi_\tau)$$

follows from the inequality (b) of Lemma 24.7.4. The first factor on the right-hand side is bounded by a constant on $M \backslash M^a_\tau$, not depending on $\tau$ (condition (b) in Definition 24.8.1), whereas the second on $M \backslash P_\tau$ is due to (18). Resorting to the inclusion $M \backslash P_\tau \subset M \backslash M^a_\tau$, we obtain that functions $e\rho(\theta_\tau)$ bounded on $M \backslash P_\tau$, which means along with (20) that the diffeotopy $\theta_\tau$ satisfies requirements (a) and (b) of Definition 24.8.1. It has been noted above that this implies the regularity of $\theta_\tau$ with respect to $M^a \cup H^n_\lambda$. Hence, the submanifolds $M^a \cup H^n_\lambda$ and $M^b$ diffeomorphic to it are regular. Since the noncritical values $a < b$ of $\mu$ were chosen arbitrarily, Lemma 24.8.4 is thus proved.

The proof of Theorem 24.6.9 is completed by combining the statements of Lemmas 24.8.3 and 24.8.4. In fact, let $M$ be a connected, closed, 2-connected, and smooth manifold of dimension $\leqslant 6$. It follows from Smale's results [395] that there exists a smooth Morse function $\mu$ on $M$, satisfying the conditions of Lemma 24.8. Further, if a connected, closed, 2-connected, and smooth manifold $M$ is of dimension $\leqslant 5$, then the standard argument involving Poincaré duality and the Hurewicz and Whitehead theorems permits us to conclude that $M$ is homotopy equivalent to the sphere. Smale showed in [394] that this implies $M$ being diffeomorphic to the sphere. For all smooth spheres $S^n$ in dimensions $n \geqslant 3$, the Morse function, the restriction to $S^n$ of any nonzero real linear funtional in the underlying space $\mathbf{R}^{n-1}$ satisfies the conditions of Lemma 24.8.4. Thus, the required function $\mu$ exists on any manifold $M$ satisfying the conditions of Theorem 24.6.9. Put $M^a = \mu^{-1}(-\infty, a)$, where $a \in \mathbf{R}$ is such a noncritical value of $\mu$ that the subset $M \backslash M^a$ contains precisely one critical point, the maximum of $\mu$. The compact submanifold $M^a$ is regular (which follows from Lemma 24.8.4), whereas the subset $M \backslash M^a$ is diffeomorphic to an open disk. Hence, the conditions of Lemma 24.8.3 are fulfilled, and the manifold $M$ is regular, i.e., there exists a diffeotopy $\varphi_t (0 \leqslant t < 1)$ of $M$, starting with the identity mapping $\varphi_0$ and such that $D(\varphi_t) \to 0$ as $t \to 1$ for any Riemannian metric on $M$; Theorem 24.6.9 is thus proved.

## 24.9. Necessary Topological Condition for the Existence of Nontrivial Globally Minimal Harmonic Mappings

We now turn to the problem of existence of homotopically nontrivial harmonic mappings realizing a global minimum of the Dirichlet functional $D$ in its homotopy class of smooth mappings. We concentrate our efforts on making out necessary topological conditions for the existence of such mappings and prove some statements from §24.6, obtained by A. I. Pluzhnikov. As is shown below, the homotopy class $[\mathrm{id}_M]$ of the identity mapping of a smooth Riemannian manifold $M$ is very important here. All the manifolds, mappings, and homotopies below are, as usual, assumed smooth (or infinitely differentiable). In particular, we identify the elements of the set $[M, N]$ of the homotopy classes of mappings of manifolds $M$ and $N$ with the corresponding connected components of the topological space $C^\infty(M, N)$. We now introduce the important variational characteristic of $M$:

$$D(M) = \inf\{D(\varphi) | \varphi \in [\mathrm{id}_M]\}.$$

In general, we will denote by $\inf D$ the infimum of the Dirichlet functional value set on a certain subset of the mapping space specified on purpose in concrete situations.

The first Theorem to be proved in the present section describes in topological terms the class of Riemannian manifolds $M$ satisfying the equality $D(M) = 0$.

**Theorem 24.6.5.** *Let $M$ a be closed, smooth Riemannian manifold. Then $D(M) = 0$ if and only if each of its connected components is 2-connected, i.e., the first and second homotopy groups are trivial.*

The theorem immediately implies the following.

**Corollary 24.6.1.** *If at least one of the groups $\pi_1(M)$ and $\pi_2(M)$ of a connected, closed, and smooth Riemannian manifold $M$ is nontrivial, then $D(M) > 0$ for any Riemannian metric on $M$.*

For the proof of Theorem 24.6.5, we need the following lemmas. Recall that, in considering smooth differential and geometric objects over a manifold with boundary, they are assumed to admit extension to a certain manifold without boundary, which contains the original manifold. The same approach is used in considering direct products, though, in general, their boundary is not a smooth manifold.

**Lemma 24.9.1.** *Let $M, P$, and $N$ be three compact, smooth Riemannian manifolds, $P$ connected, and $\inf D = 0$ in the homotopy class of a certain smooth mapping $F : M \times P \to N$. Then, for any point $z_0 \in P$, $\inf D = 0$ in the homotopy class of the mapping $f = F|M \times \{z_0\}$ from $M$ to $N$.*

**Proof.** It follows from the data that there exists a sequence $F_i (i = 1, 2, \dots)$ of smooth mappings from $M \times P$ to $N$, homotopic to $F$, such that $D(F_i) \to 0$ as $i \to \infty$. We introduce the notation

$$F_i^z = F_i|M \times \{z\} : M \to N \quad (z \in P).$$

A metric that is the direct product of those on $M$ and $P$ is assumed to be introduced on $M \times P$. Therefore, for any point $x \in M$, the integral density functions for the mappings $F_i$ and $F_i^z$ are related by $e(F_i^z)(x) \leqslant e(F_i)(x, z)$ (see formula (1) of §24.7). Therefore,

$$\int_P D(F_i^z) dv \leqslant D(F_i) \to 0 \ (i \to \infty).$$

Hence, there exists a subsequence of subscripts $i_k$ such that $D(F^z_{i_k}) \to 0$ as $k \to \infty$ almost for all $z \in P$. Fix $z$ for which this statement holds and put $f_k = F^z_{i_k}$. Then $D(f_k) \to 0$ as $k \to \infty$. Due to the connectedness of $P$, the mapping $f$ determined by the data of the lemma is homotopic to the mapping $F|M \times \{z\}$, which is, in turn, homotopic to $f_k$ for all $k$ (the latter following from $F$ and $F_{i_k}$ being homotopic).

Q.E.D.

**Lemma 24.9.2.** *Let $M$ be a connected, closed, 2-connected, and smooth Riemannian manifold. Then $D(M) = 0$.*

**Proof.** If $\dim M \geqslant 5$, then the statement of the lemma immediately follows from Theorem 24.6.9 (see §24.8). In the general case, we consider the Riemannian product $M \times P$, where $P$ is a certain connected, closed, 2-connected, and smooth Riemannian manifold of dimension $\geqslant 5$ (e.g., $S^5$). Then the manifold $M \times P$ satisfies the conditions of Theorem 24.6.9, and there exists a sequence of smooth mappings $F_i (i = 1, 2, \dots)$ of $M \times P$ into it, which are homotopic to the identity mapping, so that $D(F_i) \to 0$ as $i \to \infty$. Let $\pi$ be the projection of $M \times P$ onto the first factor. It is obvious from Lemma 24.7.2 that $D(\pi \circ F_i) \to 0$ as $i \to \infty$, i.e., $\inf D = 0$ in the homotopy class of the mapping $\pi$, since all $F_i$ are homotopic to the identity mapping of $M \times P$. Fix an arbitrary point $z \in P$, and identify $M$ with the submanifold $M \times \{z\}$ in the direct product $M \times P$. Then $\pi|M \times \{z\} = \mathrm{id}_M$. Therefore, $\inf D = 0$ in the homotopy class of the mapping $\mathrm{id}_M$ due to Lemma 24.9.1.

Q.E.D.

**Lemma 24.9.3.** *Let $N$ and $M$ by two compact, smooth Riemannian manifolds, and $D(M) = 0$. Then $\inf D = 0$ in any homotopy class of smooth mappings from $N$ to $M$.*

**Proof.** If $M$ is zero-dimensional, then the statement of the lemma is trivial. Let $\dim M > 0$. Fix an arbitrary homotopy class of smooth mappings from $N$ to $M$, and choose in it a mapping $f$ such that $f(N) \cap \partial M = \emptyset$. Consider a vector bundle $f^*TM$ over $N$ which is a tangent along $f$. For any smooth mapping $f : N \to M$, a Riemannian vector bundle $f^*TM$ is defined over the manifold $N$, induced by $f$ from the bundle $TM$, and called a *tangent bundle* along $f$. The fibre $(f^*TM)_x$ of this bundle over a point $x \in N$ is a linear space $(TM)_{f(x)}$ tangent to $M$ at the point $f(x)$, whereas a Riemannian structure, i.e., the Riemannian metric along with the corresponding connection, is carried from the tangent bundle $TM$. It is known from the theory of finite-dimensional linear bundles that $f^*TM$ is a subbundle of the trivial k-dimensional fibre bundle $N \times \mathbf{R}^k \to N$, whose fibrewise projection onto $f^*TM$ is denoted by $\pi$. We supply $N \times \mathbf{R}^k$ with the natural Riemannian metric of the product, induced by the Riemannian metric on $N$ and Euclidean on $\mathbf{R}^k$. The procedure for the identification of the fibres of the bundles $f^*TM$ and $TM$ over the points $x \in N$ and $f(x) \in M$, respectively, permits us to consider the induced family of linear submersions $\pi_x : \mathbf{R}^k \to (TM)_{f(x)}$ smoothly parametrized by the points $x \in N$. Since $N$ is compact, we can assume without loss of generality that $\|\pi_x\| \leqslant \varepsilon$ for all $x \in N$, where the value $\varepsilon > 0$ is chosen in order that $\mathrm{dist}(f(N), \partial M) \geqslant 2\varepsilon$ (if $\partial M = \emptyset$, then $\varepsilon > 0$ is arbitrary). Denote by $D^k$ a compact unit disk in $\mathbf{R}^k$ with centre in the origin, and define a mapping $F : N \times D^K \to M$ by the formula

$$F(x, v) = \mathrm{Exp}^M_{f(x)}(\pi_x v), \tag{1}$$

where $\mathrm{Exp}^M_y$ is the exponential mapping of $M$, correctly defined at points $y \in f(N)$ on the tangent vectors with norm $\leqslant \varepsilon$.

Denote by $\varphi_i (i = 1, 2, \dots)$ a sequence of smooth mappings from the class $[\mathrm{id}_M]$, existing according to the conditions of the lemma, and such that $D(\varphi_i) \to 0$ as $i \to \infty$. Put

$$F_i = \varphi_i \circ F : N \times D^k \to M.$$

The mapping $F$ introduced by formula (1) is a submersion; therefore, it follows from Lemma 24.7.3 that there exists such a positive constant $C$ that the sequence $D(F_i)$ is majorized by the sequence $C \cdot D(\varphi_i)$. Consequently, $D(F_i) \to 0$ as $i \to \infty$. It is clear that all the mappings $F_i$ are homotopic to $F$. Hence, $\inf D = 0$ in the class $[F]$. We derive from Lemma 24.9.1 that $\inf D = 0$ in the homotopy class of the mapping $F|N \times \{0\}$ coinciding with $f$ due to (1).

Q.E.D.

**Proof of Theorem 24.6.5.** Let $M$ be an arbitrary smooth Riemannian manifold. It is clear that if $M$ is compact, then $D(M) = 0$ if and only if the same equality holds for any connected component of $M$. Therefore, it suffices to prove Theorem 24.6.5 for connected manifolds. The direct statement was proved in Lemma 24.9.2, viz., $D(M) = 0$ if $M$ is a connected, closed, and 2-connected Riemannian manifold. We prove the converse statement by comparing Lemma 24.9.3 and the known theorems on the globally minimal realization of homotopy classes. Let $M$ be a connected, closed, and smooth Riemannian manifold, and $D(M) = 0$. We obtain from Lemma 24.9.3 that $\inf D = 0$ in any homotopy class from $[S^1, M]$ and $[S^2, M]$. Hence, there exist no nontrivial harmonic mappings from $S^1$ or $S^2$ in $M$, which would minimize the Dirichlet functional in their homotopy class. On the other hand, each homotopy class of mappings from $S^1$ to $M$ is representable be a globally minimal and closed geodesic. Therefore, the set $[S^1, M]$ consists of one element, and $M$ is simply-connected. The triviality of $\pi_2(M)$, and therefore the 2-connectedness of $M$, can be proved by means of another realization theorem stating that if $\pi_2(M)$ is nontrivial, then there exists a nontrivial harmonic mapping $S^2 \to M$ minimizing the Dirichlet functional in its homotopy class (Sacks and Uhlenbeck [364]), and Theorem 24.6.5 is thus proved.

A closed relation of problems related to the existence of globally minimal and stable harmonic mappings becomes clear in the class of compact, symmetric spaces of Type I, i.e., which are not Lie groups.

**Proof of Corollary 24.6.2.** Statement (1) directly follows from Theorem 24.6.5. To prove (2), we see that the equivalence of $D(M) = 0$ to the non-stability of the mapping $\mathrm{id}_M$ follows from statement (1) (since $\dim M > 0$), i.e., the stability of $\mathrm{id}_M$ is equivalent to $\mathrm{D}(M) > 0$, and the corollary is thus proved.

**Proof of Theorem 24.6.6.** The implications (b) $\Rightarrow$ (a) and (c) $\Rightarrow$ (a) are obvious, it being sufficient to make use of the substitiutions $N = M$ and $N' = M$. (a) $\Rightarrow$ (c) was proved in Lemma 24.9.3. Assume that $D(M) = 0$, and prove (b). Fix an arbitrary smooth mapping $f : M \to N$, and denote by $\varphi_i (i = 1, 2, \dots)$ a sequence of smooth mappings of the manifold $M$ into itself, so that $\varphi_i \in [\mathrm{id}_M]$ and $D(\varphi_i) \to 0$ as $i \to \infty$. We obtain from Lemma 24.9.2 that

$$D(f \circ \varphi_i) \leqslant \mathrm{const} \cdot D(\varphi_i) \to 0 \quad (i \to \infty).$$

Since all the mappings $f \circ \varphi_i$ are, obviously, homotopic to $f$, $\inf D = 0$ in the homotopy class $[f]$, and statement (b) is thus proved, for the mapping $f$ has been chosen to be arbitrary. Theorem 24.6.6 is thus proved.

**Proof of Corollary 24.6.4.** If one of the manifolds $M$ or $N$ is without boundary and 2-connected, then $\inf D = 0$ in the homotopy class of the identity mapping (Theorem 24.6.5). Meanwhile, the corresponding implication (a) $\Rightarrow$ (b) or (a) $\Rightarrow$ (c) in Theorem 24.6.6 demonstrates that $\inf D = 0$ in the homotopy class $[f]$. Therefore, $D(f) = 0$, i.e., $f$ is constant, because $M$ is connected, and Corollary 24.6.4 is thus proved.

**Remark.** Let $M$ and $N$ be 2-connected, closed, and smooth Riemannian manifolds, and let at least one of them be 2-connected. It was established in the proof of Corollary 24.6.4 that $\inf D = 0$ in any homotopy class $\alpha \in [M, N]$. Meanwhile, if $M$ and $N$ are (real) analytic manifolds, then the application of the known approximation theorem leads to the statement that there are analytic mappings with arbitrarily small values $D(f)$ of the Dirichlet functional in each homotopy class $\alpha \in [M, N]$.

Further results are related to the study of the direct products $M \times P$ of smooth manifolds. If both $M$ and $P$ are without boundary, then the smooth manifold structure on $M \times P$ is introduced canonically. Be it not so, we assume as usual that the considered smooth objects are defined on a certain smooth manifold without boundary, containing an open domain $(\mathrm{int}\, M) \times (\mathrm{int}\, P)$ and a piecewise smooth hypersurface $\partial(M \times P)$. If $M$ and $P$ are Riemannian manifolds, then the manifold $M \times P$ with the direct product canonical metric is called the *Riemannian direct product.*

**Proof of Theorem 24.6.7.** We introduce the notation $f = \pi_N \circ F$ and $\varphi = \pi_Q \circ F$. The mappings $f$ and $\varphi$ are independent and determine $F = (f, \varphi)$, with

$$D(F) = D(f) + D(\varphi) \geqslant \inf_{[f]} D. \tag{2}$$

By Corollary 24.6.3, there is a sequence $\varphi_i \in [\varphi](i = 1, 2, \dots)$ such that $D(\varphi_i) \to 0$ as $i \to \infty$. Let $f_i \in [f]$ $(i = 1, 2, \dots)$ be an arbitrary sequence minimizing $D$ on $[f]$, i.e., $D(f_i) \to \inf_{[f]} D$ as $i \to \infty$.

Set $F_i = (f_i, \varphi_i)$. Then

$$D(F_i) = D(f_i) + D(\varphi_i) \to \inf_{[f]} D \quad (i \to \infty). \tag{3}$$

It follows from (2) and (3) that

$$\inf_{[F]} D = \inf_{[f]} D.$$

If the class $[\varphi]$ is nontrivial, then $D(\varphi') > 0$ for $\varphi' \in [\varphi]$, and inequality (2) cannot turn into equality, which means the unattainability of $\inf D$ on $[F]$, and the theorem is thus proved.

**Proof of Corollary 24.6.5.** That $\inf D$ is unattainable on $[F]$ follows from Theorem 24.6.7. Furthermore, it was proved in [3, 119] that any class $[\pi_N \circ F]$ is realizable by a globally minimal harmonic mapping $\pi_N \circ F_0$, since the curvature of $N$ is nonpositive. This mapping is nonconstant, because $[\pi_N \circ F]$ is nontrivial. Therefore,

$$\inf_{[F]} D = \inf_{[\pi_N \circ F]} D = D(\pi_N \circ F_0) > 0,$$

and the corollary is thus proved.

**Proof of Theorem 24.6.8.** If $\dim P = 0$, then the statement is trivial. Assume that $\dim P > 0$. If $\pi_N$ and $\pi_Q$ are the products of the Riemannian direct product $N \times Q$ onto the first and second factors, respectively, we introduce the corresponding canonical embeddings

$$\begin{aligned} i_y &: M \to M \times \{y\} \subset P, \\ i_x &: P \to \{x\} \times P \subset M \times P \end{aligned} \tag{4}$$

for arbitrary points $x \in M, y \in P$, consider an arbitrary smooth mapping $F \in [f \times \varphi]$, and put

$$\begin{aligned} F_y &= \pi_N \circ F \circ i_y : M \to N. \\ F_x &= \pi_Q \circ F \circ i_x : P \to Q. \end{aligned} \tag{5}$$

It is clear that the Hilbert–Schmidt norms of the operators $dF, dF_x$ and $dF_y$ are related by the inequality $\|dF\|^2 \geqslant \|dF_x\|^2 + \|dF_y\|^2$. Therefore,

$$\begin{aligned} D(F) &\geqslant \int_M D(F_x)dx + \int_P D(F_y)dy \leqslant \text{vol}(M) \cdot D(F_{x_0}) + \text{vol}(P) \cdot D(F_{y_0}) \\ &= D(F_{y_0} \times F_{x_0}), \end{aligned} \tag{6}$$

where $dx$ and $dy$ are volume elements of the Riemannian manifolds $M$ and $P$, respectively, whereas $x_0$ and $y_0$ are minimum points for continuous functions $D(F_x)$ and $D(F_y)$ on the compact manifolds $M$ and $P$, respectively.

Denote by $[f] \times [\varphi]$ the set of smooth mappings of the form $f' \times \varphi' : M \times P \to N \times Q$, so that $f' \in [f]$, and $\varphi' \in [\varphi]$. It is evident that $[f] \times [\varphi] \subset [f \times \varphi]$.

**Lemma 24.9.4.** *For any $x \in M$ and $y \in P$,*

$$F_y \times F_x \in [f] \times [\varphi].$$

**Proof.** Since $F \in [F \times \varphi]$, there exists a smooth mapping $\Phi : M \times P \times I \to N \times Q$, where $I = [0,1]$, so that

$$\begin{aligned} \Phi | M \times P \times \{0\} &= F, \\ \Phi | M \times P \times \{1\} &= f \times \varphi. \end{aligned} \tag{7}$$

We introduce the following embeddings induced by embeddings (4), viz.,

$$\begin{aligned} \hat{i}_y &= i_y \times \mathrm{id}_I : M \times T \to M \times P \times I, \\ \hat{i}_x &= i_x \times \mathrm{id}_I : P \times T \to M \times P \times I. \end{aligned} \tag{8}$$

Consider the mappings

$$\begin{aligned} \Phi_y &= \pi_N \circ \Phi \circ \hat{i}_y : M \times I \to N, \\ \Phi_x &= \pi_Q \circ \Phi \circ \hat{i}_x : M \times I \to Q. \end{aligned} \tag{8}$$

From (5), (7) and (8), we obtain

$$\begin{aligned} \Phi_y | M \times \{0\} &= F_y, \quad \Phi_y | M \times \{1\} = f, \\ \Phi_x | M \times \{0\} &= F_x, \quad \Phi_x | M \times \{1\} = \varphi, \end{aligned}$$

i.e., $F_y \in [f], F_x \in [\varphi]$, and the lemma is thus proved.

Let a sequence of smooth mappings $F_i \in [f \times \varphi]$ $(i = 1, 2 \ldots)$ minimize $D$ in the class $[f \times \varphi]$. It follows from (6) and Lemma 24.9.4 that the mappings $F_i$ can be chosen in the form $F_i = f_i \times \varphi_i \in [f] \times [\varphi]$. Therefore,

$$\inf_{[f \times \varphi]} D = \inf_{[f] \times [\varphi]} D. \tag{9}$$

For any smooth mapping, $f' \times \varphi' \in [f] \times [\varphi]$, we have

$$D(f' \times \varphi') = \mathrm{vol}(P) \cdot D(f') + \mathrm{vol}(M) \cdot D(\varphi') \geqslant \mathrm{vol}(P) \cdot D(f') \geqslant \mathrm{vol}(P) \cdot \inf_{[f]} D. \tag{10}$$

By Corollary 24.6.3, there is a sequence $\varphi_i \in [\varphi]$ $(i = 1, 2, \dots)$ such that $D(\varphi_i) \to 0$ as $i \to \infty$. Let $f_i \in C^\infty(M, N)$ be an arbitrary sequence minimizing $D$ on $[f]$, i.e., $f_i \in [f]$, and $D(f_i) \to \inf_{[f]} D$ as $i \to \infty$. Taking into account equality in (10), we obtain

$$D(f_i \times \varphi_i) \to \mathrm{vol}(P) \cdot \inf_{[f]} D \quad (i \to \infty). \tag{11}$$

It follows from (9)–(11) that

$$\inf_{[f \times \varphi]} D = \inf_{[f] \times [\varphi]} D = \mathrm{vol}(P) \cdot \inf_{[f]} D.$$

If the homotopy class $[\varphi]$ is nontrivial, then $D(\varphi') > 0$ for all $\varphi' \in [\varphi]$. In particular, the first inequality in (10) cannot turn into equality, which means that $\inf D$ is unattainable on $[f] \times [\varphi]$ and, therefore, on $[f \times \varphi]$. The proof is thus completed.

**Proof of Corollary 24.6.6.** That $\inf D$ cannot be attained on $[f \times \varphi]$ follows from Theorem 24.6.8. Furthermore, let $\inf D$ be attained on a mapping $f_0$ in the class $[f]$. Then

$$\inf_{[f \times \varphi]} D = \mathrm{vol}(P) \cdot \inf_{[f]} D = \mathrm{vol}(P) \cdot D(f_)0 > 0,$$

since $f_0 \neq \mathrm{const}$ due to the nontriviality of $[f]$, thus completing the proof of the corollary.

**Proof of Corollary 24.6.7.** The positivity of $\inf D$ in the class [id] of the identity mapping of the manifold $M \times P$, i.e., the inequality $D(M \times P) > 0$, follows from Corollary 24.6.1, since one of the groups $\pi_i(M \times P)$ and $\pi_2(M \times P)$ is given to be nontrivial. The unattainability of $\inf D$ in [id] follows from Theorem 24.6.8. In fact, $\mathrm{id} = \mathrm{id}_M \times \mathrm{id}_P$, whereas the homotopy class $[\mathrm{id}_P]$ is nontrivial, since $P$ is closed, and $\dim P > 0$, thus completing the proof.

### 24.10. The Minimization of Dirichlet-Type Functionals

The results of A. I. Pluzhnikov and B. White on topological obstructions to realizable Dirichlet functional minima in homotopy classes can be extended to more general functionals and higher homotopy groups, for which we introduce a collection of functionals $D_p$,

$$D_p(f) = \int_M \|df\|^p dv, \quad 1 \leqslant p < \infty,$$

on the space of mappings of one Riemannian manifold to another. If $p = 2$, then we obtain the usual Dirichlet functional. Note that if $p = \dim M$, then the functional $D_p$ is conformally invariant.

Similarly to the usual Dirichlet functional, $D_p(f) \geqslant 0$ and $D_p(f) = 0$ if and only if $f = \mathrm{const}$.

The following two theorems due to A. V. Tyrin [417] generalize Pluzhnikov's theorems.

**Theorem 24.10.1.** *Assume that* $\inf D_p$ *in the homotopy class of the identity mapping of* $M$ *onto itself is zero. Then, for any* $N$, $\inf D_p$ *is also zero in any homotopy class of mapping of* $N$ *to* $M$ *and any homotopy class of mappings of* $M$ *to* $N$.

Therefore, the minimum is never attained in a nontrivial homotopy class in the above circumstances.

**Theorem 24.10.2.** *For a compact Riemannian manifold* $M$, *the condition* $\inf D_p = 0$ *in the homotopy class of the identity mapping of* $M$ *onto itself is equivalent to* $\pi_1 M = \cdots = \pi_{[p]} M$, *where* $[p]$ *is the greatest integer not exceeding* $p$.

*For the particular case* $N = S^n$, *we can prove the following local version of the statement regarding the unattainability of the minima* $D_p$.

**Theorem 24.10.3.** *A nonconstant mapping* $f \in C^2(M, S^n)$ *of a connected manifold* $M$ *cannot be a local minimum of the functional* $D_p$ *for* $p < n$.

For the functional $D = D_2$, it was proved in [415]. A similar theorem for $D_p$ and mappings from $C^2(S^n, M)$ is noted in [420].

Like Theorem 24.6.5, Theorem 24.10.2 also admits strengthening. Namely, the following result is due to B. White [434].

**Theorem 24.10.4.** *Consider two closed Riemannian manifolds* $M$ *and* $N$. *The mappings* $f, h \in C^\infty(M, N)$ *belong to one connected component of the closure of the space* $C^\infty(M, N)$ *in* $W_k^1(M, N)$, *where* $k \geqslant 1$ *is an integer, if and only if they are k-homotopic, i.e., their restrictions to the k-dimensional skeleton are homotopic. The connected component coincides with the closure of the set* $[f] \subset C^\infty(M, N)$. *In particular,* $\inf_{[f]} D_k = \inf_{[f]} D_k$, *where* $[f]_k$ *is the set of smooth mappings k-homotopic to a mapping* $f$.

Theorem 24.6.12 is a particular case of Theorem 24.10.4 for $k = 2$.

## 24.11. Regularity of Harmonic Mappings

Here, we consider the problem of harmonic mapping regularity minimizing the Dirichlet functional. A. V. Tyrin's theory briefly mentioned below is in a certain sense parallel to the corresponding results from volume functional theory; however, the technicalities are as a rule different.

To describe the situation in question more precisely, we assume that a manifold $N$ is isometrically embedded into Euclidean space of sufficiently large dimension, viz., $N \subset \mathbf{R}^k$. Then we can extend the definition of the Dirichlet functional to a wider class of mappings $W_2^1(M, N)$. By definition, $W_2^1(M, N) = \{f \in W_2^1(M, \mathbf{R}^k); f(x) \in N$ for almost all $x \in M\}$. Recall that extremals for $D$ in the space $C^2(M, N)$ are called harmonic mappings, for they automatically are of class $C^\infty$.

We now ask ourselves a question regarding the regularity of mappings from $W_2^1(M, N)$, minimizing $D$. We are interested in the smoothness of mappings on compact subsets of the manifold $M$. In general, they are not smooth or even continuous if $M \geqslant 2$. Nevertheless, for $m \leqslant 2$, the mappings minimizing $D$ are known to be always smooth, which already follows from the classical results of C. Morrey [289]. That the smoothness occurs if $M = 1$ is a sufficiently elementary result. Therefore, the discovery of $N$ such that the smoothness occurs in higher dimensions is of greatest interest.

Recently, R. Schoen and K. Uhlenbeck proved in [367] and [368] the following quite general criterion for regularity. Let any homogeneous mapping $F : \mathbf{R}^q \to N$ determined by a smooth harmonic mapping $f : S^{q-1} \to N$ according to the formula $F(x) = f(x/|x|)$, minimizing $D$ in $W_2^1$, and, in particular, stable on compact subsets $\mathbf{R}^q$, be constant for $3 \leqslant q \leqslant l$. Then, for any mapping $M \to N$ minimizing $D$ on compact subsets $M^m$, the Hausdorff dimension $H^{m-1-l}$ of the singular point set is finite (see [367], 368]). In particular, such mappings are always regular (smooth) for $m \leqslant l$.

What manifolds $N$ satisfy this criterion? First of all, as noted by Tyrin, if Schoen's and Uhlenbeck's criterion holds, then strong restrictions on the topology of $N$ follow.

**Theorem 24.11.1.** *Let $N$ be simply-connected and such that the Schoen–Uhlenbeck criterion is valid for homogeneous mappings $F : \mathbf{R}^q \to N$, $3 \leqslant q \leqslant 1$. Then $\pi_2 N = \cdots = \pi_{p-1} N = 0$.*

*We also indicate (which is almost obvious) that a mapping from an arbitrary $M$ to the Cartesian product of Riemannian manifolds $N_1 \times \cdots \times N_r \subset \mathbf{R}^{k_1} \times \cdots \times \mathbf{R}^{k_r}$ is harmonic (or minimizing) if and only if each coordinate mapping $M \to N_i \subset \mathbf{R}^{k_i}$ is such as this.*

As expected, this criterion holds due to the results of S. I. Alber [4], J. Eells and J. Sampson [119] for any $l$ if $N$ is of nonpositive curvature [367]. R. Schoen and K. Uhlenbeck considered the case of regularity of minimizing mappings to Euclidean spheres [369]. Below, we list A. V. Tyrin's results, who generalized the theorems of [369] to a wider class of both manifolds and metrics.

First, let $N^n \subset \mathbf{R}^{n+1}$ be a complete oriented hypersurface, let $\mu$ denote the pointwise maximum of principal curvatures on $N$, and $k(k')$ that (or minimum) of sectional curvatures on $N$.

**Theorem 24.11.2.** *If, for certain $l \geqslant 3$, the inequality*

$$\frac{l-2}{l-1}k \leqslant \frac{k'(n-1)-\mu^2}{n} \tag{1}$$

*is valid, then the mappings from $W_2^1(M,N)$, minimizing the Dirichlet functional on compact subsets of an arbitrary Riemannian manifold $M^m$, are smooth for $3 \leqslant m \leqslant \min\{l,5\}$.*

Now, let $N^n \subset S^p$ be an isometric embedding of a compact, orientable manifold in a sphere with the standard metric and second fundamental embedding form $S$, with $S$ being its length, and $k$ the same as above.

**Theorem 24.11.3.** *If, for a certain $l \geqslant 3$,*

$$\frac{l-2}{l-1}k \leqslant \frac{n-2}{n} - \frac{\max\{1, \sqrt{n-2}/2\}+1}{n}\|S\|^2, \tag{2}$$

*then the mappings from $W_2^1(M^m, N)^n$ minimizing $D$ are smooth for $3 \leqslant m \leqslant \min\{l, 5\}$.*

**Remark.** For the case of spheres $N = S^n$ with nonstandard metrics, Theorems 24.11.2 and 24.11.3 generalize the result of [369].

The above results prompt the question: Will the regularity of minimizing mappings be preserved in slightly stirring the metric on $N$? (It is clear that the theorems consider variations of metrics of quite a special form.) In general, it follows from nowhere, since the closeness of two metrics on an inverse image does not provide for that of harmonic mappings to the image. Nevertheless, it turns out that the answer is positive in a number of cases. We now formulate the following result along these lines.

**Theorem 24.11.4 (A. V. Tyrin).** *In the following results, for the Riemannian manifold $(N,g)$, there is a $C^2$-neighbourhood of the metric $g$ such that, for any metric $h$ from this neighbourhood, any mapping $M^m \to (N,h)$ minimizing $D$ on compact subsets is smooth, viz.,*

*(a) $N^n \subset \mathbf{R}^{n+1}$, and inequality (1) is strict, $m \leqslant \min\{l, 5\}$,*
*and*

*(b) $N^n \subset S^p$, and inequality (2) is strict, $m \leqslant \min\{l, 5\}$.*

Having explicitly written out the conditions for the dimensions of $M^m$ and $S^n$, we obtain that the statement of Theorem 24.11.4 (a) holds for $N = S^n$ with a metric close to the standard for $m \leqslant d(n)$, where $d(n) = [\min\{\frac{1}{2}(n+1), 5\}]$ if $n \geqslant 3, d(2) = 2$, with $[r]$ denoting the greatest integer not exceeding $r$.

**Remark.** The results of [367] and [368] show also that for manifolds $(N, g)$ due to Theorem 24.11.4, the Dirichlet problem for harmonic mappings $M^m \to (N, g)$ with an arbitrary smooth "boundary condition" $\partial M \to (N, g), W_2^1$-extendable to the whole of $M$, is always solvable.

**Remark.** To give an example of a mapping from $W_2^1(\mathbf{R}^7, S^7)$, which is not smooth but nevertheless minimizes the Dirichlet functional (see [369]), we define a mapping $F : \mathbf{R}^7\backslash\{0\} \to S^6$ by the formula $F(x) = x/|x|$. We now embed $S^6$ into $S^7$ as the equator and obtain a mapping $\mathbf{R}^7\backslash\{0\} \to S^7$. It can be proved (see [369]) that it minimizes the Dirichlet functional on compact subsets of $\mathbf{R}^7$. This is an analogue of the J. Simons cones known in volume functional theory (see [391]).

CHAPTER 6

# Solution of the Plateau Problem in Classes of Mappings of Spectra of Manifolds with Fixed Boundary. Construction of Globally Minimal Surfaces in Variational Classes $h(A, L, L')$ and $h(A, \tilde{L})$

## §25 The Cohomology Case. Computation of the Coboundary of the Pair $(X, A) = \bigcup_r (X_r, A_r)$ in Terms of Those of $(X_r, A_r)$

We now solve the Plateau problem, i.e., we give the proof of the existence theorem for globally minimal surfaces of nontrivial topological type in Riemannian manifolds.

The main geometric idea of the present section consists of gluing together a surface $X$ with boundary $A$ from separate portions $X_\alpha$ with boundaries $A_\alpha$, and then clarifying the relation between $A$ and $\bigcup_\alpha A_\alpha$. Note at first that if $X = A$, then $\nabla^*(X, A) = \emptyset$ for any $x \in A$, and, if $X \supset A$ and $X$ is contractible, then $\nabla^*(X, A) = \tilde{h}^*(A)\backslash 0$ for any $x \in A$.

**Lemma 25.1.** *Let $X = \bigvee_{r=1}^N X_r$, where $X_r \cap X_s = \emptyset$ for $r \neq s$, and $A \vee B$ denote the wedge of two pointed spaces. Furthermore, suppose $A_r$ are such compact sets that $A_r \subset X_r$ for $1 \leqslant r \leqslant N$, $x \in A_r$ for each $r$. Set $A = \bigvee_{r=1}^N A_r$. Assume that $i_r : (X_r, A_r) \to (X, A)$ are embeddings. Then the homomorphisms $i_r^* : \tilde{h}^k(X, A) \to \tilde{h}^k(X_r, A_r)$ make up the projective representation of the group $\tilde{h}^k(X, A)$ as the direct sum $\bigoplus_r \tilde{h}^k(X_r, A_r)$, i.e., for any sequence of elements $u_r \in \tilde{h}^k(X_r, A_r)$, there exists a unique element $u \in \tilde{h}^k(X, A)$ such that $i_r^* u = u_r$ for $1 \leqslant r \leqslant N$.*

**Proof.** If we consider the disjoint union $X' = \bigcup_r X_r$ i.e., $X_i \cap X_j = \emptyset$ for $i \neq j$ in the compact set $X'$, then we obtain that $h^k(X') = \bigoplus_r h^k(X_r)$. Though the mappings $j_r : (X_r, A_r) \to (X \bigcup_r A_r, A)$ are not excisions in the sense of Axiom A6 (=A6′), they are relative homeomorphisms, and since the theory $h^*$ is relatively invariant on the category $U_C$ (see Lemma 4.5.2), the homomorphisms $j_r^*$ are isomorphisms. Hence, by repeating the steps of the proof of Theorem 1.13.2 for the category of pointed spaces (see [126]), we obtain the required statement.

Note that if, within the framework of Lemma 25.1, we assume that all the compact sets $X_i$ are contractible, then $\nabla^*(X,A) = \tilde{h}^*(A)\backslash 0$.

**Lemma 25.2.** *Let* $X = \bigcup_{r=1}^N X_r$, $A = \bigcup_{r=1}^N A_r$, *where* $A_r \subset X_r$, $X_r \cap X_s = A_r \cap A_s$ *for* $r \neq s$, $x \in A_r$ *for each* $r$, *and* $i_r : (X_r, A_r) \to (X, A)$ *embeddings. Then the homomorphisms* $i_r^* : \tilde{h}^k(X,A) \to \tilde{h}^k(X_r, A_r)$ *form the projective representation of the group* $\tilde{h}^k(X,A)$ *for any* $k$ *as the direct sum* $\bigoplus_r \tilde{h}^k(X_r, A_r)$.

The statement immediately follows from the relative invariance of the theory $h^*$ on $U_C$ (which permits us to pass from the pair $(X_r, A_r)$ to $(X_r/A_r, x)$) and Lemma 25.1.

Another simple remark: Let $f : (X,A) \to (Y,B)$ be a continuous mapping, and $x \in A$, $y = f(x)$, $L \subset \nabla^k(X,A)$ be an arbitrary subset. Consider a new subset $L_1 \subset \tilde{h}^{k-1}(B)\backslash 0$, putting, by definition, $L_1 = (f^*)^{-1}L$. Then we state that $L_1 = \nabla^k(Y,B)$, i.e., $L_1 \cap \operatorname{Ker} \tilde{\delta}_B = \emptyset$.

The proof follows from the diagram

$$\begin{array}{ccc} \tilde{h}^{k-1}(A) & \xleftarrow{i_1^*} & \tilde{h}^{k-1}(X) \\ \uparrow f_2^* & & \uparrow f_1^* \\ \tilde{h}^{k-1}(B) & \xleftarrow[i_2^*]{} & \tilde{h}^{k-1}(Y) \end{array}$$

Indeed, if we assume the contrary, that there exists an element $\alpha \in L_1$ for which $\alpha = i_2^*(\omega)$, then we obtain $f_2^*(\alpha) \in L$, and $f_2^*(\alpha) = i_1^* f_1^*(\omega)$, i.e., $f_2^*(\alpha) = L \cap \operatorname{Im} i_1^*$, which contradicts the definition of $L$ due to which $L \cap \operatorname{Im} i_1^* = \emptyset$.

As a corollary, supposing $x \in A \subset X$, $L \subset \tilde{h}^{k-1}(A)\backslash 0$, $L \subset \nabla^k(X,A)$, and $Y \supset X$, we derive that $L \subset \nabla^k(Y,A)$. This clearly follows from the remark proved above, since $B = A$ and $f = i$. We denote the embedding $i : B \to A$ by $i(A,B)$, and the full inverse image of a subset $C$ in a group $K$ under a homomorphism $\alpha^* : N \to K$ by $(\alpha^*)^{-1}C$.

**Lemma 25.3.** *Let* $X = \bigcup_{r=1}^N X_r$, $A_r \subset X_r$, $A \subset X$, $x \in A_r$, *for each* $r$, $B = A \cup (\bigcup_r A_r)$, $L \subset \tilde{h}^{k-1}(A)\backslash 0$, *and* $L_r \subset \nabla^k(X_r, A_r)$. *Assume that* $i^*(B,A)^{-1}L \subset \bigcup_r i^*(B,A_r)^{-1}L_r$. *Then* $L \subset \nabla^k(X,A)$.

**Proof.** We need to show that $L \cap \operatorname{Im} i^* = \emptyset$ (where $i = i(X,A)$), or that $i^*(X,A)^{-1}L = \emptyset$.

We have

$$\begin{aligned} & i^*(X, A)^{-1}L \\ & \quad = i^*(X, B)^{-1}[i^*(B, A)^{-1}L] \subset \bigcup_r i^*(X, B)^{-1}[i^*(B, A_r)^{-1}L_r] \\ & \quad = \bigcup_r i^*(X, A_r)^{-1}L_r = \bigcup_r i^*(X, X_r)^{-1}[i^*(X_r, A_r)^{-1}L_r] = \emptyset, \end{aligned}$$

since $i^*(X_r, A_r)^{-1}L_r = \emptyset$ by assumption.

Note that Lemma 25.3 can be proved otherwise and longer than above, but more universally; the method will be repeatedly used in the following.

Consider the following commutative diagram, viz.,

$$\begin{array}{ccccc} \tilde{h}^{k-1}(X_r) & \xrightarrow{i_r^*} & \tilde{h}^{k-1}(A_r) & \xrightarrow{\delta_r} & h^k(X_r, A_r) \\ \uparrow{\scriptstyle\omega} & & \uparrow{\scriptstyle\gamma_r} & & \uparrow{\scriptstyle\tau_r} \\ \tilde{h}^{k-1}(X) & \xrightarrow{i_2^*} & \tilde{h}^{k-1}(B) & \xrightarrow{\delta_2} & h^k(X, B) \\ \downarrow{\scriptstyle\xi} & & \downarrow{\scriptstyle\alpha} & & \downarrow{\scriptstyle\beta} \\ \tilde{h}^{k-1}(X) & \xrightarrow{i_1^*} & \tilde{h}^{k-1}(A) & \xrightarrow{\delta_1} & h^k(X, A) \end{array}$$

Let $h \in L$, $h \neq 0$. We have to show that $\delta_1(h) \neq 0$. Assume the contrary, viz., let $\delta_1(h) = 0$. Then $h = i_1^*(\varphi)$, and $h = i_1^*\xi(\varphi') = \alpha i_2^*(\varphi')$, i.e., $i_2^*(\varphi') \in i^*(B, A)^{-1}h$, or $i_2^*(\varphi') \in \bigcup_r i^*(B, A_r)^{-1}L$ from the data. It follows that there exists a subscript $r$ such that $i_2^*(\varphi') \in i^*(B, A_r)^{-1}L_r$. Let just this value be taken in the upper line of the diagram. Then $\gamma_r i_2^*(\varphi') \in L$, or $\delta_r\gamma_r i_2^*(\varphi') \neq 0$. Meanwhile, we have $\delta_r\gamma_r i_2^*(\varphi') = \tau_r\delta_2 i_2^*(\varphi) = 0$ by exactness. The contradiction proves the statement.

The following lemma is important for the geometric application of our algebraic constructions in the following, since it permits us to determine the algebraic coboundary of the pair $(X, A)$ made up of separate pieces $(X_r, A_r)$ in terms of the coboundaries of $(X_r, A_r)$, which are assumed to be known.

**Lemma 25.4.** *Let $X = \bigcup_{r=1}^N X_r$, $A_r \subset X_r$, $A \subset X$, $A \cap X_r \subset A_r$, $X_r \cap X_s = A_r \cap A_s$, if $r \neq s$, $x \in A$, and $x \in A_r$ for each $r$. Put $B = \bigcup_r A_r$ (then $B \supset A$), and $\Gamma = \nabla^k(X_r, A_r) \subset \tilde{h}^{k-1}(A_r)\backslash 0$. Then*

$$\nabla^k(X, A) = \tilde{h}^{k-1}(A) \diagdown \Big\{ i^*(B, A)\Big[\tilde{h}^{k-1}(B) \diagdown \Big[\bigcup_r i^*(B, A_r)^{-1}\Gamma_r\Big]\Big]\Big\}.$$

**Proof.** Consider the commutative diagram

$$\begin{array}{ccccc} \tilde{h}^{k-1}(X_r) & \xrightarrow{i_r^*} & \tilde{h}^{k-1}(A_r) & \xrightarrow{\delta_r} & h^k(X_r, A_r) \\ \uparrow & & \uparrow{\scriptstyle\gamma_r} & & \uparrow{\scriptstyle\tau_r} \\ \tilde{h}^{k-1}(X) & \xrightarrow{i_2^*} & \tilde{h}^{k-1}(B) & \xrightarrow{\delta_2} & h^k(X, B) \\ \downarrow & & \downarrow{\scriptstyle\alpha} & & \downarrow{\scriptstyle\beta} \\ \tilde{h}^{k-1}(X) & \xrightarrow{i_1^*} & \tilde{h}^{k-1}(A) & \xrightarrow{\delta_1} & h^k(X, A) \end{array}$$

Distinguish the subset of elements (cocycles)

$$\Phi = i^*(B,A)[\tilde{h}^{k-1}(B)\backslash[\bigcup_r i^*(B,A_r)^{-1}\Gamma_r]]$$

from $\tilde{h}^{k-1}(A)$, and prove that $\nabla^k(X,A) = \tilde{h}^{k-1}(A)\backslash\Phi$. Note that, since $\Gamma_r = \tilde{h}^{k-1}(A_r)\backslash \operatorname{Ker}\delta_r$, $\Phi$ is, in fact, a subgroup in $\tilde{h}^{k-1}(A)$. We first show that $\nabla^k(X,A) \subset \tilde{h}^{k-1}(A)\backslash\Phi$. Let $a \in \nabla^k(X,A)$, which is equivalent to $\delta_1(a) \neq 0$. Assume now that $a \in \Phi$, whence $a = \alpha(\varphi)$, $\varphi \in \tilde{h}^{k-1}(B)$, and $\varphi \notin \bigcup_r i^*(B,A_r)^{-1}\Gamma_r = \bigcup_r \gamma_r^{-1}(\Gamma_r)$, i.e., $\gamma_r(\varphi) \notin \Gamma_r$ for any $r$, which means that $\delta_r\gamma_r(\varphi) = 0$, i.e., $\tau_r\delta_2(\varphi) = 0$, and then, by Lemma 25.2, we have $\delta_2(\varphi) = 0$, or $\varphi = i_2^*(m)$, $a = \alpha(\varphi) = \alpha i_2^*(m) = i_1^*\xi(m)$, with $\delta_1(a) = \delta_1 i_1^*\xi(m) = 0$, which is contrary to the choice of the element $a$. Thus, $\nabla^k(X,A) \subset \tilde{h}^{k-1}(A)\backslash\Phi$.

Conversely, we prove that $\nabla^k(X,A) \supset \tilde{h}^{k-1}(A)\backslash\Phi$. Let $a \in \tilde{h}^{k-1}(A)\backslash\Phi$. Assume that $a \notin \nabla^k(X,A)$, i.e., $\delta_1(a) = 0$. Then $a = i_1^*(m') = i_1^*\xi(m) = \alpha i_2^*(m)$. Since $\delta_2 i_2^*(m) = 0$, it follows that $\gamma_r i_2^*(m) \notin \Gamma_r$ for each $r$, because of $\delta_r\gamma_r i_2^*(m) = 0$. Therefore,

$$i_2^*(m) \in \tilde{h}^{k-1}(B)\backslash[\bigcup_r \gamma_r^{-1}(\Gamma_r)] = \tilde{h}^{k-1}(B)\backslash[\bigcup_r i^*(B,A_r)^{-1}\Gamma_r].$$

It follows that $\alpha i_2^*(m) = a \in \Phi$, which is contrary to the choice of $a$.

Q.E.D.

If the homomorphism $\alpha$ is an epimorphism, then Lemma 25.4 can be formulated in a somewhat different manner, viz., the relation

$$\nabla^k(X, A) = \delta_1^{-1}\left\{\left[\beta\left[\bigcup_r \tau_r^{-1}\delta_r(\Gamma_r)\right]\right]\backslash 0\right\}$$

holds. To prove it, denote the right-hand side by $L$, and show that $L \subset \nabla^k(X,A)$. Let $\lambda \in L$. Then $\delta_1(l) \in \beta[\bigcup_r \tau_r^{-1}\delta_r(\Gamma_r)]\backslash 0$, or $\delta_1(l) \neq 0$; therefore, $l \in \nabla^k(X,A)$. Thus, $L \subset \nabla^k(X,A)$. Conversely, we prove that $L \supset \nabla^k(X,A)$. Let $h \in \nabla^k(X,A)$, which is equivalent to $\delta_1(h) \neq 0$. Since $\alpha$ is an epimorphism, there exists an element $m$ such that $h = \alpha(m)$, i.e., $\delta_1(h) = \beta\delta_2(m)$, where $\delta_2(m) \neq 0$. Because the homomorphisms $\tau_r$ form the projective representation of the group $h^k(X,B)$ as the direct sum $\bigoplus_r h^k(X_r,A_r)$ (see Lemma 25.2), there exists a subscript $r$ such that $\tau_r d_2(m) \neq 0$. Suppose that just this value is taken in the diagram. Because $\tau_r\delta_2(m) \neq 0$, the element $\delta_r\gamma_r(m)$ is other than zero, and $\gamma_r(m) \in \nabla^k(X_r,A_r) = \Gamma_r$. Thus, we have proved the existence of $r$ such that $\varphi = \delta_2(m) \neq 0$, $\varphi \in h^k(X,B)$, $\delta_1(h) = \beta(\varphi)$, $\tau(\varphi) = \delta_r\gamma_r(m) \neq 0$, i.e., $\tau_r(\varphi) \in \delta_r(\Gamma_r)$, or $\delta_1(h) \in \beta[\tau_r^{-1}\delta_r(\Gamma_r)]\backslash 0$, and $h \in \delta_1^{-1}\{[\beta[\bigcup_r \tau_r^{-1}\delta_r(\Gamma_r)]]\backslash 0\}$.

Q.E.D.

In further applications, $\alpha$ will often be an epimorphism, and we will resort to the above statement. When studying the homology case, we shall see that the statement of Lemma 25.4 is quite far from its homology analogue, but acquires a somewhat dual character if $\alpha$ is an epimorphism.

To clarify what happens to the coboundary $\nabla^k(X, A)$ in the cases where $X = I \times Y$ and $A = (0 \times Y) \cup (1 \times Y)$, with $I$ denoting the unit interval, consider two natural embeddings $j_0 : Y \to 0 \times Y \subset A$ and $j_1 : Y \to 1 \times Y \subset A$, distinguishing the diagonal $\Delta = (h, h)$, $h \in h^{k-1}(Y)$ from the group $h^{k-1}(A)$ isomorphic to $h^{k-1}(Y) \oplus h^{k-1}(X)$. It then turns out that $(h^{k-1}(A) \backslash \Delta) \cap \operatorname{Im} i^* = \emptyset$, where $i^* : h^{k-1}(X) \to h^{k-1}(A)$, $i : A \to X$ is an embedding, i.e., $i^* h^{k-1}(X) \subset \Delta$.

Prove that if $(h_1, h_2) \notin \Delta$, then $(h_1, h_2) \notin \operatorname{Im} i^*$. Assume the contrary, viz., let $h_1 \neq h_2$, and $(h_1, h_2) = i^*(\gamma)$, $\gamma \in h^{k-1}(X)$.

But then $h_1 = j_0^*(h_1, h_2) = j_0^* i^*(\gamma) = (ij_0)^*(\gamma) = (ij_1)^*(\gamma) = j_1^*(h_1, h_2) = h_2$, i.e., $h_1 = h_2$, which is impossible. Here, we have used the fact that $ij_0$ is homotopic to $ij_1$.

Q.E.D.

Recall now that there is a base point $x$. We obtain that $\nabla^k(X, A) = (h^{k-1}(A) \backslash \Delta) \cap \tilde{h}^{k-1}(A) = \tilde{h}^{k-1}(A) \backslash (\Delta \cap \tilde{h}^{k-1}(A))$, with the subgroup $\tilde{h}^{k-1}(A)$ depending, in general, on the choice of $x$, whereas the subgroup $\Delta$ does not depend on it. In the following, while studying the homology case, we see that the homology and cohomology versions of the theory are substantially different, it being especially clearly delineated in the situation $X = I \times Y$ just considered. We shall now study the behaviour of the algebraic coboundary under continuous deformations of compact sets.

Let $f : I \times Y \to X$ be a continuous mapping, and $x \in X$ a point not in the image $f(I \times Y)$. If the whole compact set $X$ is covered with the image of $I \times Y$, then we replace $X$ by $X' = X \cup x$, where $x \notin X$. Put $A_0 = f(0 \times Y) \cup x$, $A_1 = f(1 \times Y) \cup x$, $A = A_0 \cup A_1$, where $A_0$ and $A_1$ may be intersecting, and assume that $f\big|_{0 \times Y}$ is a homeomorphism. Let $L_0 \subset \tilde{h}^{k-1}(A_0) \backslash 0$, and $\Gamma = \nabla^k(X, A)$. Then there exists a subset $L_1 \subset \tilde{h}^{k-1}(A_1) \backslash 0$ such that $\Gamma \cup i^*(A, A_0)^{-1} L_0 = \Gamma \cup i^*(A, A_1)^{-1} L_1$. To prove the equality, it suffices to select a set $L_1$ such that $(i^*(A, A_1)^{-1} L_0) \backslash \Gamma = (i^*(A, A_0)^{-1} L_1) \backslash \Gamma$. The mapping $f$ is a homotopy $\varphi_t$ of the mapping $\varphi_0 = f|_{0 \times Y}$ into $\varphi_1 = f|_{1 \times Y}$; therefore, there exists a continuous mapping $g' : f(0 \times Y) \to f(1 \times Y)$, where $g' = \varphi_1 \varphi_0^{-1}$, which can be uniquely extended to a continuous mapping $g : A_0 \to A_1$, since $\varphi_0$ is a homeomorphism of the set $Y$ onto a certain subset in $X$ by assumption and $x \notin f(0 \times Y)$. Hence, we can put $g(x) = x$. The diagram

$$\begin{array}{ccc} A_0 & \xrightarrow{\;g\;} & A_1 \\ & {\scriptstyle i_0} \searrow \quad \swarrow {\scriptstyle i_1} & \\ & X & \end{array}$$

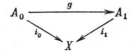

is homotopically commutative, where $i_0$ and $i_1$ are two embeddings. Put $L_1 = (g^*)^{-1}L_0 \subset \tilde{h}^{k-1}(A_1)\backslash 0$, and show that this is the required subset. Consider the diagram

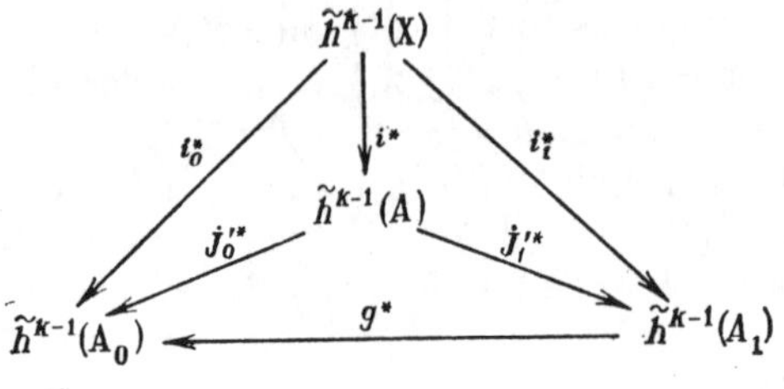


which is, in general, noncommutative, where $i : A \to X$, $j_0' : A_0 \to A$, $j_1' : A_1 \to A$ are embeddings, $i_0 = ij_0'$, $i_1 = ij_1'$, $i_0$ is homotopic to $i_1 g$, but, in the general case, $j_1'^* \neq g^* j_0'^*$, the mappings $j_0'$ and $j_1' g$ are in no way associated. We first show that $(i^*(A, A_0)^{-1}L_0)\backslash\Gamma \subset (i^*(A, A_1)^{-1}L_1)\backslash\Gamma$, i.e., $((j_0'^*)^{-1}L_0)\backslash\Gamma \subset ((j_1'^*)^{-1}L_1)\backslash\Gamma$. Let $z \in ((j_0'^*)^{-1}L_0)\backslash\Gamma$, i.e., $j_0'^*(z) \in L_0$, and $z = i^*(\alpha)$, $\alpha \in \tilde{h}^{k-1}(X)$. We then have $j_0'^*(z) = i_0^*(\alpha) = g^* i_1^*(\alpha) \in L_0$, i.e., $i_1^*(\alpha) \in L_1$ by definition of $L_1$. It follows that $j_1'^*(z) = j_1'^* i^*(\alpha) = i_1^*(\alpha) \in L_1$, or $z \in (j_1'^*)^{-1}L_1$, $z \notin \Gamma$, with $z \in ((j_1'^*)^{-1}L_1)\backslash\Gamma$.

Q.E.D.

Conversely, let $((j_0'^*)^{-1}L)\backslash\Gamma \neq \emptyset$. We need to show that $((j_0'^*)^{-1}L_0)\backslash\Gamma \supset ((j_1'^*)L_1)\backslash\Gamma$. If $((j_0'^*)^{-1}L_0)\backslash\Gamma = \emptyset$, then $(j_0'^*)^{-1}L_0 \subset \Gamma$, and we can take, e.g., the empty set as $L_1$. Let $z \in ((j_1'^*)^{-1}L_1)\backslash\Gamma$, or $g^* j_1'^* i^*(\alpha) \in L_0$. We obtain that $g^* i_1^*(\alpha) \in L_0$, or $i_0^*(\alpha) \in L_0$, $j_0'^* i^*(\alpha) = j_0'^*(z) \in L_0$, and $z \notin \Gamma$, $z \in (j_0'^*)^{-1}L_0$.

Q.E.D.

Note that if $((j_0'^*)^{-1}L_0)\backslash\Gamma \neq \emptyset$, then $L_1 \neq \emptyset$.

**Theorem 25.1.** *Let $A = \bigcup_{r=1}^{N} A_r$, $A_r \cap A_{r+1} = D_r$, $1 \leqslant r \leqslant N-1$, $A_r \cap A_s = x$ if $|r - s| > 1$, where the point $x$ belongs to each $D_r$. Put $\Gamma_t = \tilde{h}^{k-2}(D_r)\backslash 0$, and suppose that $B_r$ are such compact sets that $\nabla^{k-1}(B_r, D_r) = \Gamma_r$, $1 \leqslant r \leqslant N-1$. Consider the compact sets $C_r = A_r \cup B_r \cup B_{r-1}$, with $B_0 = B_N = x$, $D_0 = D_N = \emptyset$, $C = A \cup \left(\bigcup_{r=1}^{N-1} B_r\right) = \bigcup_{r=1}^{N} C_r$. Then the relation*

$$\bigcup_{r=1}^{N} i^*(C, C_r)^{-1}(\tilde{h}^{k-1}(C_r)\backslash 0) \supset i^*(CA)^{-1}(\tilde{h}^{k-1}(A)\backslash 0)$$

*holds.*

**Proof.** Consider the compact set $C$, and assume at first that $A \cap B_r = D_r$ for all $r$, $1 \leqslant r \leqslant N-1$, and that $B_r \cap B_s = x$ if $r \neq s$. We shall prove in the following that this restriction can be dropped. Let $B = \bigvee_{r=1}^{N-1} B_r$, $D = \bigvee_{r=1}^{N-1} D_r$ (Fig. 68).

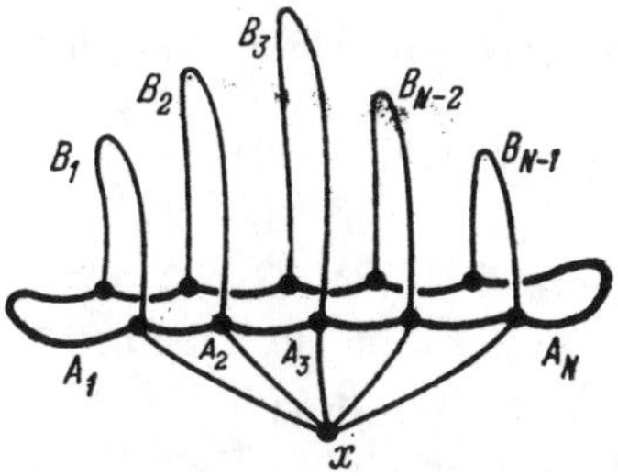


**Fig. 68**

For each $r$, $1 \leqslant r \leqslant N$, we consider the commutative diagram

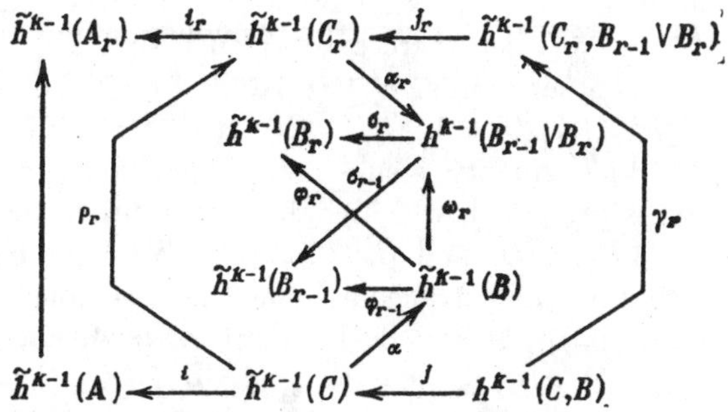

Let $z \in i^*(C,A)^{-1}(\tilde{h}^{k-1}(A)\backslash 0) = i^{-1}(\tilde{h}^{k-1}(A)\backslash 0)$; in particular, $z \neq 0$. It is requited to show that there exists a subscript $r$ such that $z \in \rho_r^{-1}(\tilde{h}^{k-1}(C_r)\backslash 0)$ i.e., $\rho_r(z) \neq 0$ in $\tilde{h}^{k-1}(C_r)$. Assume the contrary, viz., let $\rho_r(z) = 0$ for any $r$. Then, since $\alpha_r\rho_r = \omega_r\alpha$, we have $0 = \alpha_r\rho_r(z) = \omega_r\alpha(z)$, where $\alpha(z) \in \tilde{h}^{k-1}(B)$. We obtain that $\varphi_r\alpha(z) = \sigma_r\omega_r\alpha(z) = 0$, and $\varphi_{r-1}\alpha(z) = 0$. Since this is valid for any $r$, we have $\varphi_r\alpha(z) = 0$ for $1 \leqslant r \leqslant N$, and, because $B = \bigvee_{r=1}^{N-1} B_r$, where $B_r \cap B_s = x$ for $r \neq s$ by assumption, we obtain that $\alpha(z) = 0$ be Lemma 25.2. Since the sequence $\tilde{h}^{k-1}(B) \leftarrow \tilde{h}^{k-1}(C) \leftarrow h^{k-1}(C,B)$ is exact, $z = j(m)$, $m \in h^{k-1}(C,B)$. Take the elements $\gamma_r(m) \in h^{k-1}(C_r, B_{r-1} \vee B_r)$. Since $\rho_r(z) = 0$, $0 = \rho_r j(m) = j_r\gamma_r(m)$ for each $r$. Consider the commutative diagram

$$\begin{array}{ccccc}
\tilde{h}^{k-1}(C_r) & \xleftarrow{j_r} & h^{k-1}(C_r,\ B_{r-1} \vee B_r) & \xleftarrow{\delta_r} & \tilde{h}^{k-2}(B_{r-1} \vee B_r) \\
\downarrow & & \downarrow{\scriptstyle \psi_r} & & \downarrow{\scriptstyle \varepsilon_r} \\
\tilde{h}^{k-1}(A_r) & \longleftarrow & h^{k-1}(A_r,\ D_{r-1} \vee D_r) & \xleftarrow{\delta'_r} & \tilde{h}^{k-2}(D_{r-1} \vee D_r)
\end{array}$$

Since $A_r \cap A_s = x$ for $|r-s| > 1$, $D_r \cap D_s = x$ for $r \neq s$, and because $B_r \cap B_s = x$ for $r \neq s$, the diagram

$$\begin{array}{ccc} \tilde h^{k-2}(B_{r-1} \vee B_r) & \xrightarrow{\approx} & \tilde h^{k-2}(B_{r-1}) \oplus \tilde h^{k-2}(B_r) \\ \downarrow \varepsilon_r & & \downarrow \\ \tilde h^{k-2}(D_{r-1} \vee D_r) & \xrightarrow{\approx} & \tilde h^{k-2}(D_{r-1}) \oplus \tilde h^{k-2}(D_r) \end{array}$$

is also commutative. Since $\nabla^{k-1}(B_r, D_r) \supset \tilde h^{k-2}(D_r)\backslash 0$, we obtain that $\varepsilon_r \equiv 0$, and then $\psi_r\delta_r = \delta'_r\varepsilon_r \equiv 0$ on the group $\tilde h^{k-2}(B_{r-1} \vee B_r)$. Since the embedding $(A_r, D_{r-1} \vee D_r) \to (C_r, B_{r-1} \vee B_r)$ is a relative homeomorphism (recalling that $A \cap B_r = D_r$), we derive from the relative invariance of the theory $h^*$ that $\psi_r$ is an isomorphism, with the consequence that $\delta_r \equiv 0$, and $j_r$ is a monomorphism. It has been established above that $j_r\gamma_r(m) = 0$, whence $\gamma_r(m) = 0$ for any $r$. Since by Lemma 25.2, $h^{k-1}(C, B) = \bigoplus_r h^{k-1}(C_r, B_{r-1} \vee B_r)$, the latter statement means that $m = 0$, and then $z = j(m) = 0$, which is contrary to the original assumption. Thus, assuming that $B_r = B_s = x$ for $r \neq s$ and $A \cap B_r = D_r$, we have proved a strengthened result, viz., that if $z \in \tilde h^{k-1}(C)\backslash 0$, then there exists $r$ such that $z \in \rho_r^{-1}(\tilde h^{k-1}(C_r)\backslash 0)$, i.e., $\rho_r(z) \neq 0$; in other words, $\tilde h^{k-1}(C)\backslash 0 = \bigcup_r i^*(C, C_r)^{-1}(\tilde h^{k-1}(C_r)\backslash 0)$, and $i^*(C, A)^{-1}(h^{k-1}(A)\backslash 0) \subset \bigcup_r i^*(C, C_r)^{-1}(h^{k-1}(C_r)\backslash 0)$ *a fortiori*. In the generic case, the first equality is sure not to hold, but it turns out that the second inclusion is preserved, which enables us to prove the theorem. Thus, suppose that $A \cap B_r \supset D_r$ and $B_r \cap B_s = x$. To come back to the above situation, we separate the extraneous points, i.e., we construct the compact sets $\tilde C = \bigcup_r \tilde C_r$, $\tilde C_r = \tilde A_r \cup \tilde B_{r-1} \cup \tilde B_r$, where $\tilde A_r \approx A_r$, $\tilde B_r \approx B_r$ (which are homeomorphisms), $\tilde B_r \cap \tilde B_s = x$ for $r \neq s$, $A \cap \tilde B_r = \tilde D_r$, $\tilde A \approx A$, and build up a continuous mapping $f : \tilde C \to C$, a homeomorphism on $\tilde A$ and each of $\tilde B_r$, carrying out the necessary sewing to obtain the previous compact set $C$. If we put $f_r = f|_{\tilde C_r}$, then the diagram

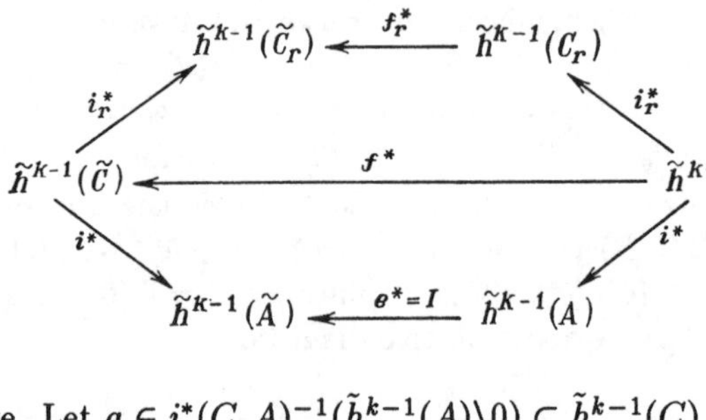


is commutative. Let $a \in i^*(C, A)^{-1}(\tilde h^{k-1}(A)\backslash 0) \subset \tilde h^{k-1}(C)$, $i^*(a) \neq 0$. Then $e^*i^*(a) \neq 0$, i.e., $i^*f^*(a) \neq 0$, $f^*(a) \in i^*(\tilde C, A)^{-1}(\tilde h^{k-1}(A)\backslash 0)$, and there exists, due to the above, such $r$ that $f^*(a) \in i^*(\tilde C, \tilde C_r)^{-1}(\tilde h^{k-1}(\tilde C_r)\backslash 0)$, or $i_r^*f^*(a) \neq 0$. Consequently, $f_r^*i_r^*(a) \neq 0$, or $i_r^*(a) \neq 0$. In other words, we have indicated a subscript $r$ such that $a \in i^*(C, C_r)^{-1}(\tilde h^{k-1}(C_r)\backslash 0)$.

Q.E.D.

We now prove an algebraic analogue of the geometric fact that sewing two or more compact sets along their common boundary leads to a compact set without boundary.

**Theorem 25.2.** *Let* $A = \bigcup_{r=1}^{N} A_r$, $A_r \cap A_{r+1} = D$, *where* $1 \leqslant r \leqslant N-1$, $A_r \cap A_s = x$ *of* $|r-s| > 1$, $\Gamma_r = \tilde{h}^{k-2}(D_r)\setminus 0$, *and* $B_r$ *be the compact sets such that* $\nabla^{k-1}(B_r, D_r) = \Gamma_r$. *Put* $C_r = A_r \cap B_{r-1} \cap B_r$, *where* $B_0 = B_N = x$, *and* $X_r$ *are such compact sets that* $\nabla^k(X_r, A) = h^{k-1}(C_r)\setminus 0$. *Then, if* $X = \bigcup_{r=1}^{N} X_r$, $\nabla^k(X, A) = \tilde{h}^{k-1}(A)\setminus 0$ *holds. (The geometric meaning of the statement is illustrated in Fig. 69.)*

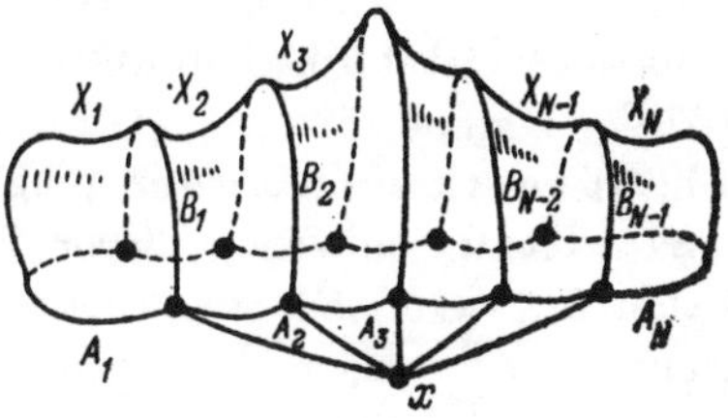


**Fig. 69**

**Proof.** Since the compact sets $A_r$, $B_r$, $D_r$ satisfy all the conditions of Theorem 25.1,

$$\bigcup_r i^*(C, C_r)^{-1}(\tilde{h}^{k-1}(C_r)\setminus 0) \supset i^*(C, A)^{-1}(\tilde{h}^{k-1}(A)\setminus 0),$$

where $C = \bigcup_r C_r$. Put $X' = X$, $X'_r = X_r$, $A'_r = C_r$, $A' = A$, $B' = A' \cup [\bigcup_r A'_r] = A \cup [\bigcup_r C_r] = C$. Furthermore, let $L = \tilde{h}^{k-1}(A')\setminus 0$, $L_r = \tilde{h}^{k-1}(A'_r)\setminus 0$. It then becomes clear that $L_r = \tilde{h}^{k-1}(C_r)\setminus 0 = \nabla^k(X_r, C_r) = \nabla^k(X'_r, A'_r)$, and the statement of Theorem 25.1 acquired the form $i^*(B', A')L \subset \bigcup_r i^*(B', A'_r)^{-1} L_r$. Since all the assumptions of Lemma 25.3 are fulfilled, we obtain $L \subset \nabla^k(X', A')$, i.e., $\tilde{h}^{k-1}(A')\setminus 0 \subset \nabla^k(X, A)$.

Q.E.D.

## §26 The Homology Case. Computation of the Boundary of the Pair $(X, A) = \bigcup_r (X_r, A_r)$ in Terms of the Boundaries of $(X_r, A_r)$

For $h_* = H_*$, certain ones of the following lemmas coincide with those obtained in [351]. They all turn out to be described by a certain very general and clear scheme embracing even the bordism Plateau problem, that of a film with parametrization.

**Lemma 26.1.** *Let* $X = \bigvee_{r=1}^{N} X_r$, *where* $X_r \cap X_s = x$ *for* $r \neq s$, *and* $A_r \subset X_r$ *be the compact sets such that* $x \in A_r$ *for each* $r$. *Then the homomorphisms* $i_{r*} : h_k(X_r, A_r) \to h_k(X, A)$, *where* $i_r$ *is an embedding, constitute an injective representation of the group* $h_k(X, A)$ *in the form of the direct sum*

*$\bigoplus_r h_k(X_r, A_r)$, i.e., any element $u \in h_k(X, A)$ is uniquely representable as the sum $u = \sum_{r=1}^N i_{r*}(u_r)$, $u_r \in h_k(X_r, A_r)$.*

The proof is similar to that of Lemma 25.1 and obtained on the basis of the argument in Theorem 1.13.2 (see [126]).

**Lemma 26.2.** *Let $X = \bigcup_{r=1}^N X_r$, $A = \bigcup_{r=1}^N A_r$, where $A_r \subset X_r$, $X_r \cap X_s = A_r \cap A_s$ for $r \neq s$, $x \in A_r$ for each $r$, and let $i_r : (X_r, A_r) \to (X, A)$ be embeddings. Then the homomorphisms $i_{r*}; h_k(X_r, A_r) \to h_k(X, A)$ form the injective representation of the group $h_k(X, A)$ as the direct sum $\bigoplus_r h_k(X_r, A_r)$.*

The proof follows from the relative invariance of the theory $h_*$ on the category $U_C$ and Lemma 26.1.

Let $f : (X, A) \to (Y, B)$ be a continuous mapping, $x \in A$, $y = f(x) \in B$, $L \subset \Delta_k(X, A)$ be an arbitrary subgroup. Consider the subgroup $L_1 = f_*(L)$. We then assert that $L \subset \Delta_k(Y, B)$, the proof obviously following from the commutative diagram

$$\begin{array}{ccc} \tilde{h}_{k-1}(A) & \longrightarrow & \tilde{h}_{k-1}(X) \\ \downarrow & & \downarrow \\ \tilde{h}_{k-1}(B) & \longrightarrow & \tilde{h}_{k-1}(Y). \end{array}$$

In particular, if $L \subset \Delta_k(X, A)$, $x \in A \subset X \subset Y$, then $L \subset \Delta_k(Y, A)$.

**Lemma 26.3.** *Let $X = \bigcup_{r=1}^N X_r$, $A_r \subset X_r$, $A \subset X$, $x \in A_r$ for each $r$. Put $B = A \cup [\bigcup_r A_r]$. Suppose that $L \subset \tilde{h}_{k-1}(A)$, $L_r \subset \Delta_k(X_r, A_r)$, and $i_*(B, A)L \subset \sum_r i_*(B, A_r)L_r$. Then $L \subset \Delta_k(X, A)$.*

**Proof.** It is necessary to show that $L \subset \operatorname{Ker} i_*$, where $i = i(X, A)$. We have

$$\begin{aligned} i_*(L) &= i_*(X, B)i_*(B, A)L \subset i_*(X, B)\sum_r i_*(B, A_r)L_r \\ &= \sum_r i_*(X, A_r)L_r = \sum_r i_*(X, X_r)i_*(X, A_r)L_r = 0, \end{aligned}$$

since $i_*(X_r, A_r)L_r = 0$. Q.E.D.

We will denote by $A + B$ the subgroup $C$ made up of elements of the form $a + b$, where $a \in A$, $b \in B$, and $A$ and $B$ are two subgroups in the Abelian group.

**Lemma 26.4.** *Let $X = \bigcup_{r=1}^N X_r, A_r \subset X_r$, $A \subset X$, $A \cap X_r \subset A_r$, $X_r \cap X_s = A_r \cap A_s$ if $r \neq s$, $x \in A$, and $x \in A_r$ for each $r$. Put $B = \bigcup_r A_r$. Then $B \subset A$, and $\Gamma_r = \Delta_k(X_r, A_r) \subset \tilde{h}_{k-1}(A_r)$. We state that the relation*

$$\Delta_k(X,\ A) = i_*(B,\ A)^{-1}\Big(\sum_r i_*(B,\ A_r)\,\Gamma_r\Big)$$

*holds.*

**Proof.** Denote the subgroup $i_*(B,A)^{-1}(\sum_r i_*(B,A_r)\Gamma_r$ by $L$. It is then clear that $i_*(B,A)L \subset \sum_r i_*(B,A_r)\Gamma_r$, equality, possibly not holding, and since all the assumptions of Lemma 26.3 are fulfilled, it follows that $L \subset \Delta_k(X,A)$. It remains to prove that $L \supset \Delta_k(X,A)$. Consider the diagram

$$\begin{array}{ccccc}
h_k(X_r,\ A_r) & \xrightarrow[\partial_r]{} & \tilde{h}_{k-1}(A_r) & \xrightarrow[\omega_r]{} & \tilde{h}_{k-1}(X_r) \\
\downarrow{\scriptstyle \tau_r} & & \downarrow{\scriptstyle \gamma_r} & & \downarrow{\scriptstyle \varphi'_r} \\
h_k(X,\ B) & \xrightarrow{\partial'} & \tilde{h}_{k-1}(B) & \xrightarrow{\omega'} & \tilde{h}_{k-1}(X) \\
\uparrow{\scriptstyle \beta} & & \uparrow{\scriptstyle \alpha} & & \uparrow{\scriptstyle \xi} \\
h_k(X,\ A) & \xrightarrow[\partial]{} & \tilde{h}_{k-1}(A) & \xrightarrow[\omega]{} & \tilde{h}_{k-1}(X)
\end{array}$$

Let $h \in \Delta_k(X,A)$, i.e., $\omega(h) = 0$, which means that $h = \partial(m)$, $m \in h_k(X,A)$. Hence, $\alpha(h) = \partial'\beta(m)$, i.e., $\omega'\alpha(h) = 0$. Since $\beta(m) \in h_k(X,B)$, the homomorphisms $\tau_r$ decompose the group $h_k(X,B)$ into a direct sum due to Lemma 26.2; therefore, $\beta(m)$ admits a unique representation in the form $\beta(m) = \sum_r \tau_r(t_r)$, where $t_r \in h_k(X_r,A_r)$. Consequently, $\partial'\beta(m) = \sum_r \partial'\tau_r(t_r) = \sum_r \gamma_r\partial_r(t_r)$, $\omega_r\partial_r(t_r) = 0$, $\varphi_r = \partial_r(t_r)$ belonging to $\Gamma_r$ by exactness, i.e, $\partial'\beta(m) = \sum_r \gamma_r(\varphi_r)$, where $\varphi_r \in \Gamma_r$, $\alpha(h) \in \sum_r \gamma_r(\Gamma)_r)$. We finally obtain that $h \in \alpha^{-1}[\sum_r \gamma_r(\Gamma_r)]$.

Q.E.D.

Note that though the statements of Lemma 25.4 and 26.4 are not dual, they can be related by a certain formal analogy, viz., if the homomorphism $\alpha : \tilde{h}^{k-1}(B) \to \tilde{h}^{k-1}(A)$ is an epimorphism, then the formula $\nabla_k(X,A) = \delta_1^{-1}\{[\beta[\bigcup_r \tau_r^{-1}\delta_r(\Gamma_r)]]\backslash 0\}$ is similar to that of Lemma 26.4.

Consider the behaviour of the boundary $\Delta_k(X,A)$ in the cases where $x = I \times Y$, $A = (0 \times Y) \cup (1 \times Y)$, and $j_0$ and $j_1$ are two embeddings of $Y$ onto the lower and upper bases of the cylinder. Distinguish the subgroup $\Delta' = \{j_{1*}(h) - j_{0*}(h)\} \cong \{j_{1*}(h), -j_{0*}(h)\}$ from $\tilde{h}_{k-1}(A)$, where $h \in \tilde{h}_{k-1}(Y)$, $\tilde{h}_{k-1}(A) = \tilde{h}_{k-1}(Y) \oplus \tilde{h}_{k-1}(Y)$, and the point $x$ is arbitrary in $A$. Recall that the subgroup $\tilde{h}_{k-1}(A)$ does not depend on the choice of $x$ at all. It then turns out that $\Delta_k(X,A) \subset \Delta'$. In fact, since the two embeddings $i(X,A)j_0$ and $i(X,A)j_1$ are homotopic, $\Delta' \subset \operatorname{Ker} i_*(X,A)$, which is just the required one. If we recall the corresponding cohomological result, then we shall see that in the cohomoloy case, the most important one is the diagonal $\Delta \subset h^{k-1}(A)$, and in the homology case, the lower estimate of the algebraic boundary of the pair $(X,A)$ is expressed in terms of the secondary diagonal $\Delta' \subset \tilde{h}_{k-1}(A)$.

Just as in the cohomological version, consider the behaviour of the boundary $\Delta_k$ under a deformation of the compact sets. Let $f : I \times Y \to X$ be a continuous mapping, and $x \in X$ be a point such that $x \notin f(I \times Y)$ (otherwise, it is always possible to replace $X$ by $X' = X \cup x$). Put $A_0 = f(0 \times Y) \cup x$, $A_1 = f(1 \times Y) \cup x$, $A = A_0 \cup A_1$, and assume that the mapping $f|_{0\times Y}$ is a homeomorphism. Let $\tilde{h}_{k-1}(A_0) \subset L$, and $\Gamma = \Delta_k(X, A)$. Then there exists a subgroup $L \subset \tilde{h}_{k-1}(A_1)$ such that $\Gamma + i_*(A, A_0)L = \Gamma + i_*(A, A_1)L_1$. Suppose that $g : A_0 \to A_1$ is a continuous mapping defined in §25, and such that the diagram

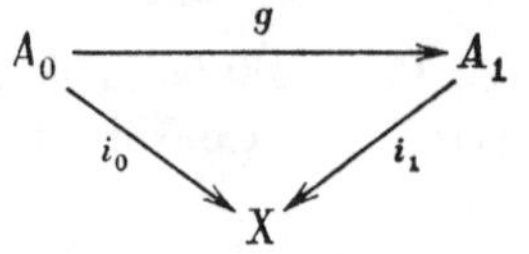


is homotopically commutative, where $i_0$ and $i_1$ are two embeddings. Put $L = g_* L_0$, and consider the diagram

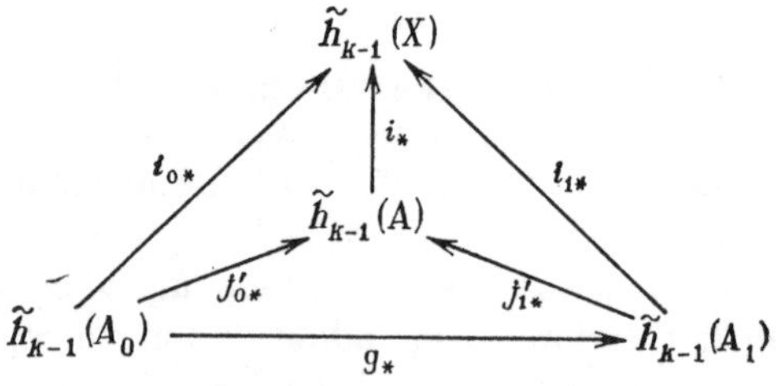


in general, to be noncommutative in the lower triangle, where $i_0 = ij_0$, $i_1 = ij_1$, $i_0 \sim i_1 g$. We first show that $\Gamma + j'_{0*}(L) \subset \Gamma + j'_{1*}(L_1)$. Let $\alpha \in \Gamma + j'_{1*}(L_1)$. Then $\alpha = \gamma + j'_{1*}g_*(m)$, where $m \in L_0$. Set $m' = j'_{0*}(m) \in \tilde{h}_{k-1}(A)$. We then obtain $i(m') = i_* j'_{0*}(m) = i_{0*}(m)$, and besides, $i_* j'_{1*} g_*(m) = i_{1*} g_*(m) = i_{0*}(m)$, i.e., $i_*(m') = i_* j'_{1*} g_*(m)$, whence $m' - j'_{1*} g_*(m) \in \operatorname{Ker} i_* = \Gamma$ by definition of $\Gamma$, or $m' - j'_{1*} g_*(m) = \gamma_1 \in \Gamma$. Hence, $\alpha = \gamma + j'_{1*} g_*(m) = \gamma + m' - \gamma_1 = \gamma_2 + m'$, where $\gamma_2 \in \Gamma$, i.e., $\alpha = \gamma_2 + j'_{0*}(m)$ with $m \in L_0$, or $\alpha \in \Gamma + j'_{0*}(L_0)$.

Q.E.D.

Conversely, we prove that $\Gamma + j'_{0*} L_0 \subset \Gamma + j'_{1*} L_1$. Let $a = \gamma + j'_{0*}(m)$, where $m \in L_0$, $\gamma \in \Gamma$. Then an element $j'_{1*} g_*(m) \in \tilde{h}_{k-1}(A)$ is defined, which can be compared with $j'_{0*}(m)$. It is clear that $i_* j'_{0*}(m) = i_{0*}(m)$, and at the same time, $i_* j'_{1*} g_*(m) = i_{1*} g_*(m) = i_{0*}(m)$, or $i_* j'_{0*}(m) = i_* j'_{1*} g_*(m)$, $j'_{0*}(m) = \gamma_1 + j'_{1*} g_*(m)$, where $\gamma_1 \in \Gamma$. It follows that $a = \gamma + j'_{0*}(m) = \gamma + \gamma_1 + j'_{1*} g_*(m) = \gamma_2 + j'_{1*}(q)$, where $\gamma_2 \in \Gamma$, $q \in L_1$.

Q.E.D.

**Theorem 26.1.** *Let* $A = \bigcup_{r=1}^{N} A_r$, $A_r \cap A_{r+1} = D_r$, $1 \leqslant r \leqslant N-1$, $A_r \cap A_s = x$ *if* $|r - s| > 1$, *where the point* $x$ *belongs to each* $D_r$. *Put* $\Gamma_r = \tilde{h}_{k-2}(D_r)$, $1 \leqslant r \leqslant N-1$. *Let* $B_r$ *be compact sets such that* $\Delta_{k-1}(B_r, D_r) = \Gamma_r$. *Consider the compact sets* $C_r = A_r \cup B_r \cup B_{r-1}$, *where* $1 \leqslant r \leqslant N$, *with* $B_0 = B_N = x$, $D_0 = D_N = \emptyset$, $C = A \cup [\bigcup_{r=1}^{N-1} B_r] = \bigcup_{r=1}^{N} C_r$. *Then* $\sum_{r=1}^{N} i_*(C, C_r)\tilde{h}_{k-1}(C_r) \subset i_*(C, A)\tilde{h}_{k-1}(A)$.

**Proof.** Consider the compact set $C$, and assume at first that $A \cap B_r = D_r$ for all $r$, $1 \leqslant r \leqslant N-1$, and that $B_r \cap B_s = x$ if $r \neq s$. In the following, we reduce the general situation to this particular case. Let $B = \bigvee_{r=1}^{N-1} B_r$, $D = \bigvee_{r=1}^{N-1} D_r$ (see Fig. 65). Construct the commutative diagram

$$\begin{array}{ccccc}
\bigoplus_r h_{k-1}(C_r, B_{r-1} \vee B_r) & \xleftarrow{\bigoplus_r \varphi_r} & \bigoplus_r \tilde{h}_{k-1}(C_r) & \xleftarrow{\bigoplus_r \gamma_r} & \bigoplus_r \tilde{h}_{k-1}(B_r \vee B_{r-1}) \\
\downarrow{\scriptstyle \omega = \bigoplus_r \omega_r} & & \downarrow{\scriptstyle \alpha = \sum_r \alpha_r} & & \downarrow{\scriptstyle \beta = \sum_r \beta_r} \\
h_{k-1}(C, B) & \xleftarrow{\quad \xi \quad} & \tilde{h}_{k-1}(C) & \xleftarrow{\psi} & \tilde{h}_{k-1}(B)
\end{array}$$

whose lines are exact sequences, and $\oplus_r \varphi_r$, $\oplus_r \gamma_r$ denote the direct sum of the homomorphisms $\varphi_r : h_{k-1}(C_r) \to h_{k-1}(C_r, B_{r-1} \bigvee B_r)$ and $\gamma_r : \tilde{h}_{k-1}(B_{r-1} \bigvee B_r) \to \tilde{h}_{k-1}(C_r)$, respectively. Note that, though $B = \bigvee_r B_r$, the homomorphism $\beta$ is not an isomorphism and possesses a kernel isomorphic to $\tilde{h}_{k-1}(B)$. The homomorphism $\beta$ is not the direct sum of the homomorphisms $\beta_r : \tilde{h}_{k-1}(B_{r-1} \bigvee B_r) \to \tilde{h}_{k-1}(B)$; that is why we have made use of the notation $\beta = \sum_r \beta_r$. Prove that, in our assumptions, i.e., $A \cap B_r = D_r$, $B_r \cap B_s = x$, $r \neq s$, a result strengthened compared with the condition of the theorem is fulfilled, viz., that $\tilde{h}_{k-1}(C) = \sum_{r=1}^{N} i_*(C, C_r) h_{k-1}(C_r)$. We establish preliminarily that $\operatorname{Ker}(\xi) \subset \sum_r i_*(C, C_r) \tilde{h}_{k-1}(C_r)$. Let $a \in \tilde{h}_{k-1}(C)$, and $\xi(a) = 0$. Then $a = \psi(b)$, $b \in \tilde{h}_{k-1}(B)$, and since $\beta$ is an epimorphism $b = \beta(b')$, where $b' = \sum_r b'_r$, $b'_r \in \tilde{h}_{k-1}(B_r \bigvee B_{r-1})$, and the decomposition $b' = \sum_r b'_r$ is unique. It follows that $a = \alpha(\bigoplus_r \gamma_r) b' = \sum_r \alpha \gamma_r (b'_r) = \sum_r \alpha(\xi_r)$, where $\xi_r \in h_{k-1}(C_r)$, i.e., $a \in \sum_r \alpha_r(\xi_r)$, which is just the required one.

Now, let $a \in \tilde{h}_{k-1}(C_r)$, and $\xi(a) \neq 0$. Since all the assumptions of Lemma 26.2 are fulfilled for the pair $(C, B) = \bigcup_r (C_r, B_{r-1} \bigvee B_r)$, the homomorphism $\omega$ establishes an isomorphism between $h_{k-1}(C, B)$ and $\bigoplus_r h_{k-1}(C_r, B_{r-1} \bigvee B_r)$, ie., $\xi(a) = \omega(c)$, where $c = \sum_r c_r$, the element $c$ and its decomposition being determined uniquely. Consider the commutative diagram

$$\begin{array}{ccccccc}
\tilde{h}_{k-1}(C_r) & \xrightarrow[\varphi_r]{} & h_{k-1}(C_r, B_{r-1} \vee B_r) & \xrightarrow[\partial_r]{} & \tilde{h}_{k-2}(B_{r-1} \vee B_r) & = & \tilde{h}_{k-2}(B_{r-1}) \oplus \tilde{h}_{k-2}(B_r) \\
\uparrow{\scriptstyle \tau_r} & & \approx \uparrow{\scriptstyle u_r} & & \uparrow{\scriptstyle \rho_r \equiv 0} & & \uparrow{\scriptstyle \nu_r \equiv 0} \\
\tilde{h}_{k-1}(A_r) & \xrightarrow{\varepsilon_r} & h_{k-1}(A_r, D_{r-1} \vee D_r) & \xrightarrow{\partial'_r} & \tilde{h}_{k-2}(D_{r-1} \vee D_r) & = & \tilde{h}_{k-2}(D_{r-1}) \oplus \tilde{h}_{k-2}(D_r)
\end{array}$$

in which the homomorphism $u_r$ is an isomorphism, since the embedding $(A_r, D_{r-1} \bigvee D_r) \to (C_r, B_{r-1} \bigvee B_r)$ is a relative homeomorphism, whereas the homomorphism $\nu_r$ is trivial, for $\Delta_{k-1}(B_r, D_r) = \Gamma_r = \tilde{h}_{k-2}(D_r)$. Therefore $\rho_r \equiv 0$ for any $r$. Since $\rho_r \partial'_r = \partial_r u_r = 0$, and $u_r$ is an isomorphism, $\partial_r \equiv 0$ on the whole group $h_{k-1}(C_r, B_{r-1} \bigvee B_r)$, i.e., the homomorphism $\varphi_r$ is an epimorphism. It follows that any element $c_r$, where $c = \sum_r c_r$ (see above), is representable as $c_r = \varphi_r(c'_r)$, and then $c = (\bigoplus_r \varphi_r) c'$, where $c' = \sum_r c'_r$. Consequently, $\xi\alpha(c') = \omega(\underset{r}{\oplus}\varphi_r)c' = \omega(c) = \xi(a)$, i.e., $\xi(a-\alpha(c')) = 0$, $a = \alpha(c') + t$ with $t \in \operatorname{Ker}(\xi)$. It follows from the above that $t = \alpha(c'')$, or $a = \alpha(c' + c'') = \sum_r \alpha_r(c'_r + c''_r)$.

Q.E.D.

We now turn to the generic case, viz., let $A \cap B_r \supset D_r$, and $x \in B_r \cap B_s$. Separating the extraneous points, we can come back to the previous situation, possessing the compact sets $\tilde{C} = \bigcup_r \tilde{C}_r$, $\tilde{D}_r = D_r$, $\tilde{B}_r = B_r$, $\tilde{A}_r = A_r$ and the mapping $f : \tilde{C} \to C$, which carries out the original sewing. We obtain the commutative diagram

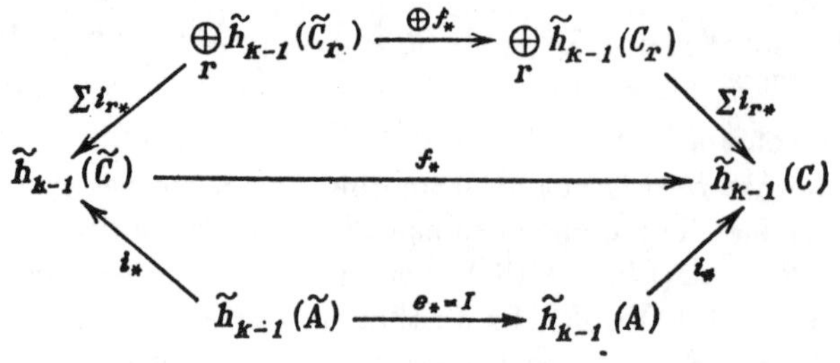


where, as follows from the above, $\tilde{h}_{k-1}(\tilde{C}) = \sum i_{r*}(\underset{r}{\oplus}\tilde{h}_{k-1}(\tilde{C}_r))$. Let $c \in \tilde{h}_{k-1}(C)$, and $c = i_*(a)$. Then

$$c = i_* e_* (\tilde{a}) = f_* i_* (\tilde{a}) = f_* (\Sigma i_{r*})(m) = \Sigma i_{r*} (\oplus f_*) m = \sum_r i_{r*} (c_r),$$

where $c_r \in h_{k-1}(C_r)$, thus completing the proof.

**Theorem 26.2.** *Let $A = \bigcup_{r=1}^N A_r$, $A_r \cap A_{r+1} = D_r$, where $1 \leqslant r \leqslant N-1$, $A_r \cap A_s = x$, if $|r-s| > 1$. Put $\Gamma_r = \tilde{h}_{k-2}(D_r)$. Suppose that $B_r$ are the compact sets such that $\Delta_{k-1}(B_r, D_r) = \Gamma_r$, $C_r = A_r \cup B_{r-1} \cup B_r$ $(1 \leqslant r \leqslant N-1)$, where $B_0 = B_N = x$, and $X_r$ are the compact sets such that $\Delta_k(X_r, C_r) = \tilde{h}_{k-1}(C_r)$. Then, if $X = \bigcup_{r=1}^N X_r$,*

$$\Delta_k(X, \; A) = \tilde{h}_{k-1}(A).$$

**Proof.** Since all the assumptions of Theorem 26.1 are fulfilled, we have $\sum_r i_*(C, C_r)\tilde{h}_{k-1}(C_r) \supset i_*(C, A)\tilde{h}_{k-1}(A)$. Put $X' = X$, $A' = A$, $X'_r = X_r$, $A'_r = C_r$, $L'_r = h_{k-1}(C_r) = \Delta_k(X'_r, A'_r)$, $B' = C = A' \cup [\bigcup_r A'_r]$. We then have $i_*(B', A')\tilde{h}_{k-1}(A') \subset \sum_r i_*(B', A'_r)L_r$ in this new notation, i.e., we come to the situation where Lemma 26.3 can be applied. In so doing, we obtain that $\tilde{h}_{k-1}(A) = \tilde{h}_{k-1}(A') \subset \Delta_k(X, A)$.

Q.E.D.

## §27 Closedness, Invariance, and Stability of Variational Classes

### 27.1. $S$-Surgery of Surfaces in a Riemannian Manifold

Until now, all the constructions have occurred in the category of compact pairs. We have now come to the situation in which we consider a Riemannian manifold $M$, and will study compact pairs $(X, A)$ embedded into it.

**Definition 27.1.1.** Let $(X, A) \subset M$ be an arbitrary compact pair $x \in A$, $\tilde{h}$ a continuous and relatively invariant (co)homology theory on $U_C$, $\sigma = \{k_i\}$ a fixed set of pairwise different integers, and $G$ an open set in $M$, so that $\overline{G} \cap A = \emptyset$, with $\overline{G}$ denoting the closure of $G$ in $M$. Consider the compact sets $X_1 = (X \cap \overline{G}) \cup x$, $A_1 = (X \cap \partial G) \cup x$ (then $A_1 \subset X_1$), $Y = Y_1 \cup Y_2$ in $M$, where $Y_2 = X \backslash G$, and $Y_1$ is a compact set in $M$ such that $\Delta_{k_i}(Y_1, A_1) \supset \Delta_{k_i}(X_1, A_1)$ for each $k_i \in \sigma$ (accordingly, the inclusion $\nabla^{k_i}(Y_1, A_1) \supset \nabla^{k_i}(X_1, A_1)$) should hold in the cohomology case). We will say that the compact set $Y = Y_1 \cup Y_2$ is *obtained by an $S$-surgery* of the compact set $X$ via an open set $G$, $Y_1 \subset M$ and the (co)homology theory $h$ in algebraic dimensions $\sigma = \{k_i\}$, denoting $Y$ by $Y_1(X)$. We will also say that the compact pair $(Y, A)$ is obtained by an $S$-surgery of $(X, A)$ (Fig. 70).

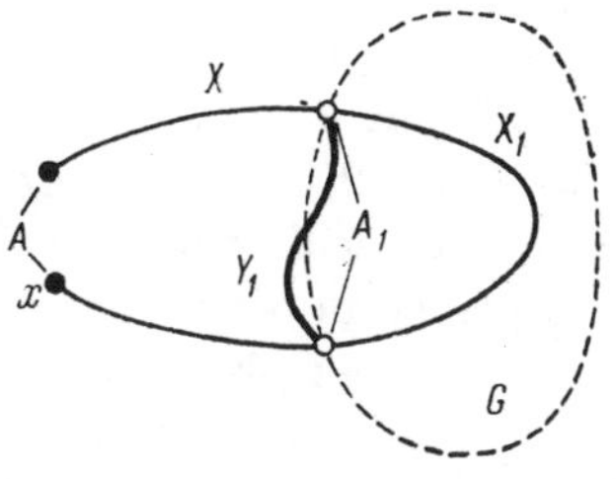


**Fig. 70**

The set of compact sets $Y_1$ with the above property is nonempty. For example, we can take $X_1$ as $Y_1$, and then $Y = X$. Since $Y_1 \supset A_1$, the point $x$ should also belong to $Y_1$, i.e., $Y \not\subset \overline{G}$. The class of surgeries possessing the important property that they wholly span the compact set $A_1$ is naturally distinguished from the set of all $S$-surgeries of the pair $(X, A)$. These special surgeries will be important in the following; therefore, we represent the situation as follows.

**Definition 27.1.2.** Let $(X, A) \subset M$ be a compact pair, and let an $S$-surgery $Y = Y_1(X)$ be given. Assume that the compact set $Y_1$ also possesses the property that $\Delta_{k_i}(Y_1, A_1) \supset \tilde{h}_{k_i - 1}(A_1)$ (accordingly, $\nabla^{k_i}(Y_1, A_1) \supset \tilde{h}^{k_i - 1}(A_1) \backslash 0$ in the cohomology case). We will call it the complete $'S$-surgery of the compact set $X$ in dimensions $\sigma = \{k_i\}$, and denote the compact set $Y$ by $'Y_1(X)$.

The compact set $Y_1$ with the above property may not exist at all (e.g., if the set $\partial G$ cuts a set $A_1$ out of $X$, which contains some nontrivial (co)cycle of the manifold $M$); however, in further applications, we will choose the set $G$ to be sufficiently small, which provides for the existence of the compact sets $Y_1$. For example, we can take $Y_1 = CA_1$.

### 27.2. The Closedness of Variational Classes Relative to the Passage to the Limit

Consider the set of all nonempty closed subsets in a Riemannian manifold $M$ and denote it by $\Theta(M)$. Let $d(x,y)$ be the distance on $M$ between two points $x$ and $y$. Then $\Theta(M)$ can be turned into a locally compact metric space (and, in the case where $M$ is compact, into a compact metric space) by introducing in $\Theta(M)$ the metric $\rho(X,Y)$, where $X, Y \in \Theta(M)$, by the formula $\rho(X,Y) = \sup_{x \in Y} d(x,Y) + \sup_{y \in Y} d(X,y), with d(z,B)$ denoting the distance from the point $z$ to the closed set $B$.

All the introduced variational classes $\mathcal{O}$ and $\tilde{\mathcal{O}}$ are subsets in the space $\Theta(M)$ and, therefore, inherit the metric $\rho(X,Y)$, turning into metric spaces. In the following, we will understand each convergence in $\Theta(M)$ as that in the metric $\rho(X,Y)$. It turns out that $\mathcal{O}$ and $\tilde{\mathcal{O}}$ are closed subsets in the metric space $\Theta(M)$, which is the important fact in applications and which is just what we are now going to prove.

**Theorem 27.2.1.** *Let $M$ be a Riemannian manifold without boundary, $A \subset M$ a fixed compact set, and $X_n$, $n = 1, 2, 3, \ldots$, a sequence of compact sets in $M$, all belonging to the same class $\mathcal{O}$ or $\tilde{\mathcal{O}}$ constructed from a certain theory $h$ continuous and relatively invariant on the category of compact pairs. Assume that there exists a compact set $X_0$, $X_0 \subset M$, $X_0 \supset A$, such that $\rho(X_0, X_n) \to 0$ as $n \to 0$. It then belongs to the same class $\mathcal{O}$ or $\tilde{\mathcal{O}}$ as $X_n$.*

**Proof.** We first consider the homology case. Let $i_0 : A \to X_0$, $j_0 : X_0 \to M$, $i_n : A \to X_n$, $j_n : X_n \to M$, $n = 1, 2, 3, \ldots$, be embeddings, and $X_n \in \mathcal{O} = h_*(A, L, L')$, where $L = \{L_p\}, L' = \{L'_q\}$. Given that $L_p \subset \operatorname{Ker}(i_{n*}) = \Delta_*(X_n, A)$, and $L'_q \subset \operatorname{Im}(j_{n*}) \subset \tilde{h}_*(M)$, it is required to establish $L_p \subset \operatorname{Ker}(i_{0*})$. Consider the new compact sets $Y_n = [\bigcup_{a \geqslant n} X_\alpha] \cup X_0$. Suppose that $i'_n : A \to Y_n$, $j'_n : Y_n \to M$ are embeddings. Since $X_n \subset Y_n$, we have $L_p \subset \operatorname{Ker} i'_{n*}$, and $L'_q \subset \operatorname{Im} j'_{n*}$, i.e., $Y_n \in \tilde{h}_*(A, L, L')$. It is clear that $Y_n \supset Y_{n+1}$ for any $n$, and the relation $X_0 = \bigcap_n Y_n = \underleftarrow{\lim} Y_n$ holds where $\underleftarrow{\lim} Y_n$ is the inverse limit of the inverse spectrum $\cdots \supset Y_n \supset Y_{n+1} \supset \ldots$. Further, it is obvious that the homomorphism $i_{0*}$ is the inverse limit of the homomorphisms $i'_{n*}$, and due to the continuity of the theory $h_*$ on the category $U_C$, we have $\tilde{h}_p(X_0) = \tilde{h}_p(\underleftarrow{\lim} Y_n) = \underleftarrow{\lim} \tilde{h}_p(Y_n)$, whence $\operatorname{Ker} i_{0*} = \underleftarrow{\lim} \operatorname{Ker} i'_{n*}$, and since $L_p \subset \operatorname{Ker} i'_{n*}$ for any $n$, we have $L_p \subset \operatorname{Ker} i_{0*}$, which is just what is required.

We now prove that $L'_q \subset \operatorname{Im} j_{0*}$. Since $h^*$ is exact on $U_C$, $L'_q \subset \operatorname{Ker} \alpha_{n*}$, where $\alpha_{n*} : \tilde{h}_*(M) \to h_*(M, Y_n)$; therefore, it suffices to establish that $L'_q \subset \operatorname{Ker} \alpha_{0*}$, where $\alpha_{0*} : \tilde{h}_*(M) \to h_*(M, X_0), \operatorname{Im} j_{0*} = \operatorname{Ker} \alpha_{0*}$. Because

$(M, X_0) = \varprojlim(M, Y_n)$, we have $\alpha_{0*} = \varprojlim \alpha_{n*}$, and then the required statement follows from the above reasoning for the set $L$. We now consider the cohomology case.

We preserve all the above notation and prove at first that $Y_n \in h^*(A, L, L')$, where $L_p \cap \operatorname{Im} i_n^* = \emptyset$, and $L'_q \cap \operatorname{Ker} j_n^* = \emptyset$. We establish that $\operatorname{Im} i_n'^* \cap L_p = \emptyset$. Indeed, suppose there exists $l \neq 0$ such that $l \in L_p$, and $l = i_n'^*(y_n)$, where $y_n \in \tilde{h}^p(Y_n)$. Let $\varphi_\alpha^\beta : X_\beta \to Y_\alpha$ be an embedding defined for $\beta \geqslant \alpha$. It then follows from the commutative diagram

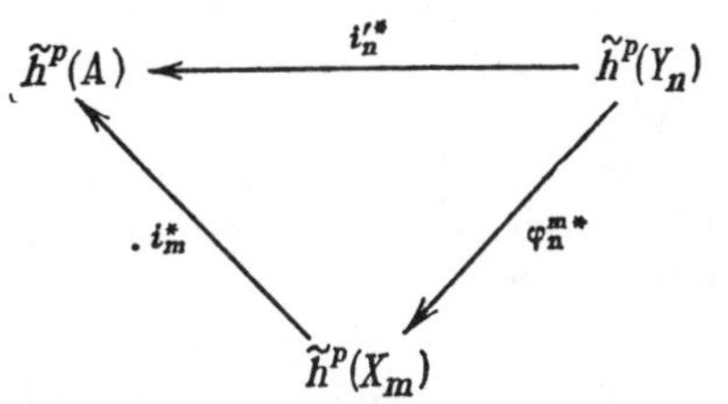


where $m$ is any integer, $m \geqslant n$, that $l = i_m^* \varphi_n^{m*}(y_n)$, i.e., $l \in \operatorname{Im} i_m^*$, which is contrary to the choice of the element $l$. We now consider all subscripts $q$ for which $L'_q \neq \emptyset$, and prove that $j_n'^*(l') = 0$. It then follows from the commutative diagram

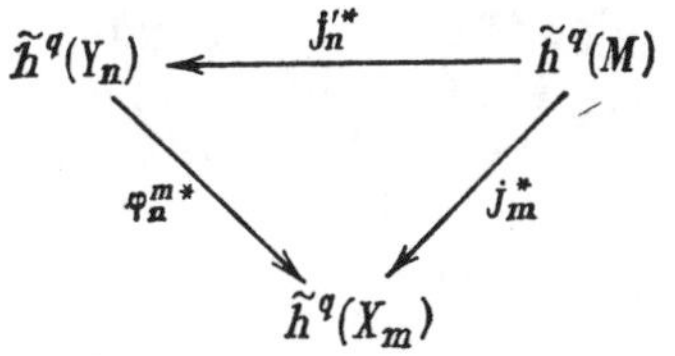


where $m \geqslant n$, that $j_m^*(l') = 0$, which is contrary to the choice of the element 1'. Thus, $Y_n \in h^*(A, L, L')$ for any $n$.

We now establish that $L_p \cap \operatorname{Im} i_0^* = \emptyset$, where $i_0 : A \to X_0$. Assume the contrary, viz., let there exist an element such that $l = i_0^*(z)$, where $z \in \tilde{h}^p(X_0)$. It then follows from Theorem 2.6, Chapter X, in [126] that there is a subscript $\alpha$ such that $z = \psi_\alpha^*(u_\alpha)$ for a certain element $u_\alpha \in \tilde{h}^p(Y_\alpha)$, where $\psi_\alpha : X_0 \to Y_\alpha$ is an embedding. Hence, we derive from the commutative diagram

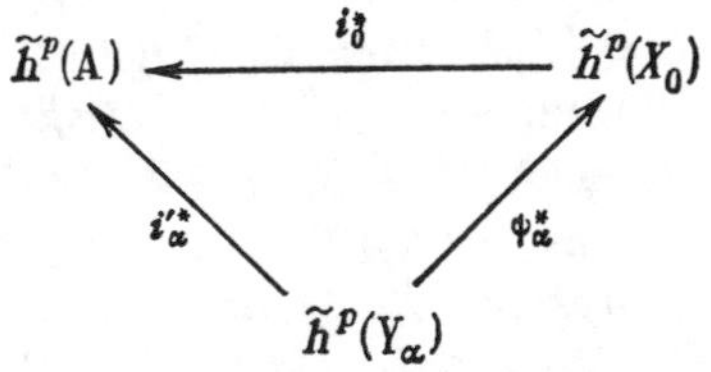


that $l = i_0^*(z) = i_0^* \psi_\alpha^*(u_\alpha) = i_\alpha'^*(u_\alpha)$, i.e., $L_p \cap \operatorname{Im} i_\alpha'^* \neq \emptyset$ due to $l \neq 0$, which is contrary to the above statement that $Y_\alpha \in h^*(A, L, L')$ for any $\alpha$.

It remains to prove that $L'_q \cap \operatorname{Ker} j_0^* = \emptyset$ for any $q$. Let $L'_q \neq \emptyset$. Assume that there exists an element $l' \neq 0, l' \in L'_q$ such that $j_0^*(l') = 0$. Consider the commutative diagram

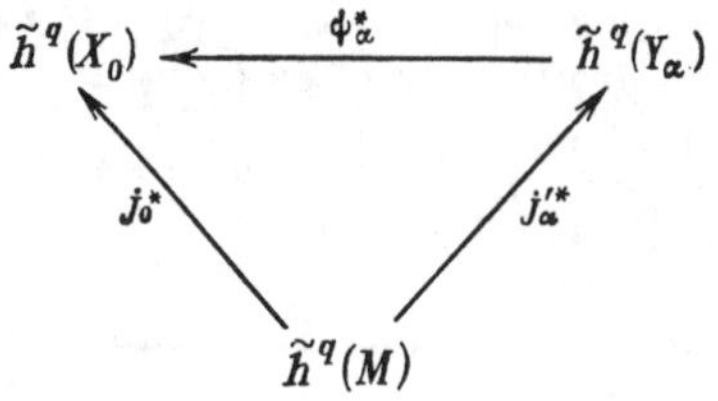


where $\alpha$ is an arbitrary integer. Then $0 = j_0^*(l') = \psi_\alpha^* j_\alpha^*(l')$, i.e., $\psi_\alpha^*(u_\alpha) = 0$, with $u_\alpha = j_\alpha'^*(l')$. Hence, it follows from Theorem 2.6, Chapter X, in [126] that there exists a subscript $\beta$ such that $\beta > \alpha$, and $\pi_\alpha^{\beta *}(u_\alpha) = 0$, where $\pi_\alpha^\beta$ is the embedding of $Y_\beta$ into $Y_\alpha$. Thereby, $j_\beta'^*(l') = \pi_\alpha^{\beta *} j_\alpha'^*(l') = \pi_\alpha^{\beta *}(u_\alpha) = 0$, and we have indicated the compact set $Y_\beta$ not realizing the element $l' \in L'_q$, $l' \neq 0$, which is contrary to $Y_\beta \in \tilde{h}^*(A, L, L')$ for any $\beta$ (see above), thus, completely proving the theorem.

Theorem 27.2.1 has been proved for the case where the functor $h$ satisfies Axioms A1–A6 and is continuous on $U_C$. This level of generality is quite convenient from the standpoint of the investigation of variational problems in the classes $\mathcal{O}$, $\tilde{\mathcal{O}}$. However, the theorem does not touch upon the problem of limit realizations, which is of special interest. We can prove a strengthened result (see [161]).

**Theorem 27.2.2.** *Let a sequence of compact sets $X_n$, $n = 1, 2, 3, \ldots$, $X_n \supset A$ be given for any $n$, let there exist a compact set $X_0 \subset M$ such that $\rho(X_n, X_0) \to 0$ as $n \to \infty$ (then $X_0 \supset A$), and $A \xrightarrow{i_0} X_0 \xrightarrow{j_0} M$, $A \xrightarrow{i_n} X_n \xrightarrow{j_n} M$ be embeddings.*

*(1) Let a contravariant continuous functor $h^* = \oplus_{k \in \mathbf{Z}} h^k$ be given on $U_C$, with values in GR satisfying Axioms A1, A2, $L = \{L_p\}$, $L' = \{L'_q\}$, where $L_p \subset h^p(A) \backslash 0$, $L'_q \subset h^q(M) \backslash 0$, and $L \cap \operatorname{Im} i_n^* = \emptyset$, $L' \cap \operatorname{Ker} j_n^* = \emptyset$ for any $n$. Then $L \cap \operatorname{Im} i_0^* = \emptyset$, and $L' \cap \operatorname{Ker} j_0^* = \emptyset$.*

*(2) Let a covariant continuous functor $h_* = \oplus_{k \in \mathbf{Z}} h_k$ satisfying Axioms A1, A2 be given on $U_C$, $L = \{L_p\}$, where $L_p \subset h_p(A)$, and $L_p \subset \operatorname{Ker} i_{n*}$ for any $n$. Then $L \subset \operatorname{Ker} i_{0*}$. Assume now that $h_*$ satisfies one of the conditions (i) $h_k(X, Y) \subset ABC$ for any $(X, Y) \in U_C, k \in \mathbf{Z}$, and all the induced homomorphisms are continuous, and (ii) $h_k(X, Y) \in GF$ for any $(X, Y) \in U_C, k \in \mathbf{Z}$, all the induced homomorphisms $X$ are linear, and, besides, $h_k(X, Y) \in GFf$ for any $(X, Y) \in P^2 \subset U_C, k \in \mathbf{Z}$. Let $L' = \{L'_q\}$, where $L'_q \subset h_q(M)$, and $L' \subset \operatorname{Im} j_{n*}$ for any $n$. Then $L' \subset \operatorname{Im} j_{0*}$.*

## 27.3. The Invariance of Variational Classes Relative to $S$-Surgeries of Surfaces

**Theorem 27.3.1.** *Let $M$ be a Riemannian manifold, $A \subset M$ a fixed compact set, $x \in A$, $G \subset M$ an open set such that $\overline{G} \cap A = \emptyset$, the class $h(A, L, L')$ be nonempty, and $X$ be an arbitrary surface (compact set) from it.*

*(1) Let a pair $(Y, A)$ be obtained from a pair $(X, A)$ by some $S$-surgery in dimensions $\sigma = \{k_i\}$, where the set of whole numbers $\{k_i\}$ contains all the values for $p$ for which $L_{p-1} \neq 0(\emptyset)$. Then we have $Y \in h(A, L, 0(\emptyset))$.*

*(2) Let $(Y, A)$ be obtained from $(X, A)$ by some complete $'S$-surgery in dimensions $\{k_i\}$, the set of integers $\sigma = \{k_i\}$ contain all the values $q, p$ for which $L_{p-1} \neq 0(\emptyset)$, and $L'_q \neq 0(\emptyset)$. Furthermore, assume that the embedding $\alpha : X_1 \cup Y_1 \to M$ (recalling that $X_1 = (X \cap \overline{G}) \cup x$) possesses the property $\tilde{h}_{k_i}(X_1 \cup Y_1) \subset \operatorname{Ker} \alpha_* = \Delta_*(M, X_1 \cup Y_1)$ for each $k$ (accordingly, we assume in the case of cohomology that $\operatorname{Im} \alpha^* = 0$ in $\tilde{h}^{k_i}(X_1 \cup Y_1)$, i.e., $\tilde{h}^{k_i}(X_1 \cup Y_1) \backslash 0 \subset \Delta^*(M, X_1 \cup Y_1)$ for each $k_i$. Then $Y \in h(A, L, L')$.*

*(3) Let the class $h(A, \tilde{L})$ be nonempty, let $X$ be an arbitrary compact set from it, $(Y, A)$ obtained from $(X, A)$ by some complete $'S$-surgery in dimensions $\sigma = \{k_i\}$, let $\sigma$ contain all integers $r$ for which $L_r \neq 0(\emptyset)$, the embedding $\alpha : X_1 \cup Y_1 \to M$ sastisfy the assumptions of (2), and $Y_1 \cap A = x$. Then $Y \in h(A, \tilde{L})$.*

**Remark.** It follows from Theorem 27.3.1 that the complete $'S$-surgeries are most important in variational problems, because they preserve the classes $h(A, L, L')$, whereas $S$-surgeries only preserve $h(A, L, 0(\emptyset))$. Apparently, the theorem about the stability of the classes $\mathcal{O}$, $\tilde{\mathcal{O}}$ can be proved also for the usual $S$-surgeries; however, all their concrete examples to be constructed in the following will turn out to be complete surgeries.

**Proof.** We first consider (1). Let $p \in \sigma = \{k_i\}$ be a number such that $L_{p-1} \neq 0(\emptyset)$. To study the cohomology version, we assume for simplicity that $Y_1 \cap Y_2 = A_1$, and subsequently come to realize that this assumption is immaterial. Recall that $X_1 = (X \cap \overline{G}) \cup x$, $A_1 = (X \cap \partial G) \cup x$, $Y = Y_1 \cup Y_2$, $Y_2 = X \backslash G$. Put $X_2 = Y_2$, $A_2 = A \cup A_1$, $A_2 = B$, $L_1 = \Delta_p(X_1, A_1)$, $L_2 = \Delta_p(X_2, A_2)$, $L'_1 = \Delta_p(Y_1, A_1)$, $L'_2 = \Delta_p(Y_2, A_2)$, where the fixed value $p$ from the set $\sigma$ determining the surgery is indicated above. Note that $L'_2 = L_2$, and, besides, all the assumptions of Lemma 26.4 are fulfilled for the triples $(X, X_1, X_2)$ and $(Y, Y_1, Y_2)$. Indeed, $X = X_1 \cup X_2 = (X \cap \overline{G}) \cup (X \backslash G)$, $A_1 \subset X_1$, $A_2 \subset X_2$, since $A \cup (X \cap \partial G) \subset X \backslash G$, $A \cap X_1 \subset A_1$, in turn, because $A \cap X_1 = x \subset A_1$, $A \cap X_2 \subset A_2$ due to $A \cap (X \backslash G) = A \subset A \cup A_1$, $X_1 \cap X_2 = A_1 \cap A_2$ following from $(X \cap \partial G) \cup x \cap (X \backslash G) = A_1 = A_1 \cap A_2$, $\Gamma_1 = L_1$, $\Gamma_2 = L_2$.

It follows, also from Lemma 26.4, that $\Delta_p(X, A) = i_*(B, A)^{-1}[i_*(B, A_1) L_1 + i_*(B, A_2) L_2]$. Consider the triple $(Y, Y_1, Y_2)$. Then $Y = Y_1 \cup Y_2$ by definition of $Y$, $A_1 \subset Y_1$, $A_2 \subset Y_2$, since $A \cup A_1 \subset X \backslash G$, $A \cap Y_1 \subset A_1$, and

$A \subset Y_2$, whereas $Y_2 \cap Y_1 = A_1$ (see the assumption above), i.e., $A \cap Y_1 = x \subset A_1$, $A \cap Y_2 \subset A_2$, because $A \cap (X \backslash G) = A \subset A \cup A_1 = A_2$, $Y_1 \cap Y_2 \subset A_1 \cap A_2$, and $Y_1 \cap Y_2 = A_1 \subset A_2$, $\Gamma_1 = L_1'$, $\Gamma_2 = L_2'$. It follows from Lemma 26.4 that $\Delta_p(Y, A) = i_*(B, A)^{-1}[i_*(B, A_1)L_1' + i_*(B, A_2)L_2']$. Since $L_2' = L_2$ and $L_1' \subset L_1$, we immediately derive that $\Delta_p(Y, A) \supset \Delta_p(X, A) \supset L_{p-1}$, where $L_{p-1} \subset L$, or $Y \in h_*(A, L, 0)$. We now reject the assumption $Y_1 \cap Y_2 = A_1$. Let $A_1 \subset Y_1 \cap Y_2$. Just as in the proof of Theorem 26.1, we separate the extraneous points in the compact set $L = X \cup Y_1$ and represent the compact set $Z$ as the continuous image of a new compact set $\widetilde{Z} = \widetilde{X} \cup \widetilde{Y}_1$, $f : \widetilde{Z} \to Z$, where $f|_{\widetilde{X}}$, $f|_{\widetilde{Y}_1}$ are homeomorphisms, and $\widetilde{Y}_1 \cap \widetilde{Y}_2 = \tilde{A}_1$, $Y_2 = \widetilde{Y}_2$, $A_1 = \tilde{A}_1$. (Note that it is not necessary to embed $\widetilde{Z}$ into the manifold $M$.) Then the above argument can be applied to $\widetilde{Z}$, and we can write $\Delta_p(\widetilde{Y}, \tilde{A}) \supset L_{p-1}$ for each $p$. Now, applying the homomorphism $f_*$, we obtain, due to the remark after Lemma 26.2, that $\Delta_p(Y, A) \subset L_{p-1}$, which is just what was required to be proved, and the homology version is studied completely.

We now consider the cohomology case. We take two triples, $(X, X_1, X_2)$ and $(Y, Y_1, Y_2)$, assuming that $Y_1 \cap Y_2 = A_1$, and see that they all satisfy all the conditions of Lemma 25.4, it being $L_1' = \nabla^p(Y_1, A_1)$, $L_1 = \nabla^p(X_1, A_1)$, $L_2' = L_2 = \nabla^p(X_2, A_2) = \nabla^p(Y_2, A_2)$.

We obtain from Lemma 25.4 that

$$\begin{aligned}
&\nabla^p(X, A) \\
&= \tilde{h}^{p-1}(A) \backslash \{i^*(B, A)[\tilde{h}^{p-1}(B) \backslash [i^*(B, A_1)^{-1}(L_1) \cup i^*(B, A_2)^{-1}(L_2)]]\}, \\
&\nabla^p(Y, A) \\
&= \tilde{h}^{p-1}(A) \backslash \{i^*(B, A)[\tilde{h}^{p-1}(B) \backslash [i^*(B, A_1)^{-1}(L_1') \cup i^*(B, A_2)^{-1}(L_2')]]\},
\end{aligned}$$

and, since $L_1' \supset L_1$, it follows that $\nabla^p(Y, A) \supset \nabla^p(X, A) \supset L_{p-1}$, where $L_{p-1} \subset L$, i.e., $Y \in h^*(A, L, \emptyset)$. As well as in the homology case, the omission of the assumption $Y_1 \cap Y_2 = A_1$ is justified by constructing a continuous mapping $f : \widetilde{Z} \to Z = X \cup Y_1$ (see above). Then $\nabla^p(\widetilde{Y}, \tilde{A}) \supset L_{p-1}$, and we derive from the remark after Lemma 25.2 that $\nabla^p(Y, A) \supset L_{p-1}$, which is just what was required, and item (1) of the present theorem is proved completely.

We now turn to item (2). First, consider the homology case. Let an $'S$-surgery of a pair $(X, A)$ be given. It is required to establish that $Y \in h_*(A, L, L')$ if $\alpha_*$ is the trivial homomorphism. Since each $'S$-surgery is an $S$-surgery, we obtain due to item (1) of the above theorem that $Y \in h_*(A, L, 0)$. It remains to be shown that $L_q' \subset i_*(M, Y)\tilde{h}_q(Y)$ for any $q \in \sigma = \{k_i\}$, where $L_q' \neq 0$. Consider the compact sets $\tilde{A}, \tilde{A}_1, \tilde{A}_2, \widetilde{D}_1, \widetilde{B}_1$, where $\widetilde{D}_1 = A_1, \tilde{A}_1 = Y_2 = X \backslash G$. Then $\tilde{A}_1 \supset \widetilde{D}_1$, because $X \backslash G \supset (X \cap \partial G) \cup x, \tilde{A}_2 = X_1$, with the consequence that $\tilde{A}_2 \subset \widetilde{D}_1$ due to $X_1 \supset A_1$, and $\widetilde{B}_1 = Y_1, \tilde{A} = X$. Accordingly, all the conditions of Theorem 26.1 are fulfilled, viz., $A = \tilde{A}_1 \cup \tilde{A}_2$, since $X = X_1 \cup (X \backslash G)$, and $\tilde{A}_1 \cap \tilde{A}_2 = \widetilde{D}_1$ due to $(X \backslash G) \cap (X \cap \overline{G}) = X \cap \partial G$, $x \in \widetilde{D}_1, \Gamma_1 = \tilde{h}_{q-1}(\widetilde{D}_1) = \tilde{h}_{q-1}(A_1) = \Delta_q(X_1, A_1) = \Delta_q(Y_1, A_1) = \Delta_q(\widetilde{B}_1, \widetilde{D}_1)$.

Consider the compact sets $\widetilde{C}_1 = \tilde{A}_1 \cup \tilde{B}_1, \widetilde{C}_1 = Y_2 \cup Y_1, \widetilde{C}_2 = \tilde{A}_2 \cup \tilde{B}_1 = X_1 \cup Y_1$, $\widetilde{C} = \widetilde{C}_1 \cup \widetilde{C}_2 = Y \cup X_1 = X \cup Y_1$. It follows from Theorem 26.1 that $i_*(X \cup Y_1, Y)\tilde{h}_q(Y) + i_*(X \cup Y_1, X_1 \cup Y_1)\tilde{h}_q(X_1 \cup Y_1) \supset i_*(X \cup Y_1, X)\tilde{h}_q(X)$.

Consider an embedding $\varphi : X \cup Y_1 \to M$. Then $i_*(M, Y)\tilde{h}_q(Y) + i_*(M, X_1 \cup Y_1) \supset i_*(M, X)\tilde{h}_q(X)$. Due to the conditions of item (2), we have

$$i_*(M, X_1 \cup Y_1)\tilde{h}_q(X_1 \cup Y_1) = \alpha_* \tilde{h}_q(X_1 \cup Y_1) = 0,$$

so that $i_*(M, Y)\tilde{h}_q(Y) \supset i_*(M, X)\tilde{h}_q(X) \supset L'_q$.

Q.E.D.

We now consider the cohomology case. Let an $'S$-surgery of a pair $(X, A)$ be given. It is required to establish that $Y \in H^*(A, L, L')$ if $\operatorname{Im} \alpha^* = 0$. Since each complete surgery is also usual, it follows from item (1) above that $Y \in h^*(A, L, \emptyset)$; therefore, it remains to show that $L'_q \cap \operatorname{Ker} i_*(M, Y) = \emptyset$ for any $q \in \sigma = \{k_i\}$, where $L'_q \neq \emptyset$. As in the homology case, consider the compact sets $\tilde{A}, \tilde{A}_1, \tilde{A}_2, \widetilde{D}_2, \widetilde{B}_1, \widetilde{C}, \widetilde{C}_1, \widetilde{C}_2$ associated with an $'S$-surgery (see above). If we put

$$\Gamma_1 = \tilde{h}^{q-1}(\tilde{D}_1) \setminus 0 = \tilde{h}^{q-1}(A_1) \setminus 0$$
$$= \nabla^q(X_1, A_1) = \nabla^q(Y_1, A_1) = \nabla^q(\tilde{B}_1, \tilde{D}_1),$$

then it is obvious that all the conditions of Theorem 25.1 are fulfilled, and we therefore obtain that

$$i^*(X \cup Y_1, Y)^{-1}[\tilde{h}^q(Y) \setminus 0]$$
$$\cup\, i^*(X \cup Y_1, X_1 \cup Y_1)^{-1}[\tilde{h}^q(X_1 \cup Y_1) \setminus 0]$$
$$\supset i^*(X \cup Y_1, X)[\tilde{h}^q(X) \setminus 0].$$

If $\varphi : X \cup Y_1 \to M$ is an embedding, then we have

$$i^*(M, Y)^{-1}[\tilde{h}^q(Y) \setminus 0]$$
$$\cup\, i^*(M, X_1 \cup Y_1)^{-1}[\tilde{h}^q(X_1 \cup Y_1) \setminus 0] \supset i^*(M, X)^{-1}[\tilde{h}^q(X) \setminus 0].$$

Due to the assumptions, we get $i^*(M, X_1 \cup Y_1)^{-1}[\tilde{h}^q(X_1 \cup Y_1) \setminus 0] = \emptyset$, i.e., $i^*(M, Y)^{-1}[\tilde{h}(Y) \setminus 0] \supset i^*(M, X)^{-1}[\tilde{h}^q(X) \setminus 0]$. Let $l' \in L'_q$. Then $i^*(M, N)l' \neq 0$, i.e., $l' \in i^*(M, Y)^{-1}[\tilde{h}^q(Y) \setminus 0]$, $i^*(M, Y)l' \neq 0$ in $\tilde{h}^q(Y)$ for any $l' \in L'_q$, which completes the proof of item (2).

We now turn to item (3). Consider the embedding $(X \cup Y_1)/A \to M/A$, and notice that, since $x = A_1 \cap A$, and $Y_1 \cap A = x$, the compact set $\widetilde{Y}_1 = Y_1/(A \cap Y_1) = Y_1$ still completely spans the compact set $A_1$ in dimensions $\{k_i\}$, i.e., we obtain the complete $'S$-surgery of the compact set $X/A = \widetilde{X}$ via $G$ and $\widetilde{Y}_1$ in the underlying space $M/A$. Since the theory $h$ is relatively invariant, $\tilde{L}_r \subset \tilde{h}_r^{(r)}(M/A)(\setminus 0) \cong \tilde{h}_r^{(r)}(M, A)(\setminus 0)$, and $\widetilde{X}$ realizes the set $\tilde{L}$ in $M/A$. We have thereby reduced the proof to the consideration of item (2) for the case where $L = 0(\emptyset), L' = \tilde{L}$. That $M/A$ is not a manifold does not alter the argument, since the embedding $X_1 \cup Y_1 = (X_1 \cup Y_1)/(X_1 \cup Y_1) \cap A \to M/A$ is still (co)homologically trivial, and the theorem is proved completely.

We have proved it for $'S$-surgeries of pairs $(X, A)$; however, it turns out that the classes $\mathcal{O}$ and $\tilde{\mathcal{O}}$ are affected by a somewhat different kind of surgery, viz., $X_1' = \overline{(X \cap G)} \cup x$, $A_1' = X_1 \backslash G$, $Y' = Y_1' \cup Y_2$. Their difference from $S$-surgeries is in $X_1 \supset X_1'$ and $A_1 \supset A_1'$. The compact set $Y'$ turns out to belong to $\mathcal{O}$, $\tilde{\mathcal{O}}$ again.

The proof is given in accordance with the above argument.

**Remark.** Let a domain $G \subset M$ be given. Put $\tilde{A}_1 = X \cap \partial G$. Let $\dot{x} \in \dot{A}_1, \dot{Y}_1$ be a compact set completely spanning the compact set $\dot{A}_1$ in dimensions $\sigma = \{k_i\}$ relative to the point $\dot{x}$. For example, in subsequent applications of Theorem 27.3.1, $\dot{Y}_1$ is contained in $\overline{G}$. Construct a compact set $Y_1$ that will completely span the compact set $A_1 = x \cup \dot{A}_1$ in dimensions $\sigma$ also relative to the point $x \in A$, for which we consider some smooth path $\gamma$ joining $x$ and $\dot{x}$, and set $Y_1 = \gamma \cup \dot{Y}_1$. It is then obvious that $Y_1$ is the required one. This remark is of technical nature; however, it is necessary, since we proceed in the category of pointed space.

## 27.4. The Stability of Variational Classes

We now take up a special class of extraordinary (co)homology theories $h$ satisfying the following.

**Condition K.** Let $h$ be a continuous and relatively invariant (co)homology theory on the category of compact pairs, and $\varphi : S^1 \to B$ an embedding of the circumference into a two-dimensional finite cell complex $B$, with the element $\varphi_*[S^1]$ either being zero in the group $H_1(B, \mathbf{Z})$ (Case 1) or the generator of finite order in $H_1(B, \mathbf{Z})$ (Case 2). We will say that $h$ satisfies Condition K on $P^2 \subset U_C$ if all the homomorphisms $\varphi_{k*} : \tilde{h}_k(S^1) \to \tilde{h}_k(B)$ are trivial in Case 1 for all such compact sets $B$ (all the homomorphisms $\varphi_k^* : \tilde{h}^k(B) \to \tilde{h}^k(S^1)$) being accordingly trivial in the cohomology case); whereas, in Case 2, the homomorphism $\varphi_{k*}$ is either trivial or is a monomorphism of all such $B$ and each $k$ (the homomorphism $\varphi_k^*$ being accordingly either trivial or an epimorphism in the cohomology case). Here, the circumference integral homology fundamental class is denoted by $[S^1]$.

Condition K can be checked easily. For example, it is obvious that the theory $h_* = N_*$ of nonoriented bordisms satisfies it. We shall show in the following that a bordism theory modulo $p$ also satisfies it.

**Proposition 27.4.1.** *Let $M$ be a compact, closed, and smooth manifold, and $\pi_1(M) = \pi_2(M) = 0$. Consider a continuous and relatively invariant (co)homology theory $h$ on $U_C$, satisfying Condition K on $P^2 \subset U_C$. Let $L' \not\equiv 0$ be an arbitrary set of subgroups (or subsets in the cohomology case) in $h_*(M)$*

*(resp. in $\tilde{h}^*(M)\backslash 0$). Then the class $h(x,0,L')$ is nonempty and 2-stable. If the class $h(A,L,L')$, or $h(A,\tilde{L})$ is nonempty, this class is also 2-stable.*

**Proof.** That $h(x,0,L')$ is nonempty is obvious. We first consider the homology case. Let $X \in U_C, X \in h_*(x,0,L')$, and $Y \subset X$ be a compact subset such that $N = \overline{X\backslash Y}$ is a finite simplicial complex in $M$, where $\dim N \leqslant 2$. Consider a sequence $\Pi_n$ of closed polyhedral neighbourhoods of the compact set $Y$ in $M$. Suppose that the $\Pi_n$ are such that $\bigcap_n \Pi_n = Y$. If we prove that $\Pi_n \in h_*(x,0,L')$ for each $n$, then the compact set $Y = \underleftarrow{\lim}\Pi_n$ also belongs to $h_*(x,0,L')$ due to Theorem 27.2.1. Since $\overline{\Pi_n \cup N} \supset X$, we have $\Pi_n \cup N \in h_*(x,0,L')$ for each $n$. Put $N_n = \overline{(\Pi_n \cup N)\backslash\Pi_n}$. Then we can assume that $N_n$, as well as $N$, is a finite simplicial complex in $M$. Fix $n$ and put $\Pi_n \cup N = X', \Pi_n = Y', N' = N_n$. If $\dim N' = 1$ then $N'$ consists of a finite number of segments; therefore, $Y' \in h_*(x,0,L')$.

Q.E.D.

Now, let $\dim N' = 2$. Then we can always construct a decomposition of $N'$ (if required) so fine that each closed 2-cell is a smooth two-dimensional disc $D^2$ embedded into $M$. For our purposes, it suffices to show that we can easily remove an arbitrary $D^2 \subset N'$ from the compact set $X' = Y' \cup N'$ without affecting the realization of the set $L'$. Thus, we can assume that $X_1 = Y_1 \cup N_1$, where $N_1 = D^2$, $N_1 = \overline{X_1\backslash Y_1}, X_1 \in h_*(x,0,L'), Z = \partial N_1 = S^1$. Consider an embedding $\overline{\varphi} : S^1 \to Y_1$. Then the following cases are possible, viz., (A) $\bar{\varphi}_*[S^1] = 0$ in the group $H_1(Y_1, \mathbb{Z})$, and (B) $\bar{\varphi}_*[S^1] \neq 0$ in the group $H_1(Y_1,\mathbb{Z})$. If we have the former case, there exists a two-dimensional subcomplex $B \subset Y_1$ such that $\varphi_*[S^1] = 0$ in $H_1(B,\mathbb{Z})$, where $\varphi : S^1 \to B$ is an embeddings. Since the theory $h_*$ satisfies Condition K, all homomorphisms $\varphi_{k*} : \tilde{h}_k(S^1) \to \tilde{h}_k(B)$ are trivial. Therefore, by setting $D = Z, B = B, A_1 = Y_1, A_2 = N_1, C_1 = Y_1, C_2 = N_1 \cup B$, we obtain $i_*(M,Y_1)\tilde{h}_k(Y_1) + i_*(M, B \cup N_1)\tilde{h}_k(B \cup N_1) \supset i_*(M,X)\tilde{h}_k(X) \supset L'_k$ by Theorem 26.1, all of whose conditions are fulfilled. Since the manifold $M$ is 2-connected, we obtain $i_*(M,X)\tilde{h}_k(B \cup N_1) = 0$, i.e., $L' \subset \operatorname{Im} i_*(M,Y_1), Y_1 \in h_*(x,0,L')$.

Q.E.D.

In case (B), since $\bar{\varphi}_*[S^1] \neq 0$, we have $\overline{\varphi}_*[S^1] = \sum_i m_i e_i$, where $e_i$ are generators in the group $H_1(Y,\mathbb{Z})$. Because $Y_1 \in P^2$, we can construct a continuous deformation $\bar{\varphi}(S^1)$ in $Y_1$ so that the new mapping (in general, not an embedding) $\varphi' : S^1 \to Y_1$ may transform the circumference $S^1$ into the one-dimensional skeleton $Y_1$, and $\varphi'(S^1) = \bigvee_\alpha S^1_\alpha$, where $S^1_\alpha$ are circumferences of $Y_1$. The deformation $\bar{\varphi}$ in $\varphi'$ can be extended to that of the disc $N_1$ in the manifold $M$. It is clear that the obtained compact set $X'_1 = Y_1 \cup N'_1$, where $N'_1$ is the deformed disc, again realizes the set $L'$. Since $\pi_1(M) = 0$, each of the circumferences $S^1_\alpha$ is contractible on $M$ to a point; therefore, we can consider a compact set $T = \bigcup_\alpha CS^1_\alpha$, where $CS^1_\alpha$ are cones over $S^1_\alpha$ in $M$. Since the compact set $T$ completely spans the compact set $\varphi'(S^1)$, we obtain from Theorem 26.1 that the compact set $Y'_1$ obtained from $Y_1$ by replacing $N_1'$ by $T$ again belongs to the class $h_*(x,0,L')$. Thus, preserving the old notation, we can assume that $X_1 = Y_1 \cup N_1$, where $N_1 = CZ$, and $Z = S^1 = \partial N_1, N_1 = \overline{X_1 \backslash Y_1}$, $X_1 \in h_*(x,0,L')$, whereas the embedding $\psi : S^1 \to Y_1$ is such that $\psi_*[S^1] \neq 0$, and $\psi$ is the generator of the group $H_1(Y_1, \mathbf{Z})$.

Let $i : X_1 \to M$, $j : Y_1 \to M$ be two embeddings, and $l' \neq 0$, $l' \in L'$. Then $l' = i_*(a)$, $a \in \tilde{h}_*(X_1)$. It is required to show that $l' \in \operatorname{Im} j_*$. Assume the contrary, viz., $l' \neq \operatorname{Im} j_*$. Consider the commutative diagrams

$$\begin{array}{ccccccc}
\tilde{h}_k(X_1) & \xrightarrow[i_*]{} & \tilde{h}_k(M) & \xrightarrow[\varphi_*]{} & \tilde{h}_k(M/X_1) & & \\
\uparrow & & \approx\uparrow\chi_* & & \uparrow\tau_* & & \\
\tilde{h}_k(Y_1) & \xrightarrow[j_*]{} & \tilde{h}_k(M) & \xrightarrow[\alpha_*]{} & \tilde{h}_k(M/Y_1) & \xrightarrow{\partial'} & \tilde{h}_{k-1}(Y_1) \\
\uparrow & & \uparrow\rho_* & & \uparrow\pi_* & & \uparrow \\
\tilde{h}_k(Z) & \longrightarrow & \tilde{h}_k(N_1) & \xrightarrow{\omega_*} & \tilde{h}_k(N_1/Z) & \xrightarrow{\partial} & \tilde{h}_{k-1}(Z),
\end{array}$$

where $\chi_*$ is an isomorphism, and $X_1/Y_1 = N_1/Z = S^2$; therefore, the homomorphism $\tau_*$ and $\pi_*$ are in the exact sequence of the triple $(M, X_1, Y_1)$. Since $l' = i_*(a)$, we have $\varphi_*(l') = 0$, and due to $l' \notin \operatorname{Im} j_*$, we obtain $\alpha_*(l') \neq 0$, $\chi_*(l') = l'$, with $\tau_*\alpha_*(l') = 0$, i.e., $\alpha_*(l') = \pi_*(\omega)$. Because $N_1 = D^2$, we derive that $\tilde{h}_k(N_1) = 0$, and $\partial$ is a monomorphism; hence, $\partial(\omega) \neq 0$ due to $\omega \neq 0$ (recalling that $\pi_*(\omega) \neq 0$). It is clear that $\psi_*\partial(\omega) = \partial'\pi_*(\omega) = \partial'\alpha_*(l') = 0$, and we have constructed an element $\xi \in \tilde{h}_{k-1}(Z)$, $\xi \neq 0$, such that $\psi_*(\xi) = 0$, where $\psi : S^1 \to Y_1$ is an embedding. Consider the element $\alpha = \psi_*[S^1] \in H_1(Y_1, \mathbf{Z})$. Recall that we are now within the framework of case (B), i.e., $\alpha \neq 0$, and $\alpha$ is a generator of the group $H_1(Y_1, \mathbf{Z})$. Two cases are possible, viz., ($B_1$) $\alpha$ is an element of infinite order, and ($B_2$) $\alpha$ is of finite order. In the former case, since $Y_1 \in P^0$, there exists a continuous mapping $f : Y_1 \to K(\mathbf{Z}, 1) = S^1$ such that $f_*(\alpha) = [S^1]$, where we have used the fact that the order of $\alpha$ is infinite. But then the composition mapping $f\psi : S^1 \to S^1$ is homotopic to the identity; therefore, all homomorphisms $\psi_{p*} : \tilde{h}_p(S^1) \to \tilde{h}_p(Y_1)$ are monomorphisms for each $p \in Z$, which is contrary to the existence of an element $\xi \in \tilde{h}_{k-1}(Z)$ proved above, so that $\psi_{k-1,*}(\xi) = 0$. Thus, the assumption that $l' \notin \operatorname{Im} j_*$ leads to a contradiction.

In case $(B_2)$, since $m\alpha = 0$ for a certain $m \in \mathbf{Z}$, there exists a two-dimensional subcomplex $B \subset Y_1$ such that $\varphi_*[S^1] \in H_1(B, \mathbf{Z})$ is of order $m$, and is a generator in the group $H_1(B, \mathbf{Z})$, where $\varphi : S^1 \to B$ is an embedding. Since the theory $h_*$ satisfies Condition K, the homomorphism $\varphi_{k-1,*} : \tilde{h}_{k-1}(Z) \to \tilde{h}_{k-1}(B)$ is either a monomorphism or trivial. On the other hand, $\xi \neq 0$ and $\xi \in \operatorname{Ker} \varphi_{k-1,*}$; therefore, $\varphi_{k-1,*}$ may not be a monomorphism and is therefore trivial. Put $D = Z$, $B = B$, $A_1 = Y_1$, $A_2 = N_1$, $C_1 = A_1 \cup B = Y_1 \cup B = Y_1$ (since $B \subset Y_1$), and $C_2 = A_2 \cup B$, $C_2 = N_1 \cup B$, $C = C_1 \cup C_2 = X_1 \cup B = X_1$, $A = A_1 \cup A_2 = Y_1 \cup N_1 = X_1$, $A_1 \cap A_2 = D$. Then the compact set $B$ wholly spans $Z = S^1$ in dimension $k$, i.e., $\Delta_k(B, D) = \tilde{h}_{k-1}(D)$, and from Theorem 26.1, all of whose conditions are fulfilled here, we derive that $i_*(C, C_1)\tilde{h}_k(C_1) + i_*(C, C_2)\tilde{h}_k(C_2) \supset i_*(C, A)\tilde{h}_k(A) \equiv \tilde{h}_k(X_1)$, since $A = C = X_1$, which means that $a = a_1 + a_2$, where $a_1 \in i_*(X, Y_1)\tilde{h}_k(Y_1)$, $a_2 \in i_*(X, B \cup N_1)\tilde{h}_k(B \cup N_1)$. Applying the homomorphism $i_*$ to the above inclusion, we obtain $l' = i_*(a) = i_*(a_1) + i_*(a_2)$, where $i_*(a_1) \in i_*(M, Y_1)\tilde{h}_k(Y_1)$, and $i_*(a_2) = 0$, since $i_*(M, B \cup N_1)\tilde{h}_k(B \cup N_1) = 0$ due to $\dim B \cup N_1 = 2$, with $\pi_1(M) = \pi_2(M) = 0$, so that the complex $B \cup N_1$ is contractible on $M$ to a point. Thus, $l' \in \operatorname{Im} j_*$, which is contrary to the original assumption, and the statement is thus proved.

We now turn to the cohomology case, the proof being similar. We can assume that the embedding $\bar{\varphi} : S^1 \to Y_1$ from the standpoint of the integral homology groups is either described by case (A) or (B). Since, in the former case, $\bar{\varphi}_*[S^1] = 0$, we have $\operatorname{Im} \bar{\varphi}_* = 0$ in the group $H^1(S^1, \mathbf{Z})$, and we can indicate a two-dimensional film $B \subset Y_1$ such that $\varphi : S^1 \to B$ is an embedding, and $\operatorname{Im} \varphi_* = 0$ in $H_1(B, \mathbf{Z})$. Then, due to Condition K, all the homomorphisms $\varphi_k^*$ are trivial, and the compact set $Y_1$ realized $L'$ because of Theorem 25.1, which is just what is required.

Now, let case (B) be realized. As before, we can assume that the embedding $\psi : S^1 \to Y_1$ realizes a generator of the group $H_1(Y_1, \mathbf{Z})$. Resorting to the prior notation for compact sets, consider the commutative diagram

$$\begin{array}{ccccccc} \tilde{h}^k(Y_1) & \xleftarrow[\alpha^*]{} & \tilde{h}^k(X_1) & \xleftarrow[\rho^*]{} & \tilde{h}^k(X_1/Y_1) & \xleftarrow[\delta']{} & \tilde{h}^{k-1}(Y_1) \\ \downarrow & & \downarrow \xi^* & & \downarrow \tau^* & & \downarrow \\ \tilde{h}^k(Z) & \longleftarrow & \tilde{h}^k(N_1) & \xleftarrow{\omega^*} & \tilde{h}^k(N_1/Z) & \xleftarrow{\delta} & \tilde{h}^{k-1}(Z) \end{array}$$

Let $l' \neq 0$, $l' \in L'_k$, and $i^*(l') \neq 0$, where $i : X_1 \to M$ is an embedding. We have to show that $j^*(l') \neq 0$, where $j : Y_1 \to M$ is an embedding. Assume the contrary. Then $i^*(X_1, Y_1)i^*(l') = 0$, i.e., $\alpha^*(\sigma) = 0$, $\sigma = i^*(l') \neq 0$. It is clear that $\sigma = \rho^*(\omega)$, and $\omega^*\tau^*(\omega) = 0$, where $\tau^*$ is an isomorphism, since $X_1/Y_1 = N_1/Z$. Then $\tau^*(\omega) = \delta(\xi)$, where $\xi \notin \operatorname{Im} \psi^*$ (otherwise, $\tau^*(\omega) = \delta\psi^*(\xi) = \tau^*\delta'(\chi)$, i.e., $\sigma = \rho^*\delta'(\chi) = 0$). Thus, we have shown that the homomorphism $\psi_{k-1}^*$ is not epimorphic.

Since we are now within the framework of case (B), both case ($B_1$) and ($B_2$) are possible. If $\alpha$ is a free generator in (B), then, having constructed a mapping $f : Y_1 \to K(\mathbf{Z}, 1)$ as before, we obtain that $\psi_{k-1}^*$ must be an epimorphism, which is contrary to the statement derived above.

Now, let ($B_2$) be realized. Then there exists a two-dimensional subcomplex $B \subset Y_1$ such that $m\varphi_*[S^1] = 0$ for certain $m \in \mathbf{Z}$, where $\varphi : S^1 \to B$ is an embedding, and we obtain from Condition K that the homomorphism $\varphi_{k-1}^* : \tilde{h}^{k-1}(B) \to \tilde{h}^{k-1}(Z)$ must either be trivial or epimorphic; however, the latter is contrary to the existence of $\xi \notin \operatorname{Im} \psi^*$. Therefore, $\varphi_{k-1}^*$ is trivial. Applying Theorem 25.1, we obtain that $Y_1$ still realizes $L'$.

We have proved Proposition 27.4.1 only for the case of realizing classes $h(x, 0(\varphi), L')$; however, it can be seen from its proof that a similar statement regarding the 2-stability is also valid for arbitrary classes $h(A, L, L')$, $h(A, \tilde{L})$.

Q.E.D.

In conclusion, one important remark: Consider the class $h(A, L_{k-1}, L'_k)$, where $L_{k-1} \subset \tilde{h}^{(k-1)}_{k-1}(A)(\backslash 0)$, $L'_k \subset \tilde{h}^{(k)}_k(M)(\backslash 0)$. It may happen that the class $\mathcal{O}$ is $(k-1)$-stable in spite of the theory $h$ being extraordinary (co)homology. The $(k-1)$-stable elements $l' \in L'_k$ and $l \in L_{k-1}$ are similar to those of the usual (co)homology theory (e.g., see Proposition 27.4.1 asserting that the 2-stability, in fact, means that $h$ behaves on two-dimensional complexes much as the usual (co)homology theory). In studying bordisms mod $p$, we shall again come back to the problem of 2-stability. Note that any nonempty class $\mathcal{O}$, $\tilde{\mathcal{O}}$ is 1-stable on a simply-connected, compact, and closed manifold $M$.

**Proposition 27.4.2.** *Let $h_* = \Omega^p_*$ be a singular bordism homology theory* mod $p$, *where $p$ is an odd prime number. Then $\Omega^p_*$ satisfies Condition $K$ on the category $P^2 \subset U_C$.*

**Proof.** Let $\varphi : S^1 \to B$ be an embedding of the circumference $S^1$ into a finite two-dimensional cell complex $B$, where $\varphi_*[S^1]$ is either zero or a generator of finite orde. in the group $H_1(B, \mathbf{Z})$. If $\varphi_*[S^1] = 0$, then all the homomorphisms $a : \tilde{\Omega}^p_k(S^1) \to \tilde{\Omega}^p_k(B)$ are obviously trivial. Now, let $\alpha = \varphi_*[S^1]$ be a generator of finite order $m$ in $H_1(B, \mathbf{Z})$, $m\alpha = 0$. Assume that $m$ and $p$ are relatively prime. Then, since the group $\tilde{\Omega}^p_k(S^1)$ is isomorphic to the group $\tilde{\Omega}^p_{k-1}(x)$, each element $\gamma \in \tilde{\Omega}^p_k(S^1)$ is of order $p$ and, therefore, representable in the form $\gamma = m\varphi'$, where $\varphi' \in \tilde{\Omega}^p_k(S^1)$. Because $m\alpha = 0$, the element $\gamma = m\varphi'$ is also carried into zero by the homomorphism $a$. Therefore, all $a$ are trivial. Suppose that $m = sp$, where s$\geqslant 1$ is an integer. Then there exists a mapping $g : B \to K(\mathbf{Z}_m, 1)$ such that the composition mapping $S^1 \xrightarrow{\varphi} B \xrightarrow{g} K(\mathbf{Z}_{ps}, 1)$ is homotopic to the identity mapping of $S^1$ into the one-dimensional skeleton of $S^1 \subset K(\mathbf{Z}_m, 1)$. Further, since the order of each element $\gamma \in \tilde{\Omega}^p_k(S^1)$ is $p$, the homomorphism $(g\varphi)_* : \tilde{\Omega}^p_m(S^1) \to \tilde{\Omega}^p_m(K(\mathbf{Z}_{ps}, 1))$ is a monomorphism, and then so is $\varphi_*$.

Q.E.D.

**Corollary 26.4.1.** *Let $M$ be a closed Riemannian manifold, and $\pi_1(M) = \pi_2(M) = 0$. Suppose that $L' \neq 0$ is an arbitrary set of subgroups in $\tilde{\Omega}^p_*(M)$. Then the class $\Omega^p_*(x, 0, L')$ is nonempty and 2-stable. If the class $h(A, L, L'$ or $h(A, \tilde{L})$ is nonempty, this class is also 2-stable.*

The proof follows directly from Proposition 27.4.1 and 27.4.2.

## §28 The General Isoperimetric Inequality

### 28.1. Choice of a Special Coordinate System

Let $M$ be a complete Riemannian manifold, either smooth or of class $C^r$, where $r \geqslant 4$. Recall that the coordinates $\omega : D^n \to M^n$, where $D^n \subset \mathbf{R}^n$ is the standard open ball with centre in the origin, of class $C^r$ are said to be normal and centered at a point $P \in M$ if $\omega(0) = P$, $g_{ij}(0) = \delta_{ij}$, where $g_{ij}$ are the components of the metric tensor with respect to $\omega$, and if the radii of $D^n$ are isometrically carried under the mapping $\omega$, into arcs of geodesics emanating from $P$.

In the following, $B(P,R)$ denotes an open $n$-dimensional ball with centre at a point $P \in M$ and radius $R$ and $l(P,A)$ the cone over $A$ with vertex at $P$, made up of all geodesic segments from $P$ to $A$, provided $C(P,A) \subset B(P,R)$). It is known that any complete Riemannian manifold $M$ of class $C^r$, $r \geqslant 4$, satisfies certain metric conditions to be called $(M)$ below (see [351]).

1. For any point $P \in M$, there exists a coordinate system $\tau = \tau(y^1,\ldots,y^n)$ centered at $P$, and such that $h_{ij}(0) = \delta_{ij}$, $h_{ij,y^k}(0) = 0$, $1 \leqslant k \leqslant n$, where $h_{ij}$ are the components of the metric tensor with respect to $\tau$, and $h_{ij,y^k}$ are partial derivatives. Furthermore, there are five constants $R_0, C_0, \eta_0, K_0, K_1$ not depending on $P$, which satisfy the relations below.

2. If $P_0 \in M$, and $\tau$ is a coordinate system centered at $P_0$ satisfying the above condition, then $\sum_{ij} |\nabla^2 h_{ij}| + |\nabla^3 h_{ij}| \leqslant K_0$ at any point $P \in B(P_0, 4R_0)$, where $\nabla z$ denotes the gradient of the vector function $z$. Besides, the constants $R_0$ and $C_0$ are related by

$$(1+C_0r^2)^{-2} \leqslant h_{ij}\lambda^i\lambda^j \leqslant (1+C_0r^2)^2, \quad 1+C_0 16R_0^2 \leqslant \frac{5}{4},$$

where $P \in B(P_0, 4R_0)$, $r = |\gamma|$, $|\gamma|$ is the length of the segment $\gamma$ of the geodesic radius joining $P_0$ to $P$, $\sum_i (\lambda^i)^2 = 1$.

3. Let $\tau$ and $B(P_0, 4R_0)$ be chosen as above. Then we can specify the normal coordinates $\omega_p$ at each point $P \in B(P_0, 4R_0)$, so that $(1+\eta_0 r)^{-2} \leqslant g_{ij}(Q,P,\omega_p)\lambda^i\lambda^j \leqslant (1+\eta_0 r)^2$, where $\sum_i (\lambda^i)^2 = 1$, $Q \in B(P,R_0)$, $r = |\gamma| \leqslant R_0$, $\gamma$ is a geodesic segment of the radius of the ball $B(P,R_0)$, joining $P$ to $Q$ in the system $\omega_p$, and $\Gamma^i_{jk}(Q,P)\lambda^i\mu^j\mu^k| \leqslant \eta_0 \cdot |\lambda| \cdot |\mu|^2$, $g_{ij}(Q,P,\omega_p)$ are the components of the metric tensor at the point $Q$, found with respect to $\omega_p$. Consider the transition function $U(Q,P,T) = \omega_T^{-1}\omega_P(Q)$. Then the second derivatives of this mapping are uniformly bounded in modulus by a constant $K_1$ for each point $P_0 \in M$ and all $P,Q,T \in B(P_0, 4R_0)$.

## 28.2. Simplicial Points of Surfaces

**Definition 28.2.1.** Let $X \in M$ be a compact set. We will say that a point $P \in X$ is a simplicial $s$-point if there exists a number $\varepsilon > 0$ such that a closed neighbourhood $\overline{U}(P,\varepsilon') = X \cap \overline{B(P,\varepsilon)}$ on $P$ on $X$ admits a representation as a finite $s$-dimensional simplicial subcomplex of class $C^r$ in the manifold $M$ for any number $\varepsilon'$, $0 < \varepsilon' \leqslant \varepsilon$ (see the definition of a subcomplex of class $C^r$ in §6). In other words, $P$ belongs to at least one closed $s$-dimensional simplex of the subcomplex $X \cap \overline{B(P,\varepsilon)}$ embedded into $M$.

Denote by $X$ the set of all simplicial points of $X_c$ for all possible $s$, and $X \backslash X_c$ by $X^*$. Then we can say that the points of $X^*$ are "bad," i.e., not simplicial. We have obtained a unique representation of $X$ as $X = X^* \cup X_c$, where $X^* \cap X_c = \emptyset$, $X_c$ is open in $X$, and $X^*$ is closed in it. Denote by $X_c^s$ the set of all simplicial $s$-points of $X$. Then $\dim X_c^s = s$ at each point $P \in X_c^s$, and a unique representation $X_c = X_c^p \cup X_c^{p-1} \cup \cdots \cup X_c^0$ arises, where certain ones of the $X_c^s$, $0 \leqslant s \leqslant p$ may be empty, $p = \dim X_c$, with $X_c^\alpha \cap X_c^\beta = \emptyset$ if $\alpha \neq \beta$.

## 28.3. Isoperimetric Inequality

Recall the definition of the Hausdorff $k$-dimensional spherical measure which we will denote by the same symbol $\mathrm{vol}_k$ as used for the Riemannian volume. Let $S$ be a subset in $M$, and $k \geqslant 1$ an integer. If $S = \emptyset$, then we put $\mathrm{vol}_k(S) = 0$. Suppose that $S \neq \emptyset$. Then we first determine the value ${}^*\Lambda_r^k(S)$, where $r > 0$ is a fixed number, putting, by definition, ${}^*\Lambda_r^k(S) = \inf \sum_{i=1} \gamma(k) r_i^k$, where $N \leqslant \infty$, $\gamma(k)$ is the $k$-dimensional volume of the unit $k$-dimensional ball $D^k$ in Euclidean space $\mathbf{R}^k$, and inf is taken for all coverings of the set $S$ with at most a countable family of open balls $B(P_i, r_i)$, with $r_i < r$ for any $i$. Since the ${}^*\Lambda_r^k(S)$ do not decrease as $r$ tends to zero, we can determine the exterior Hausdorff measure ${}^*\Lambda^k(S)$ by setting ${}^*\Lambda^k(S) = \lim_{r \to +0} {}^*\Lambda_r^k(S)$. Let $B \subset M$ be a subset. We will say that a set $S \subset M$ is Hausdorff measurable if ${}^*\Lambda^k(B) = {}^*\Lambda^k(S \cap B) + {}^*\Lambda^k(B \cap (M \backslash S))$ for any $B$, in which case we will write $\mathrm{vol}_k(S)$ instead of ${}^*\Lambda^k(S)$. If $p < q$, then $\mathrm{vol}_p(S) \geqslant \mathrm{vol}_q(S)$, and if $\mathrm{vol}_p(S) < \infty$, then $\mathrm{vol}_q(S) = 0$.

Let a (co)homology theory $h$ be fixed on $U_c$. We now prove one simple, but important, property of measure $\mathrm{vol}_k$.

**Lemma 28.3.1.** *Let $\Pi \subset \mathbf{R}^n$ be a $k$-dimensional plane passing through a point $P$, and $A = S^{k-1} = \Pi \cap \partial B(P, r)$, $\overline{D} = D^k = \Pi \cap \overline{B(P, r)}$. Suppose that $x \in A$, and $Y \supset A$ is an arbitrary compact set in $\mathbf{R}^n$, so that $\nabla^m(Y, A) \neq \emptyset$ for a certain $m \in \mathbf{Z}$ (accordingly, $\Delta_m(Y, A) \neq \emptyset$ in the homology case). Then* $\mathrm{vol}_k(Y) \geqslant \mathrm{vol}_k(D)$.

**Proof.** Consider the orthogonal projection $P : (Y, A) \to (\overline{D}, A)$, and put $Y' = P(Y)$. It then becomes clear that $\nabla^m(Y', A) \supset \nabla^m(Y, A)$ (e.g., see the remarks after Lemmas 25.2 and 26.2); $\Delta_m(Y', A) \supset \Delta_m(Y, A) \neq 0$, respectively. Hence, the compact set $Y'$ must contain the whole of the compact set $\overline{D}$ (otherwise, we could carry out a deformation retraction of $Y'$ onto $S^{k-1}$), i.e., $\mathrm{vol}_k(Y') \geqslant \mathrm{vol}_k(\overline{D})$; besides, $\mathrm{vol}_k(Y) \geqslant \mathrm{vol}_k(Y')$.

Q.E.D.

**Theorem 28.3.1. (Cap Theorem).** *Let $M^n$ be a complete Riemannian manifold of class $C^r$, $r \geqslant 4$, and $R_0$ a constant from conditions $(M)$, $A \subset M$ a compact set such that $A \subset B(P_0, R_0)$ for some point $P_0$, $x \in A$ a fixed point, $k$ an integer, $2 \leqslant k \leqslant n-1$, and $A$ representable as $A = A_c \cup A^*$ (see §28.2).*

*Assume that* $0 \leqslant \mathrm{vol}_{k-1}(A) = l^{k-1} < \infty$ *(it follows that* $A_c = A_c^s \cup \cdots \cup A_c^0$*, where* $s \leqslant k-1$*).*

*(a) If* $l > 0$*, then there exists a compact set* $X = X^* \cup X_c^k \cup \cdots \cup X_c^0$ *containing* $A$*, contained itself in the* $\varepsilon$*-neighbourhood of the convex geodesic hull of* $A$*, where* $\varepsilon$ *can be regarded as a prescribed arbitrarily small and fixed number, so that (1)* $\nabla^m(X, A) = \tilde{h}^{m-1}(A)\backslash 0$ *for all* $m \in \mathbf{Z}$ *(resp.* $\Delta_m(X, A) = \tilde{h}_{m-1}(A))$ *in the homology case), i.e.,* $X$ *completely spans* $A$ *in all dimensions, and (2) there exist two constants* $C = C(k, M)$ *and* $D = D(k, M)$ *not depending on* $A \subset B(P_0, R_0)$ *and* $P_0$*, so that* $\rho(X^*, A^*) \leqslant Cl$*,* $\rho(X^* \cup X_c^k,\ A^* \cup A_c^{k-1}) \leqslant Cl$*, the set* $X \backslash [X \cap (A^* \cup A_c^{k-1}, Cl)]$ *is a finite simplicial subcomplex of class* $C^r$ *in the manifold* $M$*, of dimension* $s \leqslant k-1$ *(note that* $X_c^k$ *may be nonempty even in the case where* $A_c^{k-1} = \emptyset$*), and, finally,* $0 \leqslant \mathrm{vol}_k(X) = \mathrm{vol}_k(X^* \cup X_c^k) \leqslant Dl^k$*;*

*(b) if* $l = \mathrm{vol}_{k-1}(A) = 0$ *(then* $A_c^{k-1} = \emptyset$*, and* $A = A_c^\alpha \cup \cdots \cup A_c^0$*,* $\alpha \leqslant k-2$*), there exists a compact set* $X = X(\delta) = X^* \cup X_c^k \cup \cdots \cup X_c^0$*,* $X \supset A$ *for any* $\delta > 0$*, contained in an* $\varepsilon$*-neighbourhood of the convex geodesic hull of the compact set* $A$ *(where* $\varepsilon$ *is an arbitrarily small number not depending on* $\delta$*) and, in general, depending on the choice of* $\delta$*, so that (1)* $\nabla^m(X, A) = \tilde{h}^{m-1}(A)\backslash 0$ *for all* $m \in \mathbf{Z}$ *(resp.* $\Delta_m(X, A) = \tilde{h}_{m-1}(A)$ *in the homology case), and (2) there exist two constants* $C = C(k, M)$ *and* $D = D(k, M)$ *not depending either on* $A \subset B(P_0, R_0)$ *or* $\delta$*, so that* $\rho(X^*, A^*) \leqslant C \cdot \delta$*,* $\rho(X^* \cup X_c^k, A^*) \leqslant C \cdot \delta$*, the set* $X \backslash [X \cap (A^*, C \cdot \delta)]$ *is a finite simplicial subcomplex of class* $C^r$ *in the manifold* $M$*, of dimension* $s \leqslant k-1$ *(recall that* $A_c^{k-1} = \emptyset$*), and* $0 \leqslant \mathrm{vol}_k(X) = \mathrm{vol}_k(X^* \cup X_c^k) \leqslant D\delta^k$*,* $X_c^k$ *possibly being empty, with* $\mathrm{vol}_k(X) = 0$*.*

In contrast to the usual homology theory $H_*$, the relation $\rho(X, A) \leqslant C(k) \cdot l$ does not hold, since, even as $l \to 0$, there may be portions of the film $X$ far off the boundary $A$, of which it is only known that their dimension is not higher than $k-1$, that they are completely made up of simplicial points, and that it is generally impossible to contract them into a small neighbourhood of $A$.

**Proof.** We can assume that $A \subset \mathbf{R}^n$, and resort to induction on $k$. Let $k = 2, l > 0$. We define the function $\varphi(a) = \sum_{s=-\infty}^{+\infty} \mathrm{vol}_0(A \cap \Pi_{a+1,1ls})$, where $\Pi_t = \{(x^1, \ldots, x^n) \in \mathbf{R}^n | x^1 = t\}$, or $\Pi_t$ is a hyperplane in $\mathbf{R}^n$. It follows from Theorem 10.2.3 in [289] that $\int_0^{1,1l} \varphi(a)da \leqslant \mathrm{vol}_1(A) = l > 0$ in case (a); therefore, there exists a value $\bar{a}$, $0 \leqslant \bar{a} \leqslant 1,1l$, such that $\varphi(\bar{a}) = 1$. Then $\varphi(\bar{a}) = 0$, i.e., $\mathrm{vol}_0(A \cap \Pi_{\bar{a}+1,1ls}) = 0$ for each $s \in \mathbf{Z}$. Repeating the construction along each axis $x^i$, $1 \leqslant i \leqslant n$, in $\mathbf{R}^n$, we break $\mathbf{R}^n$ into cubes with side $1.1l$, none containing points from $A$ on its boundary. Since $A$ is a compact set, there exists a finite number of cubes $R_1, R_2, \ldots, R_s$ such that $A \subset \bigcup_{s=1}^S R_s$. In each $R_s$, we take a point $P_s$ inside the convex geodesic hull of the compact set $A_s = A \cap R_s$, and consider $X_s = C(P_s, A_s)$, a cone with vertex at $P_s$ for a value $s_0$ such that $x \in R_{s_0}$. Put $P_{s_0} = x$. Join all $P_s$, for $s \neq s_0$ with smooth paths $\gamma_s$ to the point $x$, so that all $\gamma_s$ are contained in the convex geodesic hull of $A$. We obtain an embedding $A \to X = \bigcup_s (X_s \cup \gamma_s)$, where the compact set $X$, obviously, completely spans $A$ in all dimensions and is contained in the convex hull of $A$. It is evident that $\rho(X^*, A^*) \leqslant Cl$, and $\rho(X^* \cup X_c^2,\ A^* \cup A_c^1) \leqslant Cl$, $0 \leqslant \mathrm{vol}_2(X) = \mathrm{vol}_2(X^* \cup X_c^2) \leqslant Dl^2$.

In case (b), we assume $l = 0$, $\delta > 0$. Then $0 = \mathrm{vol}_1(A) \geqslant \int_0^{1,1\delta} \varphi(a)da$. It again follows that there exists $\bar{a}$, $0 \leqslant \bar{a} \leqslant 1,1\delta$, such that $\varphi(\bar{a}) < \delta$; therefore, the entire subsequent argument can be given by replacing $l$ by $\delta$ similarly to case (a).

Now, let $k > 2, n \geqslant k+1$, and the theorem be proved for all $k' \leqslant k-1$. Then number $n$ does not take part in the induction, i.e., any compact set $A$ for which $\mathrm{vol}_{k'-1}(A) < \infty$ can be spanned by the film $X$, being embedded into $\mathbf{R}^{n'}$, where $n'$ is arbitrary. In case (a), we define the function $\varphi^1(a^1) = \sum_{s^1=-\infty}^{+\infty} \mathrm{vol}_{k-2}(A \cap \Pi_{a^1_+ l s^1})$. Then $\int_0^l \varphi^1(a^1)da^1 \leqslant \mathrm{vol}_{k-1}(A) = l^{k-1}$ (see Theorem 10.2.3 in [289]). It follows that there exists $\bar{a}^1$, $0 \leqslant \bar{a}^1 \leqslant l$ such that $\varphi^1(\bar{a}^1) \leqslant l^{k-2}$. Put $\Pi_{s^1} = \Pi_{\bar{a}^1 + l s^1}$, $D_{s^1} = A \cap \Pi_{s^1}$, where $s^1 \in \mathbf{Z}$. Then $\sum_{s^1} \mathrm{vol}_{k-2}(D_{s^1}) = \varphi^1(\bar{a}^1) \leqslant l^{k-2}$. Let $l_{s^1}^{k-2} = \mathrm{vol}_{k-2}(D_{s^1})$. It follows that $\sum_{s^1} l_{s^1}^{k-2} \leqslant l^{k-2}$, with certain $l_{s^1}$ possibly being zero. Consider $D_{s^1} \cap A^*$, denoting this compact set by $(D_{s^1})^*$. Note that $D_{s^1}^* \subset (D_{s^1})^*$; however, there can also be simplicial points in $(D_{s^1})^*$. Further, we put $(D_{s^1})_c = D_{s^1} \backslash (D_{s^1})^*$, where again $D_{s^1,c} \supset (D_{s^1})_c$, $(D_{s^1})_c$, in general, not exhausting the whole compact set $D_{s^1,c}$. Consider an $\varepsilon'$-neighbourhood $(A^* \cup A_c^{k-1}, \varepsilon')$ of the compact set $A^* \cup A_c^{k-1}$ in $\mathbf{R}^n$. Let $F_{s^1}(\varepsilon') = \Pi_{s^1} \cap (A^* \cup A_c^{k-1}, \varepsilon')$, $N_{s^1} = \Pi_{s^1} \cap (A^* \cup A_c^{k-1})$, $F_{s^1}(\varepsilon') \supset N_{s^1} = (N_{s^1})^* \cup (N_{s^1})_c$, $(N_{s^1})^* = \Pi_{s^1} \cap A^*$, $(N_{s^1})_c = N_{s^1} \backslash (N_{s^1})^*$, $(N_{s^1})^* \supset N_{s^1}^*$, $(N_{s^1}^*)_c = N_{s^1,c}$, $\dim(N_{s^1})_c \leqslant k-2$ (see Fig. 71).

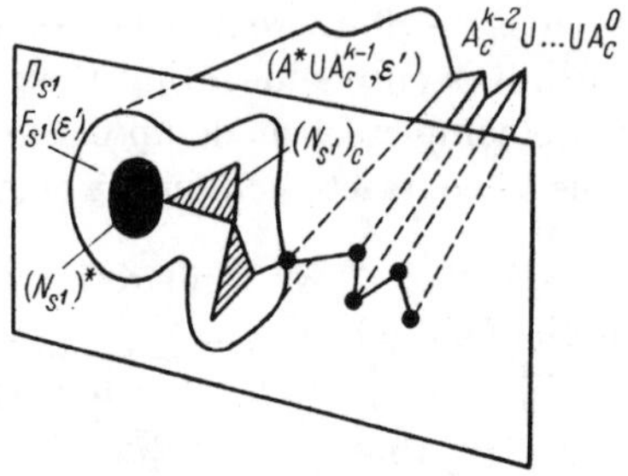


**Fig. 71**

Then $D_{s^1} \backslash F_{s^1}(\varepsilon')$ wholly consists only of simplicial points of dimension not exceeding $k-2$. It is clear that $\overline{(D_{s^1} \backslash F_{s^1}(\varepsilon'))} \cap (A^* \cup A_c^{k-1}) = \emptyset$, and we can assume that the boundary $\partial F_{s^1}(\varepsilon')$ consists of only simplicial points. Consider $\Pi_{s^1} \backslash F_{s^1}(\varepsilon')$. Then there exists an infinitesimal deformation, or diffeomorphism close to the identity, of a smooth, open manifold $\Pi_{s^1} \backslash F_{s^1}(\varepsilon')$, which is fixed on $\partial F_{s^1}(\varepsilon')$ and such that $\dim(\Pi_{s^1} \backslash F_{s^1}(\varepsilon')) \cap (A \backslash (A^* \cup A_c^{k-1}, \varepsilon')) \leqslant k-3$, where $(\Pi_{s^1} \backslash F_{s^1}(\varepsilon'))'$ denotes the deformed manifold $\Pi_{s^1} \backslash F_{s^1}(\varepsilon')$. That the plane $\Pi_{s^1}$ can be reduced to a general position relative to $A\backslash(A^* \cup A_c^{k-1}, \varepsilon')$ is based on the fact that the volume $\mathrm{vol}_{k-1} A \cap (\bigcup_{s^1} \Pi_{s^1})$ can be made finite by a convenient choice of $\bar{a}^1$. On the other hand, in general, we cannot make the value $\mathrm{vol}_{k-2}[A\backslash(A^* \cup A_c^{k-1}, \varepsilon')] \cap (\bigcup_{s^1} \Pi_{s^1})$ finite by the same procedure. Clearly, the diffeomorphism can be regarded as the restriction of that of the whole plane $\Pi_{s^1}$ replaced by a diffeomorphic submanifold $\tilde{\Pi}_{s^1} \subset \mathbf{R}^n$, with $\tilde{\Pi}_{s^1} \cap (A^* \cup A_c^{k-1}) = N_{s^1}$. The compact set $\tilde{\Pi}_{s^1} \cap [A\backslash(A^* \cup A_c^{k-1}, \varepsilon')]$ consists of simplicial points, and its dimension is not greater than $k-3$, $\tilde{\Pi}_{s^1}$ placed arbitrarily close to the plane $\Pi_{s^1}$. Put

$$\tilde{D}_{s^1} = \tilde{\Pi}_{s^1} \cap A,\ \tilde{D}_{s^1} = (\tilde{D}_{s^1})^* \cup (\tilde{D}_{s^1})_c^{k-2} \cup \ldots \cup (\tilde{D}_{s^1})_c^0,\ (\tilde{D}_{s^1})^* = \tilde{\Pi}_{s^1} \cap A^*,\ (\tilde{D}_{s^1})_c = \tilde{\Pi}_{s^1} \cap A_c,\ (\tilde{D}_{s^1})_c^{k-2} = \tilde{\Pi}_{s^1} \cap (A^* \cup A_c^{k-1}) = \Pi_{s^1} \cap (A^* \cup A_c^{k-1}).$$

It is obvious that $\mathrm{vol}_{k-2}(\tilde{D}_{s^1}) \leqslant \mathrm{vol}_{k-2}(D_{s^1}) = l_{s^1}^{k-2}$, $\mathrm{vol}_{k-2}(\tilde{D}_{s^1}) = \tilde{l}_{s^1}^{k-2}$ possibly being equal to zero, though $l_{s^1} \neq 0$. Thus, $\tilde{D}_{s^1} \subset \tilde{\Pi}_{s^1}, \mathrm{vol}_{k-2}(\tilde{D}_{s^1}) = \tilde{l}_{s^1}^{k-2} < \infty$, $k \geqslant 3$, $\tilde{D}_{s^1} = (\tilde{D}_{s^1})^* \cup (\tilde{D}_{s^1})_c$, which, in general, does not coincide with $\tilde{D}_{s^1}^* \cup \tilde{D}_{s^1,c}$. Since $\tilde{\Pi}_{s^1}$ is diffeomorphic to $\Pi_{s^1}$, we can apply the induction hypothesis to $k' = k-1$. Then there exists a compact set $B_{s^1} \supset \tilde{D}_{s^1}, B_{s^1} \subset \tilde{\Pi}_{s^1}$ such that $B_{s^1}$ is contained in the convex geodesic hull $\tilde{D}_{s^1}$ in $\tilde{\Pi}_{s^1}, \nabla^{m-1}(B_{s^1}, \tilde{D}_{s^1}) = \tilde{h}^{m-2}(\tilde{D}_{s^1})\backslash 0$ for any $m \in \mathbf{Z}$ (resp. $\Delta_{m-1}(B_{s^1}, \tilde{D}_{s^1}) = \tilde{h}_{m-2}(\tilde{D}_{s^1})), x_{s^1} \in \tilde{D}_{s^1}$; besides, there exist constants $C = C(k-1)$ and $D = D(k-1)$ for which $\rho(B_{s^1}^*, \tilde{D}_{s^1})^* \leqslant C\tilde{l}_{s^1}, \rho(B_{s^1}^* \cup B_{s^1,c}^{k-1}, (\tilde{D}_{s^1})^* \bigcup (\tilde{D}^{s^1})_c^{k-2}) \leqslant C\tilde{l}_{s^1}$ if $\tilde{l}_{s^1} > 0$, whereas if $l_{s^1} = 0$, then we have to substitute $\tilde{\delta}_{s^1}$ in both inequalities for $\tilde{l}_{s^1}$, i.e., an arbitrarily small number. Finally, $\mathrm{vol}_{k-1}(B_{s^1}) = \mathrm{vol}_{k-1}(B_{s^1}^* \cup B_{s^1,c}^{k-1}) \leqslant D\tilde{l}_{s^1}^{k-1}$ (or $\leqslant D\tilde{\delta}_{s^1}^{k-1}$ if $\tilde{l}_{s^1} = 0$).

Consider a compact set $A_{s^1} \subset A$ consisting of all points of the compact set $A$, placed between the two "hyperplanes" $\tilde{\Pi}_{s^1-1}$ and $\tilde{\Pi}_{s^1}$. Then $A = \bigcup_{s^1} \dot{A}_{s^1}$. Put $\dot{C} = A \cup (\bigcup_{s^1} B_{s^1}), \dot{C}_{s^1} = B_{s^1-1} \cup \dot{A}_{s^1} \cup B_{s^1}$. Then

$$\begin{aligned}\sum_{s^1} \mathrm{vol}_{k-1}(B_{s^1}) &\leqslant D(k-1) \sum_{s^1} (\dot{\tilde{l}}_{s^1})^{k-1} = D(k-1) \cdot \sum_{s^1} (\mathrm{vol}_{k-2}(\tilde{D}_{s^1}))^{\frac{k-1}{k-2}} \\ &\leqslant D(k-1) \sum_{s^1} (\mathrm{vol}_{k-2}(D_{s^1}))^{\frac{k-1}{k-2}} \leqslant D(k-1) \Big(\sum_{s^1} \mathrm{vol}_{k-2}(D_{s^1})\Big)^{\frac{k-1}{k-2}} \\ &= D(k-1) \cdot (\varphi^1(\bar{a}^1))^{\frac{k-1}{k-2}} \leqslant D(k-1)\, l^{k-1}.\end{aligned}$$

Note that the symbols $\dot{\tilde{l}}_{s^1}$ in the sum $\sum_{s^1}$ denote either $\tilde{l}_{s^1}$ when $\tilde{l}_{s^1} > 0$ or $\tilde{\delta}_{s^1}$ when $\tilde{l}_{s^1} = 0$. Since $\mathrm{vol}_{k-1}(\dot{C}) \leqslant \mathrm{vol}_{k-1}(A) + \mathrm{vol}_{k-1}(\bigcup_{s^1} B_{s^1})$, we have $\mathrm{vol}_{k-1}(\dot{C}) \leqslant l^{k-1} + D(k-1)l^{k-1} = (1 + D(k-1)l^{k-1}$, i.e., gluing the bridges $B_{s^1}$ to $A$ alters its volume $\mathrm{vol}_{k-1}$ only be multiplying by the same constant $1 + D(k-1)$ not depending on the original compact sets.

Consider two points $x_{s^1}$ and $x_{s^1-1}, x_{s^1} \in \tilde{\Pi}_{s^1}, x_{s^1-1} \in \tilde{\Pi}_{s^1-1}$, and join them by a segment $\Delta_{s^1}$ between the "planes" $\tilde{\Pi}_{s^1}$ and $\tilde{\Pi}_{s^1-1}$. Further, we put $C_{s^1} = \dot{C}_{s^1} \cup \Delta_{s^1}, A_{s^1} = \dot{A}_{s^1} \cup \Delta_{s^1}$. Then $x_{s^1}$ and $x_{s^1-1}$ belong to the same path component of the compact set $A_{s^1}$, which is what we shall soon make use of.

Now, we consider the coordinate $x^2$-axis in $\mathbf{R}^n$, and by replacing $A$ by $C_{s^1}$ everywhere in the above argument, we repeat the whole construction along the $x^2$-axis. We have

$$C_{s^1} = C^*_{s^1} \cup C_{s^1,c}, \ \mathrm{vol}_{k-1}(C_{s^1})$$

$$\leqslant \mathrm{vol}_{k-1}(C) \leqslant (1 + D(k-1))\, l^{k-1} < \infty, \qquad C_{s^1,c} = C^{k-1}_{s^1,c} \cup \ldots \cup C^0_{s^1,c}.$$

The compact set $C$ is obtained from $A$ by gluing the bridges $B_{s^1}$; therefore, it is important to estimate how the scattering of the "bad" points $C^*_{s^1}$ increased relative to the "bad" points of $A^*$. It is obvious that

$$\begin{aligned}\rho(C^*_{s^1}, A^*_{s^1}) &\leqslant \rho(B^*_{s^1-1} \cup A^*_{s^1} \cup B^*_{s^1}, A^*_{s^1}) \\ &\leqslant \max\,[\rho(B^*_{s^1-1}, A^*_{s^1}), \rho(B^*_{s^1}, A^*_{s^1})] \\ &\leqslant \max\,[C(k-1)\, l_{s^1-1}, C(k-1)\, l_{s^1}] \leqslant C(k-1)\, l,\end{aligned}$$

i.e., the increase of "bad" points occurs only in the neighbourhood of the radius $C(k-1)l$ of the "bad" points of $A_{s^1}$. We also have to estimate another distance

$$\begin{aligned}\rho\big(C^*_{s^1} \cup C^{k-1}_{s^1,c}, A^*_{s^1} \cup A^{k-1}_{s^1,c}\big) &\leqslant \max\big\{\rho\,[B^*_{s^1} \cup B^{k-1}_{s^1,c}, (\tilde{D}_{s^1})^* \cup (\tilde{D}_{s^1})^{k-2}_c], \\ &\qquad \rho\,[B^*_{s^1-1} \cup B^*_{s^1-1,c}, (\tilde{D}_{s^1-1})^* \cup (\tilde{D}_{s^1-1})^{k-2}_c]\big\} \\ &\leqslant \max\,(C(k-1)\,\tilde{l}_{s^1}, C(k-1)\,\tilde{l}_{s^1-1}) \leqslant C(k-1)\, l.\end{aligned}$$

Note that we have made much use of the induction hypothesis for dimension $k' = k - 1$ in both estimates. The principal conclusion to be made from the above computations is in the scattering of "bad" points occurring only in the neighbourhood of $A$ of radius $Ml$ when we pass from $A$ to $C_{s^1}$, the constant $M$ not depending on $A$, and determined from the induction hypothesis. As in the first step, we construct the function

$$\varphi^{12}(a^{12}_{s^1}) = \sum_{s^2=-\infty}^{+\infty} \mathrm{vol}_{k-2}(C_{s^1} \cap \Pi_{a^{12}+ls^2}).$$

Then

$$\int_0^l \varphi^{12}(a^{12}_{s^1})\, da^{12}_{s^1} \leqslant \mathrm{vol}_{k-1}(C_{s^1}) \leqslant (1 + D(k-1))\, l^{k-1} < \infty,$$

i.e., there exists a value $\bar{a}^{12}_{s^1}, 0 \leqslant \bar{a}^{12}_{s^1} \leqslant l$ such that $\varphi^{12}(\bar{a}^{12}_{s^1}) \leqslant (1+D(k-1))l^{k-2}$

(or $\leqslant T \cdot \delta^{k-2}$ if $\mathrm{vol}_{k-1}(C_{s^1}) = 0$). Consider the planes $\Pi_{s^1 s^2} = \Pi_{\bar{a}^{12}_{s^1} + ls^2}$, and put $D_{s^1 s^2} = C_{s^1} \cap \Pi_{s^1 s^2}$. Again by applying a sufficiently small deformation and making use of the induction, we obtain the compact sets $\tilde{D}_{s^1 s^2}, B_{s^1 s^2} \subset \tilde{\Pi}_{s^1 s^2}$, satisfying all the boundary conditions where $B_{s^1 s^2}$ completely spans $\tilde{D}_{s^1 s^2}$ in all dimensions. Further, consider part $C_{s^1}$ placed between $\tilde{\Pi}_{s^1, s^2 - 1}$ and $\tilde{\Pi}_{s^1, s^2}$. Denoting it by $\dot{A}_{s^1 s^2}$, we put $A_{s^1 s^2} = \dot{A}_{s^1 s^2} \cup \Delta_{s^1 s^2}$, where the segment $\Delta_{s^1 s^2}$ joins the points $x_{s^1, s^2 - 1}$ and $x_{s^1 s^2}$ that belong to the same path component of the compact set $A_{s^1 s^2}$. Finally, we set $C_{s^1 s^2} = B_{s^1, s^2 - 1} \cup A_{s^1 s^2} \cup B_{s^1 s^2}$. It is obvious that all estimates of "bad" point scattering are found exactly as in the first step.

Continuing the process along the axes $x^3, \ldots, x^n$, we break $\mathbf{R}^n$ into "cubes" with side 1 if $l > 0$, and side $\delta > 0$ if $l = 0$, there existing a sufficiently small diffeomorphism close to the identity for each of them onto the usual cube determined by $\bar{a}^1 + l(s^1 - 1) \leqslant x^1 \leqslant \bar{a}^1 + ls^1, \bar{a}^{12}_{s^1} + l(s^2 - 1) \leqslant x^2 \leqslant \bar{a}^{12}_{s^1} + l \cdot s^2, \ldots$ in $\mathbf{R}^n$. Each of the $R_{s^1, \ldots, s^n}$ contains a compact set $C_{s^1 \ldots s^n}$, with $A \subset \bigcup_{s^1, \ldots, s^n} C_{s^1 \ldots s^n} = C$, and $\rho(C^*, A^*) \leqslant C'(k)l, \rho(C^* \cup C_c^{k-1}, A^* \cup A_c^{k-1}) \leqslant C'(k)l, \mathrm{vol}_{k-1}(C) = \mathrm{vol}_{k-1}(C^* \cup C_c^{k-1}) \leqslant D'(k)l^{k-1}$ holding, where the constants $C'(k)$ and $D'(k)$ are obtained by the inductive process along the axes and do not depend on the compact set $A$. Our goal is to completely span the compact set $C$ with a film $X$ satisfying all the conditions of the theorem; then we will automatically span $A$. To start spanning $C$, recall that $C_{s^1 \ldots s^n} = B^1_{s^1 \ldots s^{n-1}, s^n - 1} \cup A_{s^1 \ldots s^n} \cup B_{s^1 \ldots s^n}$, where $B_{s^1 \ldots s^n} \supset \tilde{D}_{s^1 \ldots s^n} \supset x_{s^1 \ldots s^n}$, and $B_{s^1 \ldots s^n}$ completely spans $\tilde{D}_{s^1 \ldots s^n}$ relative to the point $x_{s^1 \ldots x^n}$. Two points $x_{s^1 \ldots s^{n-1}, s^n - 1}$ and $x_{s^1 \ldots s^n}$ are joined by a segment $\Delta_{s^1 \ldots s^n}$ and, therefore, belong to the same path component of the compact set $A_{s^1 \ldots x^n}$, whereas $C$ is contained in an $\varepsilon$-neighbourhood of the convex hull of $A$, because so do all the bridges $B_{s^1 \ldots s^p}$ already glued (see Fig. 72).

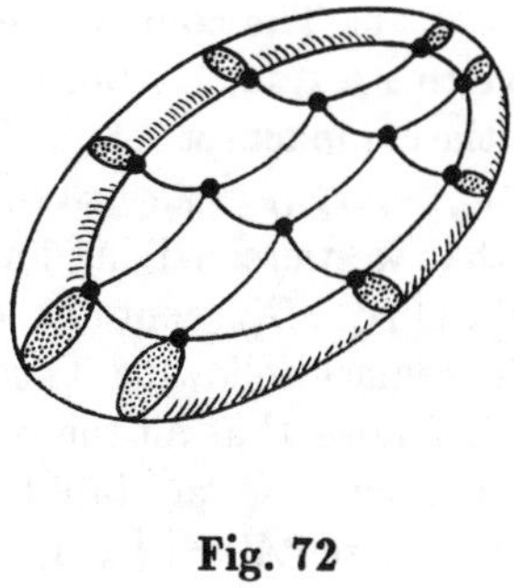

**Fig. 72**

**Remark.** It is obvious that the one-dimensional simplicial subcomplex $\Delta = \bigcup_{1 \leqslant p \leqslant n} \bigcup_{(s^1 \ldots s^p)} \Delta_{s^1 \ldots s^p}$ is contractible on itself to a point, which permits us to replace the base points in the (co)homology groups. For each compact set

$C_{s^1\dots s^n} \subset R_{s^1\dots s^n}$, we select a point $P_{s^1\dots s^n}$ in the convex hull of $C_{s^1\dots s^n}$ and consider the cones $X_{s^1\dots s^n} = C(P_{s^1\dots s^n}, C_{s^1\dots s^n})$. Then

$$\nabla^m(X_{s^1\dots s^n}, C_{s^1\dots s^n}) = \tilde{h}^{m-1}(C_{s^1\dots s^n})\setminus 0, \ x_{s^1\dots s^n} \in \tilde{D}_{s^1\dots s^n}$$

(resp. $\Delta_m(X_{s^1\dots s^n}, C_{s^1\dots s^n}) = \tilde{h}_{m-2}(C_{s^1\dots s^n})$). It is clear that $C_{s^1\dots s^n} = \bigcup_{s^n} A_{s^1\dots s^{n-1},s^n}$ (in particular, $A \cup (\bigcup_{s^1} \Delta_{s^1}) = \bigcup_{s^1} A_{s^1} = \bigcup_{s^1}(\dot{A}_{s^1} \cup \Delta_{s^1})$), $A_{s^1\dots s^{n-1},s^n} \cap A_{s^1\dots s^{n-1},s^n+1} = \tilde{D}_{s^1\dots s^{n-1},s^n}; A_{s^1\dots s^{n-1},s^n} \cap A_{s^1\dots s^{n-1},\gamma^n} = \emptyset$ if $|s^n - \gamma^n| > 1$;

$$\nabla^{m-1}(B_{s^1\dots s^{n-1},s^n}, D_{s^1\dots s^{n-1},s^n}) = \tilde{h}^{m-2}(\tilde{D}_{s^1\dots s^{n-1},s^n})\setminus 0$$

(similarly in the cohomology case),

$$C_{s^1\dots s^{n-1},s^n} = B_{s^1\dots s^{n-1},s^n-1} \cup A_{s^1\dots s^{n-1},s^n} \cup B_{s^1\dots s^{n-1},s^n},$$

and the whole picture occurring in the cylinder $(s^1,\dots,s^{n-1})$ along the $x^n$-axis. Consider all the points $x_{s^1\dots s^{n-1},s^n} \in \tilde{D}_{s^1\dots s^{n-1},s^n}$ with the variable $s^n$, and a point $x_{s^1\dots s^{n-1}} \in \tilde{D}_{s^1\dots s^{n-1}}$ possibly not belonging to $\cup\tilde{D}_{s^1\dots s^{n-1},s^n}$. Since all of them are in $\Delta$, then $x_{s^1\dots s^{n-1},s^n}$ can be joined by $\gamma_{s^1\dots s^{n-1},s^n}$ to $x_{s^1\dot{s}^{n-1}}$, where the paths $\gamma_{s^1\dots s^{n-1},s^n}$ are contained in $\Delta$ and the cylinder $(s^1\dots s^{n-1})$.

Put $A'_{s^n} = A_{s^1\dots s^{n-1},s^n} \cup x_{s^1\dots s^{n-1}}$, $A' = \bigcup_{s^n} A'_{s^n}$, $D'_{s^n} = \tilde{D}_{s^1\dots s^{n-1},s^n} \cup x_{s^1\dots s^{n-1}}, B'_{s^n} = B_{s^1\dots s^{n-1},s^n} \cup \gamma_{s^1\dots s^{n-1},s^n}$. It is obvious that $D'_{s^n} = A'_{s^n} \cap A'_{s^n+1}, A'_{s^1} \cap A'_{\gamma^n} = x_{s^1\dots s^{n-1}}$ if $|s^n - \gamma^n| > 1$. We would like to span the compact set $C'_{s^n} = B'_{s^n-1} \cup A'_{s^n} \cup B'_{s^n} = C_{s^1\dots s^{n-1},s^n} \cup \gamma_{s^1\dots s^{n-1},s^n} \cup \gamma_{s^1\dots s^{n-1},s^n-1}$ relative to $x_{s^1\dots s^{n-1}}$. The compact set $X_{s^1\dots s^{n-1},s^n}$ has wholly spanned $C_{s^1\dots s^{n-1},s^n}$ relative to $x_{s^1\dots s^{n-1},s^n}$. Since $\Delta$ is contractible on itself to a point, it is obvious that the compact set $X'_{s^n} = X_{s^1\dots s^{n-1},s^n} \cup \gamma_{s^1\dots s^{n-1},s^n} \cup \gamma_{s^1\dots s^{n-1},s^n-1}$ wholly spans $C_{s^1\dots s^{n-1},s^n} \cup \gamma_{s^1\dots s^{n-1},s^n} \cup \gamma_{s^1\dots s^{n-1},s^n-1}$ relative to $x_{s^1\dots s^{n-1}}$, which is just what we were striving for. Thus, $C' = C_{s^1\dots s^{n-1}} = \bigcup_{s^n} C'_{s^n}, C'_{s^n} = B'_{s^n-1} \cup A'_{s^n} \cup B'_{s^n}, B'_{s^n}$ contains $D'_{s^n}$ and, obviously, completely spans it (e.g., see the remark following Theorem 27.3.1). $X'_{s^n} \supset C'_{s^n}$, and completely spans $C'_{s^n}$. It follows that all the conditions of Theorem 25.2 and Theorem 26.2 in the homology case, are fulfilled; therefore, the compact set completely spans the compact set $A' = \bigcup_{s^n} A'_{s^n}$ in all dimensions relative to the point $x_{s^1\dots s^{n-1}}$, which means that $X' = X_{s^1\dots s^{n-1}}$ completely spans $A' = C_{s^1\dots s^{n-1}}$ in each cylinder $(s^1\dots s^{n-1})$. Fix $(s^1\dots s^{n-2})$.

Then $C_{s^1\dots s^{n-2}} = \bigcup_{s^{n-1}} A_{s^1\dots s^{n-2},s^{n-1}}$, and we again come back to a situation analogous to the one just investigated. Continuing the process, we finally obtain a compact set $X$ completely spanning the compact set $\bigcup_{s^1} A_{s^1} = A\cup\Delta$, and *a fortiori* spanning $A$, with all the (co)homology groups being now centered at $x \in A$.

Thus, we have constructed a certain surface $X$ completely spanning $A$ and contained in an $\varepsilon$-neighbourhood of the convex hull of $A$ in $\mathbf{R}^n$. To verify that the metric requirements are satisfied, we suppose that $P \in C$ is an $s$-simplicial point. Then the whole segment $C(Q,P)$ embedded into the cone $C(Q,C)$ consists of points which are either all $s$-simplicial or all $(s+1)$-simplicial; in other words, the cone operation transforms simplicial points into simplicial ones again. Since $\rho(C^*,A^*) \leqslant C'(k)l$, $\rho(C^* \cup C_c^{k-1}, A^* \cup A_c^{k-1}) \leqslant C'(k)l$, we have $\rho(X^*,A^*) \leqslant (1+C'(k))l$, $\rho(X^* \cup X_c^k, A^* \cup A_c^{k-1}) \leqslant (1+C'(k))l$ (recalling that $\rho(X^*,C^*) \leqslant l$, and $\rho(X^* \cup X_c^k, C^* \cup C_c^{k-1}) \leqslant l$, since the length of the generator of each cone $X_{s^1\dots s^n}$ has not exceeded $l$); therefore, we can put $C(k) = 1 + C'(k)$. In constructing $C = \bigcup C_{s^1\dots s^n}$, it was proved that $\mathrm{vol}_{k-1}(C) \leqslant \sum_{s^1\dots s^n} \mathrm{vol}_{k-1}(C_{s^1\dots s^n}) \leqslant D'(k)l^{k-1}$. Hence, it suffices to verify the inequalities for volumes $\mathrm{vol}_k$ for each cube $R_{s^1\dots s^n}$. If $Z \subset \mathbf{R}^n$, $\mathrm{vol}_{k-1}(Z) < \infty$, and $CZ$ is a cone over $Z$, whose generator length is not greater than $l$, then $\mathrm{vol}_k(CZ) \leqslant lk^{-1}\,\mathrm{vol}_{k-1}(Z)$ (e.g., see the proof in [289], Theorem 10.2.1). It follows that $\mathrm{vol}_k(X) \leqslant \sum_{s^1\dots s^n} lk^{-1}\,\mathrm{vol}_{k-1}(C_{s^1\dots s^n}) \leqslant k^{-1}D'(k)l^k$; therefore, we can put $D(k) = k^{-1}D'(k)$, and (a) has been completed. Case (b) is investigated similarly, only replacing $l$ by $\delta$, and the Cap Theorem has been completely proved.

If $l$ is sufficiently small, then the $k$-dimensional part of the compact set $X$ is concentrated in a small neighbourhood of the compact set $A^* \cup A_c^{k-1}$. We shall often make use of this circumstance.

## §29 The Minimizing Process in Variational Classes $h(A,L,L')$ and $h(A,\tilde{L})$

### 29.1. The Minimizing Sequence of Surfaces. Density Functions Related to Surfaces

We now concentrate our attention on the classes $\mathcal{O}$, since the investigation of each of $\tilde{\mathcal{O}}$ is similar. Here, we construct the minimizing process called the M-process, which will permit us to construct the surface in $\mathcal{O}$ from each minimizing sequence, which will turn out to be globally minimal. Assume that $\mathcal{O}$ is nonempty and that there exists an integer $s$, $3 \leqslant s \leqslant n-1$, such that $d_s = \inf \mathrm{vol}_s(X\backslash A)$, $x \in \mathcal{O}$, is finite. If there is no such $s$, or if $\mathrm{vol}_s(X\backslash A) = \infty$ for any $s$ and any $x \in \mathcal{O}$, then it is impossible to pose a nontrivial variational problem. Below, we will denote by $k$ the least of all such integers $s$ that $d_s < \infty$.

Since $d_k = \inf \mathrm{vol}_k(X\backslash A), X \in \mathcal{O}$, there exists an infinite sequence of compact sets $X_n^{(1)}, n = 1,2,3,\dots$, in the class $\mathcal{O}$, so that $\mathrm{vol}_k(X_n^{(1)}\backslash A) = d_k + \varepsilon_n$, where $\varepsilon_n \geqslant 0, \varepsilon_n \to 0, n \to \infty$. Put $R^0(P) = \min[d(P,A), R_0]$, where $P \in M$, and $d(P,A)$ is the distance from a point $P$ to a compact set $A$. Let $X \subset M$ be a compact set not necessarily belonging to the topological class $\mathcal{O}$ and such that $\mathrm{vol}_k(X\backslash A) < \infty$. We define the functions

$$\varphi_m(r, P, X) = \int_0^r \mathrm{vol}_{m-1}[X \cap \partial B(P, t)]\, dt,$$
$$\psi_m(r, P, X) = \mathrm{vol}_m[X \cap B(P, r)],$$

where $R^0(P) > 0$, $0 < r < R^0(P)$. Fix a countable and everywhere dense subset $\{Q_i\}$ on the manifold $M$. Let an infinite sequence of compact sets $X_\alpha, X_\alpha \supset A, \alpha = 1, 2, \ldots$, be given in $M$, so that $\mathrm{vol}_m(X_\alpha \backslash A) < \infty$ for each $\alpha$. Then we can select (in general, nonuniquely) a subsequence $X_{\alpha_j}$, $j = 1, 2, 3 \ldots$, such that the functions $\psi_m(r, Q_i, X_{\alpha_j})$ converge, as $j \to \infty$ for each $i$ and $r$, $0 < r < R^0(Q_i)$, to certain functions $\tilde{\psi}_m(r, Q_i)$. We construct the functions

$$\psi_m^+(r, Q) = \lim_{\delta \to 0} \left[ \sup_{\substack{|r - r'| < \delta \\ d(Q, Q_i) < \delta}} \tilde{\psi}_m(r', Q_i) \right],$$
$$\psi_m^-(r, Q) = \lim_{\delta \to 0} \left[ \inf_{\substack{|r - r'| < \delta \\ d(Q, Q_i) < \delta}} \psi_m(r', Q_i) \right],$$
$$\varphi_m^+(r, Q) = \lim_{j \to \infty} \left[ \sup_{p \geqslant j} \varphi_m(r, Q, X_{\alpha_p}) \right],$$
$$\varphi_m^-(r, Q) = \lim_{j \to \infty} \left[ \inf_{p \geqslant j} \varphi_m(r, Q, X_{\alpha_p}) \right].$$

from the sequence.

Denote the volume of the unit $m$-dimensional ball $D^m$ by $\gamma_m$.

Select and fix a positive number $h_m$ so that $1 + h_m \cdot r \geqslant (1 + \eta_0 r)^{2m-1}$ for $0 \leqslant r \leqslant R_0$; $h_m$ is not defined uniquely; however, this does not affect further construction. We now define the smooth function

$$h_m(r) = \gamma_m r^m (1 + h_m r)^{-m}, \qquad 0 \leqslant r \leqslant R_0.$$

**Definition 29.1.1.** Let $R^0(P) > 0$, i.e., $P \in M \backslash A$. The function

$$\Psi_m(P) = \lim_{\rho \to 0} \sup_{0 < r < \rho} [h_m^{-1}(r) \cdot \psi_m^+(r, P)]$$

is called the *m-dimensional density function* of the sequence of surfaces $X_\alpha$, $\alpha = 1, 2, \ldots$, on a manifold $M$.

In general, $\Psi_m(P)$ is not unique, and we can construct a number of density functions from the same sequence of $X_\alpha$; however, if $X_\alpha \equiv X$ for all $\alpha$, then $\Psi_m(P)$ is unique on $M$ (in particular, $\Psi_m(P) = 0$ for $P \notin X$).

### 29.2. A Rough Outline of the Minimizing Process

Consider a $k$-minimizing sequence $X_n^{(1)}$, $d_k < \infty$, in the topological variational class $\mathcal{O}$, and assume that $\mathcal{O}$ is 1-stable, which is the only condition permitting us to go on with the M-process in $\mathcal{O}$. Recall that if a manifold $M$ is simply-connected, then any nonempty topological class $\mathcal{O}$ is 1-stable (see §27.4).

Construct the density function $\Psi_k(P)$ from the sequence $X_n^{(1)}$, and put $S^k = \{P \in M\backslash A \mid \Psi_k(P) > 0\}$. As we prove below, the following is then true.

**Proposition 29.2.1.** *The set* $\widetilde{X}^k = A \cup S^k$ *is a compact set in the manifold* $M$. *There exists a new sequence* $X_n^{(1)'} \in \mathcal{O}$ *such that (1)* $\mathrm{vol}_k[X_n^{(1)'}\backslash A] \to d_k, n \to \infty$ *and, (2) for any open neighbourhood* $U$ *of* $A \cup S^k = \widetilde{X}^k$, *there exists a subscript* $N = N(U)$ *such that* $\mathrm{vol}_{k-1}[X_n^{(1)'}\backslash\overline{U}] < \infty$ *for all* $n > N(U)$.

In particular, since this means that $\mathrm{vol}_k[X^{(1)'}\backslash\overline{U}] = 0$ for $n > N$, $k$-dimensional portions of the new compact sets $X_n^{(1)'}$ are concentrated in an arbitrarily small neighbourhood of the compact set $\widetilde{X}^k$. We now consider open neighbourhoods $U_\alpha^1$ of $X^k$ so that $U_\alpha^1 \supset U_{\alpha+1}^1$, and $\bigcap_\alpha U_\alpha^1 = \widetilde{X}^k$. Put $\omega_\alpha^{k-1} = \inf \mathrm{vol}_{k-1}(X\backslash\overline{U}_\alpha^1)$, where $X \in \mathcal{O}$. By Proposition 29.2.1, we have $\omega_\alpha^{k-1} < \infty$ (see (2)). Since $\omega_{\alpha+1}^{k-1} \geqslant \omega_\alpha^{k-1}$, there exists a $\lim_{\alpha\to\infty} \omega_\alpha^{k-1} = \lambda_{k-1}$ (possibly, infinity). Note that $d_k \geqslant \lambda_k$, where $\lambda_k$ is determined similarly, i.e., $\lambda_k = \lim_{\alpha\to\infty}[\inf \mathrm{vol}_k(X\backslash\overline{U}_\alpha^0)]$, $X \in \mathcal{O}, U_{\alpha+1}^0 \subset U_\alpha^1, \bigcap_\alpha U_\alpha^0 = S^k$. It is easy to see that the values $\lambda_k, \lambda_{k-1}$ do not depend on the choice of neighbourhoods $U_\alpha^0$ and $U_\alpha^1$, respectively.

Since $\lambda_{k-1} = \lim_{\alpha\to\infty} \omega_\alpha^{k-1}$, there exists an infinite sequence $X_n^{(2)} \in \mathcal{O}$ of the compact sets $X_n^{(2)}$ such that $\mathrm{vol}_{k-1}[X_n^{(2)}\backslash\overline{U}_n^1] = \omega_n^{k-1} + \varepsilon_n, \varepsilon_n \geqslant 0, \varepsilon_n \to 0, n \to \infty$. Similarly to the first step, we construct the $(k-1)$-dimensional density function $\Psi_{k-1}(P), P \in M\backslash\widetilde{X}^k$ from it, and put

$$S^{k-1} = \{P \in M\backslash\widetilde{X}^k \mid \Psi_{k-1}(P) > 0\}.$$

**Proposition 29.2.2.** *The set* $\widetilde{X}^{k-1} = A \cup S^k \cup S^{k-1}$ *is a compact set in* $M$. *There exists a new sequence* $X_n^{(2)'} \in \mathcal{O}$ *such that (1)* $\mathrm{vol}_{k-1}[X_n^{(2)'}\backslash\overline{U}_n^1] = \omega_n^{k-1} + \varepsilon_n', \varepsilon_n \geqslant 0, \varepsilon_n' \to 0, n \to \infty$, *and (2) for any open neighbourhood* $U$ *of* $\widetilde{X}^{k-1}$, *there exists a subscript* $N = N(U)$ *such that* $\mathrm{vol}_{k-2}[X_n^{(2)'}\backslash\overline{U}] < \infty$ *for all* $n > N$.

In particular, since it means that $\mathrm{vol}_{k-1}[X_n^{(2)'}\backslash\overline{U}] = 0$, $(k-1)$-dimensional portions of new compact sets $X_n^{(2)'}$ are concentrated in an arbitrarily small neighbourhood of the compact set $\widetilde{X}^{k-1}$. Consider open neighbourhoods $U_\alpha^2$ of $\widetilde{X}^{k-1}$, so that $U_\alpha^2 \supset U_{\alpha+1}^2$, and $\bigcap_\alpha U_\alpha^2 = \widetilde{X}^{k-1}$. Put $\omega_\alpha^{k-2} = \inf \mathrm{vol}_{k-2}(X\backslash\overline{U}_\alpha^2)$, where $X \in \mathcal{O}$. By Proposition 29.2.2, we have $\omega_\alpha^{k-2} < \infty$. Since $\omega_{\alpha+1}^{k-2} \geqslant \omega_\alpha^{k-2}$, there exists $\lim_{\alpha\to\infty} \omega_\alpha^{k-2} = \lambda_{k-2}$ (possibly, infinity), not depending on the choice of the contractible system of neighbourhoods $U_\alpha^2$.

Since $\lambda_{k-2} = \lim_{\alpha\to\infty} \omega_\alpha^{k-2}$, there is an infinite sequence $X_n^{(3)} \in \mathcal{O}$ such that $\mathrm{vol}_{k-2}[X_n^{(3)}\backslash\overline{U}_n^2] = \omega_n^{k-2} + \varepsilon_n$, $\varepsilon_n \to 0, n \to \infty$. Similarly to the first step, we construct the $(k-2)$-dimensional density function $\Psi_{k-2}(P), P \in M\backslash\widetilde{X}^{k-1}$ from it, and put $S^{k-2} = \{P \in M\backslash\widetilde{X}^{k-1} | \Psi_{k-2}(P) > 0\}$, etc.

By continuing the process with respect to decreasing dimension, we obtain, eventually, a sequence of sets $S^k, S^{k-1}, S^{k-2}, \ldots, S^2, S^1$, where $S^i = \{P \in M\backslash A\backslash \bigcup_{j=i+1}^{k} S^j | \Psi_i(P) > 0\}$. It turns out that $S^1$ is empty due to the 1-stability of the class $\mathcal{O}$. At the penultimate step, we have a compact set $\widetilde{X}^4 = A \cup \bigcup_{p \geqslant 4} S^p$, $\bigcap_\alpha U_\alpha^{k-3} = \widetilde{X}^4$, $\omega_\alpha^3 = \inf \mathrm{vol}_3[X\backslash\overline{U}_\alpha^{k-3}]$, where $X \in \mathcal{O}, \omega_\alpha^3 < \infty$, $\lambda_3 = \lim \omega_\alpha^3, \lambda_3 \leqslant \infty$. Consider a sequence $X_n^{(k-2)} \in \mathcal{O}$ such that $\mathrm{vol}_3[X_n^{(k-2)}\backslash\overline{U}_n^{k-3}] = \omega_n^3 + \varepsilon_n, \varepsilon \geqslant 0, \varepsilon \to 0$, construct the function $\Psi_3(P)$ from the sequence $X_n^{(k-2)}$, and then take the set $S^3 = \{P \in M\backslash\widetilde{X}^4 | \Psi_3(P) > 0\}$.

**Proposition 29.2 ($k-2$).** *The set $\widetilde{X}^3 = A \cup \bigcup_{p=3}^{k} S^p$ is a compact set in $M$. There exists a new sequence $X_n^{(k-2)'} \in \mathcal{O}$ such that (1) $\mathrm{vol}_3[X_n^{(k-2)'}\backslash\overline{U}_\alpha^{k-3}] = \omega_n^3 + \varepsilon_n', \varepsilon_n' \geqslant 0, \varepsilon_n' \to 0, n \to \infty$, and (2) for any open neighbourhood $U$ of the compact set $\widetilde{X}^3$, there exists a subscript $N = N(U)$ such that $\mathrm{vol}_2[X_n^{(k-2)'}\backslash\overline{U}] < \infty$ for all $n > N(U)$.*

Consider the open neighbourhoods $U_\alpha^{k-2}$ of $\widetilde{X}^3$, $\bigcap_\alpha U_\alpha^{k-2} = \widetilde{X}^3$. Let $\omega_\alpha^2 = \inf \mathrm{vol}_2(X\backslash\overline{U}_\alpha^{k-2})$, $X \in \mathcal{O}$, $\lambda_2 = \lim_{\alpha\to\infty} \omega_\alpha^2$. Then it follows from the above that $\omega_\alpha^2 < \infty$. There exists a sequence $X_n^{(k-2)} \in \mathcal{O}$ for which

$$\mathrm{vol}_2[X_n^{(k-1)'}\backslash\overline{U}_n^{k-2}] = \omega_n^2 + \varepsilon_n', \varepsilon_n' \geqslant 0.$$

Construct the function $\Psi_2(P)$ and consider $S^2 = \{P \in M\backslash\widetilde{X}^3 | \Psi_2(P) > 0\}$.

**Proposition 29.2 ($k-1$).** *The set $\widetilde{X}^2 = S^2 \cup \widetilde{X}^3$ is a compact set in $M$, with $\widetilde{X}^2 \in \mathcal{O}$. There exists a new sequence $X_N^{(k-1)'} \in \mathcal{O}$ such that (1) $\mathrm{vol}_2[X_n^{(k-1)'}\backslash\overline{U}_n^{k-2}] = \omega_n^2 + \varepsilon_n', \varepsilon_n' \geqslant 0, \varepsilon_n' \to 0$, and (2) for any open neighbourhood $U$ of $\widetilde{X}^2$, there exists a subscript $N = N(U)$ such that $\mathrm{vol}_1[X_n^{(k-1)'}\backslash U] < \infty$; hence, due to the 1-stability of the class $\mathcal{O}$, we have $X_n^{(k-1)'} \subset U$ for all $n > N(U)$.*

Proposition 29.2 ($k-2$) is different from the previous Propositions 29.2(s), where $1 \leqslant s \leqslant k-2$. Namely, $\widetilde{X}^2 \in \mathcal{O}$, which cannot be generally asserted relative to the compact sets $\widetilde{X}^i$, where $3 \leqslant i \leqslant k$; moreover, if $S^2 \neq \emptyset$, then none of the compact sets $\widetilde{X}^i, 3 \leqslant i \leqslant k$ belongs to $\mathcal{O}$.

Thus, the above process takes us to a certain compact set $\widetilde{X}^2 \in \mathcal{O}$.

**Definition 29.2.1.** The above process associating each $k$-minimizing sequence of compact sets $X_n^{(1)} \in \mathcal{O}$, i.e., $\mathrm{vol}_k(X_n^{(1)}\backslash A) \to d_k$ as $n \to \infty$, with a compact set $\widetilde{X}^2 \in \mathcal{O}$, assuming that $d_k < \infty$ and that the class $\mathcal{O}$ is 1-stable, will be termed an *M-process*. We will call $\lambda = (d_k, \lambda_{k-1}, \ldots, \lambda_2)$ the $\lambda$-vector of the given M-process. We will call an M-process finite if $\lambda < \infty, 2 \leqslant i \leqslant k-1$.

An M-process actually exists in any class $\mathcal{O}$ for which $d_k < \infty$, i.e., we may not assume that $\mathcal{O}$ is 1-stable, since one-dimensional volume (length) minimization does not face difficulties. However, since we will assume 2-stability of $\mathcal{O}$ all the same to investigate the metric properties of $\widetilde{X}^2$, we will not dwell on the case of dimension one. If we assume from the beginning that the manifold is simply-connected, then we will guarantee the 1-stability of any nonempty topological variational class.

## 29.3. The Constructive Method for the Minimizing Process and the Proof for Its Convergence. First Step

We now turn to the proof of Propositions 29.2.1–29.2 $(k-1)$, where $k \geqslant 2$. In this connection, we shall need certain analytic results generalizing some constructions from E. Reifenberg [289] and C. Morrey [351].

**Lemma 29.3.1.** *Let* $X$ *be a compact set such that* $A \subset X \subset M$ $\mathrm{vol}_k(X\backslash A) < \infty$. *Then*

*(3.1.1)* $\quad \varphi_k(r_2, P, X) - \varphi_k(r_1, P, X) \leqslant \psi_k(r_2, P, X) - \psi_k(r_1, P, X)$
*if* $0 \leqslant r_1 < r_2 < R^0(P)$ *and*

*(3.1.2)* $\quad \psi_k(r', P', X) \leqslant \psi_k(r' + d(P, P'), P, X)$
*if* $r' < R^0(P')$, $r' + d(P, P') < R^0(P)$.

The proof follows from Theorem 10.2.3 of [289] and $B(P', r') \subset B(P, r' + s(P, P'))$. In the following, we will often differentiate functions of type $\psi_k(r, P, X)$ which are, in general, discontinuous; therefore, we stress that differentiation is understood as that with respect to measure (see the particulars in [289]).

**Lemma 29.3.2.** *Let* $X \in \mathcal{O}$, *and* $\mathrm{vol}_k(X\backslash A) = d_k + \varepsilon < \infty$, *where* $\varepsilon \geqslant 0$. *Let* $0 \leqslant r \leqslant R^0(P)$. *We define the functions*

$$\widehat{\varphi}(r, P, X) = \max[0, \varphi_k(r, P, X) - \varepsilon], \widehat{\psi}_k(r, P, X)$$
$$= \max[0, \psi_k(r, P, X) - \varepsilon], \ \rho_1(P, X) = \ sup(r),$$

*where the values* $r$ *are such that* $\widehat{\varphi}_k(r, P, X) = 0, \rho_1^*(P, X) = \sup(r)$, *where the* $r$ *are such that* $\widehat{\psi}_k(r, P, X) = 0$. *Then, for almost all* $r$ *such that* $0 \leqslant r \leqslant R^0(P)$,

*(3.2.1)*
$$\psi_k(r, P, X) \leqslant \begin{cases} D[\varphi_{k,r}(r, P, X)]^{k/(k-1)} + \varepsilon, \\ k^{-1} r(1 + h_k r)\varphi_{k,r}(r, P, X) + \varepsilon, \end{cases}$$

*where* $D = D(k)$ *is the constant from Theorem 28.3.1, and* $\varphi_{k,r}$ *is the partial derivative of the function* $\varphi_k$ *with respect to the argument* $r$;

*(3.2.2)* $\quad \rho_1^*(P, X) \leqslant \rho_1(P, X)$, $\widehat{\varphi}_{k,r}(r, P, X) \leqslant \widehat{\psi}_{k,r}(r, P, X)$
*for almost all* $r$, $0 \leqslant r < R^0(P)$;

*(3.2.3)* $\widehat{\varphi}_k(r_2, P, X) - \widehat{\varphi}_k(r_1, P, X) \leqslant \widehat{\psi}_k(r_2, P, X) - \widehat{\psi}_k(r, P, X)$
$0 \leqslant r_1 < r_2 < R^0(P)$;

*(3.2.4)* $r^{-k}(1 + h_k r)^k \widehat{\varphi}_k(r, P, X),\ r^{-k}(1 + h_k r)^k \cdot \psi_k(r, P, X)$
*are nondecreasing with respect to r;*

*(3.2.5)* $\widehat{\varphi}_k(r, P, X) \geqslant \kappa_k[r - \rho_1(P, X)]^k$ *if* $\rho_1(P, X) \leqslant r < R^0(P)$,
*where* $\kappa_k = k^{-k} D^{1-k} > 0$,

*(3.2.6)* $\psi_k(r, P, X) \leqslant \varepsilon + k^{-1}(1 - k')^{-1}[1 + \rho_1(P, X)]\varepsilon$
*if* $0 \leqslant r \leqslant k'\rho_1(P, X),\ 0 < k' < 1$.

**Proof.** In order to show (3.2.1), we obviously have $\varphi_{k,r}(r, P, X) = \mathrm{vol}_{k-1}[X \cap \partial B(P, r)]$ for almost all $r < R^0(P)$. To carry out the complete $'S$-surgery of the compact set $X$, we take an open ball $B(P, r)$, with $r < R^0(P)$, as an open set $G$. Then $A \cap \overline{G} = \emptyset$. Recall that $X_1 = (X \cap \overline{G}) \cup x$, $A_1 = (X \cap \partial G) \cup x$, $Y_2 = X \backslash G, Y = Y_1 \cup Y_2$, where $Y_1$ completely spans the compact set $A_1$. We take the compact set $\dot{Y}_1 \cup \gamma$ as $Y$ (see Theorem 27.3.1), where $\gamma$ is the path joining $x$ to $\dot{x} \in \dot{A}_1 = X \cap \partial G$, and $\dot{Y}_1$ is the "cap" whose existence was proved in Theorem 28.3.1. Since we have $\mathrm{vol}_{k-1}(A_1) = \mathrm{vol}_{k-1}(X \cap \partial G) = \mathrm{vol}_{k-1}(X \cap \partial B(P, r)) = \varphi_{k,r}(r, P, X) < \infty$ for almost all $r, 0 \leqslant r \leqslant R^0(P)$ (due to $\mathrm{vol}_k(X) < \infty$), we obtain $\mathrm{vol}_{k-1}(A_1) = l^{k-1} = (\varphi_{k,r})^{(k-1)/(k-1)} < \infty$, or $l = (\varphi_{k,r})^{1/(k-1)}$, in the notation of Theorem 28.3.1. Due to the same theorem, we can assume that $\dot{Y}_1$ is contained in the convex geodesic hull of $A_1$, i.e., the ball $\overline{B(P, r)}$, and also that $\mathrm{vol}_k(Y_1) = \mathrm{vol}_k(\dot{Y}_1) \leqslant Dl^k$. Since $X_1 \cup Y_1$ is contractible on the manifold $M$ to a point, we obtain by Theorem 27.3.1 that $Y = Y_1 \cup Y_2 \in \mathcal{O}$; therefore, $\mathrm{vol}_k(Y \backslash A) \geqslant d_k$. Assume that (3.2.1) does not hold, i.e., $\psi_k(r, P, X) > Dl^k + \varepsilon$, and since $(S \backslash \overline{G}) \cap (S \cap \overline{G}) = \emptyset, S = X \backslash A$, we have

$$\begin{aligned}\mathrm{vol}_k(S) &= \mathrm{vol}_k(S \setminus \bar{G}) + \mathrm{vol}_k(S \cap \bar{G}) > \mathrm{vol}_k(S \setminus \bar{G}) + Dl^k + \varepsilon \\ &\geqslant \mathrm{vol}_k(S \setminus \bar{G}) + \mathrm{vol}_k(Y_1) + \varepsilon \geqslant \mathrm{vol}_k[(S \setminus \bar{G}) \cup Y_1] + \varepsilon = \mathrm{vol}_k(Y) + \varepsilon \\ &\geqslant d_k + \varepsilon\end{aligned}$$

hence, $\mathrm{vol}_k(S) > d_k + \varepsilon$, which is contrary to the choice of $S$. Recall that $\mathrm{vol}_k(S) = d_k + \varepsilon$. Note that we have made use of the finiteness of $d_k$. The second inequality in (3.2.1) is proved similarly, since we can take the cone $\dot{Y}_1 = C(P, A_1)$ as $\dot{Y}_1 \subset \overline{B(P, r)}$. Let $w$ be normal coordinates with domain $B(0, R_0) \subset \mathbf{R}^n$, so that $w(0) = P$; we then put $\dot{A}_{10} = w^{-1}(A_1)$. Since the cone $\dot{Y}_1$ completely spans $A_1$, we again have $Y = Y_1 \cup Y_2 \in \mathcal{O}$ by Theorem 27.3.1, in which case

$$\begin{aligned}\mathrm{vol}_k(Y_1) &= \mathrm{vol}_k(\dot{Y}_1) \leqslant (1 + \eta_0 r)^k \mathrm{vol}_k[C(0, A_{10})] \\ &\leqslant (1 + \eta_0 r)^k k^{-1} r\, \mathrm{vol}_{k-1}(\dot{A}_{10}) \leqslant (1 + \eta_0 r)^{2k-1} k^{-1} r \varphi_{k,r} \\ &\leqslant (1 + h_k r) k^{-1} r \varphi_{k,r},\end{aligned}$$

as follows from Proposition 10.1.13, Theorem 10.2.1, and Lemma 10.2.1 (c) in [289], and also from the determination of the constant $h_k$, thus completing the proof of (3.2.1)

We now turn to the proof of (3.2.2). If we let $r_1$ tend to $r_2$ in (3.1.1), dividing the inequality throughout by $r_2 - r_1$, then we will have $\varphi_{k,r} \leqslant \psi_{k,r}$ for almost all $r$; therefore, $\widehat{\varphi}_{k,r} \leqslant \widehat{\psi}_{k,r}$, which is just what is required. Since $\widehat{\varphi}_k \leqslant \widehat{\psi}_k$, we have $\widehat{\varphi}_k = 0$ if $\widehat{\psi}_k = 0$, i.e., $\rho_1^* \geqslant \rho_1$ (see the definition of $\rho_1$ and $\rho_1^*$), which completes the proof.

Inequality (3.2.3) directly follows from (3.1.1) and the definition of $\widehat{\varphi}_k, \widehat{\psi}_k$. Since $\widehat{\varphi}_k \leqslant \widehat{\psi}_k$, and $\widehat{\psi}_k \geqslant \psi_k$, it follows that

$$\hat{\varphi}_k \leqslant \begin{cases} k^{-1} r(1+h_k r)\hat{\varphi}_{k,r}, \\ D(\hat{\varphi}_{k,r})^{k/(k-1)}. \end{cases} \tag{3.2.7}$$

It is clear that $\widehat{\psi} + \varepsilon \leqslant \mu(\widehat{\varphi}_k + \varepsilon)$, where $\mu = k^{-1}r(1 + h_k r)$; $\widehat{\varphi} \leqslant \widehat{\psi}_k \leqslant \mu\widehat{\varphi}_k + \varepsilon(\mu - 1)$, where $\mu < 1$. The inequality proved holds for almost all $r$.

We now show (3.2.5). It follows from (3.2.7) that $\widehat{\varphi}_k \leqslant D \times (\widehat{\varphi}_{k,r})^{k/(k-1)}$, i.e., $(k\widehat{\varphi}_k^k)'_r \geqslant D^{(1-k)/k}$. Then $\int_0^r (k\widehat{\varphi}_k^k)'_r dr \geqslant D^{(k-1)/k} \times (r - \rho_1)$, for $\int_0^r (k\widehat{\varphi}_k^k)'_r dr \equiv 0, r < \rho_1$. Hence, we find $\int_0^r (k\widehat{\varphi}_k^k)'_r dr = \int_{\rho_1}^r (k\widehat{\varphi}_k^k)'_r dr$, $k\widehat{\varphi}_k^k \geqslant D^{(k-1)/k}(r - \rho_1)$ if $\rho_1(P, X) \leqslant r < R^0(P)$, since $k\widehat{\varphi}_k^k|_{\rho_1} = 0$, and we finally derive that $\widehat{\varphi}_k \geqslant (k^{-k}D^{1-k})(r - \rho_1)^k$, which is just what was required. Recall that $\kappa_k = k^{-k}D^{1-k} > 0$.

To prove (3.2.6), we suppose that $0 < k' < 1$, $0 < r \leqslant k's$, where $s < \rho_1$. Resorting to (3.2.1), we get

$$\begin{aligned} \psi_k(r, P, X) &\leqslant \psi_k(k's, P, X) \\ &\leqslant s^{-1}(1-k')^{-1}\int_{sk'}^{s} k^{-1}t(1+h_k t)\varphi_{k,r}(t, P, X)\,dt + \varepsilon \\ &\leqslant [1 + h_k\rho_1(P, X)]k^{-1}(1-k')^{-1}\varphi_k(s, P, X) + \varepsilon \\ &\leqslant \varepsilon + k^{-1}(1-k')^{-1}(1+\rho_1)\varepsilon. \end{aligned}$$

We now prove (3.2.4). Consider the function $f(r) = r^{-k} \times (1 + h_k r)^k \widehat{\varphi}_k$. It is required to show that $f'_r$, which will imply the monotonicity, since $h_k(r)$ is a smooth and $\widehat{\varphi}_k$ is a continuous function. It is clear that

$$\begin{aligned} f'_r &= (1+h_k r)^{k-1} r^{-k}[(1+h_k r)\hat{\varphi}_{k,r} - kr^{-1}\hat{\varphi}_k] \\ &= (1+h_k r)^{k-1} k r^{-k-1}[k^{-1}r(1+h_k r)\hat{\varphi}_{k,r} - \hat{\varphi}_k] \geqslant 0 \end{aligned}$$

due to 3.2.7, and the monotonicity of $f(r)$ is proved. To prove the second part of (3.2.4), we first establish that $\frac{\partial}{\partial r}[h_k^{-1}(r)\widehat{\psi}_k] \geqslant 0$, or $[h_k^{-1}(r)]'\widehat{\psi}_k + h_k^{-1}(r)\widehat{\psi}'_{k,r} \geqslant 0$, where $h_k^{-1}(r) = \gamma_k^{-1} r^{-k}(1 + h_k r)^k$. It is clear that

$$\begin{aligned} &[h_k^{-1}(r)]'_r \hat{\psi}_k + h_k^{-1}(r)\hat{\psi}_{k,r} \\ &\qquad = \gamma_k^{-1} k(1+h_k r)^{k-1} r^{k-1}[-\hat{\psi}_k + k^{-1}r(1+h_k r)\hat{\psi}_{k,r}] \geqslant 0, \end{aligned}$$

which follows from the inequalities $\widehat{\varphi}_{k,r} \leqslant \widehat{\psi}_{k,r}, \widehat{\psi}_k \leqslant k^{-1}r(1+h_kr)\widehat{\varphi}_{k,r}$, i.e., $\widehat{\psi}_k \leqslant k^{-1}r(1+h_kr)\widehat{\psi}_{k,r}$, $-\widehat{\psi}_k + k^{-1}r(1+h_kr)\widehat{\psi}_{k,r} \geqslant 0$.

Since the function $h_k^{-1}(r)\widehat{\psi}_k$ is, in general, discontinuous, additional argument is required to show its monotonicity. We define a continuous function $\widehat{\chi} = \int_0^r \widehat{\psi}_{k,r}(t,P,X)dt$, and put $\omega = \widehat{\psi}_k - \widehat{\chi}$. Then $\widehat{\psi}_k = \widehat{\chi}+\omega$, i.e., $\widehat{\chi}$ is and $\omega$ are the regular and singular parts of the function $\widehat{\psi}$. Since $\frac{\partial}{\partial r}[h_k^{-1}(r)\widehat{\psi}_k] \geqslant 0$, we have $\frac{\partial}{\partial r}[h_k^{-1}(r)\chi] \geqslant -[-h_k^{-1}(r)]'_r\omega(r)$, because $\omega'_r = 0$ almost everywhere. Note that the function $h_k^{-1}(r)\widehat{\chi}$ is continuous. Integrating the above inequality, we obtain

$$\int_{r_1}^{r_2} \frac{\partial}{\partial r}[h_k^{-1}(r)\hat{\chi}]\,dr = h_k^{-1}(r_2)\hat{\chi}(r_2) - h_k^{-1}(r_1)\hat{\chi}(r_1)$$

$$\geqslant -\int_{r_1}^{r_2} [h_k^{-1}(r)]'_r\,\omega(r)\,dr = -\int_{r_1}^{r_2} \{d[h_k^{-1}(r)\,\omega(r)] - h_k^{-1}(r)\,d\omega(r)\}$$

$$= -h_k^{-1}(r_2)\,\omega(r_2) + h_k^{-1}(r_1)\,\omega(r_1) + \int_{r_1}^{r_2} h_k^{-1}(r)\,d\omega(r).$$

It follows that

$$h_k^{-1}(r_2)\,\hat{\psi}_k(r_2) - h_k^{-1}(r_1)\,\hat{\psi}_k(r_1) \geqslant \int_{r_1}^{r_2} h_k^{-1}(r)\,d\omega(r).$$

Note that $d\omega \neq \omega'_r dr$. We now prove that $d\omega(r) \geqslant 0$.

In fact,

$$\lim_{r'_1\to r'_2}[\omega(r'_2)-\omega(r'_1)] = \lim_{r'_1\to r'_2}[\hat{\psi}_k(r'_2)-\hat{\psi}_k(r'_1)] - \lim_{r'_1\to r'_2}\int_{r'_1}^{r'_2}\hat{\psi}_{k,r}(r)\,dr \geqslant 0,$$

since $\widehat{\psi}_k(r'_2) \geqslant \widehat{\psi}_k(r'_1)$, and the limit of the integral is zero, thus completing the proof of the lemma.

**Lemma 29.3.3.** *Let $X_n^{(1)} \in \mathcal{O}$, $\mathrm{vol}_k[X_n^{(1)}\backslash A] = d_k + \varepsilon < \infty$, $\varepsilon_n \geqslant 0$, $\varepsilon_n \to 0$, $n \to \infty$, and let $\varphi_k^{\pm}, \psi_k^{\pm}$ be the functions constructed from the sequence X (see above). Then:*

*(3.3.1) The functions $\varphi_k^{\pm}$ and $\psi_k^{\pm}$ are nondecreasing with respect to r, and, moreover, $\psi_k^{+}(r,P)$ is upper semicontinuous with respect to r at the point $(r,P)$;*

*(3.3.2)*
$$\psi_k^-(r, P) \leqslant \lim_{n\to\infty} \inf_{p\geqslant n} \psi_k(r, P, X_p^{(1')}) \leqslant \lim_{n\to\infty} \sup_{p\geqslant n} \psi_k(r, P, X_p^{(1')}) \leqslant \psi_k^+(r, P);$$

*(3.3.3) $\psi_k^-(r_2, P) \geqslant \psi_k^+(r_1, P)$ if $0 \leqslant r_1 < r_2 < R^0(P)$. Thus, $\psi_k^+(r,P) = \psi_k^-(r,P)$ if either $\psi_k^+(r,P)$ or $\psi_k^-(r,P)$ is continuous with respect to $r$ at $(r,P)$ (see (3.3.2));*

*(3.3.4) $\varphi_k^- \leqslant \varphi_k^+ \leqslant \psi_k^+$ if $0 \leqslant r < R^0(P)$;*

*(3.3.5) $\psi_k^+(r', P') \leqslant \psi_k^-(r,P)$ if $r' < R^0(P')$ and $r' + d(P,P') < r < R^0(P)$;*

*(3.3.6) At each point $P \in M\backslash A$, the functions $h_k^{-1}(r) \cdot \psi_k^{\pm}(r,P)$ and $h_k^{-1}(r)\varphi_k^{\pm}(r,P)$ are nondecreasing with respect to $r$. Furthermore, there exists $\lim_{r\to 0}[h_k^{-1}(r)\psi_k^+(r,P)]$; in particular,*

$$\Psi_k(P) = \lim_{\varepsilon\to 0} \sup_{0<r<\varepsilon} [h_k^{-1}(r)\psi_k^+(r,P)] = \lim_{\varepsilon\to 0}[h_k^{-1}(\varepsilon)\psi_k^+(\varepsilon,P];$$

*besides, $\Psi_k(P)$ is upper semicontinuous;*

*(3.3.7) There exists a positive number $\beta_k$ such that if $\Psi_k(P) > 0$, then $\Psi_k(P) \geqslant \beta_k > 0$, with the consequence that the set $S^k = \{P \in M\backslash A | \Psi_k(P) > 0\}$ is closed in $M\backslash A, \Psi_k(P) \geqslant \beta_k > 0$ on the whole set $S^k$, and $X^k = A \cup S^k$ is a compact set in $M$. In addition, $\Psi_k(P) < \infty$ for any $P \in M\backslash A$;*

*(3.3.8) $\varphi_k^-(r, P \geqslant \Psi_k(P)h_k(r)$ if $0 \leqslant r < R^0(P)$;*

*(3.3.9) if $\Psi_k(P) = 0$, then $\psi_k^+(r,P) = 0$ for all $r$ such that $0 \leqslant r < \rho(P)$, where $\rho(P)$ is a certain positive number;*

*(3.3.10) Let $P \in S^k, P_i \in S^k, 0 < r < R^0(P_i), \{B(P_i, r_i)\}$ be at most a countable family of disjoint balls such that $B(P_i,r_i) \subset B(P,r)$. Then $\psi_k^{\pm}(r,P) \geqslant \sum_i \psi_k^-(r_i, P_i)$.*

**Proof.** (3.3.1) follows from the fact that sup and inf for a countable family of nondecreasing functions are also nondecreasing functions. The upper semicontinuity is obtained similarly.

We now show (3.3.3). Let $\rho > 0$ be such that $4\rho < r_2 - r_1$. Then

$$\psi_k^-(r_2, P) \geqslant \inf_{r_2 - r' < \rho,\ d(P, Q_i) < \rho} \tilde{\psi}_k(r', Q_i),$$
$$\psi_k^+(r_1, P) \leqslant \sup_{r'' - r_1 < \rho,\ d(P, Q_j) < \rho} \tilde{\psi}_k(r'', Q_j).$$

Further, if $r' > r_2 - \rho$, $d(P,Q_i) < \rho$, $r'' < r_1 + \rho$, $d(P,Q_j) < \rho$, then $d(Q_i,Q_j) < 2\rho$, and $\tilde{\psi}_k(r',Q_i) \geqslant \tilde{\psi}_k[r' - d(Q_i,Q_j), Q_j] \geqslant \psi_k(r_2 - 3\rho, Q_j) \geqslant \tilde{\psi}_k(r'', Q_j)$, where we have used (3.1.2). The required relation is obtained by the passage to the limit. The inequalities (3.3.2) are verified exactly in the same manner.

(3.3.4) and (3.3.5) follow from (3.1.1), (3.1.2), and the passage to the limit, whereas (3.3.6) follows from (3.2.4) and (3.3.1).

To show (3.3.8), it is clear that

$$\varphi_{k,r}(r, P, X_n^{(1)}) \geqslant k \cdot r^{-1} \cdot (1 + h_k \cdot r)^{-1} \cdot [\psi_k(r, P, X_n^{(1)}) - \varepsilon_n],$$

where $\varepsilon_n \geqslant 0, \varepsilon_n \to 0$ (see (3.2.1)). By integrating with respect to $r$ from $r_0$ to $r$, where $0 < r_0 < r$, letting $n$ tend to infinity and $r_0$ to zero, we obtain

$$\varphi_k^- \geqslant \int_0^r kt^{-1}(1 + h_k t)^{-1} \psi_k^-(t, P)\, dt,$$

whence

$$\varphi_k^- \geqslant \Psi_k(P) \int_0^r k\gamma_k t^{k-1}(1 + h_k t)^{-k-1}\, dt$$

$$= \Psi_k(P) \int_0^r [h_k(t)]_t'\, dt = \Psi_k(P)\, h_k(r),$$

which is just what was required. Here, we have made use of relations (3.3.6), (3.3.3), and (3.3.2).

To prove (3.3.7), we consider two cases: (1) $\lim_{n\to\infty} \rho(P, X_n^{(1)}) = 0$ and (2) $\lim_{n\to\infty} \rho(P, X_n^{(1)}) = \rho(P) > 0$, where $n$ ranges over a certain infinite subsequence. In case (1), we derive $\varphi_k^- \geqslant \kappa_k r^k = \beta_k \gamma_k r^k$ for all $r, 0 \leqslant r < R^0(P)$, where $\beta_k = \kappa \gamma_k^{-1} = k^{-k} \gamma_k^{-1} D^{1-k} > 0$, i.e.,

$$\Psi_k(P) \cong \psi_k^+ \gamma_k^{-1} r^{-k}(1 + h_k r)^k \geqslant \varphi_k^- \gamma_k^{-1} r^{-k} \cdot (1 + h_k r)^k \geqslant \beta_k (1 + h_k r)^k,$$

and, eventually, $\Psi_k(P) \geqslant \beta_k > 0$ as $r \to 0$. In case (2), we derive from (3.2.6) that $\psi_k(r, P, X_n^{(1)}) \leqslant \varepsilon_n + k^{-1}(1 - k')^{-1}[1 + \rho_1(P, X_n^{(1)})]\varepsilon_n$, where $0 \leqslant r \leqslant k'\rho_1(P, X_n^{(1)}), 0 < k' < 1$. Letting $n$ tend to infinity, we obtain $\psi_k^+ = 0$ for $0 \leqslant r \leqslant k'\rho(P)$; therefore $\Psi_k(P) = 0$. Thus, we have proved that $\Psi_k(P) > 0$ if and only if $\lim_{n\to\infty} \rho(P, X_n^{(1)}) = 0$, in which case $\Psi_k(P) \geqslant \beta_k > 0$. Further, if $P \notin S^k$, then $\Psi_k(P) = 0 < \infty$; however, if $P \in S^k$, then $\psi_k^+ < \infty$ (see the definition of $\psi_k^+$). Hence, $h_k^{-1}(r)\psi_k^+ < \infty$ for $0 < r)R^0(P)$. Since the function $h_k^{-1}(r)\psi_k^+$ is nondecreasing with respect to $r$, $\Psi_k(P) < \infty$, which is just what is required. Since $M$ is compact and $S^k$ is closed in $M \backslash A$, $A \cup S^k$ is a compact set in $M$.

(3.3.9) now easily follows from the above. Indeed, let $\Psi_k(P) = 0$. Then it can be seen from the proof of (3.3.7) that $\rho(P) = \lim_{n\to\infty} \rho(P, X_n^{(1)}) > 0$

(otherwise, $\Psi_k(P) \geqslant \beta_k > 0$), and then $\psi_k^+ = 0$ for $0 \leqslant r \leqslant k'\rho(P), 0 \leqslant k' < 1$.

It remains to prove (3.3.10). It is clear that

$$\begin{aligned}\psi_k^+ &\geqslant \limsup_{n\to\infty\, p\geqslant n} \psi_k(r, P, X_p^{(1)}) \\ &\geqslant \liminf_{n\to\infty\, p\geqslant n} \mathrm{vol}_k[X_p^{(1)} \cap B(P, r)] \\ &\geqslant \liminf_{n\to\infty\, p\geqslant n} \sum_i \mathrm{vol}_k[X_p^{(1)} \cap B(P_i, r_i)] \\ &\geqslant \sum_i \liminf_{n\to\infty\, p\geqslant n} \mathrm{vol}_k[X_p^{(1)} \cap B(P_i, r_i)] \geqslant \sum_i \psi_k^-(r_i, P_i),\end{aligned}$$

and the lemma is thus proved.

**Lemma 29.3.4 (Surgery Lemma).** *Let* $X_n^{(1)} \in \mathcal{O}$ *be the original minimizing sequence, i.e.,* $\mathrm{vol}_k[X_n^{(1)}\backslash A] = d_k + \varepsilon_n < \infty$, $\varepsilon_n \geqslant 0$, $\varepsilon \to 0, n \to \infty$. *Let* $\widetilde{X}^k = A \cup S^k$, *and* $\varphi_k^{\pm}, \psi_k^{\pm}, \Psi_k$ *be the functions constructed above. Then there exists a new sequence of surfaces* $X_n^{(1)'} \in \mathcal{O}$, *so that* $\mathrm{vol}_k[X_n^{(1)'}\backslash A] = d_k + \varphi_n'$, $\varphi_n' \geqslant 0$, $\varepsilon_n' \to 0$, $n \to \infty$, *or the compact sets* $X_n^{(1)'}$ *still form a k-minimizing sequence, with* $'\psi_k^{\pm} \equiv \psi_k^{\pm}$, $'\Psi_k \equiv \Psi_k$, *and the new functions* $'\varphi_k^{\pm}$, *in general, not coinciding with* $\varphi_k^{\pm}$, *are still nondecreasing with respect to r, but satisfy relations (3.3.1), (3.3.6), (3.3.8). Furthermore, for any open neighbourhood* $U = U(\widetilde{X}^k)$ *of the compact set* $\widetilde{X}^k$, *there exists a subscript* $N = N(U)$ *such that* $\mathrm{vol}_{k-1}[X_n^{(1)}\backslash U] < \infty$ *for all* $n > N$; *moreover,* $X_n^{(1)'}\backslash U$ *is a finite simplicial subcomplex of dimension s, where* $s \leqslant k-1$; *in particular* $X_n^{(1)}\backslash U$ *only consists of simplicial points (see Definition 28.2.1).*

**Remark.** The new functions $'\varphi_k^{\pm}$ will possess the property of $'\varphi_k^{\pm} = 0$ if $\overline{B(P,r)} \cap \widetilde{X}^k = \emptyset$, which will follow from the compact sets $X_n^{(1)'}\backslash U$ being simplicial.

**Proof.** Let $U$ be an open neighbourhood of $\widetilde{X}^k$. Since the manifold $M$ is compact, there exists a finite number of open balls $B(P_i, r_i)$ such that: (1) $M\backslash U \subset \bigcup_i B(P_i, r_i)$ and (2) $\widetilde{X}^k \cap [\bigcup_i \overline{B(P_i, r_i)}] = \emptyset$. Consider the ball $B(P_1, r_1)$. Then either (a) there exists a subscript $N$ such that $X_n^{(1)} \cap \partial B(P_1, r_1) = \emptyset$ for all $n > N$ or (b) $X_n^{(1)} \cap \partial B(P_1, r_1) \neq \emptyset$ for a certain infinite subsequence $\{n'\} = \{n\}$. It is clear that in case (a), we have $X_n^{(1)} = {}'X_n^{(1)} \cup {}''X_n^{(1)}$ for any $n > N$, where ${}'X_n^{(1)} \cap {}''X_n^{(1)} = \emptyset$, and ${}''X_n^{(1)} \subset B(P_1, r_1)$; therefore, the compact sets $X_n^{(1)}$ can be replaced by new ones ${}'X_n^{(1)}$ such that ${}'X_n^{(1)} \in \mathcal{O}$, and ${}'X_n^{(1)} \cap \overline{B(P_1, r_1)} = \emptyset$. Carrying out this operation, we can pass to the second ball $B(P_2, r_2)$.

Now, in case (b), we still denote the subsequence $\{n'\}$ by $\{n\}$. Then $X_n^{(1)} \cap \partial B(P_1, r_1) \neq \emptyset$ for any $n > N$. We prove that $\mathrm{vol}_k[X_n^{(1)} \cap B(P_1, r_1)] \to 0$ as $n \to \infty$. Since $\overline{B(P_1, r_1)} \cap X^k = \emptyset$, we have $\Psi_k(P) \equiv 0$ for all points $P \in B(P_1, r_1 + \varepsilon')$, where $\varepsilon' > 0$ is such that $\overline{B(P_1, r_1 + \varepsilon')} \cap \widetilde{X}^k = \emptyset$. Due to

(3.3.9), there exists a number $\rho(P) > 0$ such that $\psi_k^+ \equiv 0$ for $0 \leqslant r < \rho(P)$. We obtain a finite covering of the ball $B(P_1, r_1)$ with the balls $B(P_\alpha, \rho(P_\alpha))$.

Since $\psi_k^+ = \lim_{\delta\to 0}\left[\sup_{\substack{|r-r'|<\delta \\ d(P,Q_i)<\delta}} \widetilde{\psi}_k(r', Q_i)\right] = 0$, we have

$$\mathrm{vol}_k[X_n^{(1)} \cap B(P_\alpha, \rho(P_\alpha))] \to 0,$$

whence

$$\mathrm{vol}_k[X_n^{(1)} \cap \overline{B(P_1, r_1)}] \to 0, \quad n \to \infty,$$

thus completing the proof.

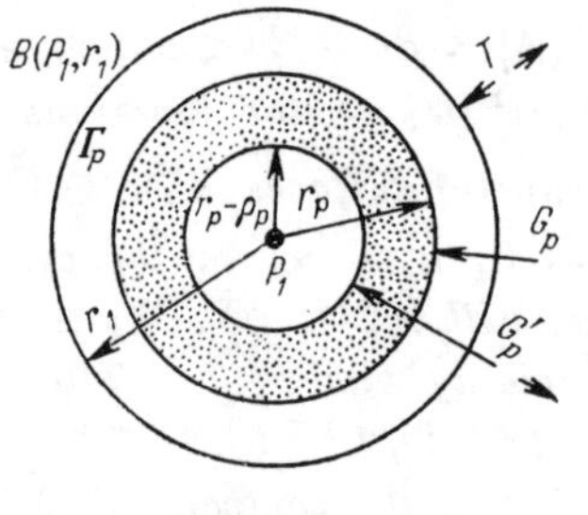


**Fig. 73**

Consider the open set $G_{p,t} = M \backslash \overline{B(P_1, r_p - t)}$, where $0 \leqslant t \leqslant \rho_p, r_1 - \frac{1}{p} < r_p < r_1 - \frac{1}{2p}$, $p$ is an arbitrary natural number, $\rho_p$ and $r_p$ are fixed for each $p$, and, moreover, the $\rho_p$ are such that $\rho_p \to 0$ as $p \to \infty$. Then $(T, \frac{1}{2p}) \subset G_{p,0} \subset (T, \frac{1}{p})$, where $T = M \backslash B(P_1, r_1)$. Put $G_p' = G_{p,\rho_p}$, $\Gamma_p = G_p' \backslash \overline{G}_{p,0}$ (see Fig. 73). We define a function $U_p(x) = t$ as on the spherical layer $\Gamma_p$, where $x \in \partial B(p_1, r_p - t)$; in particular $U_p(x) = 0$ if $x \in \partial B(P_1, r_p)$, and $U_p(x) = \rho_p$ if $x \in \partial B(P_1, r_p - \rho_p) = \partial G_p'$. It is obvious that $U_p(x) \in C^1(\overline{\Gamma}_p)$, and $\nabla U_p \neq 0$ on $\overline{\Gamma}_p$. We prove that, for each fixed $p$, we can choose the values $n_p$ and $t_p, 0 \leqslant t_p \leqslant \rho_p$ such that $\mathrm{vol}_{k-1}[\partial G_{p,t_p} \cap X_{n_p}^{(1)}] \leqslant p^{1-k}$. In fact, assume the contrary, viz., for any $t, 0 \leqslant t \leqslant \rho_p$ and $n$, let $\mathrm{vol}_{k-1}[\partial G_{p,t} \cap X_n^{(1)}] > p^{1-k}$. Since $\mathrm{vol}_k[X_n^{(1)} \backslash A] < \infty$, we have $\mathrm{vol}_k[\Gamma_p \cap X_n^{(1)}] < \infty$. Then

$$\int_0^{\rho_p} \mathrm{vol}_{k-1}[(\Gamma_p \cap X_n^{(1)}) \cap \partial B(P_1, r_p - t)]\, dt$$

$$= \int_0^{\rho_p} \mathrm{vol}_{k-1}[X_n^{(1)} \cap \partial G_{p,t}]\, dt > \rho_p p^{1-k}.$$

If we put $C_t = \{Q \in \overline{\Gamma}_p | U_p(Q) = t\}$, then $C_t = \partial G_{p,t}$, and it immediately follows from Theorem 10.2.3 in [289] that

$$M\,\mathrm{vol}_k[\Gamma_p \cap X_n^{(1)}] \geqslant \int_0^{\rho_p} \mathrm{vol}_{k-1}[X_n^{(1)} \cap C_t]\,dt > \rho_p p^{1-k},$$

where $M$ is the Lipschitz constant for $U_p(x), x \in \overline{\Gamma}_p$; therefore, $M \neq 0$. Consequently, for any natural $n$ and fixed $p$, $\mathrm{vol}_k[\overline{\Gamma}_p \cap X_n^{(1)}] > M\rho_p p^{k-1} > 0$. Since $\mathrm{vol}_l[\overline{\Gamma}_p \cap X_n^{(1)}] \to 0$ as $n \to \infty$ (see above), this inequality cannot hold. Thus, we can indicate an infinite subsequence $\{n_p\} \subset \{n\}$ and a sequence $\{t_p\}$ such that $\mathrm{vol}_{k-1}[X_{n_p}^{(1)} \cap \partial G_{p,t_p}] \leqslant p^{1-k}$, where $p^{1-k} \to 0$ as $p \to 0$. To carry out the surgery of the infinite sequence of compact sets $X_{n_p}^{(1)} \in \mathcal{O}$, we put $A_p = \partial G_{p,t_p} \cap X_{n_p}^{(1)}$. Then $\mathrm{vol}_{k-1}(A_p) = l_p^{k-l}$, $l_p \leqslant p^{-1}$, $A_p \subset B(P_1, r_1) \subset B(P_1, R_0)$. Therefore, it follows from Theorem 28.3.1 that there exists a compact set $R_p$ contained in an $\varepsilon$-neighbourhood of the geodesic hull $A_p$, where $\varepsilon$ is arbitrarily small, so that $\nabla^m(\dot{R}_p, A_p) = \tilde{h}^{m-1}(A_p)\backslash 0$ for any $m \in \mathbf{Z}, x \in A_p$ (resp. $\Delta_m(\dot{R}_p, A_p) = \tilde{h}^{m-1}(A_p)$). Besides,

$$\rho(\dot{R}_p^*, A_p) \leqslant C \cdot l_p, \ \rho(\dot{R}_p^* \cup \dot{R}_{p,c}^*, A_p^* \cup A_{p,c}^{k-1}) \leqslant C \cdot l_p,$$
$$\mathrm{vol}_k(\dot{R}_p) = \mathrm{vol}_k(\dot{R}_p^* \cup R_{p,c}^k) \leqslant D l_p^k.$$

Recall that if $l_p = 0$, then we have to replace $l_p$ by an arbitrarily small number $\delta$. Put $R_p = \dot{R}_p \cup \tilde{\gamma}_p$, where $\tilde{\gamma}_p$ is a smooth path joining $\dot{x}$ to $x \in A$. By Theorem 27.3.1 and the remark on it, the new compact set ${}'X_{n_p}^{(1)} = [X_{n_p}^{(1)} \backslash [X_{n_p}^{(1)} \cap B(P_1, r_p - t_p)]] \cup R_p$ belongs to the class $\mathcal{O}$.

Thus, we have obtained a sequence of compact sets ${}'X_{n_p}^{(1)}$, $p = 1, 2, 3, \ldots$ such that $\mathrm{vol}_k[{}'X_{n_p}^{(1)} \backslash A] \to d_k$, and ${}'X_{n_p}^{(1)} \backslash G_p$ is a finite simplicial subcomplex of class $C^r$ in $M$ of dimension $s \leqslant k-1$ (see Theorem 28.3.1), i.e., all $k$-dimensional portions of the compact sets ${}'X_n^{(1)}$ concentrate in the arbitrarily small neighbourhood of the compact set $T = M \backslash B(P_1, r_1)$; in other words, the ball $B(P_1, r_1)$ is gradually freed of them as $p \to \infty$. We now consider the next ball $B(P_2, r_2)$ and repeat the above process for it and the sequence ${}'X_{n_p}^{(1)}$. As a result, we obtain a new infinite sequence of a compact ${}''X_{n_{p_q}}^{(1)} \in \mathcal{O}$, which will still be $k$-minimizing, but whose $k$-dimensional pieces will gradually leave the interior of $B(P_2, r_2)$ as the subscript increases. It only remains to notice that the $k$-dimensional portions of the surfaces ${}''X_{n_{p_q}}^{(1)}$ cannot again get inside $B(P_1, r_1)$; therefore, they actually also leave $B(P_1, r_1) \cup B(P_2, r_2)$, and not only $B(P_2, r_2)$.

By continuing the process and by recalling that there are a finite number of $B(P_i, r_i)$, we finally derive the $k$-dimensional portions of the $k$-minimizing sequence of the compact set in a neighbourhood $U$, with only those of their portions remaining that are finally simplicial subcomplexes of dimension $s \leqslant k-1$. It follows from the above argument that ${}'\psi_k^{\pm} \equiv \psi_k^{\pm}$, and ${}'\Psi_k \equiv \Psi_k$; in general, only the functions $\varphi_k^{\pm}$ alter. It is obvious that $'\varphi_k^{\pm} \equiv 0$ on $M \backslash \widetilde{X}^k$; all other statements are also evident.

Q.E.D.

**Proof of Proposition 29.2.1.** The statement is obtained simply by combining (3.3.7) with Lemma 29.3.4.

We have actually proved a somewhat more strengthened result than that in Proposition 29.2.1, viz., that $X_n^{(1)'} \backslash U$ is a finite simplicial subcomplex of dimension $s \leqslant k-1$, and not only that $\mathrm{vol}_{k-1}[X_n^{(1)'} \backslash \overline{U}] < \infty$. We will repeatedly resort to this important remark in the following.

Thus, the first step of the M-process is completed. Consider the sequence of neighbourhoods $U_\alpha^1 \supset U_{\alpha+1}^1, \bigcap_\alpha U_\alpha^1 = \widetilde{X}^k$, and put $\omega_\alpha^{k-1} = \inf \mathrm{vol}_{k-1}(X \backslash \overline{U}_\alpha^1)$, $X \in \mathcal{O}$, $\lambda_{k-1} = \lim_{\alpha \to \infty} \omega_\alpha^{k-1}$. Recall that $\omega_\alpha^{k-1} < \infty$ for any $\alpha$, whereas $\lambda_{k-1} \leqslant \infty$, which will not present any obstacles in proceeding with the M-process even though it may turn out that $\lambda_{k-1} = \infty$.

**Lemma 29.3.5.** *The value $\lambda_{k-1}$ (possibly infinity) does not depend on the choice of a system of open contracting neighbourhoods $U_\alpha^1$, i.e., it is completely and uniquely determined only by that compact set $\widetilde{X}^k \subset M$.*

**Proof.** Given two neighbourhood systems $U_\alpha^1$ and $\widetilde{U}_\beta^1$, assume that $\tilde{\lambda}_{k-1} < \lambda_{k-1}$. Let $s$ be such that $\tilde{\lambda}_{k-1} < s < \lambda_{k-1}$. Since $\lambda_{k-1} = \lim_\alpha \omega_\alpha$, there exists a subscript $\alpha_0$ such that $\omega_{\alpha_0}^{k-1} > s$. Because $\bigcap_\beta \widetilde{U}_\beta^1 = \widetilde{X}^k$, there exists a subscript $\beta_0$ such that $\widetilde{U}_{\beta_0}^1 \subset U_{\alpha_0}^1$, and $\tilde{\omega}_{\beta_0}^{k-1} < s$; hence, there is a compact set $X \in \mathcal{O}$ for which $\mathrm{vol}_{k-1}(X \backslash \widetilde{U}_{\beta_0}^1) = \tilde{\omega}_{\beta_0}^{k-1} + \varepsilon$, $\tilde{\omega}_{\beta_0}^{k-1} + \varepsilon < s$, $\varepsilon \geqslant 0$. But since $\mathrm{vol}_{k-1}(X \backslash \widetilde{U}_{\beta_0}^1) < \infty$, we have $\mathrm{vol}_{k-1}(X \backslash \widetilde{U}_{\alpha_0}^1) < \infty$, and $\mathrm{vol}_{k-1}(X \backslash \widetilde{U}_{\alpha_0}^1) \leqslant \mathrm{vol}_{k-1}(X \backslash \widetilde{U}_{\beta_0}^1) = \tilde{\omega}_{\beta_0}^{k-1} + \varepsilon < s$. However, $\mathrm{vol}_{k-1}(X \backslash \widetilde{U}_{\alpha_0}^1) \geqslant \omega_{\alpha_0}^{k-1} > s$. The obtained contradiction proves the lemma.

If $\lambda_{k-1} = \infty$ for one neighbourhood system, then this is also true for any other system.

## 29.4. Second and Subsequent Steps in the Minimizing Process

We now turn to the study of the second step of the M-process. Let a contracting neighborhood system $U_n^1$ be fixed. Due to the definition of the values $\omega_n^{k-1} < \infty$, there exists an infinite sequence of compact sets $X_n^{(2)} \in \mathcal{O}$ such that $\mathrm{vol}_{k-1}[X_n^{(2)} \backslash U_n^1] = \omega_n^{k-1} + \varepsilon_n$, $\varepsilon_n \geqslant 0$, $\varepsilon_n \to 0$, i.e., $\mathrm{vol}_{k-1}[X_n^{(2)} \backslash U_n^1] \to \lambda_{k-1}$, and, since $\lambda_{k-1} = \sup_{(n)} \omega_n^{k-1}$, the inequality $\mathrm{vol}_{k-1}[X_n^{(2)} \backslash U_n^1] < \lambda_{k-1}$ is possible in contrast to the first step (recalling that $\mathrm{vol}_k[X_n^{(1)} \backslash A] \geqslant d_k$). However, it turns out that this difficulty can be overcome.

Since all the $k$-dimensional pieces of the compact sets $X_n^{(2)}$ concentrate around $\widetilde{X}^k$, we will now deal with the $(k-1)$-dimensional zones in $M \backslash U_n^1$.

The role of the compact set $A$ can be played by $\widetilde{X}^k$; however, there is one essential difference from the first step, viz., if $A$ is contained in each surface $X \in \mathcal{O}$, then $\widetilde{X}^k$ must not necessarily be contained in $X_n^{(2)}$. Put $R^1(P) = \min[d(P,X^k),R_0]$ and consider the functions $\varphi_{k-1}(r,P,X), \psi(r,P,X)$, where $0 \leqslant r < R^1(P), R^1(P) > 0$, and $X$ is an arbitrary compact set containing $A$, but not necessarily containing $\widetilde{X}^k$. Since the countable set $\{Q_i\}$ is everywhere dense in $M\backslash A$, and therefore in $M\backslash\widetilde{X}^k$, we can, as in the first step, consider the functions $\psi_{k-1}(r,Q_i,X_n^{(2)})$, and assume at once that they converge to $\widetilde{\psi}(r,Q_i)$, where $0 \leqslant r < R^1(P)$. Similarly to the first step, we construct the functions $\psi_{k-1}^{\pm}, \varphi_{k-1}^{\pm}$ with domains $M\backslash\widetilde{X}^k$ with respect to the argument $P$. Let $P \in M\backslash\widetilde{X}^k$. We put

$$\Psi_{k-1}(P) = \lim_{\rho\to 0}\sup_{0\leqslant r<\rho}[h_{k-1}^{-1}(r)\psi_{k-1}^{+}(r, P)],$$

$$S^{k-1} = \{P \in M\setminus\widetilde{X}^k \mid \Psi_{k-1}(P) > 0\}, \quad \widetilde{X}^{k-1} = \widetilde{X}^k \cup S^{k-1}.$$

It turns out that the properties of $\Psi_{k-1}$ are similar in many respects to those of $\Psi_k$. Let $P \in M\backslash\overline{U}_n^1$. We put $R_n^1(P) = \min[d(P,\overline{U}_n^1),R_0]$. It is clear that $R_n^1(P) < R^1(P) \leqslant R^0(P)$. We fix the subscript $n$, and reason only for the set $M\backslash\overline{U}_n^1$, and not for $M\backslash\widetilde{X}^k$, replacing by the former the prior domain $M\backslash A$ as in the first step. Further, replace $A$ by $\overline{U}_n^1$, $d_k$ by $\omega_n^{k-1}$ (recalling that $\omega_n^{k-1} < \infty$), $R^1$ by $R_n^1$, where $P \in M\backslash\overline{U}_n^1$, in the entire previous argument. We stress that the replacement does not at all throw us back to the situation in the first step, since $\mathrm{vol}_{k-1}[X_n^{(2)}\backslash\overline{U}_n^1] = \omega_n^{k-1} + \varepsilon_n$, $\varepsilon_n \geqslant 0$, $\varepsilon_n \to 0$.

**Lemma 29.4.1.** *The statement and proof of this lemma are obtained, respectively, from those of Lemma 29.3.1 by replacing $A$ by $\overline{U}_n^1$ and $R^0(P)$ by $R_n^0(P)$, where $P \in M\backslash\overline{U}_n^1$.*

**Lemma 29.4.2.** *The statement and proof of this lemma are obtained, respectively, from those of Lemma 29.3.2 by replacing $A$ by $\overline{U}_n^1$, $R^0(P)$ by $R_n^1(P)$, $d_k$ by $\omega_n^{k-1}$, and $D = D(k)$ by $D(k-1)$.*

It should be noted regarding its proof that the difference

$$\varepsilon_n = \mathrm{vol}_{k-1}(X\backslash\overline{U}_n^1) - \omega_n^{k-1},$$

where $\varepsilon_n \geqslant 0$, $\varepsilon_n \to 0$, is involved in the statement, and not the value $\mathrm{vol}_{k-1}(X\backslash\overline{U}_n^1)$ itself; therefore, that the inequality $\mathrm{vol}_{k-1}(X\backslash\overline{U}_n^1) < \lambda_{k-1}$ is possible does not at all affect the M-process construction (see Fig. 74). The constant $\kappa_{k-1}$ is of the form $(k-1)^{1-k}[D(k-1)]^{2-k}$ and, therefore, does not depend on the subscript $n$, which will be important for the following.

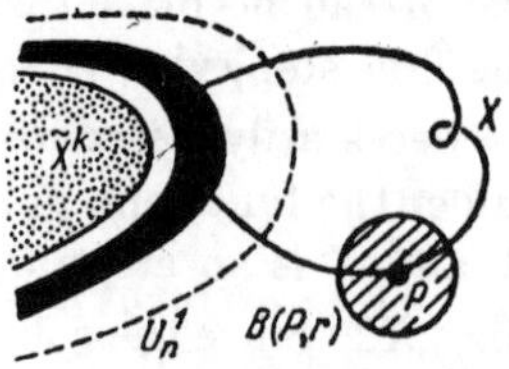


**Fig. 74**

**Lemma 29.4.3.** *Let $X_n^{(2)} \in \mathcal{O}$ be the compact set sequence fixed above, $\mathrm{vol}_{k-1}[X_n^{(2)} \backslash \overline{U}_n^1] = \omega_n^{k-1} + \varepsilon_n < \infty$, $\varepsilon_n \geqslant 0, \varepsilon \to 0$, and $\varphi_{k-1}^{\pm}, \psi_{k-1}^{\pm}$ the functions constructed from $X_n^{(2)}$ with domain $M\backslash\widetilde{X}^k$ (consequently, they are also defined on $M\backslash U_n^1$). Then the statements are valid, which are similar to (3.3.1)–(3.3.10) of Lemma 29.3.2 and are obtained by replacing $R^0(P)$ by $R^1(P)$, $A$ by $\widetilde{X}^k$, $X_n^{(1)}$ by $X_n^{(2)}$, and $\widetilde{X}^k$ by $\widetilde{X}^{k-1}$.*

**Proof.** Let $P \in M\backslash\widetilde{X}^k, 0 \leqslant r < R^1(P)$. We have to prove at the point $(r, P)$ relations only referring to the functions $\psi_{k-1}^{\pm}, \varphi_{k-1}^{\pm}, \Psi_{k-1}$. It is clear that there exists a subscript $N = N(r, P)$ such that $\overline{U}_n^1 \cap B(P, r) = \emptyset$ for all $n > N$, i.e., $r < R_n^1(P)$, and we can apply Lemma 29.4.2 as in the proof of Lemma 29.3.3, also involving Lemma 29.3.2. We stress that $\beta_{k-1} = (k-1)^{1-k}\gamma_{k-1}^{-1} \times [D(k-1)]^{1-k} > 0$, with $\beta_{k-1}$ not depending on the subscript $n$; therefore, if $\Psi_{k-1}(P) > 0$, where $P \in M\backslash\widetilde{X}^k$ similarly to the first step, then $\Psi_{k-1} \geqslant \beta_{k-1} > 0$.

Q.E.D.

Note that the closedness of the set $S^{k-1}$ in $M\backslash\widetilde{X}^k$ follows from Lemma 29.4.3; therefore, $\widetilde{X}^{k-1} = S^{k-1} \cup \widetilde{X}^k$ is a compact set. Further, if $P \in M\backslash\widetilde{X}^{k-1}$, then $\Psi_{k-1}(P) = 0$, with $\psi_{k-1}^{+} \equiv 0$ for $0 \leqslant r < k'\rho(P)$, where $0 < k^1, \rho(P) > 0$. We will make use of this fact in restructuring the sequence similarly to the first step.

**Lemma 29.4.4.** *Let $X_n^{(2)} \in \mathcal{O}$ be the original sequence such that $\mathrm{vol}_{k-1}[X_n^{(2)} \backslash \overline{U}_n^1] = \omega_n^{k-1} + \varepsilon_n$, $\varepsilon_n \geqslant 0$, $\varepsilon_n \to 0$, $n \to \infty$, $\omega_n^{k-1} < \infty$ and $\varphi_{k-1}^{\pm}, \psi_{k-1}^{\pm}$, $\Psi_{k-1}$ the functions constructed above from $X_n^{(2)}$. Then there exists a new sequence of compact sets $X_n^{(2)'} \in \mathcal{O}$ such that $\mathrm{vol}_{k-1}[X_n^{(2)'} \backslash \overline{U}_n'] = \omega_n^{k-1} + \varepsilon_n'$, $\varepsilon_n \geqslant 0$, $\varepsilon_n \to 0$, $n \to \infty$, with $'\psi_{k-1}^{\pm} \equiv \psi_{k-1}^{\pm}, '\Psi_{k-1} \equiv \Psi_{k-1}$, whereas the new functions $\varphi_{k-1}^{\pm}$ (in general, nonconsistent with $\varphi_{k-1}^{\pm}$) are still nondecreasing with respect to $r$ for $0 \leqslant r < R^1(P)$ and satisfying relations similar to (3.3.1), (3.3.6), and (3.3.8). Further, for any open neighbourhood $U = U(\widetilde{X}^{k-1})$ of the compact set $\widetilde{X}^{k-1}$, there exists a subscript $N = N(U)$ such that $\mathrm{vol}_{k-2}(X_n^{(2)'} \backslash U] < \infty$ for all $n > N$, and, moreover, $X_n^{(2)'} \backslash U$ is a finite simplicial subcomplex of dimension $s \leqslant k-2$. In particular, $X_n^{(2)'} \backslash U$ only consists of simplicial points.*

The proof is obtained as in Lemma 29.3.2, and we will not dwell on it.

At the first step of the M-process, we operated with the number $d_k$, and not with $\lambda_k = \lim_{\alpha\to\infty} \inf \mathrm{vol}_k(X\backslash\overline{U}^0_\alpha)$ (see above), which is just what made the first and second steps different.

**Proof of Proposition 29.2.2.** The statement is obtained simply by combining an analogue of (3.3.7) with Lemma 29.4.4.

Thus, the second step of the M-process is also completed. Consider a neighbourhood system $U^2_\alpha \supset U^2_{\alpha+1}, \bigcap_\alpha U^2_\alpha = \widetilde{X}^{k-1}$, and put $\omega^{k-2}_\alpha = \inf \mathrm{vol}_{k-2}(X\backslash\overline{U}^2_n), X \in \mathcal{O}, \lambda_{k-2} = \lim_{\alpha\to\infty}(\omega^{k-2}_\alpha)$. Then $\omega^{k-2}_\alpha < \infty$.

**Lemma 29.4.5.** *The number $\lambda_{k-2}$ (possibly infinity) does not depend on the choice of an open neighbourhood $U^2_\alpha$ system, i.e., it is completely determined only by the compact set $\widetilde{X}^{k-1} \subset M$.*

The proof is quite similar to that of Lemma 29.3.5.

Thus, we can continue the M-process into smaller dimensions, which yields a sequence of compact sets $A \subset \widetilde{X}^k \subset \widetilde{X}^{k-1} \subset \widetilde{X}^{k-2} \subset \ldots$, stopping only at dimension one. We now describe this last step.

**Proof of Proposition 29.2 $(k-1)$.** We have the compact set $\widetilde{X}^3$ and a neighbourhood system $U^{k-2}_\alpha, \bigcap_\alpha U^{k-2}_\alpha = \widetilde{X}^3$. In full conformity with the above scheme, we construct $S^2 = \{P \in M\backslash\widetilde{X}^3 | \Psi_2(P) > 0\}$, $\widetilde{X}^2 = S^2 \cup \widetilde{X}^3$. Then there is a sequence of compact sets $X^{(k-1)'}_n$ for which $\mathrm{vol}_2[X^{(k-1)'}_n\backslash\overline{U}^{k-2}_n] = \omega^2_n + \varepsilon'_n$, $\varepsilon'_n \geqslant 0$, $\varepsilon'_n \to 0-$, and there exists a subscript $N = N(U)$ for any open neighbourhood $U$ of the compact set $\widetilde{X}^2$, so that the compact set $X^{(k-1)'}_n\backslash Y^{k-1}_n$ is generally a one-dimensional finite simplicial subcomplex for all $n > N$; therefore, due to the 1-stability of the class $\mathcal{O}$, $X^{(k-1)'}_n \cap \overline{U}^{k-1}_n$ also belongs to $\mathcal{O}$ for $n > N$, i.e., we can assume that $X^{(k-1)'}_n \subset \bar{U}^{k-1}_n$. To prove the statement, it suffices to establish that $\widetilde{X}^2 \in \mathcal{O}$. Indeed, consider a sequence $X^{(k-1)'}_n$. Let $X$ be the set of all limit points for the sequence $X^{(k-1)'}_n$. It is then obvious that $\widetilde{X} \subset \widetilde{X}^2$. (Note that it is not necessary that $\rho(X^{(k-1)'}_n, \widetilde{X}) \to 0$ as $n \to \infty$.) Since $\mathcal{O}$ is closed under the passage to the limit in metric $\rho$, with it being clear that $Z_p = \widetilde{X} \cup [\bigcup_{n\geqslant p} X^{(k-1)'}_n] \in \mathcal{O}$ and $\widetilde{X} = \bigcap_p Z_p$, $\rho(Z_p, \widetilde{X}) \to 0$ as $p \to \infty$, it follows, due to Theorem 27.2.1, that $\widetilde{X} \in \mathcal{O}$.

Q.E.D.

Summarizing, we have constructed (in general, nonunique) the compact set $\widetilde{X}^2 \in \mathcal{O}$ from each $k$-minimizing sequence $X^{(1)}_n \in \mathcal{O}$, $\mathrm{vol}_k(X^{(1)}_n\backslash A) \to d_k < \infty$, to which the sequence of numbers $d_k, \lambda_{k-1}, \lambda_{k-2}, \ldots, \lambda_3, \lambda_2$ is related, and in which the natural stratification is determined by the compact subsets $\widetilde{X}^2 \supset \widetilde{X}^3 \supset \widetilde{X}^4 \cdots \supset \widetilde{X}^k \supset A$, where $\widetilde{X}^{m-1} = \widetilde{X}^m \cup S^{m-1}$,

$S^{m-1} = \{P \in M \backslash \widetilde{X}^m | \Psi_{m-1}(P) > 0\}, 3 \leqslant m \leqslant k+1, \widetilde{X}^{k+1} = A$. For each compact set $\widetilde{X}^m$, there exists a sequence of compact sets $X_n^{(k-m+2)'}$ such that $\mathrm{vol}_{m-1}[X_n^{(k-m+2)'} \backslash U_n^{k-m+1}(\widetilde{X}^m)] = \omega_n^{m-1} + \varepsilon_n$, $\varepsilon_n \geqslant 0, \varepsilon \to 0$, i.e., $\mathrm{vol}_{m-1}[X_n^{(k-m+2)'} \backslash U_n^{k-m+1}(\widetilde{X}^m)] \to \lambda_{m-1} < \infty$. Further, $\widetilde{X}^{m-1} = \widetilde{X}^m \cup S^{m-1}$, where $\omega_n^{m-1} < \infty$, $X_n^{(k-m+2)'} \backslash U_n^{k-m+2}(\widetilde{X}^{m-1})$ is an $s$-dimensional finite-simplicial subcomplex, and $s \leqslant m-2$ (see Fig. 75).

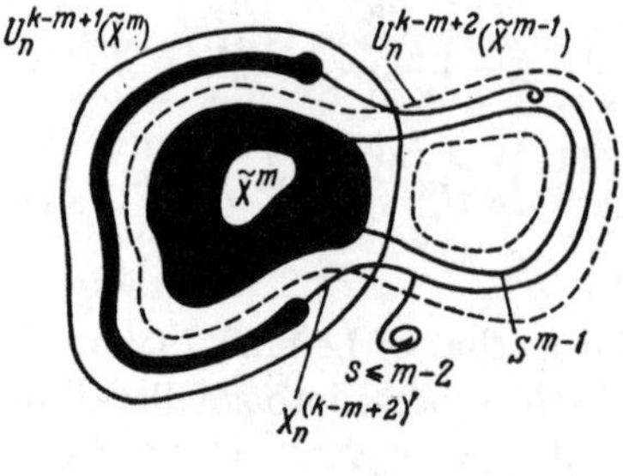


**Fig. 75**

Thus, we have proved that the M-process is convergent. The compact set $\widetilde{X}^2 \in \mathcal{O}$ constructed by us admits a representation $\widetilde{X}^2 = A \cup S^k \cup S^{k-1} \cup \cdots \cup S^2$; however, we still cannot say anything regarding its metric properties and do not have a geometric interpretation of the vector $(d_k, \lambda_{k-1}, \ldots, \lambda_2)$, which will be done in the subsequent section.

## 29.5. The Theorem on the Coincidence of the Least Stratified Volume with Least $\lambda$-Vector in a Variational Class

We now turn to the problem of interpreting the vector SV $= (d_k, d_{k-1}, \ldots, d_3)$. We can construct a number of M-processes from a given $k$-minimizing sequence $X_n^{(1)}$ and, therefore, obtain many $\lambda$-vectors, in general, associated with different compact sets $\widetilde{X}^3(M)$, which can be different from each other, starting with the component $\lambda_{k-1}$. Consider the set $T$ of all $\lambda$-vectors obtained in the class $\mathcal{O}(\widetilde{\mathcal{O}})$ in the course of all possible M-processes, and introduce lexicographic order. The question arises: Can we find in $T$ a $\lambda$-vector that is the least one with respect to this ordering?

**Theorem 29.5.1.** *Let all the assumptions of Theorem 7.2.1 be fulfilled. Consider a class $\{X\}_3$ whose existence and nonemptiness are asserted in the theorem. It then turns out that each compact set $X \in \{X\}_3$ is a result of a certain M-process in $\mathcal{O}(\widetilde{\mathcal{O}})$, i.e., it can be represented in the form $X = \widetilde{X}^3(M)$, with the $\lambda$-vector $(d_k, \lambda_{k-1}, \lambda_{k-2}, \ldots, \lambda_3)$ coinciding with the least stratified volume vector $SV = (d, d_{k-1}, d_{k-2}, \ldots, d_3)$. Therefore, the $\lambda$-vector only depends on* *the class $\mathcal{O}(\widetilde{\mathcal{O}})$ and does not depend on the compact set $X \in \{X\}_3$, $X = \widetilde{X}^3(M)$. Finally, being common to all $X \in \{X\}_3$, it is the least $\lambda$-vector with respect to lexicographic order in $T$.*

## §30 Properties of Density Functions. The Minimality of Each Stratum of the Surface Obtained in the Minimization Process

### 30.1. The Value of the Density Function is Always not Less than Unity on Each Stratum, and Unity only at Regular Points

Consider the compact set $\widetilde{X}^3 \in \mathcal{O}(\widetilde{\mathcal{O}})$ obtained in a certain M-process, and study its metric properties. If the process is finite, then we prove that $\Psi_\alpha(P) \geqslant 1$ holds for $P \in S^\alpha$, $3 \leqslant \alpha \leqslant k$, whence the surfaces $S^\alpha$ are relatively minimal in dimension $\alpha$.

In constructing the compact set $\widetilde{X}^3$, we have proceeded from the greatest dimension $k$ towards 3, whereas we move in the reverse direction from $\mathrm{vol}_3$ to $\mathrm{vol}_k$ in proving the relative minimality of the films $S^\alpha$. It is important that $\widetilde{X}^3$ obtained in the course of the M-process must not at all belong to the class $\{X\}_3$ (see Theorem 7.2.1), whose existence is not yet proved. In particular, the values $\lambda_m$ and $d_m$ may not coincide. Consider the $(k-s+1)$th step of the M-process. We have $\widetilde{X}^s = \widetilde{X}^{s+1} \cup S^s$, $S^s = \{P \in M \backslash \widetilde{X}^{s+1} | \Psi_s(P) > 0\}$, $U_n^{k-s+1}(\widetilde{X}^s) \supseteq U_n^{k-s}(\widetilde{X}^{s+1})$. We purposefully select a neighbourhood system $U_n^{k-s+1}(\widetilde{X}^s)$ such that $\mathrm{vol}_s(X_n^{k-s+1)'} \backslash U_n^{k-s}(\widetilde{X}^{s+1})] = \omega_n^s + \varepsilon_n'$, $\varepsilon_n' \geqslant 0$, $\varepsilon_n' \to 0$, $\mathrm{vol}_{s-1}[X_n^{k-s+1)'} \backslash U_n^{k-s+1}(\widetilde{X}^{s+1})] < \infty$. Recall that the functions $\varphi_s^\pm, \psi_s^\pm, \Psi_s$ are related to the sequence $X_n^{(k-s+1)'}$. For brevity, we will write $X_n$ instead of $X_n^{(k-s+1)'}$. Put $\check{X}_n = X_n \cap \overline{U}_n^{k-s+1}(\widetilde{X}^s)$. Then $\mathrm{vol}_{s-1}[X_n \backslash \check{X}_n] < \infty$, i.e., the compact set $\check{X}_n$ is the "massive" part of the compact set $X_n$, and $\mathrm{vol}_s[X_n \backslash U_n^{k-s}] = \mathrm{vol}_s[\check{X}_n \backslash U_n^{k-s}]$. Note that, in general, the distance $\rho(\check{X}_n, \widetilde{X}^s)$ must not tend to zero; we can only assert that $\rho[\check{X}_n \backslash U_n^{k-s}(\widetilde{X}^{s+1}), S^s] \to 0$.

**Lemma 31.1.** *Consider the set $S^s = \{P \in M \backslash \widetilde{X}^{s+1} | \Psi_s(P) > 0\}$, where $s \geqslant 3$, and the compact set $\widetilde{X}^3 \in \mathcal{O}$ is obtained in a certain M-process. Let $\beta_0 > 0$ be a fixed number, and $P' \in S^s$, $0 \leqslant r' < R^{k-s}(P')$. Then there exist two constants $\varepsilon_0 = \varepsilon_0(\xi, \beta_0, s, M) > 0$ and $v = v(\xi, \beta_0, x, M) > 0$ for each $\xi > 0$, so that if*

$$(1.1.1) \quad \beta_0 \leqslant h_s^{-1}(r)\varphi_s^-(r, P) \leqslant h_s^{-1}(r)\psi_s^+(r, P) \leqslant \beta_0 + \varepsilon$$

*for each pair $(r, P)$, where $P \in S^s$, and $B(P, r) \subset B(P', r')$, then there exists a point $P^* \in S^s$, and a geodesic $s$-plane $\Pi$, i.e., $\dim \Pi = s$, filled with geodesics emanating from $P^*$, with centre at $P^*$, so that*

(1.1.2) $B(P^*, vr) \subset B(P, r)$ and $K(P^*, vr) \subset (\Pi, \xi vr)$,

*where* $K(P^*, vr) = S^s \cap B(P^*, vr)$, *and* $(\Pi, \xi vr)$ *is an open neighbourhood of radius* $\xi vr$ *of* $\Pi$. *Moreover, the values* $\varepsilon_0$ *and* $v$ *can be selected, so that* $\varepsilon_0 = C(\beta_0, s, M)\xi^h$, *with* $h = h(\beta_0, s, M)$, $v = D(\beta_0, s, M)\xi^h$.

The proof of this technicality is as in [289] for Lemma 10.4.3 and the usual homology theory, $s = k$.

Until now, we could only consider any $s$ such that $2 \leqslant s \leqslant n-1$. Lemma 30.1.1 is that unique statement for whose proof it is necessary to assume that $s \geqslant 3$, which is just what takes us to the study of 2-stable topological variational classes.

**Lemma 30.1.2.** *Let* $G$ *be an open set in* $M$, *so that* $G \cap \widetilde{X}^{s+1} = \emptyset$, $G \cap S^s \neq \emptyset$, *where* $S^s \subset \widetilde{X}^3$, $\widetilde{X}^3$ *is obtained in a certain M-process, and* $\beta_G = \inf \Psi_s(P)$, *where inf is taken for points* $P \in G \cap S^s$, *i.e.,* $\beta_G \geqslant \beta_s > 0$ *(see (3.3.7) in §29.3 for the definition of* $\beta_s$*). Then there exists a ball* $B(P_1, r_1) \subset G$ *for an arbitrary constant* $\varepsilon_0' > 0$, *in general, depending on* $\varepsilon_0'$, *so that* $P_1 \in S^s \cap G$. *Furthermore,*

*(1.2.1)* $\psi_s^+ \leqslant (\beta_G + \varepsilon_0')h_s(r)$ *for any ball* $B(P, r)$, *where* $P \in S^s, 0 \leqslant r < R^{k-s}(P)$, $B(P, r) \subset B(P_1, r_1)$. *Besides,* $\varphi_s^- \geqslant \beta_G h_s(r)$ *holds for all* $B(P, r) \subset B(P_1, r_1) \subset G$, *where* $P \in S^s$, $0 \leqslant r < R^{k-s}(P)$.

The proof is as in [289].

**Lemma 30.1.3.** $\Psi_s(P) \geqslant 1$ *if* $S^s \neq 0$ *for any point* $P \in S^s$, *where* $S^2 \subset \widetilde{X}^3$ *is the result of certain M-process, and any* $s$, $3 \leqslant s \leqslant k$.

**Remark.** The proof is similar to the argument of [351] for the density lemma in certain respects of analytic character and also to Theorem 10.4.3 in [289]. In general, this inequality for the density function only holds for minimal surfaces. As has been repeatedly demonstrated above (e.g., in Chapter 3), this statement leads to far-reaching geometric consequences.

**Proof.** Let $P \in S^s$, $\beta = \inf \Psi_s(P)$. Then $\beta \geqslant \beta_s > 0$ (see (3.3.7) in §29.3). Suppose that $\xi > 0$ is an arbitrary number, whereas $\varepsilon_0 = \varepsilon_0(\xi, \beta, s, M)$ and $v = v(\xi, \beta, s, M)$ are the values whose existence is stated in Lemma 30.1.1. Meanwhile, we have to put $\beta_0 = \beta = \inf \Psi_s(P)$. It follows from (3.3.8) in §29.3 that $\varphi_s^- \geqslant \Psi_s(P)h_s(r)$, i.e., $h_s^{-1}(r)\varphi_s^- \geqslant \beta$. Since $\varphi_s^- \leqslant \psi_s^+$ due to (3.3.4) in §29.3, $\beta \leqslant h_s^{-1}(r)\varphi_s^- \leqslant h_s^{-1}(r)\psi_s^+$. Consider an open set $G$ such that $G \cap \widetilde{X}^{s+1} = \emptyset$, $G \cap S^s \neq \emptyset$, $P \in G \cap S^s$. Let $\beta_G = \inf \Psi_s(P)$, $P \in G \cap S^s$. It is then clear that $\beta \leqslant \beta_G$. Put $\varepsilon_0' = \varepsilon_0$, where $\varepsilon_0$ is taken from Lemma 30.1.1. We are then in the situation where Lemma 30.1.2 can be applied, and derive that there is a ball $B(P_1, r_1) \subset G$ such that $P_1 \in S^s \cap G$ and $\psi_s^+(r, P) \leqslant (\beta_G + \varepsilon_0)h_s(r)$ for any ball $B(P, r)$, where $P \in S^s$ and $B(P, r) \subset B(P_1, r_1)$. Since $h_s^{-1}(r)\varphi_s^- \geqslant \beta_G$ if $P \in G \cap S^s$, we finally obtain that $\beta_G \leqslant h_s^{-1}(r)\varphi_s^- \leqslant h_s^{-1}(r)\psi_s^+ \leqslant \beta_G + \varepsilon_0$ for any ball $B(P, r) \subset B(P_1, r_1) \subset G$. Having proved this fact, we find ourselves in a situation where Lemma 30.1.1 is applicable, i.e., where there exists a point $P^* \in S^s$ and a geodesic $s$-plane $\Pi$ with the origin of coordinates at $P^*$, for any ball $B(P,r) \subset B(P_1,r_1)$, $P \in G \cap S^s$ if we put $\beta_0 = \beta_G$, so that $B(P^*,vr) \subset B(P,r)$ and $K(P^*,vr) \subset (\Pi,\xi vr)$. Consider a sequence $X_n^{(k-s+1)'}$, where $\rho[\check{X}_n^{(k-s+1)'} \cap B(P^*,vr),\ S^s \cap B(P^*,vr)] \to 0$ as $n \to \infty$ (see above), which means that we can select a subscript $N$ such that

$$\check{K}_n(P^*,\ vr) \subset (\Pi,\ 2\xi vr), \tag{1.3.1}$$

for all $n > N$, where $\check{K}_n(P^*,vr) = \check{X}_n^{(k-s+1)'} \cap B(P^*,vr)$, and $\mathrm{vol}_s[\check{K}_n(P^*, vr/2)] = \mathrm{vol}_s[K_n,P^*,vr)] \geqslant \beta_G h_s(vr/2) > 1/2\beta_G h_s(vr/2)$. Since $\varphi_s^- \leqslant (\beta_G+\varepsilon_0)h_s(r)$, we have $\varphi_s(vr,P^*,X_n^{(k-s+1)'} < (\beta_G+2\varepsilon_0)h_s(vr)$ for all $n > N$. Therefore, there exists a value $\rho_n$, $vr/2 \leqslant \rho_n \leqslant vr$ for each $n > N$, so that

$$\mathrm{vol}_{s-1}[X_n \cap \partial B(P^*,\ \rho_n)] \leqslant 2(\beta_G + 2\varepsilon_0)(vr)^{-1}h_s(vr). \tag{1.3.2}$$

Indeed, assume the contrary (here and henceforward, we omit the index $(k-s+1)'$ in the notation $X_n^{(k-s+1)'}$). Let $\mathrm{vol}_{s-1}[X_n \cap \partial B(P^*,\rho)] > 2(\beta_G + 2\varepsilon_0)(vr)^{-1}h_s(vr)$ for any $\rho, vr/2 \leqslant \rho \leqslant vr$. Then

$$\int_{vr/2}^{vr} \mathrm{vol}_{s-1}[X_n \cap \partial B(P^*,\ \rho)]\,d\rho > \frac{vr}{2}\, 2(\beta_G + 2\varepsilon_0)(vr)^{-1}h_s(vr) = (\beta_G + 2\varepsilon_0)\,h_s(vr),$$

i.e., we obtain

$$\begin{aligned}\varphi_s(vr,\ P^*,\ X_n) &= \int_0^{vr} \mathrm{vol}_{s-1}[X_n \cap \partial B(P^*,\ \rho)]\,d\rho \\ &\geqslant \int_{vr/2}^{vr} \mathrm{vol}_{s-1}[X_n \cap \partial B(P^*,\ \rho)]\,d\rho > (\beta_G + 2\varepsilon_0)\,h_s(vr),\end{aligned}$$

which is contrary to (1.3.1), thus completing the proof of (1.3.2).

Since $\check{K}_n(P^*,vr) \subset (\Pi, 2\xi vr)$, we can assume that the intersection $X_n \cap [\partial B(P^*,\rho_n)\backslash U_n^{k-s+1}]$ only consists of $q$-simplicial points, where $q \leqslant s-2$ (see Fig. 76). Recall that $X_n\backslash U_n^{k-s+1}$ is a finite simplicial subcomplex of dimension $q+1 \leqslant s-1$. This intersection structure can be obtained by an arbitrarily small stirring of the simplicial subcomplex $X_n\backslash U_n^{k-s+1}$, with the immediate consequence that

$$\begin{aligned}&\mathrm{vol}_{s-1}[X_n \cap \partial B(P^*,\ \rho_n)] \\ &\quad = \mathrm{vol}_{s-1}[\check{X}_n \cap \partial B(P^*,\ \rho_n)] \leqslant 2(\beta_G + 2\varepsilon_0)(vr)^{-1}h_s(vr).\end{aligned} \tag{1.3.3}$$

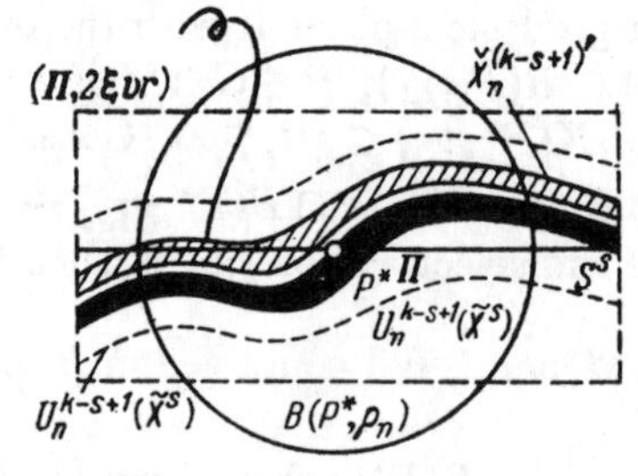


**Fig. 76**

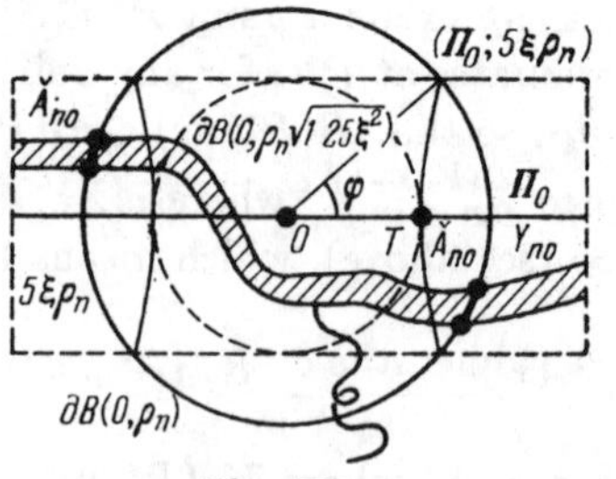


**Fig. 77**

Select the normal coordinates $\tau$ in the ball $B(P^*, vr)$, which are centered at the point $P^*$, and define the compact sets

$$
\begin{aligned}
&Y_{n0}=\tau^{-1}[X_n\cap\overline{B(P^*,\ \rho_n)}],\quad \check{Y}_{n0}=\tau^{-1}[\check{X}_n\cap\overline{B(P^*,\ \rho_n)}],\\
&A_{n0}=\tau^{-1}[X_n\cap\partial B(B^*,\ \rho_n)],\quad \check{A}_{n0}=\tau^{-1}[\check{X}_n\cap\partial B(P^*,\ \rho_n)],\\
&\Pi_0=\tau^{-1}(\Pi),\ \Gamma_{n0}=f_n[C(\Pi_0,\ A_{n0})],\quad \check{\Gamma}_{n0}=f_n[C(\Pi_0,\ \check{A}_{n0})],\\
&A^*_{n0}=f_n\pi A_{n0}=\Gamma_{n0}\cap\Pi_0,\quad \check{A}^*_{n0}=f_n\pi\check{A}_{n0}=\check{\Gamma}_{n0}\cap\Pi_0,\\
&A'_{n0}=A_{n0}\cup A^*_{n0},\quad \check{A}'_{n0}=\check{A}_{n0}\cup\check{A}^*_{n0},\\
&Y_n=\tau(Y_{n0}),\quad \check{Y}_n=\tau(\check{Y}_{n0}),\\
&A_n=\tau(A_{n0}),\quad \check{A}_n=\tau(\check{A}_{n0}),\quad \Gamma_n=\tau(\Gamma_{n0}),\quad \check{\Gamma}_n=\tau(\check{\Gamma}_{n0}),\\
&\check{A}^*_n=\tau(\check{A}^*_{n0}),\quad A^*_n=\tau(A^*_{n0}),\quad A'_n=\tau(A'_{n0}),\\
&\check{A}'_n=\tau(\check{A}'_{n0}),
\end{aligned}
$$

where $\pi$ denotes the orthogonal projection $\mathbf{R}^n\to\Pi$, $C(\Pi_0, A_{n0})$, the cylinder of the mapping $\pi: A_{n0}\to\pi(A_{n0})$, $\pi(A_{n0})\subset\Pi_0$, $f_n$ is the radial projection onto the boundary $\partial B(0,\rho_n)$ of the exterior of the ball $B(0,\rho_n\sqrt{1-25\xi^2})$ in $B(0,\rho_n)$, or the projection of the spherical layer, or, which is equivalent, its homothety onto the whole ball $B(0,\rho_n)$. Since $\check{K}_n(P^*, vr)\subset(\Pi, 2\xi vr)$, we have $\check{Y}_{n0}\subset(\Pi_0, 2(1+\eta_0 vr)\xi vr)\subset(\Pi_0, 5\xi\rho_n)$ (see Conditions (M)). Note that an analogous inclusion does not hold for the compact set $Y_{n0}$. It follows from Lemma 10.2., Theorem 10.2.2 in [289], and relation (1.3.1), as well as from the proof of Theorem 10.4.3 in [289] (see above) that

$$(1.3.4)\qquad \mathrm{vol}_s(\Gamma_n)\leqslant Z_1(\beta_\sigma+2\varepsilon_0)h_s(vr)\xi,$$

where $Z_1=8C_1\rho_n(vr)^{-1}(1+\eta_0\rho_n)^s(1-25\xi^2)^{-s/2}$, assuming that $1+\eta_0\rho_n\leqslant 5/4$ (see §28.1) and $C_1=C_1(s)$. Further, due to (1.3.3), we have $\mathrm{vol}_s(\Gamma_n)=\mathrm{vol}_s(\check{\Gamma}_n)$, since $\dim(\Gamma_n\backslash\check{\Gamma}_n)\leqslant s-1$, and $\dim(A_{n0}\backslash\check{A}_{n0})\leqslant s-2$. We are going to show that the projection $\pi(Y_{n0})$ completely covers the ball $\Pi_0\cap B(0,\rho_n\sqrt{1-25\xi^2})$. Assume the contrary, viz., let $\pi(Y_{n0})$ contain no points

$Q_n \in \Pi_0 \cap B(0, \rho_n\sqrt{1-25\xi^2})$. Put $Q'_n = f_n(Q_n) \in \Pi_0 \cap B(0,\rho_n)$. Let $g_n$ denote the radial projection from the point $Q'_n$ of the domain $B(0,\rho_n)\backslash Q'_n$ onto $\partial B(0,\rho_n)$. Let $\varphi_n = g_n f_n \pi$, and $Z_{n0} = \varphi_n(Y_{n0}), \check{Z}_{n0} = \varphi_n(\check{Y}_{n0})$. Note that $\check{A}^*_{n0} = \varphi(\check{A}_{n0})$, since $\pi(\check{A}_{n0}) \subset \overline{B(0,\rho_n)\backslash B(0,\rho_n\sqrt{1-25\xi^2})}$. Indeed, $0T = \rho_n \cos\varphi = \rho_n\sqrt{1-25\xi^2}$, or $T \in \partial B(0,\rho_n\sqrt{1-25\xi^2})$ (see Fig. 77).

Q.E.D.

Meanwhile, it is clear that, in general, $\pi(A_{n0})$ is not contained in $B(0,\rho_n)\backslash B(0,\rho_n\sqrt{1-25\xi^2})$ in contrast to $\pi(\check{A}_{n0})$. We prove that the compact set $(X_n\backslash Y_n) \cup (Z_n \cup g_n\Gamma_n)$ belongs to the class $\mathcal{O}$. We have already noted that the cylinder of the mapping $\pi : A_{n0} \to \Pi_0 \cap B(0,\rho_n)$ is the compact set $C(\Pi_0, A_{n0}) \cup [\Pi_0 \cap B(0,\rho_n)]$. Construct a homotopy $F_n$ contracting the compact set $A_{n0}$ into the plane $\Pi_0$ along the projection $\pi : F_n(t,x) = (1-t)x + t\pi(x)$, where the vector $x \in A_{n0}$, $x$ is the radius vector emanating from the point 0, and the vector $\pi(x) \in \Pi_0$, $F_n(0,x) = x$, $F_n(1,x) = \pi(x)$. It is evident that $F_n(1,\check{A}_{n0}) = \pi(\check{A}_{n0}) \subset B(0,\rho_n)\backslash B(0,\rho_n \cdot \sqrt{1-25\xi^2})$.

We prove that $(X_n\backslash Y_n) \cup [C(\Pi, A_n) \cup \pi(Y_n)] \in \mathcal{O}$. Carry out the homotopy $F_n(t,x)$ only until the value $t = \frac{1}{2}$, and consider the compact set $(X_n\backslash Y_n) \cup \bar{A}_n \cup F_n(1/2, Y_n) = \overline{X}_n$, where $\bar{A}_n = \{F(t, A_n), 0 \leqslant t \leqslant 1/2\}$. It is obvious that $F_n(1/2, Y_n)$ is homeomorphic to $Y_n$, and in particular, $F_n(1/2, A_n) \cong A_n$, i.e., $\bar{A}_n \cong A_n \times [0, 1/2]$. Let $\overline{Y}_n = \bar{A}_n \cup F(1/2, Y_n)$, $\overline{X}_n = (X_n\backslash Y_n) \cup \overline{Y}_n$. Then the embedding $i_1 : Y_n \to \overline{Y}_n$ and the projection $i_2 : \overline{Y}_n \to Y_n$ are determined, with $i_2 i_1 = 1_{Y_n}$, $i_1 i_2$ homotopic to $1_{\overline{Y}_n}$, or $i_2$ is a homotopy equivalence. Construct a continuous mapping $f : \overline{X}_n \to X_n$ by setting

$$f(x) = \begin{cases} x, & x \in [(X_n\backslash Y_n) \cup \overline{Y}_n]\backslash \overline{Y}_n, \\ i_2(x), & x \in \overline{Y}_n. \end{cases}$$

Then the mapping $f' : (\overline{X}_n, \overline{Y}_n) \to (X_n, Y_n)$ is a relative homeomorphism. It immediately follows from the commutative diagram

$$\begin{array}{ccccccccc} \tilde{h}_{q+1}(X_n, Y_n) & \to & \tilde{h}_q(Y_n) & \to & \tilde{h}_q(Y_n) & \to & \tilde{h}_q(X_n, Y_n) & \to & \tilde{h}_{q-1}(Y_n) \\ \uparrow\cong & & \uparrow\cong & & \uparrow & & \uparrow\cong & & \uparrow\cong \\ \tilde{h}_{q+1}(\overline{X}_n, \overline{Y}_n) & \to & \tilde{h}_q(\overline{Y}_n) & \to & \tilde{h}_q(\overline{X}_n) & \to & \tilde{h}_q(\overline{X}_n, \overline{Y}_n) & \to & \tilde{h}_{q-1}(\overline{Y}_n) \end{array}$$

(a similar diagram is also constructed in the cohomology case) that the homomorphisms $f_* : \tilde{h}_q(\overline{X}_n) \to \tilde{h}_q(X_n)$ are isomorphisms for all $q \in \mathbf{Z}$. Since the embedding $\bar{j} : \overline{X}_n \to M$ is homotopic to the mapping $f\bar{j} : \overline{X}_n \to M$, we have $f_* j_* \tilde{h}_*(\overline{X}_n) = i_* \tilde{h}_*(X_n)$, where $i : X_n \to M$ is an embedding, i.e., $\overline{X}_n \in \mathcal{O}$, which is just what was required.

Further, the compact set $(X_n\backslash Y_n) \cup [C(\Pi, A_n) \cup \pi(Y_n)]$ is obtained from $\overline{X}_n$ by a homotpy of the embedding $\bar{j}$, i.e., we have to complete the homotopy

$F_n(t,x)$; therefore, $(X_n\backslash Y_n)\cup[C(\Pi, A_n)]\cup\pi(Y_n) \in \mathcal{O}$. Applying the homotopy $f_n$ to this compact set, we obtain that the compact set $(X_n\backslash Y_n)\cup\Gamma_n\cup[f_n\cdot\pi(Y)]$ is in the class $\mathcal{O}$. Finally, by performing the last homotopy, i.e., by pushing $Q'_n$ onto the boundary $\partial B(0,\rho_n)$, we obtain

$$\begin{aligned}&(X_n\backslash Y_n)\cup(g_n\Gamma_n)\cup[g_nf_n\pi(Y_n)]\\&\qquad=(X_n\backslash Y_n)\cup(g_n\Gamma_n)\cup\varphi_n(Y_n)=(X_n\backslash Y_n)\cup(g_n\Gamma_n)\cap Z_n\in\mathcal{O}.\end{aligned}$$

It is clear that $(g_n\Gamma_n)\cup Z_n \supset (g_n\check{\Gamma}_n)\cup\check{Z}_n$. However, since $\check{\Gamma}_n \subset \partial B(P^*,\rho_n)$, we have $g_n\check{\Gamma}_n \equiv \check{\Gamma}_n$. Because $\check{Z}_{n0} \subset Z_{n0} \subset \Pi\cap\partial B(0,\rho_n)$, $\mathrm{vol}_s(Z_{n0}) = 0$. Therefore,

$$\begin{aligned}\mathrm{vol}_s(Z_n\cup g_n\Gamma_n)&=\mathrm{vol}_s(g_n\Gamma_n)=\mathrm{vol}_s(g_n\check{\Gamma}_n)=\mathrm{vol}_s(\check{\Gamma}_n)\\&=\mathrm{vol}_s(\Gamma_n)\leqslant Z_1(\beta_G+2\varepsilon_0)\,h_s(vr)\,\xi\end{aligned}$$

(see (1.3.4)). Let

$$\mathrm{vol}_s[X_n\backslash U_n^{k-s}(\hat{X}^{s+1})]=\omega_n^s+\varepsilon'_n,\qquad \varepsilon'_n\geqslant 0,\ \varepsilon'_n\to 0.$$

Then

$$\begin{aligned}\mathrm{vol}_s(\Omega_n)&=\mathrm{vol}_s\{[(X_n\backslash Y_n)\cup(g_n\Gamma_n)\cup Z_n]\backslash U_n^{k-s}(\tilde{X}^{s+1})\}\\&\leqslant\omega_n^s+\varepsilon'_n-2^{-s-1}\beta_Gh_s(vr)+Z_1\ (\beta_G+2\varepsilon_0)\,h_s(vr)\,\xi,\end{aligned}$$

where we have made use of relation (1.3.1), viz.,

$$\begin{aligned}&\mathrm{vol}_s(Y_n)=\mathrm{vol}_s(\check{Y}_n)=\mathrm{vol}_s[\check{K}_n(P^*,\ \rho_n)]\\&\leqslant\mathrm{vol}_s[\check{K}_n(P^*,\ vr/2)]>{}^1\!/_2\beta_Gh_s(vr/2)=\beta_G(vr)^s\,2^{-s-1}[1+(h_svr)/2]^{-s}\cdot\\&\qquad\geqslant\beta_G\cdot(vr)^s\,2^{-s-1}(1+h_svr)^{-s}=\beta_G2^{-s-1}h_s(vr).\end{aligned}$$

Hence,

$$\begin{aligned}\mathrm{vol}_s(\Omega_n)\leqslant\omega_n^s+\varepsilon'_n-h_s(vr)\,[2^{-s-1}\beta_G-\xi(\beta_G+2\varepsilon_0)\\ \times 8C_1(s)\,\rho_n\,(vr)^{-1}(1+\eta_0\rho_n)^s(1-25\xi^2)^{-s/2}],\end{aligned}$$

where $\varepsilon_0 = \varepsilon_0(\xi)$, $v = v(\xi)$, $\rho_n(vr)^{-1} < 1$; therefore, the expression in the square bracket is approximately equal to $2^{-s-1}\beta_G$ for sufficiently small $\xi$, with the consequence that it determines a positive number not depending on $n$. It immediately follows that $\mathrm{vol}_s(\Omega_n) \leqslant \omega_n^s + \varepsilon'_n - \tau$, where $\tau > 0$ and does not depend on $n$, i.e., there exists a subscript $N_1$ such that we have $\mathrm{vol}_s(\Omega_n) \leqslant \omega_n^s - (\tau - \varepsilon'_n) < \omega_n^s - \frac{\tau}{2} < \infty$ for all $n > N_1$, which is impossible, since $(X_n\backslash Y_n) \cup (g_n\Gamma_n) \cup Z_n \in \mathcal{O}$, and the number $\omega_n^s$ is constructed from the neighbourhood $U_n^{k-s}(\widetilde{X}^{s+1})$. The obtained contradiction means that the projection $\pi(Y_{n0})$ completely covers the ball $B(0, \rho_n\sqrt{1-25\xi^2})$; therefore, we have from Lemma 28.3.1 that $\text{vol}_s(Y_{n0}) = \text{vol}_s(\check{Y}_{n0}) \geqslant \gamma_s\rho_n^s(1-25\xi^2)^{-s/2}$, i.e., $\text{vol}_s(Y_n) \geqslant \gamma_s\rho_n^s(1-25\xi^2)^{-s/2}(1+\eta_0 vr)^{-s}$. Letting $\xi \to 0$, $n \to \infty$, $\rho_n \to \rho$, and then resorting to the arbitrariness of $r$, we obtain that $\Psi_x(P) \geqslant 1$.

Q.E.D.

As an important corollary, we derive that the system of inequalities $\Psi_s(P) \geqslant 1$, $P \in S^s$, $3 \leqslant s \leqslant k$ turns out to provide for the minimality of all surfaces $S^s$, viz., $\text{vol}_s(S^s) = \lambda_s$.

**Lemma 30.1.4.** *Let the compact set $X^3 = X^3(M)$ be obtained in a certain M-process, and $S^s \subset \widetilde{X}^3$, $3 \leqslant s \leqslant k$, $\widetilde{X}^3 \supset \widetilde{X}^s = \widetilde{X}^{s+1} \cup S^s$. Then the equalities $\text{vol}_s(S^s) = \text{vol}_s(\widetilde{X}^s\backslash\widetilde{X}^{s+1}) = \text{vol}_s(\widetilde{X}^3\backslash\widetilde{X}^{s+1}) = \lambda_s$ hold, where $\lambda_s$ is the sth component of the $\lambda$-vector of the M-process for $3 \leqslant s \leqslant k-1$, and $\text{vol}_k(\widetilde{X}^k\backslash A) = \text{vol}_k(\widetilde{X}^3\backslash A) = d_k$. Moreover, $\Psi_s(P) = 1$ almost everywhere on the set $S^s$, and $\psi_s^- = \text{vol}_s[S^s \cap B(P,r)]$ for almost all $r < R^{k-s}(P)$. If $\lambda_s = 0$, then $S^s = \emptyset$.*

**Remark.** Here, we have for the first time made use of M-process finiteness. All the constructions until now have been carried out for an arbitrary M-process. Besides, for the first time, we also resort to $\widetilde{X}^3 \in \mathcal{O}$.

**Proof.** We carry it out by induction on $s$, starting with $s = 2$. Since $S^2 = \emptyset$, and $\lambda_2 = 0$, due to the 2-stability of the variational class, the first induction step has already been made due to the trivial argument. Thus, let the lemma be proved for all subsets $S^{s'}$, where $3 \leqslant s' \leqslant s-1$; in particular, this means that $\text{vol}_{s-1}(\widetilde{X}^3\backslash\widetilde{X}^s) = \text{vol}_{s-1}(S^{s-1}) = \lambda_{s-1} < \infty$. It follows that $\text{vol}_s(\widetilde{X}^3\backslash\widetilde{X}^{s+1}) = \text{vol}_s(\widetilde{X}^s\backslash\widetilde{X}^{s+1}) = \text{vol}_s(S^s)$.

To prove that $\text{vol}_s(X^3\backslash X^{s+1}) \geqslant \lambda_s$, we assume the contrary, viz., let $\text{vol}_s(\widetilde{X}^3\backslash\widetilde{X}^{s+1}) < \lambda_s$. Since $\lambda_s = \lim_{n\to\infty}\omega_n^s$, where $\omega_n^s \leqslant \omega_{n+1}^s$, there exists a subscript $n_0$ such that $\text{vol}_s(\widetilde{X}^3\backslash\widetilde{X}^{s+1}) < \omega_{n_0}^s$. Consider a neighbourhood system $U_n^{k-1}(\widetilde{X}^{s+1})$ determining the numbers $\omega_n^s$ in the M-process. Then $\text{vol}_s[\widetilde{X}^3\backslash U_{n_0}^{k-s}] \leqslant \text{vol}_s(\widetilde{X}^3\backslash\widetilde{X}^{s+1}) < \omega_{n_0}^s$, which is impossible, $\widetilde{X}^3 \in \mathcal{O}$. Due to Lemma 10.2.3 in [289], we can find a countable family of disjoint balls $B(P_i, r_i)$ for each $\rho > 0$, where $5r_i < \rho$, $P_i \in S^s$, $B(P_i, r_i) \cap \widetilde{X}^{s+1} = \emptyset$ for any $i$, so that

$$S^s \subset \left[\bigcup_{i=1}^{\infty} B(P_i,\ r_i)\right] \cup \left[\bigcup_{i=p+1}^{\infty} B(P_i,\ 5r_i)\right], \tag{1.4.1}$$

where $p = 1, 2, 3, \ldots$. It follows from Lemma 30.1.3 that

$$\sum_{i=1}^{\infty} \gamma_s r_i^s = \sum_{i=1}^{\infty} \gamma_s r_i^s (1+h_s r_i)^{-s} (1+h_s r_i)^s \leqslant (1+h_s\rho)^s \sum_{i=1}^{\infty} h_s(r_i)$$

(since $\Psi_s(P_i) \geqslant 1) \leqslant (1+h_s\rho)^s \sum_{i=1}^{\infty} \psi_s^-(r_i, P_i)$

$$= (1+h_s\rho)^s \sum_{i=1}^{\infty} \lim_{n\to\infty} \left[ \inf_{q \geqslant n} \psi_s(r_i, P_i, X_q^{(k-s+1)'}) \right]$$
$$\leqslant (1+h_s\rho)^s \lim_{n\to\infty} \inf_{q \geqslant n} \sum_{i=1}^{\infty} \psi_s(r_i, P_i, X_q^{(k-s+1)'})$$

(since the balls $B(P_i, r_i)$ are disjoint)

$$\leqslant (1+h_s\rho)^s \lim_{n\to\infty} [\mathrm{vol}_s(X_n^{(k-s+1)'} \setminus \tilde{X}^{s+1})]$$
$$= (1+h_s\rho)^s \lim_{n\to\infty} (\omega_n^s + \varepsilon_n') = (1+h_s\rho)^s \lambda_s.$$

Thus, we have $\sum_{i=1}^{\infty} \gamma_s r_i^s \leqslant (1+h_s\rho)^s \lambda_s < \infty$ by M-process finiteness, i.e., the series $\sum_{i=1}^{\infty} \gamma_s r_i^s$ converges. Recall that, in general, we have found its own family of balls $B(P_i, r_i)$ for each $\rho > 0$. Making use of special balls (see (1.4.1)), we obtain $\mathrm{vol}_s^\rho(S^s) \leqslant \sum_{i=1}^{\infty} \gamma_s r_s^i \leqslant (1+h_s\rho)^s \cdot \lambda_s$, whence $\mathrm{vol}_s(S^s) \leqslant \lambda_s$ if we let $\rho$ tend to zero. Comparing this inequality with the above, we have $\mathrm{vol}_s(S^s) = \lambda_s$.

Now, consider the intersection $S^s \cap B(P, r)$, where $r < R^{k-s}(P)$, and take a countable family of balls $\{B(P_i, r_i)\}$ as in (1.4.1), but so that $B(P_i, r_i) \subset B(P, r)$ for each $i$. Since $\Psi_s(P) \geqslant 1$ on $S^s \cap B(P, r)$, by repeating the above argument for $S^s \cap B(P, r)$, we obtain

(1.4.2) $$\mathrm{vol}_s[S^s \cap B(P, r)] \leqslant \psi_s^- \leqslant \lim_{n\to\infty} \inf_{q\geqslant n} \mathrm{vol}_s[X_q^{(k-s+1)'} \cap B(P, r)].$$

Suppose now that $P \in S^s, r < R^{k-s}(P)$, and $r$ is chosen so that $\mathrm{vol}_s[S^s \cap \partial B(P, r)] = 0, \mathrm{vol}_s[X_n^{(k-s+1)'} \cap \partial B(P, r)] = 0$ for each $n$ from an infinite subsequence in $\{n\}$. Such values of $r$ are everywhere dense in the half-interval $[0, R^{k-s}(P)]$. Besides, let $\overline{B(P, r)} \cap U_n^{k-s}(\tilde{X}^{s+1}) = \emptyset$. Since $\Psi_s(P) \geqslant 1$ on the set $S^s \setminus \overline{B(P, r)}$, we obtain

(1.4.3) $$\mathrm{vol}_s[S^s \setminus \overline{B(P, r)}] \leqslant \lim_{n\to\infty} \inf_{q\geqslant n} \mathrm{vol}_s[(X_q^{(k-s+1)'} \setminus U_q^{k-s}) \setminus \overline{B(P, r)}],$$

repeating the above argument for $S^s \setminus \overline{B(P, r)}$, whence

$$\text{vol}_s[S^s \setminus B(P, r)] \leqslant \lim_{n\to\infty} \inf_{q \geqslant n} \text{vol}_s[(X_q^{(k-s+1)'} \setminus U_q^{k-s}) \setminus B(P, r)],$$

where we have used the choice of $r$. Thus,

$$\begin{aligned}\lambda_s = \text{vol}_s(S^s) &= \text{vol}_s[S^s \setminus B(P, r)] + \text{vol}_s[S^s \cap B(P, r)] \\ &\leqslant \lim_{n\to\infty} \inf_{q \geqslant n} \text{vol}_s[(X_q^{(k-s+1)'} \setminus U_q^{k-s}) \setminus B(P, r)] \\ &+ \lim_{n\to\infty} \inf_{q \geqslant n} \text{vol}_s[(X_q^{(k-s+1)'} \setminus U_q^{k-s}) \cap B(P, r)] \\ &\leqslant \lim_{n\to\infty} \inf_{q \geqslant n} \text{vol}_s[X_q^{(k-s+1)'} \setminus U_q^{k-s}] = \lim_{n\to\infty} (\omega_n^s) = \lambda_s,\end{aligned}$$

whence inequalities (1.3.2) and (1.4.3) are actually equalities. Therefore, $\psi_s^- = \text{vol}_s[S^s \cap B(P,r)]$ for almost all $r < R^{k-s}(P)$. In particular, $\Psi_s(P) = \lim_{r\to 0}\{\gamma_s^{-1} r^{-s} \text{vol}_s[S^s \cap B(P,r)]\}$. It remains to prove that $\Psi_s(P) = 1$ almost everywhere on $S^s$.

Denote by $Z_m$ a subset in $S^s$ on which $\Psi_s(P) \geqslant 1 + \frac{1}{m}, m = 1,2,3,\ldots$. Then $Z_m$ is closed in $M \setminus \widetilde{X}^{s+1}$. Assume $\text{vol}_s(Z_m) > 0$, choosing $\varepsilon$ so that $0 < \varepsilon < \text{vol}_s(Z_m)/2m$. Cover $Z_m$ with a set $G$ which is open in the manifold $M$, where $G \subset M \setminus \widetilde{X}^{s+1}$, so that $\text{vol}_s[G \cap (S^s \setminus Z_m)] < \varepsilon$. Due to Lemma 10.2.3 in [289], we can find a countable family of disjoint balls $B(P_i, r_i)$ for any $\rho > 0$, where $P_i \in Z_m$, $r_i < R^{k-s}(P_i), 5r_i < \rho$, $B(P_i, r_i) \subset G$, so that $Z_m \subset [\bigcup_{i=1}^{\infty} B(P_i, r_i)] \cup [\bigcup_{i=p+1}^{\infty} B(P_i, 5r_i)]$, $p = 1,2,3,\ldots$. It is clear that $B(P_i, r_i)$ covers the whole of $Z_m$, possibly, except a set $\xi_m \subset Z_m$ with $\text{vol}_s(\xi_m) = 0$. Then

$$\text{vol}_s(Z_m) + \varepsilon \geqslant \sum_{i=1}^{\infty} \text{vol}_s[S^s \cap B(P_i, r_i)] = \sum_{i=1}^{\infty} \psi_s(r_i, P_i, S^s)$$

(since $\Psi_s(P) \geqslant 1 + m^{-1}$) $\geqslant \sum_{i=1}^{\infty}(1 + m^{-1}) h_s(r_i)$

$$\begin{aligned}&= (1 + m^{-1}) \sum_{i=1}^{\infty} \gamma_s r_i^s (1 + h_s r_i)^{-s} \geqslant (1 + m^{-1})(1 + h_s\rho)^{-s} \sum_{i=1}^{\infty} \gamma_s r_i^s \\ &\qquad \geqslant (1 + m^{-1})(1 + h_s\rho)^{-s} \text{vol}_s^{\rho}(Z_m).\end{aligned}$$

Since this inequality must hold for any $\varepsilon > 0$ and any $\rho > 0$, $\text{vol}_s(Z_m) \geqslant (1 + m^{-1}) \text{vol}_s(Z_m)$, whence $\text{vol}_s(Z_m) = 0$ for any $m$. Since $\Psi_s(P) \geqslant 1$ on $S^s$, $\Psi_s(P) \geqslant 1$ almost everywhere $S^s$. For $s = k$, the argument is similar to the above in the last step; however, we do not have to involve the neighbourhoods $U_n^0, \bigcap_n U_n^0 = A$. Though, we can assume that $U_n^0$ are closed, and $U_n^0 \equiv A$.

Q.E.D.

### 30.2. Each Stratum Is a Smooth Minimal Submanifold, Except Possibly a Set of Singular Points of Measure Zero

As in the usual homology theory case, we can prove the following (see [289]).

**Lemma 30.2.1.** *Let $S^3 \subset \widetilde{X}^3 = \widetilde{X}^3(M)$, where the M-process is finite. Then there exists a subset $Z_s \subset S^s$ such that $\mathrm{vol}_s(Z_s) = 0$, and $S^s \backslash Z_s$ is an s-dimensional topological submanifold in the manifold $M$. The $\lambda$-vector introduced above can thereby be simply interpreted geometrically, viz., it is made up of the volumes of submanifolds $S^s \backslash Z_s \subset M$.*

**Lemma 30.2.2.** *Let $S^i \subset \widetilde{X}^3(M), 3 \leqslant i \leqslant k$, where the M-process is finite, and $K^i = S^i \backslash Z_i \backslash [\bigcup_{\alpha=3}^{i-1} \overline{S}^\alpha]$, where $\overline{S}^\alpha$ is the closure of $S^\alpha$ in $M$, and the set $Z_i$ is determined by Lemma 30.2.1. Then each set $K^i$ is a differentiable submanifold in the manifold $M$. Besides, $K^i$ is a minimal submanifold in the sense of classical differential geometry, i.e., its mean curvature is zero.*

The proof is based on the results of [351]–[353] and obtained separately for each dimension similarly to that of Theorem 10.7.1 in [289]; therefore, we will not dwell on it at length.

## §31 Proof of Global Minimality for Constructed Stratified Surfaces

### 31.1. Proof of the Basic Existence Theorem for a Globally Minimal Surface

Thus, let all the assumptions of Theorem 7.2.1 hold. Consider the class of surfaces $\{X\}_k$ and prove that it is nonempty. Indeed, let $\widetilde{X}^3 = \widetilde{X}^3(M)$ be the compact set obtained as a result of some M-process. If the variational problem makes sense, i.e., if the M-processes are finite, it then immediately follows from Lemma 30.1.4 that $\mathrm{vol}_k(\widetilde{X}^3 \backslash A) = \mathrm{vol}_k(\widetilde{X}^k \backslash A) = \mathrm{vol}_k(S^k) = d_k < \infty$, where $d_k = \inf \mathrm{vol}_k(Y \backslash A)$, $Y \in \mathcal{O}$. Thus, $\{X\}_k \supset \{\widetilde{X}^3(M)\}$ for all M-processes.

Now, let $X \in \{X\}_k$ be an arbitrary surface from this class. We have to find a subset $S^k \subset X$ satisfying the conditions of Theorem 7.2.1. Consider an infinite $k$-minimizing sequence $X_n^{(1)}$, where $X_n^{(1)} \equiv X$ for any $n$. Then $\mathrm{vol}_k[X_n^{(1)} \backslash A] = d_k$, and the M-process can be applied to the sequence. As the first step, we have to construct the density function $\Psi_k(P)$ which is, obviously, determined uniquely, since all the passages to the limit involved turn into a chain of the identity transformations. Therefore, the subset $S^k = \{P \in M \backslash A | \Psi_k(P) > 0\}$ is determined uniquely. Recall that $S^k = \emptyset$ if $d_k = 0$, and $S^k \neq \emptyset$ if $d_k > 0$. It is clear that $S^k \subset X$, for $\Psi_k(P) \equiv 0$ if $P \in X$.

In general, the second step has already been made nonunique, as in the general situation. *A fortiori*, the M-process is determined nonuniquely in lesser dimensions, which is just what leads to the appearance of many a globally minimal solution, with the absolute minimum uniqueness theorem not holding in the general case. It follows from Propositions 29.2.1–29.2 $(k-2)$ that we obtain as a result a certain compact set $\widetilde{X}^3 = \widetilde{X}^3(M)$ with the decomposition

$\widetilde{X}^3(M) = A \cup S^k \cup S^{k-1} \cup \cdots \cup S^3$, it being $X \cap \widetilde{X}^3(M) \supset A \cup S^k = \widetilde{X}^k(M)$. Thus, we have distinguished a uniquely determined subset $S^k$ from the compact set $X$ and included it into a certain compact set of the form $\widetilde{X}^3(M)$, $\Psi_k$ having truncated all pieces of lesser dimensions off $X$, and left only its $k$-dimensional part, which is nonempty if $d_k > 0$. We stress that the above M-process, possibly, has changed $X$ in dimensions $s \leqslant k-1$, which is, however, immaterial, since we stay within the framework of the prior variational class. Because $S^k \subset \widetilde{X}^3(M)$, we have $\mathrm{vol}_k(S^k) = \mathrm{vol}_k[\widetilde{X}^3(M)\backslash A] = vol_k[\widetilde{X}(M)\backslash A] = \mathrm{vol}_k(X\backslash A)$ due to Lemma 30.1.4, i.e., if $s = k$, and due to Lemmas 30.2.1 and 30.2.2, the subset $S^k$ satisfies all local properties stated in Theorem 7.2.1. Thus, we obtained a complete description of the class $\{X\}_k$, which is only made up of such $X$ that, for each of them, there exists an M-process possessing the property $X \cap \widetilde{X}^3(M) \supset A \cup S^k$, i.e., the $k$-dimensional part of $X$ is involved in a certain M-process, and (1) of Theorem 7.2.1 is thereby proved.

We now turn to (2). Starting with this point, the proof gets extremely complicated. Consider the class $\{X\}_k$. Let $d_{k-1} = \inf \mathrm{vol}_{k-1}(Y\backslash A\backslash S^k)$, $Y \in \{X\}_k$, $S^k = S^k(Y)$. The existence of a uniquely determined subset $S^k(Y) \subset Y$ has been proved in (1). Then there is a sequence $X_n$ such that $X_n \in \{X\}_k$, and $\mathrm{vol}_{k-1}(X_n\backslash S^k_n) = d_{k-1} + \varepsilon_n \to d_{k-1}$, where $\varepsilon_n \geqslant 0$, $\varepsilon_n \to 0$. Since all M-processes in the class $\mathcal{O}$ are finite, we have $\mathrm{vol}_{k-1}[\widetilde{X}^3(M)\backslash S^k] = \lambda_{k-1}(M) < \infty$ for any $\widetilde{X}^3(M) \in \{X\}_k$ (see Lemma 30.1.4, second step). It follows that $d_{k-1} \leqslant \lambda_{k-1}(M) < \infty$. It is required to show that $\{X\}_{k-1} \neq \emptyset$. Certainly, we possess the powerful means to "smooth" volumes, i.e., the M-process; however, to apply it here is impossible, as will be seen from the following. We shall only apply it to "smooth" $\mathrm{vol}_i$ for $i \leqslant k-1$. Construct the function $\Psi_k|_{\{X_n\}}$ from the sequence $X_n$. We can put $X_n = X^{(1)}_n$, in which case the construction is nonunique either, since we can select different subsequences in $\{X_n\}$ for the convergence. Consider a compact set $A \cup S^k$, where $S^k = \{P \in M\backslash A | \Psi|_{\{X_n\}}(P) > 0\}$. That $A \cup S^k$ is compact follows from relation (3.3.7) in § 29.3. Let $U^1_\alpha$ be a contracting system of open neighbourhoods such that $U^1_\alpha \supset U^1_{\alpha+1}$, and $\bigcap_\alpha U^1_\alpha = A \cup S^k$. We prove that, for any $\alpha$, there is a subscript $N(\alpha)$ such that the compact sets $A \cup S^k_{n'} = A \cup S^k(X_{n'})$ are contained in a neighbourhood $U^1_\alpha$ for all $n' > N(\alpha)$ (recalling that $S^k(Y)$ is defined for any $Y \in \{X\}_k$), where $n'$ ranges over a certain infinite subsequence in $\{n\}$. We stress that earlier we could guarantee a similar inclusion for $k$-dimensional portions of the $k$-minimizing sequence, e.g., in constructing the M-process, but only after restructuring the sequence $X^{(1)}_n$ to the sequence $X^{(1)'}_n$ (see Lemma 29.3.4). Here, we cannot apply Lemma 29.3.4, since surgeries in dimension $k$ may violate the fixed convergence of volumes to a minimum in dimension $k-1$. However, as will now be proved, the necessity of a topological surgery for compact sets in dimension $k$, in fact, falls away due to the minimality of surfaces in this maximal dimension, already proved.

Thus, we assume the contrary, viz., let there exist a subscript $\alpha$ such that

$S_n^k \cap (M\backslash U_\alpha^1) \neq \emptyset$ for all $n$, starting with a certain subscript $n = n_0$. Then there exists a point $P_0 \in M\backslash U_\alpha^1$ which is a limit one for a certain sequence $P_n \in S_n^k \cap (M\backslash U_\alpha^1)$. Furthermore, there is $R > 0$ such that $\overline{B(P_0, R)} \cap \overline{U}_m^1 = \emptyset$ for all subscripts $m > m_0$, where $m_0 \geqslant \alpha + 1$, since $\bigcap_m U_m^1 = A \cup S^k, \overline{U}_{m_0}^1 \subset U_\alpha^1$. Because $\overline{B(P_0, R)} \cap (A \cup S^k) = \emptyset$, $\Psi_k|_{\{X_n\}}(P) = 0$ for all points $P \in B(P_0, R)$. Hence, due to (3.3.9) in § 29.3, we derive that there exists $R_1 > 0$, $R_1 < \rho(P_0)$ (see Lemma 29.3.3) such that $\psi_k^+(R_1, P_0) \equiv 0$, where $\psi_k^+ = \psi_k^+|_{\{X_n\}}$, i.e., due to the definition of $\psi_k^+$, we have

$$0 = \lim_{\delta \to 0} \Big[ \sup_{|R_1 - r| < \delta,\ d(P_0, Q_i) < \delta} \tilde{\psi}_k(r', Q_i) \Big],$$

with

$$\tilde{\psi}_k(r', Q_i) = \lim_{n_s \to \infty} \mathrm{vol}_k [B(Q_i, r') \cap X_{n_s}] = \lim_{n_s \to \infty} \mathrm{vol}_k [B(Q_i, r') \cap S_{n_s}^k].$$

We stress that the equality

$$\mathrm{vol}_s [B(Q_i, r') \cap X_{n_s}] = \mathrm{vol}_k [B(Q_i, r') \cap S_{n_s}^k]$$

follows from $X_{n_s} \in \{X\}_k$ (see (1)), i.e., $\mathrm{vol}_k(X_{n_s} \backslash A) = \mathrm{vol}_k(S_{n_s}^k)$. Thus, there exist sequences $Q_i$ and $r_i$ such that $Q_i \to P_0$, $r_i \to R_1$, and $\mathrm{vol}_k[B(Q_i, r_i) \cap S_{n_i}^k] \to 0, i \to \infty$. Since $r_i \to R_1 > 0$, we can assume that $Q_i \in S_{n_i}^k$ (if necessary, we can replace $Q_i$ by a nearby point $\widetilde{Q}_i$ and decrease $r_i$).

We now fix a subscript $i$ and the compact set $X_{n_i} \in \{X\}_k$. As well as in the proof of (1), we can apply the M-process to the $k$-minimizing sequence $Y_\beta$, where $Y_\beta \equiv X_{n_i}$, $i$ is fixed, and $\beta = 1, 2, 3, \ldots$. As a result, we obtain a certain compact set $X'_{n_i} = \widetilde{X}^3(M)$ such that $A \cup S_{n_i}^k \subset X_{n_i} \cap X'_{n_i}$, and $\Psi_k|_{\{Y_\beta \equiv X_{n_i}\}}(P) \geqslant 1$ on the set $S^k(X'_{n_i}) = S^k(X_{n_i}) = S_{n_i}^k$ (note that, for each $i$, there is certainly its own M-process; see also Lemma 30.1.3). For brevity, we denote the density function $\Psi_k|_{\{Y_\beta \equiv X_{n_i}\}}$ by $\Psi_k^{(i)}$. Since $Q_i \in S^k(X_{n_i}) = S^k(X'_{n_i})$ (see above), we have $1 \leqslant \Psi_k^{(i)}(Q_i) = \lim_{r \to 0}[h_k^{-1}(r) \cdot \psi_k^{+(i)}(r, Q_i)]$ due to Lemma 30.1.3, where $\psi_k^{+(i)} = \psi_k^+|_{\{Y_\beta \equiv X_{n_i}\}}$ (see (3.3.6) in § 29.3). Because the function $h_k^{-1}(r)\psi_k^+(r, Q_i)$ is nondecreasing as $r$ grows from 0 to $R^0(Q_i)$ (see (3.3.6) in §29.3), $1 \leqslant h_k^{-1}(r)\psi_k^{+(i)}(r, Q_i)$ for all $r$, and $0 \leqslant r < R^0(Q_i)$. At the same time, $Q_i \to P_0$; therefore, $R^0(Q_i) \to R^0(P_0)$, i.e., there exists a $r_0$, $0 < r_0 < \min[R_1, R^0(P_0)]$ such that $0 < r_0 < R^0(Q_i)$ for all $i \geqslant i_0$, where $i$ is a certain, possibly large subscript, and, which is especially important, $r_0$ does not depend on $i$. In the following, we will again assume that $i$ is fixed, but $i \geqslant i_0$. Hence $1 \leqslant h_k^{-1}(r_0)\psi_k^{+(i)}(r_0, Q_i)$, or

$$h_k(r_0) \leqslant \psi_k^{+(i)}(r_0, Q_i) = \lim_{\delta\to 0}\left[\sup_{|r_0-r'|<\delta,\, d(Q_i, Q_j)<\delta} \tilde{\psi}_k^{(i)}(r', Q_j)\right],$$

which means that there exists a sequence of points $Q_j$ and of radii $r'_j$ such that $\lim_{j\to\infty} Q_j = Q_i$, $\lim_{j\to\infty}(r'_j) = r_0 > 0$, and $\psi_k^{(i)}(r'_j, Q_j) \geqslant h_k(r_0) - \gamma_j$, where $\lim_{j\to\infty}\gamma_j = 0$. Recall that $Y_\beta \equiv X_{n_i}$, $\beta = 1, 2, 3, \ldots$, i.e.,

$$\tilde{\psi}_k^{(i)}(r'_j, Q_j) \equiv \mathrm{vol}_k[B(Q_j, r'_j) \cap X_{n_i}]$$
$$= \mathrm{vol}_k\left[B(Q_j, r'_j) \cap S^k_{n_i}\right] \geqslant h_k(r_0) - \gamma_j.$$

Here, we have again made use of the fact that $Y_\beta \in \{X\}_k$; therefore, $\mathrm{vol}_k[B(P,r) \cap Y_\beta] = \mathrm{vol}_k[B(P,r) \cap S^k(Y_\beta)]$ (see (1)). Passing to the limit with respect to $j$, we obtain that $\mathrm{vol}_k[B(Q_i, r_0) \cap S^k_{n_i}] \geqslant h_k(r_0)$ due to the local Euclideaness almost everywhere of the set $S^k_{n_i}$.

From Lemmas 30.1.3, 30.2.1, and 30.2.2, and since $r_0 < R_1$, we have

$$0 < h_k(r_0) \leqslant \mathrm{vol}_k\left[B(Q_i, r_0) \cap S^k_{n_i}\right] \leqslant \mathrm{vol}_k\left[B(Q_i, r_i) \cap S^k_{n_i}\right].$$

Recall that $i \geqslant i_0$, and since $\lim_{i\to\infty} r_i = R_1$, $r_0 < R_1$, we can assume $r_i \geqslant r_0$ for $i \geqslant i_0$. However, proceeding from the assumption that $P_0 \in M\backslash U^1_\alpha$, we have obtained above that $\mathrm{vol}_k[B(Q_i, r_i) \cap S^k_{n_i}] \to 0$ as $i \to \infty$. The obtained contradiction proves the existence of the subscript $N(\alpha)$ for any $\alpha$. Note that the geometric meaning of the above reasoning is in the fact that the sequence of minimal surfaces $S^k_{n_i}$ cannot converge to one from which long and thin "whiskers" would grow, not affecting in the limit the behaviour of the density function $\Psi_k$.

Thus, we can choose a subsequence $X_n$ from $X_\alpha$ so that $S^k_\alpha = S^k(X_\alpha) \subset U^1_\alpha$ for any subscript $\alpha$, $\alpha \to \infty$. Denote the subsequence $\{\alpha\} \subset \{n\}$ again by $\{n\}$. One circumstance is important, viz., since the original sequence was not subject to '$S$-surgeries in dimension $k$ in contrast to the M-process, the basic relation $\mathrm{vol}_{k-1}(X_n\backslash S^k_n) \to d_{k-1} < \infty$ is preserved. Apply an M-process to the sequence $X_n$, so as to repeat the construction of the function $\Psi_k|_{\{X_n\}}$ at the first step. We then obtain a set $A \cup S^k$ around which $S^k_n$ are concentrated. In the second and all subsequent steps, the M-process is arbitrary. As a result, we obtain a compact set $\widetilde{X}^3(M) \in \mathcal{O}$ such that $A \cup S^k = \widetilde{X}^k(M)$, $\mathrm{vol}_k(\widetilde{X}^3\backslash A) = \mathrm{vol}_k(\widetilde{X}^k\backslash A) = \mathrm{vol}_k(S^k) = d_k < \infty$. Further, $\widetilde{X}^3(M) \supset S^{k-1}(M)$, and $\mathrm{vol}_{k-1}(\widetilde{X}^3\backslash\widetilde{X}^k) = \mathrm{vol}_{k-1}(\widetilde{X}^{k-1}\backslash\widetilde{X}^k) = \mathrm{vol}_{k-1}(S^{k-1}) = \lambda_{k-1} < \infty$, where $\lambda_{k-1}$ is determined by the M-process, and we can assume that it is constructed from the system of neighbourhoods $U^1_n$. It immediately follows from the proof of Lemmas 30.2.1 and 30.2.2, second step, that the subset $S^{k-1}(M) \subset \widetilde{X}^3(M)$ satisfies all the conditions of (2) of Theorem 7.2.1. Since $\mathrm{vol}_{k-1}(X_n\backslash S^k_n) \geqslant \mathrm{vol}_{k-1}(X_n\backslash U^1_n)$, we have, $\mathrm{vol}_{k-1}(X_n\backslash S^k_n) \geqslant \omega^{k-1}_n$, and $\lim_{n\to\infty}\mathrm{vol}_{k-1}(X_n\backslash S^k_n) \geqslant \lim_{n\to\infty}\omega^{k-1}_n$ i.e., $d_{k-1} \geqslant \lambda_{k-1}$. On the other hand, since $\widetilde{X}^3 \in \{X\}_k$, we have $\lambda_{k-1} = \mathrm{vol}_{k-1}(\widetilde{X}^3 \backslash \widetilde{X}^k) \geqslant d_{k-1}$, or $d_{k-1} = \lambda_{k-1}$. Thus, the compact set $\widetilde{X}^3(M) \in \{X\}_{k-1}$, and we have proved that $\{X\}_{k-1} \neq \emptyset$. Now, let $X \in \{X\}_{k-1}$ be an arbitrary compact set from this class. Similarly to the proof of the first part, we distinguish a compact set $A \cup S^k$ from $X$, which is determined uniquely. Then $\mathrm{vol}_{k-1}(X \backslash A \backslash S^k) = d_{k-1} < \infty$, and we can consider a sequence of compact sets $X_n^{(2)} \equiv X$, $n = 1, 2, 3, \dots$, constructing the density function $\Psi_{k-1}|_{\{X_n^{(2)}\}}$ on $M \backslash A \backslash S^k$. Further, consider $S^{k-1} \subset M \backslash A \backslash S^k$, $S^{k-1} = \{P \in M \backslash A \backslash S^k | \Psi_{k-1}(P) > 0\}$, and apply the M-process to the sequence $X_n^{(2)}$, which yields a certain compact set $\widetilde{X}^3(M)$ such that $A \cup S^k \cup S^{k-1} \subset X \cap \widetilde{X}^3(M)$, with the sets $A, S^k, S^{k-1}$ being involved in the decomposition of $\widetilde{X}^3(M)$, which is determined by the $M$-process. It follows from the proof of Lemma 30.1.4, second step, that $\mathrm{vol}_{k-1}(S^{k-1}) = \lambda_{k-1} = \lambda_{k-1}(M) < \infty$. Since $\widetilde{X}^3(M) \in \mathcal{O}$, we have $\mathrm{vol}_{k-1}(\widetilde{X}^3 \backslash A \backslash S^k) = \mathrm{vol}_{k-1}(S^{k-1}) \leqslant d_{k-1}$, i.e., $\lambda_{k-1} \geqslant d_{k-1}$. On the other hand, if $U$ is an open neighbourhood $A \cup S^k$, then $\mathrm{vol}_{k-1}(\widetilde{X}^3 \backslash U) \leqslant \mathrm{vol}_{k-1}(\widetilde{X}^3 \backslash A \backslash S^k)$, or $\omega_n^{k-1} \leqslant \mathrm{vol}_{k-1}(\widetilde{X}^3 \backslash A \backslash S^k) = \mathrm{vol}_{k-1}(X \backslash A \backslash S^k) = d_{k-1}$. Hence, $\lambda_{k-1} \leqslant d_{k-1}$, or $d_{k-1} = \lambda_{k-1}$. Since $S^{k-1} \subset \widetilde{X}^3(M)$, we obtain from Lemmas 30.2.1 and 30.2.2 that $S^{k-1} \subset X$ satisfies all the conditions of (2) of Theorem 7.2.1, and the second part is proved.

By extending the above argument to lesser dimensions, we obtain the proof of Theorem 7.2.1.

## 31.2. The Proof of the Theorem on the Coincidence of the Least Stratified Volume with the Least $\lambda$-Vector in a Variational Class

In fact, we have already proved Theorem 29.5.1 in Theorem 7.2.1. In particular, we have shown that $d_s = \lambda_s$ for $3 \leqslant s \leqslant k$, and that each compact set $X \in \{X\}_3$ is a result of a certain M-process. That the vector $(d_k, d_{k-1}, \dots, d_3)$ is least in lexicographic ordering also follows from the proof of Theorem 7.2.1. Thus, Theorem 29.5.1 is proved.

Theorems 7.2.1 and 29.5.1 state the existence of a globally minimal surface in all dimensions and each 2-stable class $\mathcal{O}(\widetilde{\mathcal{O}})$. However, all the compact sets $\widetilde{X}^3(M)$ obtained in an arbitrary M-process also possess relative minimality properties. More precisely, $\widetilde{X}^3(M) = A \cup S^k \cup S^{k-1} \cup \cdots \cup S^3$, where $\mathrm{vol}_s(S^s) = \lambda_s$, and $\lambda_s = \mathrm{infty\,vol}_s(X \backslash \widetilde{X}^{s+1})$, $X \in \mathcal{O}(\widetilde{\mathcal{O}})$, with $X \subset \widetilde{X}^{s+1}$. Indeed, if there existed a compact set $X$ such that $X \subset \widetilde{X}^{s+1}$, and $\mathrm{vol}_s(X \backslash \widetilde{X}^{s+1}) < \lambda_s - \varepsilon, \varepsilon > 0$, then the inequality $\mathrm{vol}_s(X \backslash U) < \mathrm{vol}_s(X \backslash \widetilde{X}^{s+1}) < \lambda_s - \varepsilon$ would hold for any open neighbourhood $U = U(\widetilde{X}^{s+1})$, or $\omega_n^s \leqslant \lambda_s - \varepsilon$, $\lambda_s \leqslant \lambda_s - \varepsilon$, which is impossible. Thus, it means from the geometric point of view that the film $S^s$ is a globally minimal surface in the absolute sense if we consider only such compact sets $X$ that contain $\widetilde{X}^{s+1}$, i.e., provided that the compact contour $\widetilde{X}^{s+1}$ is fixed in the manifold $M$.

In the case of usual homological Plateau problem, we always assume that variational problem makes sense, i.e., that there exists the film $X$ that $\partial X = A$ and $\text{vol}_k X \leqslant \infty$. In the general case, the assumption that $M$-process is finite has the same sense, i.e., variational problem in the class $\mathcal{O}$ makes sense. This condition is the generalization on all dimensions of the usual condition $\text{vol}_k X \leqslant \infty$.

## §32 The Fundamental (Co)Cycles of Globally Minimal Surfaces. Exact Realization and Exact Spanning

### 32.1. Fundamental (Co)Cycle Theorem

Let $H_*^{(*)}$ be the usual (co)homology theory, which is continuous and relatively invariant on the category of compact pairs (see above for the construction of such theories). Consider an arbitrary variational class $\mathcal{O} = H(A, L, L')$, where $3 \leqslant k \leqslant n-1$, $n = \dim M$, $L \subset H_{k-1}^{(k-1)}(A)$, $L' \subset H_k^{(k)}(M)$. As was proved above, $\mathcal{O}$ is $(k-1)$-stable. It is also clear that all the M-processes are reduced here only to implementing the first step in dimension $k$, i.e., $A \cup S^k = \widetilde{S}^3$, $S^\alpha = \emptyset$ for $3 \leqslant \alpha \leqslant k-1$. Besides, as $d_k < \infty$, any M-process is finite in $\mathcal{O}$. Let $X_0 \in H(A, L, L')$ be a globally minimal surface. Then $\dim X_0 = k$.

**Theorem 32.1.1.** *Let $M$ be a compact, smooth Riemannian manifold, $A \subset M$, a fixed compact contour, $G$ the coefficient group of a homology theory $H_*$, $G$ a vector space over a certain field $F$, and $\dim_F G = 1$. Assume that there are no $k$-cycles in $A$, i.e., that $H_k(A, G) = 0$. For example, let $\dim A = k-1$, where $3 \leqslant k \leqslant n-1$. We state that*

*(1) if $X_0 \in H_*(A, L, 0)$, $\dim X_0 = k$, is a globally minimal surface, where $L \subset H_{k-1}(A, G)$, and $L \neq 0$, then $H_k(X_0, G) = 0$;*

*(2) if $X_0 \in H_*(A, L, L')$, $\dim X_0 = k$ is a globally minimal surface, where $L' \subset H_k(M, G), L' \neq 0$, then the homomorphism $i_* : H_k(X_0, G) \to H_k(M, G)$, where $i : X_0 \to M$ is an embedding, is a monomorphism.*

**Remark.** The theorem has a clear geometric meaning. In (1), we state that if the boundary curve $A$ contains no $k$-cycles, then a minimal film of spanning type for $A$ contains no $k$-cycles either, i.e., it is arranged similarly to "a manifold with boundary $A$." It turns out in (2) that the minimality of a film of realizing type entails that of its $k$-homology, and none of the $k$-cycles is annihilated under the embedding $X \to M$, i.e., each is involved in minimizing the $k$-dimensional volume. The proof is based on the following.

**Lemma 32.1.1.** *Let $X_0$ be a metric compact set of dimension $k$, $X_0 \supset R$, $R$ a certain compact subset, and the homology theory coefficient group $G$ a vector space over a certain field $F$, $\dim_F G = 1$, $H_k(R, G) = 0$, $X_0 \backslash R$ an open topological $k$-dimensional manifold, and $l' \in H_k(X_0, G)$, $l' \neq 0$. Then there* *exists an open disk* $D = D^k(l') \subset X_0 \backslash R$ *such that if* $i_* : H_k(X_0\backslash R) \to H_k(X_0)$ *is a homomorphism induced by the embedding, then (1)* $i_*$ *is a monomorphism,* $\mathrm{Codim}(\mathrm{Im}\, i_*) = 1$ *in the group* $H_k(X_0, G)$, *and* $l' \in \mathrm{Im}\, i_*$.

**Proof.** Consider the diagram

$$\begin{array}{ccccc}
 & H_{k-1}(X_0\backslash D) & & & \\
 & \uparrow & & & \\
 & H_k(S^k) & & & \\
 & \uparrow j_* & & & \\
0 \to & H_k(X_0) & \xrightarrow{\varphi_*} H_k(X_0, R) & \xrightarrow{f_*} & H_k(X_0/R) \\
\uparrow & \uparrow i_* & \uparrow \alpha_* & & \uparrow \omega_* \\
0 \to & H_k(X_0\backslash D) & \xrightarrow{\psi} H_k(X_0\backslash D, R) & \xrightarrow{f'_*} & H_k((X_0/R)\backslash f(D)) \\
 & \uparrow & & & \\
 & 0 & & &
\end{array}$$

where $D = D^k$ is a certain open disk in $X_0\backslash R$, $f : X_0 \to X_0/R$ is a projection, $x = f(R)$, $S^k = X_0/(X_0\backslash D)$ (recalling that $H_*(A, B) = H_*(A/B)$), and $f_*$, $f'_*$ are isomorphisms. Put $l = f_*\varphi_*(l')$. Then $l \neq 0$. Let $T = X_0/R$. Assume that we have successfully found a disk $f(D) \subset T\backslash x$ such that $l \notin \mathrm{Im}\,\omega_*$. Then $l' \notin \mathrm{Im}\,\omega_*$, and we have thereby reduced the problem to the study of the compact set $T$ which is a topological manifold everywhere except at one point $x$. We introduce some metric $d$ on $T$. The group $H_k(T)$ is the inverse limit of the spectrum of groups $H_k(T_\alpha)$, where $\alpha \in \mathrm{Cov}^f\, T$, and $T_\alpha$ is a nerve of the covering $\alpha$. Since $k = \dim T$, we can select a cofinal subset $\overline{\mathrm{Cov}}^f\, T$ in the set $\mathrm{Cov}^f\, T$, consisting of finite coverings of multiplicity not exceeding $k$, and whose diameters tend to zero, which means that the nerves $T_\alpha$, $\alpha \in \overline{\mathrm{Cov}}^f\, T$ are finite simplicial complexes, $\dim T_\alpha \leqslant k$. Since $l \neq 0$, there exists an element $\alpha_0 \in \overline{\mathrm{Cov}}^f\, T$ such that the representative $l$, an element $l_{\alpha_0} \in H_k(T_{\alpha_0})$, is other than zero. Let $U \in \alpha_0$ be an element of the covering $\alpha_0$, where $x \in U$, and $T\backslash\overline{U} \neq \emptyset$ (we can assume that $\alpha_0$ has already been taken to be sufficiently fine). Let $U \in T_{\alpha_0}$ be the vertex associated with $U \in \alpha_0$. Consider an open ball $B(x, r)$ such that $\overline{B(x, r)} \subset U$. Let $\varepsilon > 0$ be such that $\overline{B(x, r+\varepsilon)} \subset U$. Construct a finite covering $\kappa$ of the compact set $T\backslash B(x, r)$ with open sets of $f(D)$, homeomorphic to the disk $D = D^k$. Let $\eta > 0$ be the Lebesgue number of the covering $\kappa$, $\xi = \min(\varepsilon, \eta)$, $\xi > 0$. Let a covering $\alpha_1$ be a refinement of $\alpha_0$, so that $\alpha_1 \in \overline{\mathrm{Cov}}^f\, T$, and $\mathrm{diam}(\alpha) < \xi$. Then there exists a homomorphism $\pi^{\alpha_1}_{\alpha_0} : H_k(T_{\alpha_1}) \to H_k(T_{\alpha_0})$, and $l_{\alpha_1} \in H_k(T_{\alpha_1})$, $\lambda_{\alpha_1} \neq 0$, so that $\pi^{\alpha_1}_{\alpha_0}(l_{\alpha_1}) = l_{\alpha_0}$. Let $\mathrm{Car}\, \hat{l}_{\alpha_1}$ be the carrier of an element $l_{\alpha_1}$ in the nerve in $T_{\alpha_1}$. Since $\dim T_{\alpha_1} = k$, the homology in $k$-chains is fixed; therefore, there is a uniquely determined set of simplexes $\Delta^k$ in $T_{\alpha_1}$ onto which the chain $\hat{l}_{\alpha_1}$ is projected.

There are two possibilities for the position of vertices $s$ from $\mathrm{Car}\, \hat{l}_{\alpha_1}$ in $T$, viz., (1) $S \cap B(x, r) \neq \emptyset$ for any vertex $s \in \mathrm{Car}\, \hat{l}_{\alpha_1}$, and (2) there exists a vertex $s_0 \in \operatorname{Car} \hat{l}_{\alpha_1}$ such that $S_0 \subset T \backslash B(x,r)$, where $S = S(s)$. In (1), since $S \cap B(x,r) \neq \emptyset$, we have $S \subset B(x, r+\varepsilon)$, for $\operatorname{diam}(\alpha_1) < \xi \leqslant \varepsilon$, and then *a fortiori* $S \subset U$, which means that all $s \in \operatorname{Car} \hat{l}_{\alpha_1}$ are contained in $U$. We map $T_{\alpha_1}$ into $T_{\alpha_0}$ by sending all $s \in \operatorname{Car} \hat{l}_{\alpha_1}$ into $u \in T_{\alpha_0}$, and the other $s' \in \operatorname{Car} \hat{l}_{\alpha_1}$ arbitrarily with respect to the inclusion. Since the homomorphism $\varphi : H_k(T_{\alpha_1}) \to H_k(T_{\alpha_0})$ does not depend on the choice of a simplicial mapping with respect to the inclusion $T_{\alpha_1} \to T_{\alpha_0}$, we have $\varphi(l_{\alpha_1}) = l_{\alpha_0}$, $\varphi = \pi_{\alpha_0}^{\alpha_1}$. On the other hand, $\varphi(l_{\alpha_1}) = 0$, or $l_{\alpha_0} = 0$, which is contrary to the choice of this element.

Thus, (2) is, in fact, realized. Since $S_0 \subset T \backslash B(x,r)$, and $\operatorname{diam} S_0 \leqslant \operatorname{diam}(\alpha_1) < \xi \leqslant \eta$, $S_0$ is wholly contained in a certain open disk $f(D^k)$ of the covering $\kappa$ (see above). Let $\omega : T \backslash f(D^k) \to T$ be an embedding. Construct the spectrum mapping $H_k(\widetilde{T}_\alpha) \to H_k(T_{\alpha'})$, where $\widetilde{T} = T \backslash f(D^k)$, $\alpha = \omega^{-1}(\alpha')$, and $\omega^{-1} : \overline{\operatorname{Cov}}^f T \to \overline{\operatorname{Cov}}^f \widetilde{T}$ associates each $\alpha$ with its inverse image on $\widetilde{T}$ under the embedding $\omega$. Consider the embedding $\omega_{\alpha_1}^{\omega^{-1}(\alpha_1)} : \widetilde{T}_{\omega^{-1}(\alpha_1)} \to T_{\alpha_1}$. Then the star $\operatorname{St}(s_0) \subset T_{\alpha_1}$ does not belong to the image $\widetilde{T}_{\omega^{-1}(\alpha_1)}$, i.e., $l_{\alpha_1} \notin \operatorname{Im} H_k(\widetilde{T}_{\omega^{-1}(\alpha_1)})$, whence $l \notin \operatorname{Im} \omega_*$, and thereby $l' \in \operatorname{Im} i_*$.

Thus, $l' \in \operatorname{Im} i_*$. It follows from the diagram that $\operatorname{Im} j_* = 0$, and since $\dim G = 1$, we have $\operatorname{Im} j_* = G$, i.e., $H_k(X_0) = G \oplus H_k(X_0 \backslash D)$, which is just what is required.

Q.E.D.

**Proof of Theorem 32.1.1.** We first consider (1). Put $R = A \cup Z$, where $Z = Z_k$ is the set of singular points in $S^k$. By Theorem 7.2.1, we have $\operatorname{vol}_k(Z) = 0$. Since $H_k(A) = 0$ by assumption, it is clear that $H_k(R) = 0$, which follows, e.g., from Lemma 19.A in [351] and its corollaries. Let $X_0 \in H_*(A, L, 0), L \neq 0, l \subset H_{k-1}(A)$. Assume that there exists $l' \in H_k(X_0), l' \neq 0$. By Lemma 32.1.1, there is a disk $D = D^k(l')$ such that $l' \notin \operatorname{Im} i_*$, where $i : X_0 \backslash D \to X_0$ is an embedding. Consider the commutative diagram

$$
\begin{array}{ccccccccccc}
0 & \longrightarrow & H_k(X_0\backslash D) & \xrightarrow{i_*} & H_k(X_0) & \xrightarrow{j_*} & G & \xrightarrow{\varphi_*} & H_{k-1}(X_0\backslash D) & \xrightarrow{\varphi_*} & H_{k-1}(X_0) \longrightarrow 0 \\
 & & & & & & & & \nwarrow{\scriptstyle \alpha_*} & & \nearrow{\scriptstyle \beta_*} \\
 & & & & & & & & & H_{k-1}(A) &
\end{array}
$$

It is clear that $\operatorname{Im} j_* \neq 0$. Since $\dim G = 1$, we have that $\operatorname{Im} j_* = G$, or $\psi_*$ is a monomorphism. Because $\operatorname{vol}_k(X_0) > \operatorname{vol}_k(X_0 \backslash D)$, we have $X_0 \backslash D \in H_*(A, L, 0)$, i.e., there exists $l \in H_{k-1}(A), l \neq 0$, such that $\alpha_*(l) \neq 0$; however, since $\beta_*(l) = 0$, we have $\psi_* \alpha_*(l) = 0$, which is contrary to $\psi_*$ being a monomorphism, and (1) is thus proved.

Consider (2). Let $X_0 \in H_*(A, L, L')$, where $L \subset H_{k-1}(A), L' \neq 0$. Assume that there exists $l' \in H_k(X_0), l' \neq 0$, such that $i_*(l') = 0$, where $i_* : H_k(X_0) \to H_k(M)$ is induced by the embedding. Due to Lemma 32.1.1, there is a disk

$D = D^k(l') \subset X_0 \backslash (A \cup Z_k)$ such that $l' \in \operatorname{Im} j_*$, and $\operatorname{Codim}(\operatorname{Im} j_*) = 1$ in $H_k(X_0)$, $j : X_0 \backslash D \to X_0$ being an embedding. Then there exists a decomposition into the direct sum of vector subspaces $H_k(X_0) = G \oplus \operatorname{Im} j_*$, where the one-dimensional subspace $G$ is spanned by the vector $l'$. Let $\alpha$ be an arbitrary element from $L'$, $\alpha \neq 0$. Then $\alpha = i_*(\varphi)$, where $\varphi \in H_k(X_0)$, $\varphi = \lambda l' + \varphi'$, $\varphi' \neq 0$ (otherwise, $\alpha = \lambda i_*(l') = 0$), and $\varphi' \in \operatorname{Im} j_*$. Hence, $\alpha = i_*(\varphi) = i_*(\varphi') = (ij)_*(\varphi'')$, $\varphi'' \in H_k(X_0 \backslash D)$, i.e., $L' \subset \operatorname{Im}(ij)_*$, and $X_0 \backslash D \in H_*(A, 0, L')$. Moreover, to prove that $X_0 \backslash D \in H_*(A, L, L')$, let us consider the commutative diagram

$$
\begin{array}{ccccccccccccc}
0 & \longrightarrow & H_k(X_0\backslash D) & \xrightarrow{j_*} & H_k(X_0) & \xrightarrow{\tau_*} & G & \xrightarrow{\xi_*} & H_{k-1}(X_0\backslash D) & \xrightarrow{\rho_*} & H_{k-1}(X_0) & \longrightarrow & 0 \\
 & & {\scriptstyle\varphi_*}\searrow & & \swarrow{\scriptstyle i_*} & & & & {\scriptstyle\psi_*}\nwarrow & & \nearrow{\scriptstyle\omega_*} & & \\
 & & & H_k(M) & & & & & & H_{k-1}(A) & & &
\end{array}
$$

where $L' \subset \operatorname{Im} \varphi_*$. Similarly to the proof of (1), we obtain that $\rho_*$ is a monomorphism, and then $\psi_*(L) = 0$, i.e., $X_0 \backslash D \in H_*(A, L, L')$, which is impossible because of the inequality $\operatorname{vol}_k(X_0 \backslash D) < \operatorname{vol}_k(X_0)$.

Q.E.D.

It is obvious that we cannot drop the assumption that $H_k(A) = 0$ in the statement of Theorem 32.1.1, since, otherwise, nontrivial cycles could be projected onto the boundary $A$. We cannot either omit the assumption that the homology theory coefficient group is a one-dimensional vector space over a certain field. In fact, we put, e.g., $G = S^1 = \mathbf{R}^1(\bmod 1)$ (i.e., the circle) and prove the existence of the minimal surface $X_0 \in H_*(A, L, 0)$ possessing not only a nontrivial group $H_k(X_0, S^1)$ but also realizing a nontrivial cycle in the manifold $M$. Consider $M = S^1 \times RP^{2s}$, where $\mathbf{R}P^{2s}$ is a real projective space, and put $A = RP^{s-1} \subset RP^{2s} = X_0$. It is clear that $X_0 \in H_*(A, S^1, 0)$, and $X_0$ is a globally minimal surface in $M$, but realizes the subgroup $L' \neq 0$, $L' = \mathbf{Z}_2 \subset H_{2s}(M, S^1)$.

## 32.2. Exact Minimal Realization and Exact Minimal Spanning

From the standpoint of minimal realizations, the following question is of interest: In which cases can a fixed subgroup be realized by means of a globally minimal surface whose homology group is isomorphic to the subgroup?

In other words, in what cases does a minimal surface realize more cycles than there are in a prescribed subgroup, and when does the surface realize precisely as many cycles as chosen *a priori*?

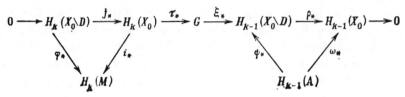

Let $\mathcal{O}_* = H_*(\emptyset, 0, L')$, $L' \neq 0$, $L' \in H_k(M)$. Put $\operatorname{vol}(L') = d_k(\emptyset, 0, L')$, and $L_1' = L_2'$, $L_2'/L_1' \neq 0$. Is it true that $\operatorname{vol}_k(L_1') < \operatorname{vol}_k(L_2')$, i.e., that the subgroup volume also is strictly decreasing as the number of cycles decreases? Elementary examples demonstrate that this statement generally does not hold, i.e., that there are cases where $\mathrm{vol}_k(L'_1) = \mathrm{vol}_k(L'_2)$. However, the following can be proved easily.

**Proposition 32.2.1.** *Let $M$ be a compact, smooth Riemannian manifold, and $G$ a one-dimensional vector space over a certain field $F$. Then any subgroup $L' \neq 0$, $L' \subset H_k(M,G)$ contains a subgroup $\tilde{L}'$ such that* $\dim L' - \dim \tilde{L}' = 1$, *and* $\mathrm{vol}_k(\tilde{L}') < \mathrm{vol}_k(L')$. *Moreover, if $\mathcal{O}_* = H_*(A,L,L')$, where $L' \neq 0$, $H_k(A,G) = 0$, $\dim_F G = 1$, then there always exists a subgroup $\tilde{L}' \subset L'$,* $\dim L' - \dim \tilde{L}' = 1$ *such that $d_k(A,L,L') > d_k(A,L\tilde{L}')$.*

The proof is similar to that of Theorem 32.1.1; therefore, we shall not repeat the argument. It follows that if all the assumptions are fulfilled and $L' \subset H_k(M,G)$ is an arbitrary subgroup, then there exists a descending series of subgroups $L' = L'_0 \supset L'_1 \supset \cdots \supset L'_N$, so that $\dim L'_p - \dim L'_{p+1} = 1$ for $0 \leqslant p \leqslant N-1$, $\dim L'_N = 1$, and $\mathrm{vol}_k(L'_p) > \mathrm{vol}_k(L'_{p+1})$.

**Definition 32.2.1.** We will say that a subgroup $L' \subset H_k(M)$ admits an exact minimal realization if there is a minimal surface $X_0 \in H_*(\emptyset, 0, L')$ such that the embedding $i : X_0 \to M$ induces the isomorphism $i_* : H_k(X_0) \simeq L'$.

Simple examples show that there are such subgroups $L' \in H_k(M)$ that do not admit an exact minimal realization in the given Riemannian metric. Moreover, such subgroups form the "majority" among all possible subgroups in $H_k(M)$. Meanwhile, we can always make up an additive basis in $H_k(M)$ of subgroups $L'$ admitting an exact minimal realization, and not only indicate them. It turns out that this is a reflection of a certain general fact to be now proved. We denote by $\{a_1, \ldots, a_m\}$ the subgroup generated by the elements $a_1, \ldots, a_m \in H_k(M)$.

**Proposition 32.2.2.** *Let $M$ be a compact, smooth, and closed Riemannian manifold, and the homology theory coefficient group $G$ be a group $Z_p$, where $p \neq 0$, $p$ is a prime number, and $r = \dim H_k(M, \mathbb{Z}_p)$. If $r > 1$, then there exists a basis $l_1, \ldots, l_r$ in the group $H_k(M, \mathbb{Z}_p)$, so that (1) $\mathrm{vol}_k\{l_i\} < \mathrm{vol}_k(H_k(M))$ for any $i$, and $\mathrm{vol}_k\{l_i\} < \mathrm{vol}_k\{l_1, \ldots, l_r\}$ for $i > 1$, (2) all one-dimensional subgroups $\{l_i\}$, $1 \leqslant i \leqslant r$, admit an exact minimal realization, and (3) if an element $a \in H_k(M, \mathbb{Z}_p)$ is of the form $\alpha = \sum_{p=1}^{s} a_{i_p} e_{i_p}$, where all $a_{i_p}$ are different from zero, then*

$$\mathrm{vol}_k\{\alpha\} \geqslant \max_{(p)} \mathrm{vol}_k\{e_{i_p}\}, \qquad 1 \leqslant p \leqslant s.$$

The proof is similar to that of Proposition 32.2.1., whereas a more general statement holds, viz., let $M$ be a compact, closed Riemannian manifold, $G = \mathbb{Z}_p$, $p$ is prime, $A \subset M$ is a fixed compact boundary, so that $H_k(A,G) = 0$, and $H_*(A,L,L') \neq \emptyset$, where $L' \neq 0$, $L \subset H_{k-1}(A,G)$, $r = \dim_{\mathbb{Z}_p} L'$. If $r > 1$, then we can choose a basis $l_1, \ldots, l_r$ in $L'$, so that (1) $d_k(A,L,\{e_j\}) < d_k(A,L,L')$, $1 \leqslant j \leqslant r$; $d_k(A,L,\{e_j\}) < d_k(A,L,\{e_1, \ldots, e_r\}) \leqslant d_k(A,L,L')$

for $j < 1$, (2) there exist minimal surfaces $X_{0j} \in H_*(A, L, \{e_j\})$ such that $(\operatorname{Im} H_k(X_{0j})) \cap L' = \{e_j\}$ under the embedding of surfaces $X_{0j}$ in $M$, (3) if $\alpha \in L'$ and $\alpha = \sum_{p=1}^{s} a_{i_p} e_{i_p}$, where $a_{i_p} \neq 0, 1 \leqslant p \leqslant s$, then $d_k(A, L, \{\alpha\}) \geqslant \max_{1 \leqslant p \leqslant s} d_k(A, L, \{e_{ip}\})$.

**Definition 32.2.2.** Let $A \subset M$ be a fixed compact set (or contour). We will say that a subgroup $L \neq 0$, $L \subset H_{k-1}(A)$ admits an exact minimal spanning if there exists a minimal surface $X_0 \in H_*(A, L, 0)$ such that $\Delta_k(X_0, A) = L$.

As well as in the case of a realization, an exact minimal spanning is possible for a small class of subgroups; however, it turns out that we can select an additive basis in the homology group from them. Denote $d_k(A, L, 0)$ by $\operatorname{vol}_k(L)$ and formulate the following general statement.

**Proposition 32.2.3.** *Let $M$ be a compact, smooth, and closed Riemannian manifold $G = \mathbb{Z}_p$, $p \neq 0$, and a prime number, $A \subset M$ a fixed boundary (or contour) such that $H_k(A, G) = 0$, and $L \neq 0$, $L \subset H_{k-1}(A, G)$, $1 < \dim L = r < \infty$, $H_*(A, L, 0) \neq \emptyset$. Then we can choose a basis $e_1, \ldots, e_r$ for the subgroup $L$, so that (1) $\operatorname{vol}_k\{e_j\} < \operatorname{vol}_k(L)$ for each $j$, and $\operatorname{vol}_k\{e_j\} < \operatorname{vol}_k\{e_1, \ldots, e_j\} \leqslant \operatorname{vol}_k(L)$, $1 < j \leqslant r$, (2) there exist minimal surfaces $X_{0j} \in H_*(A_1\{e_j\}, 0)$ such that $\Delta_k(X_{0j}, A) \cap L = \{e_j\}$, $1 \leqslant j \leqslant r$, and (3) if $\alpha \in L$, $\alpha \neq 0$, and $\alpha = \sum_{p=1}^{s} a_{i_p} e_{i_p}$, $a_{i_p} \neq 0$, $1 \leqslant p \leqslant s$, then $\operatorname{vol}_k\{\alpha\} \geqslant \max_{1 \leqslant p \leqslant s} \operatorname{vol}_k\{e_{i_p}\}$. If we set $L = H_{k-1}(A)$, then $e_1, \ldots, e_r$ admits an exact minimal spanning.*

The proof is similar to that of Proposition 32.2.2, and we are not going to dwell on it.

Any element $e_j$ in the homology group $H_{k-1}(A)$ can thereby be spanned by a minimal film only annihilating this one and its multiples.

## 32.3. Minimal Surfaces with Boundaries Homeomorphic to the Sphere

In the two-dimensional Plateau problem, the boundary of a minimal surface always decomposes into the union of circles in the case where it is a smooth one-dimensional manifold. Consider multidimensional minimal surfaces whose boundary is also a sphere smoothly embedded into the ambient Riemannian manifold. This special arrangement of a minimal surface boundary permits us to derive interesting statements of purely metric character, which relate the topology of the minimal surface to its volume. Note that the values $\tilde{d}$ and $d'$ characterizing the geometric position of a sphere $A$ in a manifold $M$ are naturally related to the pair $(M, A)$, viz., $d' = \inf \operatorname{vol}_k(X \backslash A)$, where $A \subset X \subset M$, $\operatorname{Im} H_k(X, G) \neq 0$ in $H_k(M, G)$ under the embedding $X \to M$, $G$ is a certain fixed homology theory coefficient group, and $\tilde{d} = \inf \operatorname{vol}_k(X \backslash A)$, $A \subset X \subset M$, $\Delta_k(X, A) \neq 0$ in the group $H_{k-1}(A, G)$. In other words, $d'$ is the $k$-dimensional volume of the least $k$-cycle in $H_k(M)$ and does not depend on the boundary sphere $A$ at all, whereas $d$ is the $k$-dimensional volume of the least $k$-dimensional film which can span any nontrivial cycle in $H_{k-1}(A)$ in the manifold $M$.

Consider an arbitrary variational class $H_*(A, L, L')$. As before, let $d = \inf \mathrm{vol}_k(X \backslash A)$, $X \in H_*(A, L, L')$. Then $d \geqslant d'$, $d \geqslant \tilde{d}$. An arbitrary minimal surface $X_0$ from this class may be of a sufficiently complex topological structure; therefore, in particular, the problem of the greatest volume of one connected component $\Pi_i$ only of regular points of $X_0$ is of considerable interest. In other words, what is the greatest volume of the "nonsingular" part of the minimal surface? Note that it easily follows from Theorem 7.2.1 that $\tilde{d} > 0$, and $d' > 0$.

**Theorem 32.3.1.** *Let $M$ be a compact, closed Riemannian manifold, and $A = S^{k-1}$, $G = S^1 = \mathbf{R}^1(\bmod 1)$ be the coefficient group of a homology theory $H_*$. Assume that $\tilde{d} < \infty$, noting that the inequality $d^1 < \infty$ is always true, and put $\Delta = \max(d - \tilde{d}, d - d')$. Let $X_0 \in H_*(A, L, L')$ be a minimal surface, where $L \neq 0, L' \supset S^1$, and $Z \subset X_0$ be the set of singular points of $X_0$. Suppose that $X_0 \backslash (A \cup Z) = \bigcup_i \Pi_i$ is the decomposition of the regular part of $X_0$ into the connected, smooth, and minimal submanifolds $\Pi_i \subset M$ which are open in $X_0$ (there may be an infinite set of them). Then $0 < \sup_i \mathrm{vol}_k(\Pi_i) \leqslant \Delta$.*

**Proof.** Obviously, it suffices to show that $\sup \mathrm{vol}_k(\Pi_i) \leqslant \Delta$, since the inequality $\sup \mathrm{vol}_k(\Pi_i) > 0$ is an immediate corollary of Theorem 7.2.1. Thus, assume the contrary, viz., let there exist a subscript $i_0$ such that $\mathrm{vol}_k(\Pi_{i_0}) > \Delta$. Consider an open, smooth disk $D = D^k \subset \Pi_{i_0}$ such that $\mathrm{vol}_k(\Pi_{i_0} \backslash D) = \varepsilon$, where $\varepsilon > 0$ is a fixed arbitrarily small number, with $D = D(\varepsilon)$. Then, $vol_k(\Pi_{i_0}) = \varepsilon + \mathrm{vol}_k(D)$. The disk existence follows from Theorem 7.2.1. Denote the boundary of $D$ by $S = S^{k-1}$. Then

$$\begin{aligned}\mathrm{vol}_k(X_0 \diagdown (A \cup D)) &= \mathrm{vol}_k(X_0 \diagdown A) - \mathrm{vol}_k(D) \\ &= \mathrm{vol}_k(X_0 \diagdown A) - \mathrm{vol}_k(\Pi_{i_0}) + \varepsilon \\ &= \mathrm{vol}_k(X_0 \diagdown (A \cup \Pi_{i_0})) + \varepsilon < d - \Delta,\end{aligned}$$

if $\varepsilon$ is sufficiently small. Hence,

$$\mathrm{vol}_k(X_0 \diagdown (A \cup D)) < d - (d - d') = d',$$
$$\mathrm{vol}_k(X_0 \diagdown (A \cup D)) < d - (d - \tilde{d}) = \tilde{d},$$

i.e., $\mathrm{Im}\, H_k(X_0 \backslash D) = 0$ in the group $H_k(M)$, and $\Delta(X_0 \backslash D, A) = 0$ in $H_{k-1}(A)$. Consider the commutative diagram

$$
\begin{array}{ccccccc}
0 & \longrightarrow & H_k(M) & \xrightarrow{p_*} & H_k(M,\, S) & \xrightarrow{\tau_*} & H_{k-1}(S) \\
\downarrow & & \downarrow{\scriptstyle \alpha_1^{-1}} & & \downarrow{\scriptstyle \psi_*} & & \downarrow{\scriptstyle \varphi_*} \\
H_k(X_0\setminus D) & \longrightarrow & H_k(M) & \xrightarrow{i_*} & H_k(M,\, X_0\setminus D) & \xrightarrow{\partial} & H_{k-1}(X_0\setminus D) \\
\downarrow & & \downarrow{\scriptstyle \alpha_2^{-1}} & & \downarrow & & \downarrow \\
H_k(X_0) & \xrightarrow{j_*} & H_k(M) & \longrightarrow & H_k(M,\, X_0) & \longrightarrow & H_{k-1}(X_0)
\end{array}
$$

where $\alpha_1, \alpha_2$ are isomorphisms, $i_*$ is a monomorphism (see above), $H_k(M,S) = H_k(M) \oplus H_k(\overline{D},S)$, since the sphere $S$ is contractible on the disk to a point; therefore, the homomorphism $\psi_*$ monomorphically maps the subgroup $H_k(\overline{D},S) \subset H_k(M,S)$ into $H_k(M, X_0\setminus D)$. We prove that $i_*\alpha_2 j_* H_k(X_0) \subset \psi_* X_k(D,S)$ (geometrically, this is obvious). Put $\xi_* = p_*\alpha_1\alpha_2 j_*$. Then $\psi_*\xi_* = i_*\alpha_2 j_*$. Let $x \in H_k(X_0)$. Then $\xi_*(x) = y_1 + y_2$, $y_1 \in p_* H_k(M), y_2 \in H_k(D,S)$, $\psi_*\xi_*(x) = \psi_*(y_1) + \psi_*(y_2)$. The homomorphism $\xi_*$ admits a decomposition

$$\xi_* = \chi_*\rho_* \colon\ H_k(X_0) \to H_k(X_0,\, S) \to H_k(M,\, S),$$

where $H_k(X_0, S) \sim H_k(X_0\setminus D, S) \oplus H_k(\overline{D},S)$, i.e., $\rho_*(x) = z_1 + z_2$, and $\psi_*\xi_*(x) = \psi_*\chi_*(z_1) + \psi_*\chi_*(z_2)$. It is clear that $\psi_*\chi_*(z_1) = 0$, or $\psi_*\chi_*(x) = \psi_*\chi_*(z_2) \subset \psi_* H_k(\overline{D},S)$, which is just what is required. Since $\psi_*$ is a monomorphism onto $H_k(\overline{D},S)$, for $S^1 \subset L' \subset j_* H_k(X_0)$, and $i_*$ is a monomorphism, we have $\psi_* H_k(\overline{D},S) \supseteq S^1$, and because $H_k(\overline{D},S) = S^1$, we derive that $\psi_* H_k(\overline{D},S) = S^1 \supset \operatorname{Im} i_*$, with $\partial\psi_* H_k(\overline{D},S) = 0$.

Since the homomorphism $H_{k-1}(A) \to H_{k-1}(X_0\setminus C)$ possesses no kernel, and $\dim X_0 = k$, the sphere $A$ is a retract of the compact set $X_0\setminus D$ due to the Hopf theorem (recalling that $S^1 = G$), i.e., there exists a continuous mapping $g : X_0\setminus D \to A$, which is the identity on $A$. Consider $t = g|_s : S = \partial D \to A$. It is evident that the degree of the mapping $t$ is other than zero; otherwise, $t$ could be extended to a continuous mapping $t' : \overline{D} \to A$, generating a new mapping $t'' : X_0 \to A$, the identity on $A$, which is impossible due to $L \neq 0$ from the data. Thus, since $m = \deg(t) \neq 0$, the homomorphism $t_* : H_{k-1}(S) \to H_{k-1}(A)$ is multiplication by $m$. It follows from the diagram

$$
\begin{array}{ccc}
H_{k-1}(S) & \xrightarrow{\quad \varphi_* \quad} & H_{k-1}(X_0\setminus D) \\
 & {\scriptstyle t_*}\searrow \quad \swarrow{\scriptstyle g_*} & \\
 & H_{k-1}(A) &
\end{array}
$$

that $\varphi_* \not\equiv 0$. Returning to the original one, we see that the homomorphism $\varphi_*\tau_*$ on the subgroup $H_k(\overline{D},S) \subset H_k(M,S)$ is nontrivial, and then the homomorphism $\partial\psi_*$ equal to $\varphi_*\tau_*$ cannot be trivial on the subgroup $H_k(\overline{D},S)$ either, which is contrary to the relation $\partial\psi_* H_k(\overline{D},S)$ established above. The obtained contradiction proves the theorem.

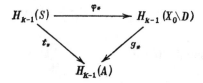

Certainly, this theorem can be formulated for the case of a cohomology theory with integral coefficients because of the duality. We cannot drop any of the conditions $L \neq 0, L' \not\supset S^1$ in the theorem statement. Further, if $A$ is not the sphere $S^{k-1}$, then, generally following, the statement, or the inequality $\mathrm{vol}_k(\Pi_i) \leqslant \Delta$, is also violated. In fact, consider the manifold $M = S^1 \times \mathbf{R}P^{2s}$. Let $A = \mathbf{R}P^{2s-1} \subset \mathbf{R}P^{2s} = X_0, H_{2s}(M, S^1) = H_{2s}(\mathbf{R}P^{2s}, S^1) = \mathbf{Z}_2$.

Suppose that $L' = \mathbf{Z}_2$ (noting that $L' \not\supset S^1$), $H_{2s}(M) \supset L'$, $L = S^1 = H_{2s-1}(A, S^1)$. Then $X_0 \in H_*(\mathbf{R}P^{2s-1}, S^1, \mathbf{Z}_2)$, $X_0$ is a globally minimal surface in the variational class $\mathcal{O}$, the space $\mathbf{R}P^{2s}$ being the least surface in $M$, which spans some nontrivial subgroup $L' \subset S^1 = H_{k-1}(A, S^k)$, $k = 2s$; hence, $d = \tilde{d}, d = d'$ i.e., $\Delta = 0$, whereas $\sup \mathrm{vol}_k(\Pi_i) = \mathrm{vol}_k(\mathbf{R}P^{2s}) = d > 0$.

In the case where the boundary of a minimal surface is a sphere, many homology characteristics can be reduced to homotopy ones, and vice versa. As an example, we give the existence theorem for a minimal surface in the class of films whose boundary is a sphere, and which do not retract onto it. It then turns out that we can derive a homotopy existence theorem from the homology minimal solution existence theorem (see [351]). More precisely, let a compact boundary $A$ be homeomorphic to the sphere $S^{k-1}$ and embedded into $\mathbf{R}^n$. Consider the class $K^*$ of all compact sets $X$, $A \subset X \subset \mathbf{R}^n$ that are not retracted onto their boundary $A$, meaning the usual retraction, and not a deformation one. We call the compact set $X \in K^*$ *simple* if it does not contain a proper subset $Q$ that would have the sphere $A$ as its boundary, i.e., which is not retracted on it either. It then turns out that (1) the minimum volume in the variational class $K^*$ is attained on a certain compact set $X_0 \in K^*$, (2) each compact set $X \in K^*$ contains a simple compact subset, and (3) each simple compact subset $X_0 \in K^*$ realizing the minimum of $\mathrm{vol}_k$ is locally Euclidean almost at all its points, i.e., on the set of complete measure (see [351]).

This theorem is stated in nonstable homotopy terms (retraction). The stable homotopy situation is completely contained in our Theorem 7.2.1 for the case of an extraordinary homology theory $h_* = \pi_*^s$. However, the "nonstable" homotopy theorem is here reduced to the homology existence theorem and follows from it. In fact, due to the Hopf theorem, we have the identity

$$K^* = \bigcup_L H_*(S^{k-1}, L, 0), \qquad L \neq 0,$$

where $G = S^1 = \mathbf{R}^1(\bmod 1)$. Indeed, let $X \in K^*$. We have to prove that $\operatorname{Ker} i_* \neq 0$, where the homomorphism $i_* : H_{k-1}(A) \to H_{k-1}(X)$ is embedding induced. If $i_*$ is a monomorphism, then, by the Hopf theorem, the sphere $A$ is a retract of the compact set $X$, which is impossible. Conversely, let $X \in H_*(S^{k-1}, L, 0), L \neq 0$. We have to prove that the sphere $A$ is not a retract of the surface $X$. Since $L \neq 0$, it is obvious.

It is interesting to determine which other unstable homotopy situations in the Plateau problem are reduced to homology ones. A hypothesis was offered in [351], which, if true, would generalize the above existence theorem in the class $K^*$. More precisely, consider an arbitrary compact, closed, and smooth submanifold $V^{k-1}$ as the compact boundary $A \subset \mathbf{R}^n$. Let $K^*(V^{k-1})$ be the class of all compact sets $X$, $V \subset X \subset \mathbf{R}^n$, so that each of them cannot be retracted on any of its closed subsets $\widetilde{V}$, $V \subset \widetilde{V} \subset X$, of dimension $k-1$. Meanwhile, the retraction is not assumed to be a deformation. The conjecture formulated in [351] consisted of equality $K^*(V) = \bigcup_L H_*(V, L, 0)$, where $L \neq 0$, $G = S^1 = \mathbf{R}^1(\mod 1)$, and we employ the notation from the previous sections. In the case $V = S^{k-1}$, we obtain the class $K^*$ described above. However, this hypothesis is not valid. We give a simple counterexample. Preliminarily, note that the inclusion $K^*(V^{k-1}) \supset \bigcup_L H_*(V, L, 0)$ certainly holds, but the converse turns out not to be valid any more than in the general case. Put $A = V^{k-1} = S^1 \times S^{k-2}$, where $k \geqslant 5$. Then $\pi_{k-1}(A) = \mathbf{Z}_2$. Let $\alpha : S^{k-1} \to A$ be a representative of the element $1 \in \mathbf{Z}_2$. Consider the subcomplex $X = A \bigcup_\alpha D$, gluing the disk with respect to the mapping of the boundary $\alpha$. Since the image of the sphere $S^{k-1}$ in $A$ under the mapping $\alpha$ can be homotopically pressed out into the sphere $S^{k-2}$, the embedding $i : A \to X$ induces a monomorphism in the $(k-1)$-dimensional homology groups, i.e., $X \notin \bigcup_L H_*(V^{k-1}, L, 0)$, $L \neq 0$. On the other hand, $X \in K^*(V)$. In fact, assume that there exists a compact set $\widetilde{V} \subset U$ such that there is a continuous mapping $f : X \to \widetilde{V}$, which is constant on $\widetilde{V}$. Since $\widetilde{V}$ is a compact set, and $\dim \widetilde{V} = k - 1 < k = \dim D^k$, the mapping $f$ is homotopic to the mapping $f' : X \to A$, for $\widetilde{V}$ is pressed out into $A = V$. Meanwhile, we can assume that $f'$ is the identity on $A$, which means that the mapping $\alpha : S^{k-1} \to A$ can be extended to that of the disk $D^k$ into $A$, or $\alpha = 0$, which is contrary to its choice. We have thereby exhibited a counterexample in contrast to the conjecture suggested in [351].

# APPENDIX I

# Minimality Test for Lagrangian Submanifolds in Kähler Manifolds.
Submanifolds in Kähler Manifolds.
Maslov Index in Minimal Surface Theory

While studying globally minimal surfaces in $R^{2n}$, R. Harvey and H. Lawson proved that any locally minimal Lagrangian submanifold in $R^{2n} = C^n$ with the standard Kähler metric is special Lagrangian and, therefore, globally minimal [196]. The question of describing the topological classes of minimal Lagrangian submanifolds arises for an arbitrary Kählerian manifold. Meanwhile, we assume that the Riemannian metric and symplectic structure are naturally determined by a Kählerian structure on the manifold. We formulate and prove, together with Lê Hông̃ Vân, a *general minimality test for the maximally isotropic submanifolds* $L$ (which we will call $\Phi$-Lagrangian) in Hermitian manifolds $M$, stated in terms of a certain differential $I$-form of $L$. Applying it, we prove the minimality of certain Lagrangian surfaces in Kähler manifolds. A Lagrangian submanifold in the symplectic space $R^{2n}$, and also in certain other Kählerian manifolds $M^{2n}$, is known to possess a topological invariant, i.e., the Maslov index, and, more generally, the characteristic Maslov–Arnold classes $\alpha_i$. The author has formulated the hypothesis that *minimal Lagrangian submanifolds possess in many cases trivial characteristic Maslov–Arnold classes.* It turns out that this conjecture is, in fact, valid for the case of $M^{2n} = R^{2n}$. The theorem follows from the general minimality criterion for $\Phi$-Lagrangian submanifolds and [196].

## §1 Definitions

**1.1.** The Riemannian metric on a complex manifold $M$ is said to be *Hermitian* if $g(JX, JY) = g(X, Y), \forall X, Y \in TM$, where $J$ is a complex structure on $M$.

**1.2.** The fundamental 2-form $\Phi$ on a Hermitian manifold is defined as $\Phi(X,Y) = g(X,JY)$. If $\Phi$ is closed, then the Hermitian manifold is said to be *Kähler*. Each Kähler manifold is a symplectic with the symplectic form $\Phi$.

**1.3.** We will call an $n$-dimensional plane $l$ in $T_xM^{2n}$ *$\Phi$-Lagrangian* if $l$ is the maximally isotropic plane of the restriction of $\Phi$ to $T_xM^{2n}$. In other words, $Jl \perp l$, $Jl \oplus l = T_xM^{2n}$. Denote by $G_L(M^{2n})$ the fibration of orientable $\Phi$-Lagrangian planes over $M$. $G_L(M^{2n})$ is formed by the pairs $(x,l)$, where $x \in M^{2n}$ and $l$ is a $\Phi$-Lagrangian plane in $T_xM$. We call a submanifold $L$ in $M^{2n}$ $\Phi$-Lagrangian if all of its tangent planes are $\Phi$-Lagrangian. Then, for each $\Phi$-Lagrangian submanifold $L$ in $M^{2n}$, a mapping $p : L \to G_L(M^{2n}) : X \mapsto (x, T_xL)$ is defined, where $T_xl$ is a $\Phi$-Lagrangian tangent plane to $L$, which is an analogue of a Gauss map.

**1.4.** Let the standard complex structure and Hermitian metric be given on $R^{2n} = C^n$. Then a form $\varphi = \mathrm{Re}(e^{i\theta}dz_1 \wedge \cdots \wedge dz_n)$, where $z_i$ is a certain unitary basis for $R^{2n}$, is said to be a *special Lagrangian form* $\varphi \in \wedge^n R^{2n}$ (see [196]). Harvey and Lawson have proved that being special Lagrangian does not depend on the choice of a unitary basis. On a Hermitian manifold, $\varphi$ is said to be *SL-form* if, $\forall_x \in M^{2n}$, the restriction of $\varphi$ to $T_xM^{2n}$ is special Lagrangian.

**1.5.** A Hermitian manifold $M^{2n}$ is said to be *locally calibrated* if, for any point $x \in M^{2n}$, there is a neighbourhood $0(x)$ with a closed $SL$-form on it.

## §2 The Local Minimality Test for $\Phi$-Lagrangian Submanifolds $L$ in the Hermitian Manifold $M^{2n}$

Let $\Sigma g_{\alpha\bar{\beta}}dz_\alpha d\bar{z}_\beta$ be a Hermitian metric with respect to local coordinates, and $G(z) = \det(g_{\alpha\bar{\beta}})$. We define the function $\bar{f}(x) = \ln\sqrt{G}$ and the complex form $w(x) = \sqrt{G}dz_1 \wedge \cdots \wedge dz_n$ on $M^{2n}$. Then we can (locally) define a function $f$ on the fibration $G_L(M^{2n})$, so that $f(x,l_x) = \bar{f}(x)$, where the mapping $\pi : (x,l_x) \to x$ is the natural projection from $G_L(M^{2n})$ onto $M^{2n}$. We can also define a function $\theta$ with period $2\pi$, i.e., a mapping into a circumference, so that $\theta(x,l_x) = (-i)\ln w(x), \vec{l}_x)$, where $\vec{l}_x$ is the unit polyvector determining the Lagrangian plane $l_x$ and $(w(x), \vec{l}_x)$ is the value of $w(x)$ on $\vec{l}_x$.

**Theorem 1 (due to Lê Hông̃ Vân and the author).** *The I-form $\psi = Jdf + d\theta$, where $J$ is the complex structure operator, is defined correctly on the whole fibration $G_L(M^{2n})$, i.e., it is independent of the choice of local complex coordinates.*

**Theorem 2 (Minimality Test for $\Phi$-Lagrangian Submanifolds; due to Lê Hông̃ Vân and the author).** *A $\Phi$-Lagrangian submanifold $L$ in $M^{2n}$ is locally minimal if and only if the induced form $p^*(\psi)$ vanisher on $L$.*

For the proof, we shall need the following lemmas. Let $v_1, \ldots, v_n$ be an orthonormal basis in $T_xL$, and $H$ the mean curvature of the submanifold $L$ in $M^{2n}$. By definition, we put

$$(X \rfloor d\varphi_\theta)(v_1, \ldots, v_n) = d\varphi_\theta(X, v_1, \ldots, v_n).$$

Recall that $H = H(x)$ is the normal bundle vector of $L$. If $\langle\ ,\ \rangle$ is the scalar product on $T_xM^{2n}$ determined by the Riemannian metric, then we can compute $\langle H, Y\rangle = \langle H(x), Y(x)\rangle$ for any normal vector field $Y(x)$.

**Lemma 2.1.** *Let $L$ be a $\Phi$-Lagrangian submanifold in $M^{2n}$ and $\varphi_\theta$ be a certain $SL$-form in the neighbourhood of $L$ with the property that $\varphi_\theta(\overrightarrow{T_xL}) = 1$, $\forall x \in L$. Then, for any vector $X$ in the normal bundle $N_xL$, the equality*

$$(X \rfloor d\varphi_\theta)(v_1, \ldots, v_n) = \langle -H, X\rangle$$

*is valid.*

**Lemma 2.2.** *The form $\varphi = \operatorname{Re} w$ is an $SL$-form on the given domain with coordinates $z_i$ in $M^{2n}$. Let $L$ be a $\Phi$-Lagrangian submanifold in the same domain. Then $(we^{-i\theta\cdot(x)}, \overrightarrow{T_xL}) = 1$, $\forall x \in L$.*

*The function $\theta \cdot p$ can be extended to a certain neighbourhood $L$ in $M^{2n}$, so that the form $\varphi_\theta = \operatorname{Re}(we^{-i\theta\cdot p})$ satisfies the conditions of Lemma 2.1.*

**Lemma 2.3.** *With the notation of Lemmas 2.2. and 2.1, we have*

$$\begin{aligned}(X \rfloor d\varphi_\theta)_{|L}(v_1, \ldots, v_n) &= (d\bar{f}_{|L} - (d\theta.p)_{|L})(X)\\ &= J(d\bar{f}_{|L} + d(\theta.p)_{|L})(JX),\end{aligned}$$

*where, by identifying the tangent and cotangent spaces by means of the Riemannian metric, we may assume that the equalities*

$$(d\bar{f}, X) = (Jd\bar{f}, JX), \qquad (J(d\theta.p), X) = (-d\theta.p, JX)$$

*hold.*

**Proof of Lemma 2.1.** We first establish that if $v_i$ is an orthonormal basis in $T_xL$, then

$$\begin{aligned}&\text{(i) } \varphi_\theta(v_1 \wedge \cdots \wedge v_n) = 1;\\ &\text{(ii) } \varphi_\theta(v_1 \wedge \cdots \wedge \tilde{v}_i \wedge \cdots \wedge v_n \wedge w_j) = 0; \quad \forall w_j \in N_xL.\end{aligned}$$

In fact, (i) means that $\varphi_\theta(\overrightarrow{T}_xL) = 1$, and (ii) that $\varphi_\theta$ attains a maximum or simply extremum on the $n$-vector $v_1 \wedge \cdots \wedge v_n$ as a function on $G_{2n,n}(T_xM^{2n})$, where $G_{p,q}$ denotes the usual Grassmannian manifold of $q$-dimensional planes in $p$-dimensional space. This fact follows from the comass of $\varphi_\theta$ being equal to unity. The latter statement was established in [196].

We now compute the value of $X\rfloor d\varphi_\theta(v_1,\dots,v_n)$, where the vector fields $v_i$ are assumed to be given in the neighbourhood of $L$, so that $\langle v_i, v_j\rangle = \delta ij$.

We have

$$(X\rfloor d\varphi_\theta)|_L(v_1,\dots,v_n) = \sum_{i=1}^{n}(-1)^{i+1}v_i(\varphi_\theta(x_1,v_1,\dots,\hat{v}_i,\dots,v_n))|_L$$
$$+X(\varphi_\theta(v_1,\dots,v_n))|_L + \sum_{1\leqslant i<j\leqslant n}(-1)^{i+j}\varphi_\theta([v_i,v_j],$$
$$X,\dots,\hat{v}_i,\dots,\hat{v}_j,\dots,v_n)|_L + \sum_{i=1}^{n}(-1)^i\varphi_\theta([X,v_i],\dots,\hat{v}_i,\dots,v_n)|_L.$$

The first and third sums vanish due to (ii), whereas the second due to (i). Because of (i) and (ii), we obtain

$$\varphi_\theta([X,v_i],\dots,\hat{v}_i,\dots,v_n) = (-1)^i\langle[X,v_i],v_i\rangle.$$

Finally, we have

$$(X\rfloor d\varphi_\theta)(v_1,\dots,v_n) = \sum_i\langle[X,v_i],v_i\rangle = \sum\langle\nabla_x v_i - \nabla_{v_i}X, v_i\rangle,$$

and $\langle\nabla_x v_i, v_i\rangle = \frac{1}{2}\nabla_x\langle v_i,v_i\rangle = O$.

Therefore,

$$(X\rfloor d\varphi_\theta)|_L(v_1,\dots,v_n) = (-1)\sum_i\langle\nabla_{v_i}X,v_i\rangle|_L = \langle -H, X\rangle,$$

and the lemma is thus proved.

**Proof of Lemma 2.2.** Let a unitary basis $\partial z_i^0$ in $T_xM^{2n}$ exist at a point $x\in M^{2n}$ and let $dz_i^0$ be the dual basis in $T_x^*M^{2n}$. Then the Hermitian metric on $T_xM^{2n}$ is of the form $g=\sum_\alpha dz_\alpha^0 d\bar{z}_\alpha^0$. Suppose that $A(x)$ is the transition matrix $dz_i^0 = a_{ij}dz_j$ in the space $T_x^*M^{2n}$. Then $d\bar{z}_i^0 = \bar{a}_{ij}d\bar{z}j$. We know that the metric on $M^{2n}$ is given by

$$g = dz_j g_{jk} d\bar{z}_k. \tag{1}$$

On the other hand,

$$g = \sum dz_\alpha^0 d\bar{x}_\alpha^0 = \sum a_{ij}dz_j\cdot\bar{a}_{ik}d\bar{z}_k = \sum dz_j a_{ij}\bar{a}_{ik}d\bar{z}_k. \tag{2}$$

By comparing (1) and (2), we obtain $g_{jk} = \sum a_{ij}\bar{a}_{ik}$; therefore, $G = A^T\bar{A}$, and $\det G = \det A^T\cdot\det\bar{A} = |\det A|^2$, hence $|\det A| = \sqrt{\det G}$.

Let $\det A = |\det A| \cdot e^{i\varphi}$. Then $|\det A| dz_1 \wedge \cdots \wedge dz_n = e^{-i\varphi} \det A\, dz, \wedge \cdots \wedge dz_n = e^{-i\varphi} dz_1^0 \wedge \cdots \wedge dz_n^0$. By definition, the form $\mathrm{Re}(e^{-i\varphi} dz_1^0 \wedge \cdots \wedge dz_n^0)$ is an $SL$-form; therefore, the form $\mathrm{Re}(\sqrt{\det G}\, dz_1 \wedge \cdots \wedge dz_n)$ is an $SL$-form, too, and the first part of the lemma is thus proved.

The function $\theta$ on $G_L(M^{2n})$ is determined by

$$\theta(x, l_x) = (-i) \ln(w(x), l_x),$$

which means that $(w(x), l_x) = e^{i\theta(x,l_x)}$, or $(e^{-i\theta(x,l_n)} \cdot w_x, l_x) = 1$. Hence, the second statement of our lemma follows that, for any $\Phi$-Lagrangian submanifold, $L \to M^{2n}$: $(e^{i\theta \cdot p(x)}, \overrightarrow{T_x L}) = 1$.

The statement regarding the extension of the function $\theta \cdot p$ follows from the existence of a normal neighbourhood of $L$ in $M^{2n}$, and the lemma is thus proved completely.

**Proof of Lemma 2.3.** Recall that $\bar{f} = \ln \sqrt{G} = \frac{1}{2} \ln G, \quad \theta = \theta \cdot p$. Then

$$w = e\bar{f} \cdot dz_1 \wedge \cdots \wedge dz_n, \qquad \varphi = \mathrm{Re}(e^{\bar{f}} \cdot dz_1 \wedge \cdots \wedge dz_n)$$
$$w_\theta = e^{\bar{f}+i\theta} dz_1 \wedge \cdots \wedge dz_n, \qquad \varphi_\theta = \mathrm{Re}(e^{\bar{f}+i\theta} \cdot dz_1 \wedge \cdots \wedge dz_n).$$

For convenience, we select the new coordinate system $(z_i')$ such that at the fixed point $x$, we have $w_\theta = dz_1' \wedge \cdots \wedge dz_n'$. It can be done by choosing $\partial \overrightarrow{x}_i' = \partial \overrightarrow{z_i} + \partial \overrightarrow{\bar{z}_i}$ in $T_x L$, since $T_x L \perp J(T_x L)$ and $T_x L + J(T_x L) = T_x M$. First, we observe that

$$de^{\bar{f}+i\theta} = e^{\bar{f}+i\theta} \cdot d(\bar{f} + i\theta) = e^{\bar{f}+i\theta} \cdot \sum_i (\partial/\partial z_i (f + i\theta) dz_i + \partial/\partial \bar{z}_i (f + i\theta) d\bar{z}_i).$$

Therefore,

$$de^{\bar{f}+i\theta} \wedge dz_1 \wedge \cdots \wedge dz_n = \sum_i (\partial/\partial z_i' (\bar{f} + i\theta) dz_i' + \partial/\partial \bar{z}_i' (\bar{f} + i\theta) d\bar{z}_i')$$
$$\wedge dz_i' \wedge \cdots \wedge dz_n' = \sum_i \partial/\partial \bar{z}_i' (\bar{f} + i\theta) d\bar{z}_i' \wedge dz_1' \wedge \cdots \wedge dz_n'.$$

Then

$$d\varphi_\theta = \sum_i \mathrm{Re}(\partial/\partial \bar{z}_i' (\bar{f} + i\theta) d\bar{z}_i') \cdot \mathrm{Re}(dz_1' \wedge \cdots \wedge dz_n')$$
$$- \sum_i \mathrm{Im}(\partial/\partial \bar{z}_i' (\bar{f} + i\theta) d\bar{z}_i') \cdot \mathrm{Im}(dz_1' \wedge \cdots \wedge dz_n'). \qquad (3)$$

Recall the formula for differentiation

$$\partial/\partial \bar{z}_i' (\bar{f} + i\theta) = \tfrac{1}{2} (\partial/\partial x_i' + i\partial/\partial y_i')(\bar{f} + i\theta); \quad d\bar{z}_i' = dx_i' = i dy_i'.$$

Then

$$\begin{aligned}&\partial/\partial\bar{z}_i'(\bar{f}+i\theta)d\bar{z}_i'\\&=\frac{1}{2}\Big(\frac{\partial(\bar{f}+i\theta)}{\partial x_i'}dx_i'+\frac{\partial(\bar{f}+i\theta)}{\partial y_i'}dy_i'-i\frac{\partial(\bar{f}+i\theta)}{dx_i'}dy_i'+i\frac{\partial(\bar{f}+i\theta)}{dy_i'}dx_i'.\end{aligned} \quad (4)$$

Substituting (4) in (3), we obtain

$$\begin{aligned}d\varphi_\theta&=\frac{1}{2}\sum(\bar{f}_{x_i'}dx_i'+\bar{f}_{y_i'}dy_i'-\theta_{y_i'}dx_i'+\theta_{x_i'}dy_i')\cdot\varphi_\theta\\&-\frac{1}{2}\sum(\bar{f}_{y_i'}dx_i'-\bar{f}_{x_i'}dy_i'+\theta_{x_i'}dx_i'-\theta_{y_i'}dy_i')\cdot\operatorname{Im}w_\theta.\end{aligned} \quad (5)$$

We now calculate $X\rfloor d\varphi_\theta$ via (5). Note that the statement of our lemma is of linear character; it therefore suffices to verify the lemma only for $X = \partial y_i' \in N_xL$.

**Proposition.** *The following equalities hold, viz.,*

*(1)* $(dx_i' \wedge \varphi_\theta)(X, \partial x_1', \ldots, \partial x_n') = 0,$

*(2)* $(dy_i' \wedge \bar{\varphi}_\theta)(X, \partial x_1', \ldots, \partial x_n') = 0,$

*(3)* $(dy_i' \wedge \varphi_\theta)(X, \partial x_1', \ldots, \partial x_n') = (dy_i', X),$

*(4)* $(dx_i' \wedge \bar{\varphi}_\theta)(X, \partial x_1, \ldots, \partial x_n') = (dy_i', X),$

*where* $\bar{\varphi}_\theta = \operatorname{Im} w_\theta, X \in N_xL$.

*Before the proof, we see that certain necessary corollaries are valid. Assume that (1), (2), (3), and (4) have been proved. Then, substituting the relations in (2.3.3), we immediately see that the statement* $X\rfloor d\varphi_\theta(\partial x_1', \ldots, \partial x_n') = X\rfloor d\varphi_\theta(v_1, \ldots, v_n) = \langle d\bar{f} + Jd\theta, X\rangle$ *is valid.*

*Thus, it remains to be verified that (1), (2), (3), and (4) are valid. In fact, (1) follows from (ii) (see also the proof of Lemma 2.1).*

*(ii)* $\varphi_\theta(\partial x_1' \wedge \cdots \wedge \partial \hat{x}_i \cdots \wedge \partial y_j') = 0;$

*(i)* $\partial_\theta(\partial x_1' \wedge \cdots \wedge \partial x_n') = 1.$

*However, relations (2) and (4) follow from (i') and (ii') regarding the SL-form* $\operatorname{Im} w_\theta = \bar{\varphi}_\theta$.

*(i')* $\bar{\varphi}_\theta(\partial x_i' \wedge \cdots \wedge \partial_{x_{i-1}}' \wedge \partial y_i' \wedge \partial x_{i+1}' \wedge \cdots \wedge \partial x_n') = 1$

*(ii')* $\bar{\varphi}_\theta(\partial x_1' \wedge \cdots \wedge \partial x_n') = 0.$

*Formula (3) follows from (i), and our lemma is proved.*

**Proof of Theorem 1.** Let $z_i' = z_i'(z_i)$ be another complex coordinate system on $M^{2n}$, and $(\frac{\partial z_i'}{\partial z_i}) = A$ be the transition matrix. Then $\det A$ is a holomorphic function of $(z_i)$, $\det A = e^{\alpha+i\beta}$. We prove that

$$df' = df - d\alpha \qquad \text{(i)} \qquad (6)$$

$$d\theta' = d\theta + d\beta, \tag{ii}$$

$$Jd\alpha = d\beta. \tag{7}$$

Let (6) and (7) hold. Then, because of (6),

$$\begin{aligned} Jdf' + d\theta' &= J(df - d\alpha) + (d\theta + d\beta) \\ &= J(df + d\theta) - (Jd\alpha - d\beta) \\ &= Jdf + d\theta \end{aligned}$$

due to (7).

The equalities (6i) and (6ii) are easily verifiable. In fact, since $z_i' = z_i'(z_j)$, we have $dz_i' = \frac{\partial z_i'}{\partial z_i} dz_j$. Hence,

$$dz_1' \wedge \cdots \wedge dz_n' = \det(\frac{\partial z_i'}{\partial z_j}) dz_1, \wedge \cdots \wedge dz_n. \tag{8}$$

With our notation, $\det(\frac{\partial z_i'}{\partial z_j}) = e^{\alpha + i\beta}$. Let us rewrite (8) as

$$dz_1' \wedge \cdots \wedge dz_n' = e^{\alpha + i\beta} dz_1 \wedge \cdots \wedge dz_n. \tag{9}$$

By definition, $G = \det(g_{\alpha\bar{\beta}})$, and we have in the new coordinate system that $G' = \bar{A}^{T(-1)} G \bar{A}^{-1}$ (actually, $(dz)G(d\bar{z}) = (dz' A^{T(-1)}) \times G(\bar{A}^{-1} d\bar{z}')$, since $A = \frac{\partial z_i'}{\partial z_1}$).

Hence, $\det G' = \det G / |\det A|^2$.

But,

$$f' = \tfrac{1}{2} \ln(\det G') = \tfrac{1}{2} \ln(\det G) - \ln|\det A|. \tag{10}$$

Therefore, $f' = f - \alpha$, and (i) of (6) is thus proved.

By definition, $w^1 = \sqrt{\det G} dz_1' \wedge \cdots \wedge dz_n'$. Then, in view of (9) and (10), we have

$$w' = (\sqrt{\det G} / |\det A|) e^{\alpha + i\beta} dz_1 \wedge \cdots \wedge dz_n.$$

However, $|\det A| = e^{\alpha}$; hence,

$$w^1 = \sqrt{\det G} \cdot e^{i\beta} dz_1 \wedge \cdots \wedge dz_n = e^{i\beta} w.$$

Recall that the function $\theta$ is defined as $(w(x), l_x) = e^{i\theta(l_x)}$, where $l$ is the Lagrangian $p$-vector in $T_x M$,

$$\begin{aligned} (w'(x), l_x) = e^{i\theta'(l_x)} &= e^{i\beta} w(x), l_x) = e^{i\beta} \cdot (w(x), l_x) \\ &= e^{i\beta + i\theta(l_x)}. \end{aligned}$$

Hence, $\theta' = \theta + \beta$, and (ii) of (7) is thus proved. It remains to prove (7), viz., $Jd\alpha = d\beta$. As we have already noticed, the function $\det A$ is holomorphic, and therefore, the function $(\alpha+i\beta)(x)$ is also holomorphic. The Cauchy–Riemann relation is then written as

$$\alpha_{x_i} = \beta_{y_i}, \qquad \alpha_{y_i} = -\beta_{x_i}. \tag{11}$$

However, the operator $J$ is determined with respect to the coordinates $x_i, y_i$ as

$$\begin{aligned} J dx_i &= dy_i \\ J dy_i &= -dx_i, \end{aligned}$$

and $J d\alpha = \alpha_{x_i} dy_i - \alpha_{y_i} dx_i$ due to (11); hence

$$J d\alpha = \beta_{y_i} dy_i + \beta_{x_i} dx_i = d\beta.$$

Equality (7) is thus proved, thereby completing the theorem.

**Proof of Theorem 2.** Lemmas 2.1 and 2.3 show that $L$ is locally minimal if and only if $\forall_x - H_x = [d\bar{f}_x - Jd(\theta \cdot p)]^* = 0$, where $^*$ denotes the projection onto $N_x L$). But, $N_x L = J(T_x L)$; therefore, $[d\bar{f} - Jd(\theta \cdot p)]^* = 0$ if and only if $Jd\bar{f} + d^*(\theta \cdot p)\mid_{T_x L} = 0$. By definition,

$$\begin{aligned} d(\theta \cdot p) &= p^*(d\theta) \\ d\bar{f}\mid_L &= p^*(df). \end{aligned}$$

Finally, the operator $J$: $T^*G_L(M^{2n}) \to T^*G_L(M^{2n})$ defined by $(Jf^*, v) = (f^*, -Jv)\forall v \in TG_L(M^{2n})$,

$$\begin{aligned} J_v &= J^v \text{ if } v \in T_x M^{2n} \\ J^v &= v \text{ if } v \in T_l G_L^n. \end{aligned}$$

Therefore, $Jd\bar{f}|_L = p^*(Jdf)$, and $Jd\bar{f} + d(\theta \cdot p)|_L = p^*(Jdf + d\theta)$.

Finally, the vanishing of the mean curvature is equivalent to that of the form $p^*(Jdf + d\theta)$ on $L$, and Theorem 2 is thus proved.

## §3 Certain Corollaries. New Examples of Minimal Surfaces. The Maslov Index for Minimal Lagrangian Submanifolds

**3.1.** Consider the case $M^{2n} = R^{2n}$ with the standard Kählerian structure. Then $G_L(R^{2n})$ is the direct product $R^{2n} \times U(n)/\mathrm{SO}(n)$. Note that $U(n)/\mathrm{SO}(n)$ is a two-sheeted covering over $U(n)/O(n)$, a Grassmannian Lagrangian manifold. Our function $\theta : G_L(R^{2n}) \to S^1$ is defined by $\theta(x, l_x) = \det A(l_x)$, where $A(l_x) \in U(n)$, $A$ is the matrix carrying the plane $R^n$ into the Lagrangian plane $l_x$.

Consider the diagram

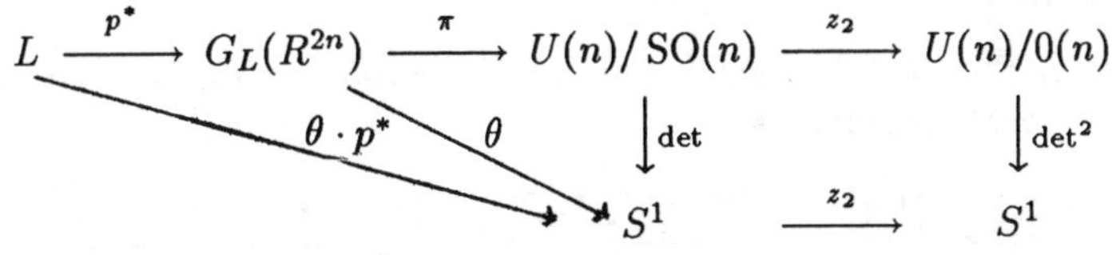


The element $[\alpha] = \deg(\det^2)$ has been proved to be a generator of the group $H^1(U(n)/O(n); \mathbf{Z})$. It is easy to see that $[\tilde{\alpha}] = \deg(\det)$ is a generator of $H^1(U(n)/\operatorname{SO}(n), \mathbf{Z})$, with $\mathbf{Z}_2^*[\alpha] = 2[\tilde{\alpha}]$.

Since the function $\det(g\alpha_{\bar{\beta}})$ is constant on $R^{2n}$, the form $\psi = Jdf + d\theta$ equals $d\theta$. We see from the diagram that the form $\frac{k(\det)}{2_\pi}$ is a generator of $H^1(U(n)/\operatorname{SO}(n), R)$, i.e., the class $[\tilde{\alpha}] \deg = (\det)$ under the embedding $H^1(U(n)/\operatorname{SO}(n), \mathbf{Z}) \to H^1(U_n/\operatorname{SO}(n), R)$ coincides with the class $[\frac{d(\det)}{2_\pi}]$.

**Theorem 3.** *The image of the Maslov class* $[\alpha] = \deg(\det)$ *in the group* $H^1(U(n)/\operatorname{SO}(n), R)$ *coincides with the class* $[\frac{\psi}{\pi}]$. *Therefore, any minimal Lagrangian submanifold L in* $R^{2n}$ *has the Maslov index zero and trivial characteristic Maslov-Arnold classes with coefficient group in* $\mathbf{Z}_2$ *or* $\mathbf{Z}$.

**Proof.** We have

$$H^1(U(n)/O(n), \mathbf{Z}) \xrightarrow{\mathbf{Z}_2^*} H^1(U(n)/\operatorname{SO}(n), \mathbf{Z})$$
$$\xrightarrow{i_*} H^1(U(n)/\operatorname{SO}(n), \mathbf{R}) \xrightarrow{\pi^*} H^1(G_L(\mathbf{R}^{2n}), \mathbf{R}).$$

It has been proved that $i_* \mathbf{Z}_2^*([\alpha]) = [\frac{d(\det)}{\pi}]$. However, $\theta = \pi \cdot \det$, therefore

$$\pi^* i_* \cdot \mathbf{Z}_2^*([\alpha]) = [\frac{\psi}{\pi}], \tag{3.1.1}$$

and the first part of the theorem is thus proved. The second part follows from Theorem 2. If $L$ is locally minimal, then $p^*[\psi] = 0$. Due to (3.1.1), $[p^*(\psi)] = p^* \pi^* i_* \mathbf{Z}_2^*([\alpha]) = \alpha(L)$, where $\alpha(L)$ is the Maslov index for $L$. Fuchs proved that if the Maslov index is zero, then the other characteristic Maslov–Arnold classes with coefficient group in $Z$ or $Z$ also vanish, and our theorem is thus proved.

**3.2.** Let $G/H$ be a real homogeneous space (e.g., symmetric canonically embedded into its complexification $G^c/H^c$), where $G^c$ and $H^c$ are complexifications of $G$ and $H$), respectively, assuming that $G^c$ is a semisimple Lie group, and $H$ a closed subgroup in $G$. Let the Hermitian metric on $G^c/H^c$ be induced by the Killing form. Then $G/H$ is a Lagrangian minimal (even totally geodesic) submanifold in $G^c/H^c$.

**Proof.** As locally complex coordinates on $G^c/H^c$, we take the mapping exp : $V \to O_\Sigma(e) \subset G^c/H^c$, where $V$ is the direct complement of $lH$ in the algebra

$lG$. That $G/H$ is totally geodesic follows from this submanifold being the set of fixed points under the involution.

**3.3.** Let $M^{2n} = CP^n$. Recall that the Kählerian metric is given here with respect to local coordinates as

$$ds^2 = \frac{(1+\Sigma_\alpha t^\alpha \bar{t}^\alpha)(\Sigma_\alpha dt^\alpha d\bar{t}^\alpha) - (\Sigma_\alpha \bar{t}^\alpha dt^\alpha)(\Sigma_\alpha t^\alpha d\bar{t}^\alpha)}{(1+\Sigma_\alpha t^\alpha \bar{t}^\alpha)^2},$$

where $t^\alpha = t_0^\alpha = z^\alpha / z_0$ are the coordinates on the chart $\{z_0 = 1\}$.

**Lemma 3.1.** *We have* $G = \det(g_{\alpha\bar{\beta}}) = (1+\Sigma_\alpha t^\alpha \bar{t}^\alpha)$.

**Proof.** Perform elementary transformations on the matrix $M = (1+\Sigma_\alpha t^\alpha \bar{t}^\alpha)\cdot(g_{\alpha_{\bar{\beta}}})$, writing it explicitly, viz.,

$$g_{\alpha_{\bar{\beta}}} = -\bar{t}^\alpha t^\beta \text{ if } \alpha \neq \beta,$$
$$g_{\beta\bar{\beta}} = 1 + \Sigma_\alpha t^\alpha \bar{t}^\alpha - t^\beta t^{\bar{\beta}}.$$

$$M = \left|\begin{array}{c|c|c} 1+\sum_\alpha -t^{\alpha_1}\overline{t^{\alpha_1}} & -\overline{t^{\alpha_1}}t^{\alpha_2} & -t^{\alpha_1}\overline{t^{\alpha_n}} \\ \hline -t^{\alpha_1}\overline{t^{\alpha_2}} & & \\ \hline & & \\ \hline t^{\alpha_1}\overline{t^{\alpha_n}} & & 1+\sum_\alpha -t^{\alpha_n}\overline{t^{\alpha_n}} \end{array}\right|.$$

Then $\det M = \det(\operatorname{diag}(1/t^\alpha, \ldots, 1/t^{\alpha_n}) \cdot M \cdot \operatorname{diag}(t^{\alpha_1}, \ldots, t^{\alpha_n}) = \det M'$, where $M'$ is of the form

$$\begin{cases} M'_{i_i} &= g_{ii}, \\ M'_{ij} &= t^{\alpha_i} t. \end{cases}$$

Further, $M' = \det M'_n$, where $M'_n$ is obtained from $M'$ by adding to the $n$th row the sum of the first $n-1$ rows

$$M'_n = \left|\begin{array}{c|c|c} 1+\sum_\alpha -t^{\alpha_1}\overline{t^{\alpha_1}} & -t^{\alpha_1}\overline{t^{\alpha_1}} & -t^{\alpha_1}\overline{t^{\alpha_1}} \\ \hline & & \\ \hline -t^{\alpha_{n-1}}\overline{t^{\alpha_{n-1}}} & & -t^{\alpha_{n-1}}\overline{t^{\alpha_{n-1}}} \\ \hline 1 & 1 & 1 \end{array}\right|.$$

Then $\det M'_n = \det M'_{n,n}$, where $M'_{n,n}$, is obtained from $M'_n$ by subtracting the $n$th column from each $k$th column, and consequently, $M'_{n,n}$ is of the form

$$M'_{n,n} = \begin{array}{|c|c|c|c|c|} \hline 1+\sum_\alpha & 0 & & 0 & -t^{\alpha_1}\overline{t^{\alpha_1}} \\ \hline 0 & 1+\sum_\alpha & & & \\ \hline & & \ddots & & \\ \hline & & & 1+\sum_\alpha & -t^{\alpha_{n-1}}\overline{t^{\alpha_{n-1}}} \\ \hline 0 & 0 & & & 1 \\ \hline \end{array}.$$

Thus, $\det m'_{n,n} = (1+\sum_\alpha)^{n-1}_{(n+1)} = \det M$.

Furthermore, $\det G = (1+\Sigma_\alpha)^{-2n} \cdot \det M = (1+\Sigma_\alpha)$, and the lemma is thus proved.

Replace the coordinates $(z_i, \bar{z}_i)$ on $M$ and pass to the almost polar coordinates $(r_i, \theta_i)$

$$z_i = e^{r_i + i\theta_i},$$
$$\overline{z_i} = e^{r_i - i\theta_i},$$

i.e., $(r_i + i\theta i) = \ln z_i$ is a new complex coordinate system on $M^{2n}$. Then

$$dz_i = e^{ri+i\theta_i} \cdot (dr_i + i\theta_i),$$
$$d\bar{z}_i = e^{ri-i\theta_i}(dr_i - id\theta_i).$$

Thus,

$$J(dr_i + id\theta_i) = -i(dr_i + id\theta_i),$$
$$J(dr_i - id\theta_i) = i(dr_i - id\theta_i).$$

Finally,

$$\begin{cases} Jd\theta i &= -dr_i, \\ Jdr_i &= d\theta i. \end{cases}$$

Then the form $\Phi(X, Y) = g(X, JY)$ with respect to $(r_i, \theta i)$ is

$$\Phi = \frac{4}{(1+\Sigma_i e^{2r_i})^2} \cdot ((1+\Sigma_i e^{2r_i})(\Sigma_i dr_i \wedge d\theta_i \cdot e^{2r_i})$$
$$- (\Sigma_i e^{2r_i}(dr_i + id\theta_i)) \wedge (\Sigma_i e^{2r_i}(dr_i - id\theta_i))). \qquad (3.3.1)$$

**Proposition 3.2.** *If $L$ is a Lagrangian submanifold in $CP^n$, $dp^*(\theta)_{|L} = 0$, and $L$ is a cone in the chart $\{z_0 = 1\}$, then it is locally minimal.*

**Proof.** By Lemma 1, we have $G = (1+\Sigma_\alpha t^\alpha \bar{t}^\alpha)^{-(n+1)}$. Then

$$df = \frac{1}{2} d\ln G = \frac{-(n+1)}{2}\,\frac{d(1+\Sigma_\alpha t^\alpha \bar{t}^\alpha)}{(1+\Sigma_\alpha t^\alpha t^{\bar{\alpha}})}$$

$$= \frac{-(n+1)}{2(\Sigma_\alpha t^\alpha t^{\bar{\alpha}}+1)} \cdot \Sigma_i dt^{\alpha_i} d\bar{t}^{\alpha_i} = (-(n+1)/2(1+\Sigma t_\alpha \bar{t}^\alpha)) \cdot e^{2r_i} \cdot 2dr_i.$$

Let $g : T^N M^{2n} \to TM^{2n} : \langle gf, \nu\rangle = f(v)$. We prove that $g(df) \in T_x L$, and then $g(Jdf) = Jg(df) \in N_x L$ with the consequence that $Jdf_{|T_x L} = 0$.

**Lemma 3.3.** $g(e^{2r_i}dr_i + \cdots + e^{2r_n}dr_n) = -(1+\Sigma_i e^{2r_i})^2(\partial r_i + \cdots + \partial r_n))$.

**Proof.** We proceed from the fact that, for any $i$,

$$g^{-1}(\partial r_i) = [(1+\sum_{j\neq i} e^{2r_i})e^{2r_i}\cdot dr_i - 2\sum_{j\neq i} e^{2(r_i+r_j)}dr_j]\cdot \frac{1}{(1+\sum_i e^{2r_i})^2},$$

$$ds^2 = 4\left[\frac{(1+\Sigma e^{2r_i})(\Sigma e^{2r_i}(dr_i^2 + d\theta_i^2) - 2\Sigma e^{2(r_i+r_j)}(dr_i dr_j + d\theta_i d\theta_j)}{(1+\Sigma_i e^{2r_i})^2}\right].$$

We have

$$g[(1+\sum_{j\neq i} e^{2r_j})e^{2r_i}dr_i - \sum_{j\neq i} e^{2(r_i+r_j)}dr_j] = -\partial r_i \cdot (1+\Sigma_i e^{2r_i})^2.$$

Summing up $n$ equations, we obtain

$$g(e^{2r_i}dr_i) = -(1+\Sigma e^{2r_i})^2 (\sum_i \partial r_i).$$

Q.E.D.

**Lemma 3.4.** *If $L$ is a cone, i.e., $\forall x \in L$, $\forall k \in R$, $kx \in L$, then $\Sigma \partial r_i \in T_x L$.*

**Proof.** The vector $\frac{d}{dk}_{|k=0}(k\overrightarrow{x}) = x_i\partial\overrightarrow{x_i} \in T_x L$ if $L$ is a cone. We prove that $\Sigma x_i \partial \overrightarrow{x}_i = \Sigma \partial \overrightarrow{r}_i$, for which it suffices to establish that $x_i \rfloor \partial \overrightarrow{x_i} + y_i \partial \overrightarrow{y}_i = \partial \overrightarrow{r}_i$, equivalently to (*)

$$\begin{cases} (x_i\partial x_i + y_i \partial y_i, dr_i) = 1, & (i) \\ (x_i \partial x_i + y_i \partial y_i, d\theta_i) = 0. & (ii) \end{cases}$$

However, $d(x_i^2 + y_i^2) = d(e^{2r_i}) = 2e^{2r_i}dr_i$. Therefore,

$$dr_i = \frac{d(x_i^2+y_i^2)}{2e^{2r_i}} = \frac{x_i dx_i + y_i dy_i}{e^{2r_i}}.$$

Thus, $(dr_i, x_i\partial x_i + y_i\partial y_i) = \frac{1}{x_i^2+y_i^2}(x_i^2+y_i^2) = 1$, and part (i) of (*) is proved. We have

$$\theta = \tan^{-1}(x/y) \Rightarrow d\theta$$
$$= \left(1/\left(\frac{x^2}{y^2}+1\right)\right) \cdot \left(\frac{dx}{y} - \frac{xdy}{y^2}\right) = \frac{ydx - xdy}{x^2+y^2}.$$

Therefore, $(x_i\partial x_i + y_i\partial y_i, d\theta_i) = \frac{x_i y_i - y_i x_i}{x_i^2+y_i^2} = 0$, and part (ii) is also proved.

Lemma 3.3 demonstrates that $g(df) = A(n)\cdot(\partial r_1 + \cdots + \partial r_n)$ if $L$ is a cone. Since $L$ is Lagrangian, $J(dr_1 + \cdots + \partial r_n) \in N_x L$; hence, $g(Jdf)\ |_{T_x L}$, and Proposition 3.2 is proved completely.

**Proposition 3.5.** *The cone* $L = \{z_0 = 1, z_j = ke_j^{i\theta}, \sum_j \theta_j = \text{const}\}$, *is special Lagrangian and therefore, locally minimal.*

**Proof.** (i) We first establish that the cone $L$ is Lagrangian. The restriction of the form $\Phi$ to it is

$$\Phi = \frac{4}{(1+ne^{2r})^2}[(\Sigma_i e^{2r} dr_i \wedge (\Sigma d\theta_i)) - e^{2r}(ndr$$
$$+ i(\Sigma d\theta_i)) \wedge e^{2r}(ndr - i(\Sigma d\theta_i))] = 0$$

(see formula (3.3.1)), since $\sum d\theta_i = 0$, with $r_i = r = \ln|k|$.

We prove that $p^*(d\theta)\ |_L = 0$, i.e., $L$ is special Lagrangian. By definition, $e^{i\theta(l_x)} = (\sqrt{G}dz_1 \wedge \cdots \wedge dz_n, \vec{l}_x) = \sqrt{G}(\bigwedge_{i=1}^n (dr_i + id\theta_i)e^{2r_i}, \overrightarrow{l_x})$. Replacing $\vec{l}_x = \overrightarrow{T_x L} = \frac{\partial r \wedge \partial\theta_1 \wedge \cdots \wedge \partial\theta_{n-1}}{|1\partial r \wedge \partial\theta_1 \wedge \cdots \wedge \partial\theta_{n-1}|}$, we obtain

$$\forall x \in L : e^{i\theta p(x)} = \sqrt{G}(e^{2r} \bigwedge_{i=1}^{n} (dr + id\theta_i),$$
$$\partial r_1 \wedge \cdots \wedge \partial\theta_{n+1}|\partial r \wedge \cdots \wedge \partial\theta_{n-1}|) = \frac{\sqrt{G}\cdot e^{2r}\cdot (i)^{(n-1)}}{|\partial r_1 \wedge \cdots \wedge \partial\theta_{n-1}|}$$
$$\cdot(dr \wedge d\theta_1 \wedge \cdots \wedge d\theta_{n-1}, \partial r_1 \wedge \cdots \wedge \partial\theta_{n-1})$$
$$= \begin{cases} k \in \text{Re} & \text{if } n \text{ is odd,} \\ k \in \text{Im} & \text{if } n \text{ is even.} \end{cases}$$

Anyway, the function $\theta \cdot p$ is constant; therefore, $d(\theta \cdot p) = 0$.

**Proposition 3.6.** *Let* $M^{4n} = CP^{2n}$. *Then the surface* $L = \{z_0 = 1, z_0 = \bar{z}_{n+i}, i = 1, n\}$ *is minimal in the manifold* $M^{4n}$.

We prove (i), first establishing that $L$ is Lagrangian. The restriction of the form $\Phi$ (see (3.3.1)) to $L$ takes the form

$$\Phi = \frac{4}{(1+2\sum_{i=1}^n e^{2r_i})^2}\left[\left(1+\sum_{i=1}^n e^{-2r_i}\right)(\Sigma dr_i \wedge (d\theta_i + d\theta_{n+i})\cdot e^{2r_i})\right.$$
$$\left. - (\Sigma e^{2r_i}(2dr_i + i(d\theta_i + d\theta_{n+i}))) \wedge \Sigma(e^{2r_i}(2dr_i - i(d\theta_i + d\theta_{n+i})))\right],$$

where $dr_i = dr_{n+i}$, $d\theta_i + d\theta_{n+i} = 0$; therefore, $\Phi_{|_L} = 0$.

We now prove (ii). We show that $p^*(d\theta)_{|_L} = 0$, i.e., $L$ is a special Lagrangian cone. Since $dz_i = e^{r_i}(dr_i = id\theta_i)$, the restriction of the form $\sqrt{G}dz, \wedge \cdots \wedge dz_n$ to $L$ is

$$\begin{aligned} w &= \sqrt{G}e^{2(r_1+\cdots+r_n)} \bigwedge_{i=1}^{n} (dr_i + id\theta_i) \wedge (dr_i - id\theta_i) \\ &= (-i)^n \cdot 2\sqrt{G} \cdot e^{2(r_1+\cdots+r_n)} \cdot \bigwedge_{i=1}^{n} dr_i \wedge d\theta_i. \end{aligned}$$

By definition,

$$\begin{aligned} \forall x \in L, e^{i\theta.p(x)} &= (\sqrt{G}dz_1 \wedge \cdots \wedge dz_n, \overrightarrow{T_xL}) \\ &= \frac{(-i)^n.2\sqrt{G}e^{2(r_1+\cdots+r_n)}}{|\partial r_1 \wedge \partial\theta_1 \wedge \cdots \wedge \partial r_n \wedge \partial\theta_n|} \cdot (\bigwedge_{i=1}^{n} dr_i \wedge d\theta_i, \partial r_1 \wedge \partial\theta_1 \wedge \ldots \\ &\qquad \cdots \wedge \partial r_n \wedge \partial\theta_n); \begin{cases} k \in \mathrm{Re} & \text{if } n \text{ is even}, \\ k \in \mathrm{Im} & \text{if } n \text{ is odd}. \end{cases} \end{aligned}$$

Due to the continuity of the function $\theta p$, $k$ must be constant; therefore, $d(\theta \cdot p)_{|_L} = 0$. Applying Proposition 3.2, we conclude that the surface $L$ is minimal.

## §4 The Integrability Condition for the Form $\psi$

**Theorem 4.** *If a form $\psi$ is integrable, i.e., if it determines a foliation of codimension one, then it is closed. Further, $\psi$ is closed if and only if the manifold $M$ is locally calibrated. A Kählerian manifold is locally calibrated if and only if its Ricci tensor is identically zero.*

To prove (i), we first find a condition for $\psi$ to be integrable, consisting of $0 \equiv \psi \wedge d\psi$, and

$$\psi = Jdf + d\theta, \quad d\psi = dJdf.$$

Therefore,

$$\psi \wedge d\psi = (Jdf + d\theta) \wedge dJdf. \tag{4.1.1}$$

Since $f(x) = \bar{f}(\pi x)$, we have $dJdf = dJ\bar{f}(\pi x)$, or the form $dJdf \in \Lambda^2(T^*M) \subset \Lambda^2(T^*G_L(M))$.

By definition, the function $\theta$ is locally independent of $x$. Thus, at each point $(x, l_x) \in G_L(M)$, we have

$$\psi \wedge d\psi = Jdf \wedge dJdf + d\theta \wedge dJdf.$$

The first addend is in space $\Lambda^3(T^*M)$, whereas the second in $\Lambda^1(T^*G_L^n) \times \Lambda^2(T^*M)$, where $G_L^n$ denotes the Grassmannian manifold of oriented Lagrangian planes. Therefore, $\psi \wedge d\psi \equiv 0$ if and only if both addends vanish. But $d\theta \neq 0$, and hence, the second is zero if and only if $dJdf = 0$. But $Jdf = \Sigma_\alpha(-i)(f_{z_\alpha} dz_\alpha - f_{\bar{z}_\alpha} d\bar{z}_\alpha)$, and the form $dJdf = z_i \sum_{\alpha,\beta}(f_{\bar{z}_\alpha z_\beta} \cdot dz_\beta \wedge d\bar{z}_\alpha)$.

Substituting $f = \ln\sqrt{G}$, we obtain that the form $dJdf$ is zero if and only if

$$\frac{\partial^2 \ln\sqrt{G}}{\partial\bar{z}_\alpha \partial z_\beta} = 0, \tag{4.1.2}$$

or, which is tantamount, $(\partial^2/\partial\bar{z}_\alpha\partial_{z_\beta})(\ln G) = 0$.

(ii) We show that (4.1.2) holds if and only if $M$ is locally calibrated. We prove the following.

**Lemma 4.1.** *Let $w = \operatorname{Re}\hat{w}$ be a closed SL-form. Then the form $\hat{w}$ is also closed.*

**Proof.** We write $\hat{w}$ with respect to the local coordinates, viz.,

$$\hat{w} = g(z)dz_1 \wedge \cdots \wedge dz_n.$$

Then

$$\begin{aligned} d\operatorname{Re}\hat{w} &= \operatorname{Re} d\hat{w} = \operatorname{Re}(\partial/\partial\bar{z}_i(g(z)) \cdot d\bar{z}_i \wedge dz_1 \wedge \cdots \wedge dz_n) \\ &= \sum_i \frac{1}{2} \operatorname{Re}\left[\left(\frac{\partial(g)z}{\partial x_i} + i\frac{\partial g(z)}{\partial y_i}\right)(-1)^i dx_i \wedge dy_i \wedge dz_1 \wedge \cdots \wedge_i \cdots \wedge dz_n\right] \\ &= \frac{1}{2}\sum_i \operatorname{Re}\left[i\left(\frac{\partial g^R}{\partial x_i} - \frac{\partial g^I}{\partial y_i}\right) - \left(\frac{\partial g^R}{\partial y_i} + \frac{\partial g^I}{\partial x_i}\right)\right] \\ &\quad (-1)^i dx_i \wedge dy_i \wedge (dz_1 \wedge \cdots \wedge_i \cdots \wedge dz_n), \end{aligned}$$

where $g^R = \operatorname{Re} g(z), g^I = \operatorname{Im} g(z)$.

Due to (4.1.3), $d\operatorname{Re}\hat{w} = 0$ if and only if

$$\forall i \operatorname{Re}\left[i\left(\frac{\partial g^R}{\partial x_i} - \frac{\partial g^I}{\partial y_i}\right) - \left(\frac{\partial g^R}{\partial y^i} + \frac{\partial g^I}{\partial x^i}\right)\right] dx_i \wedge dy_i \wedge dz_1 \wedge \cdots \wedge_i \cdots \wedge dz_n = 0,$$

which is equivalent to

$$\left(\frac{\partial g^R}{\partial y_i} + \frac{\partial g^I}{\partial x_i}\right) \cdot \operatorname{Re}(dz^{v_i}) - \left(\frac{\partial g^R}{\partial x_i} - \frac{\partial g^I}{\partial y_i}\right)\operatorname{Im}(dz^{v_i}) = 0, \tag{4.1.5}$$

where $dz^{v_i}$ denotes the $(n-1)$-form $dz_i \rfloor (dz, \wedge \cdots \wedge dz_n)$. The forms $\operatorname{Re}(dz^{v_i})$ and $\operatorname{Im}(d_z v_i)$ are linearly independent; therefore, (4.1.5) holds if and only if

$$\begin{cases} \frac{\partial g^R}{\partial y_i} + \frac{\partial g^J}{\partial x_i} &= 0, \\ \frac{\partial g^R}{\partial x_i} - \frac{\partial g^J}{\partial y_i} &= 0, \end{cases}$$

which means that $\frac{\partial g(z)}{\partial \bar{z}}$. Hence, $d\hat{w} = 0$, and the lemma is thus proved.

Returning back to the proof of the theorem, we see that our lemma shows that $\hat{w} = g(z) \cdot dz_1 \wedge \cdots \wedge dz_n$, where $g(z)$ is a holomorphic function. Put $g(z) = e^{f(z)+i\varphi(z)}$, where $f, \varphi \in R$. Since the form $\operatorname{Re}\hat{w}$ is special Lagrangian by assumption, $\operatorname{Re}(e^{\ln\sqrt{G}} \cdot dz_1 \wedge \cdots \wedge dz_n)$ is also special Lagrangian by Lemma 2.2, we have

$$\begin{aligned} e^{\ln\sqrt{G}} dz_1 \wedge \cdots \wedge dz_n &= e^{i\theta_1(z)} \cdot dz_1^0 \wedge \cdots \wedge dz_n^0, \\ e^{f+i\varphi} dz_1 \wedge \cdots \wedge dz_n &= e^{i\theta_2(z)} \cdot dz_1^0 \wedge \cdots \wedge dz_n^0, \end{aligned} \tag{4.1.7}$$

where $dz_i^0$ is a unitary basis at the point $z$.

It follows from (4.1.7) that

$$e^{\ln\sqrt{G}} dz_1 \wedge \cdots \wedge dz_n = e^{i(\theta_1-\theta_2)} \cdot e^{f+i\theta} \cdot dz_1 \wedge \cdots \wedge dz_n.$$

Hence,

$$\ln\sqrt{G} = f + i[(\varphi + \theta_i - \theta_2) + 2k\pi]. \tag{4.1.8}$$

Since $f + i\varphi$ is a holomorphic function, we have

$$\frac{\partial^2}{\partial z_\alpha \partial z_{\bar{\beta}}} f = 0 \Rightarrow \frac{\partial^2 \ln G}{\partial z_\alpha \partial z_\beta} = 0, \tag{4.1.9}$$

which is a necessary condition for the manifold $M$ to be locally calibrated. Conversely, if $\frac{\partial^2}{\partial z_\alpha \partial \bar{z}_\beta} \ln G = 0$, then $\ln\sqrt{G}$ is locally the real part of the holomorphic function $g(z) = \ln\sqrt{G} + i\theta$. Then the form $\hat{w} = e^{\ln\sqrt{G}+i\theta} dz_1 \wedge \cdots \wedge dz_n$ is closed and $\operatorname{Re}\hat{w}$ is an $SL$-form. Therefore, (4.1.9) is also sufficient to be locally calibrated.

(iii) Comparing (4.1.2) and (4.1.9), we see that $\psi$ is closed if and only if $M^{2n}$ is locally calibrated. If it is Kählerian then the Ricci tensor is given on it by $K_{\alpha\bar{\beta}} = \frac{\partial^2 \ln G}{\partial z_\alpha \partial z_\beta}$ (see [198]). In view of (4.1.2) and (4.1.9), the remaining part of our theorem is thus proved.

# APPENDIX II

## Calibrations, Minimal Surface Indices, Minimal Cones of Large Codimensional and the One-Dimensional Plateau Problem

Lê Hông Vân has found new minimality criteria for currents in [244]–[245], [451] and [452]. Here, we give a rough outline of the results. Recall that a differential $k$-form $\varphi$ on a Riemannian manifold $M$ is called a *calibration* if it is closed and is of comass 1 (for details see [195] and [196]). A current $S^k$ on a manifold $M$ is called a $\varphi$*-current* if, almost for all $x \in M$ in the sense of measure $\|S^k\|$, the equality $\langle \varphi, \overrightarrow{S_x} \rangle = \|\overrightarrow{S_x}\| = 1$ holds true. It is known that any $\varphi$-current is globally minimal (see [195], [196]).

**Definition (see [244], [245]).** A calibration $\varphi$ is said to be *efficient* if there is at least one $\varphi$-current. We will say that a calibration $\varphi'$ is weaker than $\varphi$ if any $\varphi'$-current is also a $\varphi$-current.

Let $\varphi$ be efficient on a compact, homogeneous Riemannian manifold $G/H$. According to [244], [245], there exists a $G$-invariant calibration $\varphi_G$ for any $\varphi$, so that $\varphi$ is weaker than $\varphi_G$. If $\varphi$ is efficient, then $\varphi_G$ is also efficient, and the comass of the form $\varphi_G$ is minimal for all $G$-invariant differential forms cohomologous to it. The Lê Hông Vân criterion for a normal current on a compact, homogeneous Riemannian space $G/H$ to be globally minimal generalizes that of Đao Chông Thi on a symmetric space.

Let $V$ be the tangent space to $G/H$, $\Lambda^*_H(V)$ the space of all $H$-invariant forms on $V$, and the invariant measure $\int_g d\mu_g = 1$ be given on the group $G$. Denote the current $\int_G S d\mu g$ by $\pi_G S$. It is representable as $\overrightarrow{\pi_G S} \times d\|\pi_G S\|$.

Let the $k$-vector $M(\pi_G S)/\operatorname{vol}(G/H)) \cdot \overrightarrow{\pi_G S_e} \in \Lambda^k_H(V)$ be $\overrightarrow{\pi_H \int_e}$.

**Theorem 1 (Lê Hông Vân).** *A closed current $S^k$ on a compact, homogeneous Riemannian space is globally minimal if and only if $M(S) = M(\pi_G S) =$ $\operatorname{vol}(G/H) \min\{\|\overrightarrow{\pi_H S_e} + \overrightarrow{\partial\theta}\|, \theta \in \Lambda^{k+1}_H(V)\}$.*

This criterion permits us to complete and substantially develop the author's and Dao Chong Thi's results regarding the minimality of submanifolds in symmetric and homogeneous spaces. For example, a real Grassmannian manifold $G_{2,2k}(R)$ is globally minimal in the manifold $G_{2,n}(C)$ for any number $k \leqslant [n/2]$. The four-dimensional quaternion projective space $HP^1$ is a globally minimal submanifold (current) in the quaternion Grassmannian manifold $G_{p,q}(H)$.

**Theorem 2 (Lê Hông̃ Vân).** *For any classical group $G$ from* $\mathrm{SU}_n$, $\mathrm{SO}_n$, $\mathrm{Sp}_n$, *the following submanifolds are globally minimal in the orbit $G/T$ of the adjoint representation of $G$ in its Lie algebra (the metric on the orbit being induced by the Killing form from the Lie algebra):*

*(1) two-di 2-spheres* $S^2 = \mathrm{SU}_2(\alpha)/S^1, \alpha \in \Delta$;

*(2) suborbits* $\mathrm{SU}_{m+1}/T^m$, *where* $m+1 \leqslant n = rkG$ *and the embedding* $\mathrm{SU}_{m+1} \to G$ *is standard;*

*(3) suborbits* $\mathrm{SO}_{2m}/T^m$, *where the embedding* $\mathrm{SO}_{2m} \to G \in \{\mathrm{SO}_n, \mathrm{Sp}_n\}$ *is standard;*

*(4) when* $G = \mathrm{Sp}_n$, *there is an additional series* $S_{p_m}/T^m$, *where the embedding* $S_{p_m} \to S_{p_n}$ *is standard.*

*In the generic orbit $G/T$, the current $S_{D_\alpha}$ is globally minimal in each homology class $[\alpha]$ of codimension 2. Moreover, any globally minimal current of codimension 2 in $G/T$ in positive relative to a certain invariant complex structure on the orbit. A current $S$ of dimension 2 is globally minimal if and only if $M(S) = \rho^*(0, V_{[S]})$ (see [244], [245]). The following submanifolds are globally minimal in the singular orbit $G/Z(x)$, where $Z(x) = U_{n_1}x \ldots xU_{n_k}XG_p$ if $G = \mathrm{SO}_n(\mathrm{Sp}_n)$ (and then $G_p = \mathrm{SO}_p(\mathrm{Sp}_p)$, $Z(x) = S(U_{n_1}x \ldots xU_{n_k}j)$ if $G = \mathrm{SU}_n$:*

*(1) suborbits* $\mathrm{SU}_{m_1+m_2}/S(U_{m_1}xU_{m_2})x \ldots x\,\mathrm{SU}_{m_{2r-1}+m_{2r}}/S(U_{m_{2r-1}} \times U_{m_{2r}})$;*where* $2r \leqslant k$ *and the embeddings* $U_{m_i} \to U_{n_i}$ *are standard;*

*(2) suborbits* $\mathrm{SU}_{m+1}/S(U_{n_1}x \ldots xU_{n_r}$, $\Sigma n_i = m+1, R \leqslant k$ *and the embedding* $\mathrm{SU}_{m+1} \to G$ *is standard;*

*(3) suborbits* $\mathrm{SO}_{2m}/(U_{n_1}x \ldots xU_{n_r})$, $\Sigma n_i = m, r \leqslant k$, *and the embeddings* $\mathrm{SO}_{2m} \to G$ *is standard.*

An important characteristic of a minimal surface is its index defined in analogy with the index of a geodesic whose many properties are carried to the index of a minimal surface without change. For example, this index is used in the stability of a minimal surface and the theorem on the relation of the index of a geodesic with the multiplicity of conjugate points on it, and so on (see [396], [391]).

We illustrate this by giving examples of indices of certain classical noncompact minimal surfaces in $R^3$, in particular, of the catenoid and Enneper surface (see [39], [414]).

**Definition.** We call the supremum of the indices of compact subdomains with smooth boundaries the *index* of a noncompact minimal submanifold.

It is convenient to specify minimal surfaces in $R^3$ by a Weierstrass representation.

Let $U \subset C = R^2$ be a domain.

**Definition.** We call a pair of complex-valued functions $f, g$ on a domain $U$, so that $f$ is holomorphic, $g$ is meromorphic, and $fg^2$ is holomorphic, a *Weierstrass representation.* If $\varphi_1 = \frac{1}{2}f(1-g^2), \varphi_2 = \frac{i}{2}f(1+g^2), \varphi_3 = fg$, then we define the mapping $x : U \to R^3$ by $x^k(z) = \mathrm{Re} \int_p^z \varphi_k(w)dw$, where $p \in U$ is a fixed point, $z \in U$, and the integral is along a certain path in $U$. We assume that $x^k(z)$ are one-valued functions, i.e., the integral is independent of a path. The obtained mapping determines a generalized minimal surface [312].

We consider surfaces determined by a Weierstrass representation globally. In this case, the domain $U$ is replaced by the Riemannian surface $M$, and the function $f$ is replaced by the holomorphic 1-form $w$

**Theorem 3 (A. A. Tuzhilin).** *Let an immersed minimal surface $M$ in $R^3$ be given by the Weierstrass representation $(f(w), w^m)$ on (a) the $w$-plane $C$, where $m \in Z, m > 0$, or (b) $C \backslash 0$, where $m \in Z \backslash 0$. Then the index of $M$ equals* $2|m| - 1$.

Here, the immersion of $M$ is equivalent to $f(w) \neq 0$. The Weierstrass representation of the Enneper surface is $(1, w)$ on the plane $C$, of the catenoid it is $(-\frac{1}{2W^2}, w)$ on the plane $C$ without the point 0, and of the Richmond surface, it is $(w^2, \frac{1}{w^2})$ on $C \backslash 0$.

**Corollary.** *The index of the Enneper surface is* 1, *of the catenoid* 1, *and of the Richmond surface* 3.

D. Fischer-Colbrie [467] was the first who calculated the index of the catenoid and the Enneper surface. Tuzhilin's results were obtained independently.

**Statement 1 (Fischer-Colbrie, [467]).** The index of the complete minimal surface in $\mathbf{R}^3$ is finite iff the total curvature of this surface is finite.

Tuzhilin's theorem below is the generalization of Fischer-Colbrie's result.

**Definition.** The meromorphic function $g$ on the Riemannian surface $M$ is called *nice* iff there exists a holomorphic 1-form $w$ on $M$, i.e., $(M, w, g)$ is the global Weierstrass representation of some complete minimal surface.

**Statement 2 (Tuzhilin).** The index of the minimal surface with a nice function $g$ in the Weierstrass representation $(M, w, g)$ is finite iff the total curvature of this surface is finite.

**Corollary.** *Let $(\mathbf{C}, f dw, g)$ be a Weierstrass representation with nice $g$. Then the index of the corresponding minimal surface is finite iff $g$ is a rational function.*

For example, the meromorphic function $g$ on $\mathbf{C}$ is nice when $g = P/h + c/Q$ or $g = h/P$, where $P$ and $Q$ are polynomials, $c =$ const, and $h$ is a holomorphic function.

**Corollary.** *The index of a helicoid, of all periodic minimal surfaces, for example, Schwarz–Riemann surfaces, and Scherk's surfaces, is equal to infinity.*

For a compact minimal submanifold $M$ with boundary $\partial M$ in a Riemannian manifold considered as a critical point of the volume functional $A$, the second variation $\delta^2 A$ is a symmetric bilinear form correctly defined on $C_0^\infty(NM)$, i.e., on the space of all smooth sections of the normal bundle $NM$ over $M$, vanishing on $\partial M$.

**Definition.** The index (resp. null index) of $\delta^2 A$ as of a bilinear form is called the *index* (resp. *null index*) of the compact minimal submanifold $M$. We call $\partial M$ conjugate if the null index of $M$ does not vanish, in which case the null index is called the multiplicity of $\partial M$.

Similarly to $\varepsilon$-type contraction (see [391]), we define the exhaustion of a noncompact minimal surface as follows.

**Definition.** The *exhaustion* of a connected minimal submanifold $M$ in a Riemannian manifold is the family of its subdomains $K_t, t \in R_t$, with smooth boundaries, so that

(1) $\partial K_t$ depends on $t$ smoothly, with $K_{t_1} \subset K_{t_2}, K_{t_1} \neq K_{t_2}$ for $t_1 \neq t_2$;

(2) the index and null index of $K_t$ are zero for sufficiently small $t$;

and

(3) for any compact subset $K$, there exists $K_t$ such that $K \subset K_t$. It is clear that $\bigcup K_t = M$.

Propositions 1, 2, and 3 have been proved by A. A. Tuzhilin [437].

**Proposition 1 (generalizing the Smale and Simons theorems).** *Let $\{K_t, t \in R_+\}$ be the exhaustion of a noncompact, minimal, and connected submanifold $M$ in a Riemannian manifold.*

We define

$$\alpha(t) = \begin{cases} 0 \text{ if } \partial K_t \text{ is not a conjugate boundary,} \\ \text{multiplicity of } \partial K_t \text{ if } \partial K_t \text{ is a conjugate boundary.} \end{cases}$$

Then $\operatorname{ind} M = \sum_{t \in R_+} \alpha(t)$.

Proposition 1 reduces the computation of the minimal surface indices to that of the conjugate boundary $\partial K_t$ multiplicities, i.e., the Jacobi equation solution (see [391] for the case of a minimal submanifold in a Riemannian manifold). For a minimal surface in $R^3$, the equation assumes a simpler form.

**Proposition 2.** *Let $M$ be a minimal, immersed, and orientable surface in $R^3$, let $n$ be the field of unit normals to $M$, and $V = T \cdot n$ be an arbitrary section of the normal bundle $NM$, where $T$ is a function on $M$. Then the Jacobi equation has the form $\Delta T - 2KT = 0$, where $\Delta$ is the metric Laplacian on $M$, and $K$ is the Gaussian curvature of $M$. With respect to the conformal coordinates $u, v$ on $M$ (i.e., such that $ds^2(M) = \lambda^2(du^2 + dv^2)$), the Jacobi equation is*

$$\frac{\partial^2 T}{\partial u^2} + \frac{\partial^2 T}{\partial v^2} - 2K\lambda^2 T = 0.$$

If a surface $M$ is given by a Weiestrass representation $(f(w), g(w))$, where $w = u + iv$, then the Jacobi equation is rewritten as

$$\frac{\partial^2 T}{\partial u^2} + \frac{\partial^2 T}{\partial v^2} + \frac{8|g'|^2}{(1+|g|^2)^2} T = 0,$$

where $g' = \frac{dg}{dw}$. It has been remarked in [38] that the Jacobi equation turns into an equation on the sphere $S^2$, not depending on a minimal surface, under a Gauss map of the minimal surface in $R^3$.

**Proposition 3.** *Let $M_1$ and $M_2$ be two orientable, minimal and immersed surfaces in $R^3$, and $n_1 : M_1 \to S^2$ and $n_2 : M_2 \to S^2$ be Gauss maps. Let $M_1$ and $M_2$ be isomorphic in the sense that there exists a diffeomorphism $F : M_1 \to M_2$ associated with the maps, i.e., the diagram*

$$\begin{array}{ccc} M_1 & \xrightarrow{F} & M_2 \\ {\scriptstyle n_1}\downarrow & & \downarrow{\scriptstyle n_2} \\ S^2 & \xrightarrow{\varphi} & S^2 \end{array}$$

*is commutative, where $\varphi$ is an isometry $(S^2, h_{ij})$ and $h_{ij}$ is the metric induced on $S^2$. Then* $\operatorname{ind} M_1 = \operatorname{ind} M_2$.

*The Jacobi equation on the sphere is of the form $\Delta_{S^2}(T) + 2T = 0$, where $\Delta_{S^2}$ is the metric Laplacian on $S^2, h_{ij}$). We omit the details.*

Note that A. Yu. Borisovich has carried out the reduction of the problem of searching for minimal surface bifurcations to operator equations in investigating the Plateau operator and bifurcations in the neighbourhood of the catenoid, helicoid, Scherk and Enneper surfaces [444]–[446]. These works clarify the functional nature of the Plateau operator.

An attempt to calculate the two-dimensional minimal surface indices was made in [457]. However, its author based himself on the wrong equation for the second variation of the areal functional (considering the Jacobi equation Laplacian as metric, it being Euclidean).

Above, we have discussed the global minimality of cones of small codimensions. There is no such classification in large codimensions. However, partial results have already been obtained.

**Theorem 4 (A. O. Ivanov [436], [442]).** *The cones CA of codimension two in Euclidean space $R^N$ over the following locally minimal submanifolds $A \subset S^{N-1}$ are unique globally minimal surfaces with boundary A in $R^N$, viz.,*

*(1)* $S^{r-1} \times S^{r-1} \times S^{r-1} = \frac{\mathrm{SO}(r)\times\mathrm{SO}(r)\times\mathrm{SO}(r)}{\mathrm{SO}(r-1)\times\mathrm{SO}(r-1)\times\mathrm{SO}(r-1)}$ *in* $\mathbf{R}^{3r}$, $r \geqslant 7$ ,

*(2)* $\frac{\mathrm{SO}(r)\times\mathrm{SO}(3)}{\mathrm{SO}(r-1)\times\mathbf{Z}_2\times\mathbf{Z}_2}$ *in* $\mathbf{R}^r \times \mathbf{R}^5$, $r \geqslant 53$,

*(3)* $\frac{\mathrm{SO}(r)\times\mathrm{SU}(3)}{\mathrm{SO}(r-1)\times T^2}$ *in* $\mathbf{R}^r \times \mathbf{R}^8$, $r \geqslant 39$,

*(4)* $\frac{\mathrm{SO}(r)\times\mathrm{Sp}(3)}{\mathrm{SO}(r-1)\times\mathrm{Sp}(1))^3}$ *in* $\mathbf{R}^r \times \mathbf{R}^{14}$, $r \geqslant 74$,

*(5)* $\frac{\mathrm{SO}(r)\times\mathrm{Sp}(2)}{\mathrm{SO}(r-1)\times T^2}$ *in* $\mathbf{R}^r \times \mathbf{R}^{10}$, $r \geqslant 51$,

*(6)* $\frac{\mathrm{SO}(r)\times G_2}{\mathrm{SO}(r-1)\times T^2}$ *in* $\mathbf{R}^r \times \mathbf{R}^{14}$, $r \geqslant 75$,

*(7)* $\frac{\mathrm{SO}(r)\times F_4}{\mathrm{SO}(r-1)\times\mathrm{Spin}(8)}$ *in* $\mathbf{R}^r \times \mathbf{R}^{26}$, $r \geqslant 74$,

*(8)* $\frac{\mathrm{SO}(r)\times\mathrm{Spin}(10)\times U(1)}{\mathrm{SO}(r-1)\times\mathrm{SU}(4)\times T^1}$ *in* $\mathbf{R}^r \times \mathbf{R}^{32}$, $r \geqslant 136$.

The estimation obtained for $r$ is not optimal, but making them more concise faces serious computational difficulties. The proof that these surfaces are globally minimal is based on the construction of calibration forms in [94] and [196]. Some other interesting examples of symmetric, locally minimal cones in Lie algebras have been obtained by I. S. Balinskaya [439], [440], [443] (see below).

Let an exterior differential $p$-form $\varphi$ be given on a Riemannian manifold $X^N$, so that (1) $\varphi$ is closed (i.e., $d\varphi = 0$) and (2) $\varphi|_\xi \leqslant \mathrm{vol}_\xi$ for any oriented $p$-plane $\xi$ at any point $x \in X$. As we have noted, this form is called a calibration. If we take a $p$-dimensional submanifold $M^p \subset X$ such that $\varphi|_M = \mathrm{vol}_M$, then $M$ is a minimal surface in $X$ in the sense that $\mathrm{vol}\, M \leqslant \mathrm{vol}\, M'$ for any $M' \subset X$ with the same boundary $\partial M = \partial M'$.

Let $M \subset X$ be submanifold, let $i : M \to X$ be an embedding, and let there exist a retraction $j : X \to M$ (where all points of $M$ remain fixed). Then we can consider the form $\omega = j^*(*1_M)$ which is closed and coincident with the volume form on $i(M)$. We only need a convenient Riemannian metric for property (2) to be fulfilled. The construction has been noted by Lê Hông̅ Vân. It turns out that such a metric can be found in some cases. Let $M^p$ be a Riemannian manifold with metric $g_{ij}$, $\pi : V^{p+k} \to M^p$ be a real vector bundle locally trivial with fibre $R^k$ over $M^p$, and the affine connection $\Gamma$, the family of "horizontal" planes $\Gamma(v)$ depending smoothly on $v \in V$, be given in the fibre bundle. That they are horizontal means the nondegeneracy of the differential of the projection on $\Gamma(v)$ at each point. We require that the zero section $f_0 : M \to V$ should be horizontal, i.e., $df_0(m)(T_m M) = \Gamma(m, 0)$.

We introduce the metric on the total space of the fibre bundle $V$ as follows. At each point $v \in V$, the tangent space $T_v V$ to the total space is decomposed in to the direct sum of two subspaces $\Gamma(v)$ and $F(v)$, where $\Gamma(v)$ is a horizontal plane and $F(v)$ the tangent plane to the fibre at the point $v$. We specify the family of metrics $\tilde{g}_{ij}(m), 1 \leqslant i, j \leqslant k$, depending on $m \in M$ smoothly, and the metric on $V$ by

$$\left(\begin{array}{c|c} g_{ij}(m) & 0 \\ \hline 0 & \tilde{g}_{ij}(m) \end{array}\right).$$

The tangent plane to the fibre and the horizontal plane are assumed orthogonal. The metric on $M$ is in the upper left-hand corner (giving the scalar product on the horizontal planes), and the above metric $\tilde{g}_{ij}$ in the lower upper-hand corner (giving the scalar product on the tangent planes to the fibres).

**Example.** Consider the tangent bundle $TM^p \to M^p$ to a Riemannian manifold $M^p$, and take the linear connection generated by the metric on $M$ as the connection $\Gamma$, taking the metric on the corresponding tangent space $T_m M$ to $M$ as the metric $\tilde{g}_{ij}$ on the fibres. We then obtain a metric of the form

$$\left(\begin{array}{c|c} g_{ij}(m) & 0 \\ \hline 0 & g_{ij}(m) \end{array}\right).$$

**Theorem 5 (A. O. Ivanov).** *Let $M^p$ be a Riemannian manifold, $\pi : V^{k+p} \to M^p$ a vector bundle, locally trivial with the above metric, $f_0 : M^p \to V^{k+p}$ the zero horizontal section of the fibre space $V$. Then $f_0(M^p)$ is a globally minimal surface in $V$.*

Symmetric minimal cones of large codimension are of considerable interest. Let $\mathfrak{G}$ be a compact Lie group, and $G$ its Lie algebra. Consider the adjoint action of the group on the algebra and distinguish generic orbits. Suppose the Killing metric is fixed on the Lie algebra, and $S^{N-1}$ is the sphere of unit radius in the Lie algebra. Then the sphere is mapped onto itself under the adjoint action of the group, and therefore, fibres into the orbits of adjoint action. The problem is to describe locally minimal orbits in the sphere and to determine when the cone over them with vertex at the origin is a locally or globally minimal surface. Locally minimal invariant cones turn out to be describable for classical simple Lie algebras (see I. S. Balinskaya). Let $H$ be a Cartan subalgebra in $G = \mathrm{so}(2r)$. Then the element (matrix) $a \in H$ can be written in canonical block-diagonal form, each block being $\begin{pmatrix} 0 & a_i \\ -a_i & 0 \end{pmatrix}$. Let $a_1, \dots, a_n$ be the set of numbers specifying the matrix $a$. If $G = \mathrm{so}(2n+1)$, then the canonical form of $a \in H$ differs from the above one only in the zero eigenvalue in the diagonal form. If $G = \mathrm{su}(n)$, then the canonical form of $a$ from the Cartan subalgebra is $a = \mathrm{diag}(ia_1, \dots, ia_n)$, $\sum_k a_k = 0$. If $G = \mathrm{sp}(n)$, then the matrix $a \in H$ can be written as $a = \mathrm{diag}(ia_1, \dots, ia_n, -ia_1, \dots, -ia_n)$.

Denote the orbit of $a$ by $\mathcal{O}(a)$. If $a$ is in general position (i.e., is a regular element of the algebra), then $\mathcal{O}(a)$ is diffeomorphic to the quotient space of the group with respect to its maximal torus. The above parameters $a_1, \ldots, a_n$ can be regarded as coordinates on the Cartan subalgebra $H$ in $G$. Let $f(A) = \operatorname{vol} \mathcal{O}(a)$ be the volume of the orbit $\mathcal{O}(a)$, where $f(a) = f(a_1, \ldots, a_n)$ and $a = (a_1, \ldots, a_n)$.

The following theorem is a (local) solution of the spherical Bernstein problem for the case where the sphere is embedded into a Lie algebra (for the earlier results, see W. Hsiang [213]).

We call a locally minimal cone $CA$ over a manifold $A \subset S^{N-1}$ *stable* if any of its isotopic variations, not displacing its vertex and "base" $A$, and sufficiently small in amplitude, does not decrease its volume.

**Theorem 6 (I. S. Balinskava [439], [440], [443]).** *In the sphere $S^{N-1}$ (standardly embedded into a classical simple Lie algebra), there exists one, and only one, generic orbit $A = \mathcal{O}(a)$ which is a locally minimal submanifold in the sphere. The orbit is of maximal volume in the generic orbit class. For $n \leqslant 4, n \leqslant 8, n \leqslant 3$ and the groups $\mathrm{SU}(n), \mathrm{SO}(n)$ and $\mathrm{Sp}(n)$, respectively, the cone $CA$ is locally minimal, but not stable; it is stable for the remaining $n$.*

Therefore, there exists one, and only one, locally minimal generic orbit $A$ in the sphere, with maximal volume for all regular orbits. In proving Theorem 6, explicit expressions for the volumes of adjoint action orbits were found in terms of the roots of the corresponding Lie algebra. The formulas are given in the following theorem (see [443]), viz.,

**Proposition 4.** *Let $a = (a_1, \ldots, a_n) \in H$, and $A = \mathcal{O}(a)$ the orbit of $a$. Then the volume of the orbit, $\operatorname{vol} \mathcal{O}(a)$, is found by the formulas*

*(1)* $G = \mathrm{so}(2n), \operatorname{vol} \mathcal{O}(a) = \prod_{i<j} (a_i^2 - a_j^2)^2$;

*(2)* $G = \mathrm{so}(2n+1)$ *or* $\mathrm{sp}(n), \operatorname{vol} \mathcal{O}(a) = \prod_{i<j} (a_i^2 - a_j^2) \prod_i a_i^2$;

*and*

*(3)* $G = \mathrm{su}(n), \operatorname{vol} \mathcal{O}(a) = \prod_{i<j} (a_i - a_j)^2$,

*where $a_1, \ldots, a_n$ are canonical coordinates on $H$.*

A. O. Ivanov obtained the following effective sufficient condition of symmetric minimal cones of arbitrary codimension to be stable.

Let $G$ be a compact connected Lie group acting on $\mathbf{R}^N$ as a subgroup of $\mathrm{SO}(N)$. Let an orbit space $\mathbf{R}^N/G \equiv w \subset \mathbf{R}^n$, and let $w$ be a subset of $\mathbf{R}^n$ bounded by some finite collection of hyperplanes $P$, with origin $0 \in P$.

Let the function $\tilde{f} : w \to \mathbf{R}^1$ be equal to the volume of the corresponding orbit and "$M$-homogenious with repect to the radius," i.e., if $(r, \tilde{\varphi}_1, \ldots, \tilde{\varphi}_{n-1})$ are some regular coordinates in $w \subset \mathbf{R}^n$, where $r$ is a distance to the origin 0, then

$$f(r, \tilde{\varphi}_1, \ldots, \tilde{\varphi}_{n-1}) = r^M \cdot f(1, \tilde{\varphi}_1, \ldots, \tilde{\varphi}_{n-1}) \equiv r^M \cdot f(\tilde{\varphi}_1, \ldots, \tilde{\varphi}_{n-1}),$$

where $f$ is a restriction of $\tilde{f}$ to the standard sphere $S^{n-1}(1)$.

Let $a \in S^{n-1}(1) \cap w$ be a critical point of $f$, let $l$ be an arbitrary tangent vector to the sphere $S^{n-1}(1)$ in point $a$, and let $A$ be an orbit, corresponding to $a$.

**Statement (A. O. Ivanov).** If $D_l^2 f(a) < f(a) \cdot (M+1)2/4$ (here $D_l$ means a derivative with respect to direction $l$), then the cone $CA$ is stable.

As seen from Theorem 6, a "critical" dimension has been discovered for cones of large codimensions (like the above), separating the stability and instability zones (relative to the dimension).

Along with the classical multidimensional Plateau problem, of considerable interest is its "one-dimensional" version, viz., of finding a thread (one-dimensional continuum) of least length, joining a finite number of fixed points on a Riemannian manifold, or a thread without boundary, of (locally) least length. Meanwhile, the thread may be branched. For simplicity, we consider threads on two-dimensional manifolds. Then a locally minimal thread should (1) contain ramification points of multiplicity no more than three, (2) possess smooth arcs converging at 120° at each singular point, and (3) have smooth segments of the arcs as geodesics in the given metric. In [438] and [441], I. V. Shklyanko has given a complete classification of all closed one-dimensional minimal threads (networks) on the flat torus $T^2$. Recall that the classification of minimal nets on the standard 2-sphere is trivial, and there are just ten such networks (see the details in [20]). The case of the torus proves to be much more complicated. Shklyanko then derived necessary conditions for a closed minimal network to exist, with singularities on surfaces $M_g^2$ of genus $g$, where $g > 1$ (the sphere with handles being endowed with a metric with constant negative curvature, i.e., $M_g^2$ represented as quotient spaces of the Lobachevsky plane relative to the action of the corresponding discrete isometry group). All globally minimal (and, therefore, stable) networks with any fixed number of nodes have been found on the torus. In other words, all networks of absolutely minimal length among networks with a fixed number of nodes were described.

We illustrate this by a flat torus. The minimal network breaks $T^2$ into polygons with 120° at the vertices, all of them being hexagons, the number of nodes (singularities) being even and twice greater than that of the polygons. We call a network *regular* if it consists of congruent hexagons, and two nets *equivalent* if there is a homeomorphism of one onto the other, sending the sides of the polygons into parallel (on the torus), and preserving their oriented normals. Shklyanko has proved that any minimal network on the flat torus is equivalent to a regular network. Equivalent networks are of the same length (see [438], [441]). There is always a minimal network with $2n$ nodes on the flat torus for any fixed integer $n$. Note that flat tori are given by integral unimodular matrices of order two. It turns out that each minimal network on the flat torus determines uniquely, and is determined uniquely by, a certain quadruple of integers $(k_1, k_2, km_1, m_2)$, which can be regarded as an integral point in $R^4$. Therefore, the description of minimal networks is equivalent to that of a certain set of integral points in $R^4$. The problem was solved in [438] and [441].

**Theorem 7 (I. V. Shklyanko).** *Let $T^2$ be an arbitrary flat torus with the matrix $\begin{pmatrix} a & b \\ c & d \end{pmatrix}$. Then the number of nonequivalent minimal (locally stable) nets with $2n$ nodes on the torus equals the number of decompositions of the integer as $n = |k_1k_2 + m_1m_2|$, where $k_1, k_2, m_1$, and $m_2$ satisfy some other conditions explicitly given as a system of algebraic inequalitites (we omit the actual formulas).*

Thus, the set of minimal nets on a flat torus is given by integral points of a certain domain in $R^4$, its boundary described by explicit equations. This strengthens the algebraic theorems by A. Edmonds, J. Ewing, and R. Kulkarni [460], proved within the framework of the theory of discrete transformation groups (with the networks associated with these groups being only part of the general class of minimal networks described above). Recently, for a more detailed description of minimal networks on a torus, computer geometry methods were employed. A. Melnikov constructed and realized on a computer an algorithm producing all minimal networks with a fixed number of nodes. Tables were obtained, listing all minimal networks where the number of nodes is not too large, and curious number-theoretic dependencies in the distribution of minimal networks were discovered.

In conclusion, we note the following interesting fact. The known example by J. Adams (see its description in [343], and, for more details, [170], [171], [455]) is a stable minimal surface (with fixed boundary and singularities, embedded into $R^3$), nevertheless, retractable onto its boundary. Moreover, even deformation retraction can be carried out in one of the example modifications. Soboleva has proved, in fact, that this circumstance is a reflection of the following statement. It turns out that if we remove a small disc from a 2-complex, the known Bing house, then we obtain a surface with singularities, which can be realized in $R^3$ as a stable minimal surface whose boundary is a circle (boundary of the discarded 2-disc). Furthermore, it turns out that the whole Bing house (as a complex "without boundary") can be immersed into $R^3$, so that a two-dimensional minimal stable surface with singularities, retracting into a point, is obtained. Recall that the Bing house is homotopically, but not combinatorially contractible to a point. The minimal realization of the Bing house, constructed by O. Yu. Soboleva in $R^3$, turned out to be closely related to the Adams example (though not coinciding with the minimal film indicated by him; see the details in [459]).

# BIBLIOGRAPHY


1. Adams, J. "Vector fields on spheres," *Ann. Math.* **75** (1962), 603–632.
2. Alber, S. I. "Higher-dimensional problems in global variational calculus," *DAN SSSR* **156** (1962), 727–730 (Russian).
3. Alber, S. I. "Spaces of mappings into a manifold of negative curvature," *DAN SSSR* **178** (1968), 13–16 (Russian).
4. Alber, S. I. "The topology of functional manifolds and the calculus of variations in the large," *UMN* **25** (1970), 57–123 (Russian).
5. Alexander, H. and Osserman, R. "Area bounds for various classes of surfaces," *Amer. J. Math.* **97** (1975), 753–769.
6. Allard, W. "On boundary regularity for Plateau's problem," *Bull. Amer. Math. Soc.* **75** (1969), 522–523.
7. Allard, W. " On the first variation of a varifold. Boundary behavior," *Ann. of Math.* **95** (1972), 417–491.
8. Allard, W. "On the first variation of a varifold. Boundary behavior," *Ann. of Math.* **101** (1975), 418–446.
9. Allard, W. and Almgren, F. "The structure of stationary one-dimensional varifolds with positive density," *Invent. Math.* **34** (1976), 83–97.
10. Allard, W. and Almgren, F. "On the radial behavior of minimal surfaces and the uniqueness of their tangent cones," *Ann. of Math.* **113** (1981), 215–265.
11. Almgren, F. " The homotopy groups of the integral cycle groups," *Topology* **1** (1962), 257–299.
12. Almgren, F. "Some interior regularity theorems for minimal surfaces and an extension of Bernstein's theorem," *Ann. of Math.* **84** (1966), 277–293.

13 Almgren, F. *Plateau's Problem. An Invitation to Varifold Geometry.* New York, 1966.

14. Almgren, F. "Existence and regularity almost everywhere of solutions to elliptic variational problems among surfaces of varying topological type and singularity structure," *Ann. of Math., Ser. 2*, **87** (1968), 321–391.
15. Almgren, F. "Measure-theoretic geometry and elliptic variational problems," *Bull. Amer. Math. Soc.* **75** (1969), 285–304.
16. Almgren, F. "Existence and regularity almost everywhere of solutions to elliptic variational problems with constraints," *Mem. Amer. Math. Soc.* **4** (1976), 1–199.

17. Almgren, F. "$Q$-valued functions minimizing Dirichlet's integral and the regularity of area minimizing rectifiable currents up to codimension two," *Bull. Amer. Math. Soc.* **8** (1983), 327–328.

18. Almgren, F., Schoen, I., and Simon, L. "Regularity and singularity estimates on hypersurfaces minimizing parametric elliptic variational integrals," *Acta Math.* **139** (1977), 217–265.

19. Almgren, F. and Simon, L. "Existence of embedded solutions of Plateau's problem," *Ann. Scuola Norm. Sup. Pisa* **6** (1979), 447–495.

20. Almgren, F. and Taylor, J. "The geometry of soap films and soap bubbles," *Scientific American*, July 1976, 82–83.

21. Almgren, F. and Thurston, W. P. "Examples of unknotted curves which bound only surfaces of high genus within their convex hulls," *Ann. of Math.* **105** (1977), 527–538.

22. Alt, H. "Verzweigungspunkte von $H$-Flächen. I," *Math. Z.* **127** (1972), 333–362.

23. Alt, H. "Verzweigungspunkte von $H$-Flächen. II," *Math. Annal.* **201** (1973), 33–55.

24. Aminov, Yu. A. "The metric of approximately minimal surfaces," *Sibirsky Matematichesky Zhurnal* **8** (1967), 483–493 (Russian).

25. Aminov, Yu. A. "The outer diameter of an immersed Riemannian manifold," *Matematichesky Sbornik* **92** (1973), 456–460 (Russian).

26. Aminov, Yu. A. "On the problem of the stability of a minimal surface in a Riemannian space of positive curvature," *DAN SSSR* **224** (1975), 745–747 (Russian).

27. Aminov, Yu. A. "An analogue of Ricci's intrinsic condition for a minimal variety in a Riemannian space," *Ukrainsky Geometrichesky Sbornik* **17** (1975), 15–22 (Russian).

28. Aminov, Yu. A. "On the problem of the stability of a minimal surface in a Riemannian space of positive curvature," *Matematichesky Sbornik* **100** (1976), 400–419 (Russian).

29. Aminov, Yu. A. "The unboundedness of a minimal surface in a Riemannian space of nonpositive curvature," *Ukrainsky Geometrichesky Sbornik* **19** (1976), 3–9 (Russian).

30. Aminov, Yu. A. "Determination of a surface in a four-dimensional Euclidean space from its image in a Grassmannian," *Matematichesky Sbornik* **117** (1982), 147–160 (Russian).

31. Anosov, D. *Cellular Decompositions and Whitehead's theorem. Poincare's Duality and Gluing Handles.* Moscow, 1965 (Russian). (A supplement to the Russian edition of *Morse Theory* by J. Milnor, Mir Publishers, 1965.)

32. Arnold, V. *Mathematical Methods in Classical Mechanics.* Moscow, Nauka, 1974 (Russian).

33. Arnold, V., Varchenko, A. N., and Gusein-Zade, S. M. *Singularities of Differentiable Mappings.* Moscow, Nauka, 1982 (Russian).

34. Atiyah, M. "Algebraic topology and elliptic operators," *Comm. Pure Appl. Math.* **20** (1967), 237–249.

35. Bakelman, I. Ja. "Mean curvature and quasilinear elliptic equations," *Sibirsky Matematichesky Zhurnal* **9** (1968), 1014–1040 (Russian).

36. Barbosa, J. and Do Carmo, M. "On the size of a stable minimal surface in $R^3$," *Amer. J. Math.* **98** (1976), 515–528.

37. Beer, A. *Einl. in die Math. Theorie der Elasticität und Capillarität.* Leipzig, 1869.

38. Beeson, M. "Some results on finiteness in Plateau's problem. I," *Math. Z.* **175** (1980), 103–123; II. *Math. Z.* **181** (1982), 1–30.

39. Beeson, M. and Tromba, A. " The cusp catastrophe of Thom in the bifurcation of minimal surfaces," *Manusc. Mathem.* **46** (1984), 273–307.

40. Beklemishev, D. V. "On strongly minimal surfaces on the Riemannian surface," *DAN SSSR* **114** (1957), 256–258.

41. Beklemishev, D. V. "Strongly minimal surfaces," *Izvestiya Vyshykh Uchebnykh Zavedeniy, Matematika* **3** (1958), 13–23 (Russian).

42. Berndtsson, B. "Zeros of analytic functions of several variables," *Ark. Mat.* **16** (1978), 251–262.

43. Bernshtein, S. N. *Collected Works.* Vol. III: *Differential Equations, Calculus of Variations and Geometry.* Moscow, AN SSSR, 1960 (Russian).

44. Bers, L. "Abelian minimal surfaces," *J. Anal. Math.* **1** (1951), 43–58.

45. Bers, L. "Isolated singularities of minimal surfaces," *Ann. of Math.* **53** (1951), 364–386.

46. Bishop, R. and Crittenden, R. J. *Geometry of Manifolds.* Academic Press, New York, London, 1964.

47. Bindschadler, D. "Invariant solutions to the oriented Plateau problem of maximal codimension," *Trans. Amer. Math.* **261** (1980).

48. Böhme, R. "Stability of minimal surfaces," *Lecture Notes in Math.* **456** (1976), 123–137.

49. Böhme, R., Hildebrandt, S., and Tausch, E. "The two-dimensional analogue of the catenary," *Pacific J. Math.* **88** (1980), 245–278.

50. Böhme, R. and Tromba, A. "The index theorem for classical minimal surfaces," *Ann. Math.* **113** (1981), 447–499.

51. Bombieri, E., De Giorgi, E., and Giusti, E. "Minimal cones and the Bernstein problem," *Invent. Math.* **7** (1969), 243–268.

52. Bombieri, E. and Miranda, M. "Una maggiorazione a priori relativa alle superfice minimali non parametriche," *Arch. R. Mech.* **32** (1969), 255–267.

53. Borel, A. "Sur la cohomologie des espaces fibres principaux et des espaces homogenes de groupes de Lie compacts," *Ann. of Math.* **57** (1953), 115–207.

54. Borel, A. "Sur l'homologie et la cohomologie des groupes de Lie compacts connexes," *Amer. J. Math* **76** (1954), 273–342.

55. Borisenko, A. A. "Saddle surfaces in spherical space," *Ukrainsky Geometrichesky Sbornik* **18** (1975), 25–27 (Russian).

56. Borisenko, A. A. "Explicitly represented minimal surfaces," *Ukrainsky Geometrichesky Sbornik* **25** (1982), 6–7 (Russian).
57. Borisovich, Yu. G., Bliznyakov, N. M., Izrailevich, Ya. A., and Fomenko, T. N. *Introduction to Topology.* Mir Publishers, Moscow, 1985.
58. Bott, R. "Nondegenerate critical manifolds," *Ann. of Math. Ser. 2*, **60** (1954), 248–261.
59. Bourbaki, N. *Eléments de Mathématique.* Vol. V: *Espaces Vectoriels Topologiques*, Hermann, Paris, 1958.
60. Bourbaki, N. *Eléments de Mathématique.* Vol. VI: *Intégration.* Hermann, Paris, 1965.
61. Bourbaki, N. *Eléments de Mathématique.* Vol. III: *Topologie Générale.* Hermann, Paris, 1967.
62. Bourguignon, J.-P. "Harmonic curvature for gravitational and Yang-Mills fields," *Lect. Notes in Math.* **949** (1982), 35–47.
63. Brothers, J. "Invariance of solutions to invariant parametric variational problems," *Trans. Amer. Math. Soc.* **262** (1980), 159–180.
64. Caccioppoli, R. "Misura ed integrazione sugli insiemi dimensionalmente orientati," *Atti. Accad. Naz. Lincei, Rend. Cl. fis. mat., Ser. 8*, **13** (1952), I: 13–83: II: 137–146.
65. Calabi, E. "Minimal immersions of surfaces in Euclidean spheres," *J. Diff. Geom.* **1** (1967), 111–125.
66. Calabi, E. "Quelques applications de l'analyse complexe aux surfaces d'aire minimal," (with Rossi, H., in *Topics in Complex Manifolds.* Les Presses de l'Univ. de Montréal, Montréal, Que., 1968).
67. Carleman, T. "Zur Theorie der Minimalflächen," *Math. Z.* **9** (1921).
68. do Carmo, M. and Peng, C. "Stable complete minimal surfaces in $R^3$ are planes," *Bull. Amer. Math. Soc.* **1** (1979), 903–906.
69. do Carmo, M. and Wallach, N. "Representations of compact groups and minimal immersions into spheres," *J. Differential Geometry* **4** (1970), 91–104.
70. do Carmo, M. and Wallach, N. "Minimal immersion spheres into spheres," *Proc. Nat. Acad. Sci. USA* **63** (1969), 640–642.
71. Cartan, E. "Sur une classe remarquable d'espaces de Riemann," *Bull. Soc. Math. France* **54** (1926), 214- 264; **55** (1927), 114–134.
72. Cartan, E. *Leçons sur la Théorie des Spineurs.* Hermann, Paris, 1938.
73. Cartan, H. "La transgression dans un groupe de Lie et dans un espace fibre principal," *Colloque de Topologie* (espaces fibres). Bruxelles, 1950, pp. 57–71.
74. Chen, Y. W. "Branch points, poles and planar points of minimal surfaces in $R^3$," *Ann. Math.* **49** (1948), 790–806.
75. Chern, S.-S. *Complex Manifolds.* The University of Chicago, 1955–1956.
76. Chern, S.-S. "Minimal surfaces in an Euclidean space of $N$ dimensions," (in: *Differential and Combinatorial Topology.* Princeton University Press, Princeton, N. J., 1965, pp. 187–199).

77. Chern, S.-S. "Brief survey of minimal submanifolds," *Tagungsbricht*, Oberwolfach, 1969, pp. 43–60.

78. Chern, S.-S. "Differential geometry, its applications past and its future," *Actes. Cong. Intern. Math.* **1** (1970), 41–53.

79. Chern, S.-S., and Osserman, R. "Complete minimal surfaces in Euclidean $n$-space," *J. Analyse Math.* **19** (1967), 15–34.

80. Conner, P. E. and Floyd, E. E. *Differentiable Periodic Maps.* Springer, Berlin, 1964.

81. Courant, R. "Plateau's problem and Dirichlet's principle," *Ann. of Math.* **38** 679–724.

82. Courant, R. "The existence of minimal surfaces of given topological structures under prescribed boundary conditions," *Acta Math.* **72** (1940), 51–98.

83. Courant, R. " On a generalized form of Plateau's problem," *Trans. Amer. Math. Soc.* **50** (1941), 40–47.

84. Courant, R. "On the Plateau problem with free boundaries," *Proc. Nat. Acad. Sci. USA* **31** (1945), 242–246.

85. Courant, R. *Dirichlet's Principle, Conformal Mapping, and Minimal Surfaces.* Interscience, New York, 1950.

86. Courant, R. and Davids, N. "Minimal surfaces spanning closed manifolds," *Proc. Nat. Acad. Sci. USA* **26** (1940), 194–199.

87. Crittenden, R. "Minimum and conjugate point in symmetric spaces," *Canad. J. Math.* **14** (1962), 320–328.

88. Đao Chông Thi. "Algebraic problems of the realization of cycles in symmetric spaces," *Vestnik MGU* **2** (1976), 62–66 (Russian).

89. Đao Chông Thi. "On minimal currents and surfaces in Riemannian manifolds," *DAN SSSR* **253** (1977), 21–22 (Russian).

90. Đao Chông Thi. "On minimal real currents on compact Riemannian manifolds," *Izvestiya AN SSSR, Ser. mat.* **41** (1977), 853–867 (Russian).

91. Đao Chông Thi. "On the stability of the homology of compact Riemannian manifolds," *Izvestiya AN SSSR, Ser. mat.* **42** (1978), 500–505 (Russian).

92. Đao Chông Thi. "Multidimensional variational problem in symmetric spaces," *Funktsionalny analis* **12** (1978), 72–73 (Russian).

93. Đao Chông Thi. "Minimal surfaces in compact Lie groups," *UMN* **33** (1978), 163–164 (Russian).

94. Đao Chông Thi. "Minimal real currents in compact Lie groups," in: *Trudy seminara po vektornomu i tenzornomu analizu*, Moscow University Press, Moscow, Vol. 19, 1979, pp. 112–129 (Russian).

95. Đao Chông Thi. "Multivarifolds and classical multidimensional problems of Plateau," *Izvestiya AN SSSR* **44** (1980), 1031–1065 (Russian).

96. Đao Chông Thi. "Isoperimetric inequalities for multivarifolds," *Acta Math. Vietnamica* **6** (1981), 88–94 (Russian).

97. Đao Chông Thi. "Isoperimetric inequalities for multivarifolds," *Izvestiya AN SSSR* **48** (1984), 1031–1065 (Russian).

98. Đao Chông Thi. "Multivarifolds and problems of minimizing functionals of multidimensional volume type," *DAN SSSR* **276** (1984), 1074–1075 (Russian).
99. Đao Chông Thi. "Parametrization and parametrized multivarifold spaces," in *Trudy seminara po vectornomu i tenzornomu analizu*, Moscow University Press, Moscow, Vol. 22, 1985, pp. 31–59 (Russian).
100. Đao Chông Thi and Fomenko, A. T. "Topology of absolute minima of volume-type functionals and Dirichlet functionals in modern problems of real and complex analysis," *AN USSR*, Kiev, 1984, pp. 40–64 (Russian).
101. Darboux, G. *Leçons sur la Théorie Générale des Surfaces et les Applications Géométriques du Calcul Infinitésimal*, Part I. Gauthier-Villars, Paris, 1914.
102. Davids, N. "Minimal surfaces spanning closed manifolds and having prescribed topological position," *Amer. J. Math.* **64** (1942), 348-362.
103. De Giorgio, E. " Su una teoria generale della misura $(r-1)$-dimensionale in uno spazio ad $r$ dimensioni," *Ann. Mat. Pura Appl.* **36** (1954) 191–213.
104. De Giorgi, E. "Nuovi teoremi relativi alle misure $(r-1)$-dimensionali in uno spazio ad $r$ dimensioni," *Ricerche Mat.* **4** (1955), 95–113.
105. De Giorgi, E. "Frontiere orientate di misura minima," *Sem. Mat. Scu. Norm. Sup.* Pisa, 1960–61.
106. De Giorgi, E. "Complementi alla teoria della misura $(n-1)$-dimensionale in uno spazio $n$-dimensionale," *Sem. Mat. Scu. Norm. Sup.* Pisa, 1960–61.
107. De Giorgi, E. "Una estensione del teorema di Bernstein," *Ann. Scuola Norm. Sup. Pisa* **19** (1965), 79–85.
108. Đoàn Quỳnh "Poincaré polynomials of compact, homogeneous Riemannian spaces with irreducible stability group," in: *Trudy seminara po vektornomu i tenzornomu analizu*, Vol. 14, 1968, pp. 33–93 (Russian).
109. Douglas, J. "Solution of the problem of Plateau," *Trans. Amer. Math. Soc.* **3** (1931), 263–321.
110. Douglas, J. "Minimal surfaces of general topological structure," *J. Math. Phys.* **15** (1936), 105–123.
111. Douglas, J. "The most general form of the problem of Plateau," *Proc. Nat. Acad. Sci. USA* **24** (1938), 360–364.
112. Douglas, J. "Minimal surfaces of higher topological structure," *Proc. Nat. Acad. Sci. USA* **24** (1938), 343–353.
113. Douglas, J. "Minimal surfaces of higher topological structure," *Ann. Math.* **40** (1939), 205–298.
114. Douglas, J. "The higher topological form of Plateau's problem," *Ann. Scuola Norm. Sup. Pisa* **8** (1939), 195–218.
115. Dubrovin, B. A., Novikov, S. P., and Fomenko, A. T. *Modern Geometry.* Part 1, 2, Nauka, Moscow, 1979; Part 3, Nauka Moscow, 1984 (Russian). Engl. transl.: B. A. Dubrovin, A. T. Fomenko, S. P. Novikov. *Modern Geometry-Methods and Applications.* Springer-Verlag. Part 1, GTM, v. 93, 1984; Part 2, GTM, v. 704, 1985.

116. Dynkin, E. B. "Homologies of compact Lie groups," *UMN* **8** (1953), 73–120 (Russian).
117. Edgar, H. and Brown, H. "Cohomology theories," *Ann. Math.* **75** (1962), 467–484.
118. Eells, J. and Lemaire, L. "A report on harmonic maps," *Bull. London Math. Soc.* **10** (1978), 1–68.
119. Eells, J. and Sampson, J. "Harmonic mappings of Riemannian manifolds," *Amer. J. Math.* **86** (1964), 109–160.
120. Eells, J. and Wood, J. "Maps of minimum energy," *J. London Math. Soc., Ser. 2*, **23** (1981), 303–310.
121. Eells, J. and Wood, J. "Harmonic maps from surface to complex projective spaces," *Advances in Math* **49** (1983), 217–263.
122. Efimov, N. V. "The impossibility of Euclidean 3-space of a complete regular surface with a negative upper bound of the Gaussian curvature," *DAN SSSR* **150** (1963), 1206–1209 (Russian).
123. Efimov, N. V. "Surfaces with slowly varying negative curvature," *UMN* **21** (1966), 3–58.
124. Efimov, N. V. *Higher Geometry.* Nauka, Moscow, 1978 (Russian).
125. Efimov, N. V. and Poznyak, E. G. "Generalization of Hilbert's theorem on surfaces of constant negative curvature," *DAN SSSR* **137** (1961), 509–512 (Russian).
126. Eilenberg, S. and Steenrod, N. *Foundations of Algebraic Topology.* Princeton University Press, Princeton, N. J., 1952.
127. Eisenhart, L. *An Introduction to Differential Geometry.* Princeton University Press, Princeton, N. J., 1949.
128. Eisenhart, L. *Riemannian Geometry.* Princeton University Press, Princeton, N. J., 1949.
129. Ehresmann, C. "Sur la topologie de certaines variétés algébriques réeles," *J. Math. Pur. Appl.* **16** (1937), 69–100.
130. Erdem, S. and Wood, J. C. "On the construction of harmonic maps into a Grassmannian," *J. London Math. Soc.* **28** (1983), 161–174.
131. Federer, H. "Hausdorff measure and Lebesque area," *Proc. Nat. Acad. Sci. USA* **37** (1951), 90–94.
132. Federer, H. "Measure and area," *Bull. Amer. Math. Soc.* **58** (1952), 306–378.
133. Federer, H. "Currents and area," *Trans. Amer. Math. Soc.* **98** (1961), 204–233.
134. Federer, H. "Some theorems on integral currents," *Trans. Amer. Math. Soc.* **117** (1965), 43–67.
135. Federer, H. *Geometric Measure Theory.* Springer, Berlin, Bd. 153, 1969.
136. Federer, H. "The singular sets of area minimizing rectifiable currents with codimension one and of area minimizing flat chains modulo two with arbitrary codimensions," *Bull. Amer. Math. Soc.* **76** (1970), 767–771.

137. Federer, H. "Real flat chains, cochains and variational problems," *Indiana Univ. Math* **24** (1974–1975), 351–407.

138. Federer, H. and Fleming, W. H. "Normal and integral currents," *Ann. Math.* **72** (1960), 458–520.

139. Finn, R. "Isolated singularities of solutions of nonlinear partial differential equations," *Trans. Amer. Math. Soc.* **75** (1953), 387–404.

140. Finn, R. "On equations of minimal surface type," *Ann. of Math.* **60** (1954), 397–416.

141. Finn, R. "New estimates for equations of minimal surface type," *Arch. Rational Mech. Anal.* **14** (1963), 337–375.

142. Finn, R. "Remarks relevant to minimal surfaces, and to surfaces of prescribed mean curvature," *J. Anal. Math.* **14** (1965), 139–160.

143. Finn, R. and Osserman, R. " On the Gauss curvature of nonparametric minimal surfaces," *J. Anal. Math.* **12** (1964), 351–364.

144. Fleming, W. "An example in the problem of least area," *Proc. Amer. Math. Soc.* **7** (1956), 1063–1074.

145. Fleming, W. "On the oriented Plateau problem," *Rend. Circ. Mat. Palermo* **11** (1962), 69–90.

146. Fleming, W. *Functions of Several Variables.* Springer, New York-Hiedelberg-Berlin, 1977.

147. Fleming, W. and Young, L. "A generalized notion on boundary," *Trans. Amer. Math. Soc.* **76** (1954), 457–484.

148. Fomenko, A. T. "Existence and regularity almost everywhere of minimal compact with given homologycal properties," *DAN SSSR* **187** (1969), 747–749 (in Russian).

149. Fomenko, A. T. "The Poincaré polynomials of certain homogeneous spaces," in *Trudy seminara po tenzornomu i vectornomu analizu*, Moscow University Press, Vol. 15, 1970, pp. 128–152 (Russian).

150. Fomenko, A. T. "Some cases of the realization of elements of homotopy groups of homogeneous spaces by totally geodesic spheres," *DAN SSSR* **190** (1970), 492–795 (Russian).

151. Fomenko, A. T. "Homological properties of minimal compacta in the multidimensional Plateau problem," *DAN SSSR* **192** (1970), 38–41 (Russian).

152. Fomenko, A. T. "The multidimensional Plateau problem and singular points of minimal compacta," *DAN SSSR* **192** (1970), 293–296 (Russian).

153. Fomenko, A. T. "The realization of cycles in compact symmetric spaces by totally geodesic submanifolds," *DAN SSSR* **195** (1970), 789–792 (Russian).

154. Fomenko, A. T. "Bott periodicity from the point of view of the multidimensional Dirichlet functional," *Izvestiya AN SSSR* **35** (1971), 667–681 (Russian).

155. Fomenko, A. T. "The multidimensional Plateau problem in extraordinary theories of homology and cohomology," *DAN SSSR* **200** (1971), 797–800

(Russian).
156. Fomenko, A. T. "The multidimensional Plateau Problem in Riemannian manifolds," *Matematichesky sbornik* **89** (131), 3(1972), 475–519 (Russian).
157. Fomenko, A. T. "Totally geodesic models of cycles," in: *Trudy seminara po vektornomu i tenzornomu analizu*, Moscow University Press, Moscow, Vol. 16, (1972), pp. 14–98 (Russian).
158. Fomenko, A. T. "Minimal compacta in Riemannian manifolds and Reifenberg's conjecture," *Izvestiya AN SSSR* **36** (1972), 1049–1080 (Russian).
159. Fomenko, A. T. "Geometric variational problems," *Itogi nauki i tekhniki* **1** (1973), 39–59 (Russian).
160. Fomenko, A. T. "Multidimensional Plateau problem on Riemannian manifolds. On the problem of the algorithmical recognizability of the standard three-dimensional sphere," *Proc. of the Intern. Congress of Math.*, Vancouver, Vol. 1, pp. 515–525.
161. Fomenko, A. T., "Multidimensional Plateau problems on Riemannian manifolds and extraordinary homology and cohomology theories," in: *Part 1. Trudy seminara po vektornomu i tenzornomu analizu*, Vol. 17. Moscow University Press, Moscow, 1974, pp. 3–176 (Russian).
162. Fomenko, A. T. "Multidimensional Plateau problems on Riemannian manifolds and extraordinary homology and cohomology theories," in: *Part 2. Trudy seminara po vektornomu i tenzornomu analizu*, Vol. 18. Moscow University Press, Moscow, 1978, pp. 4–93 (Russian).
163. Fomenko, A. T. "Universal lower bound of the rate of globally minimal solutions," *DAN SSSR* **251** (1980), 295–299 (Russian).
164. Fomenko, A. T. "On lower volume bounds for surfaces which are globally volume minimizing with respect to a cobordism constraint," *Izvestiya AN SSSR* **45** (1981), 187–212 (Russian).
165. Fomenko, A. T. "Multidimensional variational problems in the topology of extremals," *UMN* **36** (1981), 105–135 (Russian).
166. Fomenko, A. T. "On absolute minima of a volume functional on Riemannian manifolds. Integration of Hamiltonian systemes on Lie algebras," in: *Short Communication Abstracts of Inter. Congress of Math.*, Warszawa, Vol. 3 (1983), p. 17.
167. Fomenko, A. T. *Differential Geometry and Topology. Additional Chapters.* Moscow University Press, Moscow, 1983 (Russian). English transl. Plenum Publishing Corporation, 1987.
168. Fomenko, A. T. "Indices of minimal and harmonic surfaces," in: *Topological Methods for Mathematical Physics.* Vilnius University Press, Vilnius, 1984, pp. 95–108 (Russian).
169. Fomenko, A. T. "On certain properties of extremals in variational problems," *Lecture Notes in Math.* **1108** (1984), 209–217.
170. Fomenko, A. T. *Topological Variational Problems.* Moscow University Press, Moscow, 1985 (Russian). English transl. Gordon and Breach (to

appear).

171. Fomenko, T. N. "Boundaries of surfaces and Hopf retraction theorem," in *Abstracts of the V-th Symposium on General Topology and Application.* Kishinev, 1985, pp. 237–239 (Russian).

172 Frankel, T. "On the fundamental group of a compact minimal submanifold," *Ann. of Math.* **83** (1966), 68–73.

173. Freed, D. and Uhlenbeck, K. *Instantons and Four-Manifolds.* Springer, N. Y., 1984.

174. Fuchs, D. B., Fomenko, A. T., and Gutenmakher, V. L. *Homotopic Topology.* Moscow University Press, Moscow, 1969 (Russian). English transl. Akadémiai Kiado, Budapest, 1986.

175. Giaquinta, M. and Pepe, L. "Esistenza e regolaritá per il problema dell'area minima con ostacoli in $n$ variabili," *Ann. Scuola Norm. Sup.* Pisa **25** (1971), 481–507.

176. Guisti, E. "Superfici minime cartesiane con ostacoli discontinui," *Arch. Rational Mech. and Anal.* **40** (1971), 251–267.

177. Golubitsky, M. and Guillemin, V. *Stable Mappings and Their Singularities.* Springer-Verlag, N. Y., 1973.

178. Gorokh, V. P. "Hypersurfaces that contain minimal submanifolds," *Ukrainsky Geometrichesky sbornik* **24** (1981), 18–26 (Russian).

179. Gray, A. "Minimal varieties and almost Hermitian submanifolds," *Michigan Math J.* **12** (1965), 273–287.

180. Griffiths, P. and King, J. "Nevanlinna theory and holomorphic mappings between algebraic varieties," *Acta Mathematica* **130** (1973), 145–220.

181. Gromoll, D., Klingenberg, W., and Meyer, W. *Riemannsche Geometry im Grossen.* Springer, Berlin-Heidelberg-New York, 1968.

182. Gromov, M. L. "Volume and bounded cohomology," *Inst. Hautes Sci. Publ. Math.* **56** (1982), 5–99.

183. Gromov, M. L. and Rohlin, V. A. "Imbeddings and immersions in Riemannian geometry," *Uspekhi Mat. Nauk* **25** (1970), 3–62 (Russian).

184. Grüter, M., Hilderbrandt, S., and Nitsche, J. C. "On the boundary behavior of minimal surfaces with a free boundary, which are not minima of area," *Manusc. Math.* **35** (1981), 387–410.

185. Gulliver, R. D. "Regularity of minimizing surfaces of prescribed mean curvature," *Ann. Math.* **97** (1973), 275–305.

186. Gulliver, R. and Lesley, F. D. "On the boundary branch points of minimal surfaces," *Arch. R. Mech* **52** (1973), 20–25.

187. Gulliver, R. D., Osserman, R., and Royden, H. L. "A theory of branched immersions of surfaces," *Amer. J. Math.* **95** (1973), 750–812.

188. Gulliver, R. and Spruck, J. "On embedded minimal surfaces," *Ann. Math.* **103** (1976), 331–347.

189. Haar, A. "Über das Plateausche Problem," *Math. Ann.* **97** (1927), 124–158.

190. Hamilton, R. S. *Harmonic maps of manifolds with boundary.* Springer, Berlin, 1975.

191. Hartman, P. "On homotopic harmonic maps," *Canad. J. Math.* **19** (1967), 673–687.

192. Harvey, R. and Lawson, B. "On boundaries of complex analytic varieties," *Ann. of Math* **102** (1975), 223–290.

193. Harvey, R. and Lawson, H. "Extending minimal varieties," *Invent. Math.* **28** (1975), 209–226.

194. Harvey, R. and Lawson, H. "On boundaries of complex analytic varieties II," *Ann. of Math.* **106** (1977), 213–238.

195. Harvey, R. and Lawson, H. "Calibrated foliations (foliations and mass-minimizing currents)," *Amer. J. Math.* **104** (1982), 607–633.

196. Harvey, R. and Lawson, H. "Calibrated geometries," *Acta Math.* **148** (1982), 47–157.

197. Heinz, E. and Hildebrandt, S. "Some remarks on minimal surfaces in Riemannian manifolds," *Com. PA Math.* **23** (1970), 371–377.

198. Helgason, S. *Differential Geometry and Symmetric Spaces.* Academic Press, New York-London, 1962.

199. Helgason, S. "Totally geodesic spheres in compact symmetric spaces," *Math. Ann.* **165** (1966), 309–317.

200. Hildebrandt, S. "Boundary behavior of minimal surfaces," *Arch. R. Mech.* **35** (1969), 47–82.

201. Hildebrandt, S. and Kaul, H. "Two-dimensional variational problems with obstructions, and Plateau's problem for $H$-surfaces in Riemannian inmanifold," *Com. PA Math.* **25** (1972), 187–223.

202. Hildebrandt, S., Kaul, H., and Widman, K. "An existence theorem for harmonic mappings of Riemannian manifolds," *Acta Math.* **138** (1977), 1–16.

203. Hildebrandt, S. and Nitsche, J. C. C. "Minimal surfaces with free boundaries," *Acta Math.* **143** (1979), 251–272.

204. Hildebrandt, S. and Nitsche, J. C. "Optimal boundary regularity for minimal surfaces with a free boundary," *Manusc. Math.* **33** (1981), 357–364.

205. Hildebrandt, S. and Nitsche, J. C. "A uniqueness theorem for surfaces of least area with partially free boundaries on obstacles," *Arch. R. Mech.* **79** (1982), 189–218.

206. Hoffman, D. and Osserman, R. "The geometry of the generalized Gauss map," *Mem. Amer. Math. Soc.* **28** (1980), 1-105.

207. Hoffman, D. and Osserman, R. "The area of the generalized Gaussian image and the stability of minimal surfaces in $S^n$ and $R^n$," *Math. Annal.* **260** (1982), 437–452.

208. Hopf, H. and Rinow, W. "Über den Begriff der vollständigen differentialgeometrischen Flächen," *Comment. Math. Helv.* **3** (1981), 209–225.

209. Horkova, N. G. *Minimal Cones in Riemannian Manifolds in Application of Topology to Modern Analysis.* Voronezh University Press, Voronezh, 1985, pp. 167–171.

210. Hsiang, W. and Tomter, P. "On the construction of infinitely many, mutually noncongruent examples of minimal imbeddings of $S^{2n-1}$ into $CP^n$, $n \geqslant 2$," *Bull. Amer. Math. Soc.* **8** (1983), 463–465.

211. Hsiang, W. "On the compact homogeneous minimal submanifolds," *Proc. Nat. Acad. Sci. USA* **56** (1966), 5–6.

212. Hsiang, W. "Remarks on closed minimal submanifolds in the standard Riemannian $m$-sphere," *J. Differential Geometry* **1** (1967), 257–267.

213. Hsiang, W. "New examples of minimal imbeddings of $S^{n-1}$ into $S^n(1)$—the spherical Bernstein problem for $n = 4, 5, 6$," *Bull. Amer. Math. Soc.* **7** (1982), 377–379.

214. Hsiang, W. and Lawson, H. "Minimal submanifolds of low cohomogeneity," *J. Diff. Geometry* **5** (1971), 1–38.

215. Ioffe, A. D. and Tihomirov, V. M. "Extension of variational problems," in: *Trudy Moskovskogo Matematicheskogo Obschestwa* **18** 1968, 187–246.

216. Jacobson, N. "Some groups of transformations defined by Jordan algebras," I: *J. Reine Angew. Math.*, 3/4 (1959), 201, pp. 178–195; II: *J. Reine Angew. Math.*, 1/4 (1960), 294, pp. 74–98, III: *J. Reine Angew. Math.*, 1/2 (1961), 207, pp. 61–85.

217. Jacobson, N. *Lie Algebras.* Interscience. New York, 1962.

218. Jenkins, H. and Serrin, J. "Variational problems of minimal surface area type," *I. Arch. R. Mech.* **12** (1963), 185–212.

219. Jenkins, H. and Serrin, J. "Variational problems of minimal surface type. II. Boundary value problems for the minimal surface equation," *Arch. R. Mech.* **21** (1966), 321–342.

220. Jenkins, H. and Serrin, J. "Variational problems of minimal surface type. III. The Dirichlet problem with infinite data," *Arch. Rational Mech. Anal.* **29** (1968), 304–322.

221. Jenkins, H. and Serrin, J. "The Dirichlet problem for the minimal surface equation in higher dimension," *J. Reine Angew. Math.* **229** (1968), 170–187.

222. Kamke, E. *Differentialgleichungen. Lösungsmethoden und Lösungen.* Leipzig, Akademishe Verlagsgesellschaft, Gees & Portig, 1967.

223. Katznelson, V. E. and Ronkin, L. I. "On minimal volume of analytic set," *Sibirsky matematichesky jurnal* **15** (1974), 516–528.

224. Karcher, H. and Wood, J. C. "Nonexistence results and growth properties for harmonic maps and forms," *J. Reine Angew. Math.* **353** (1984), 165–180.

225. Kesselman, V. M. "On the Bernstein theorem for surfaces with quasiconformal mapping," *Matematicheskiye zametki* **35** (1984), 445–453 (Russian).

226. Klingenberg, W. *Lectures on Closed Geodesics.* Springer, Berlin, 1978.

227. Klotz, T. and Sario, L. "Existence of complete minimal surfaces of arbitrary connectivity and genus," *Proc. Nat. Acad. Sci. USA* **54** (1965),

42–44.

228. Korolev, E. A. and Fomina, T. N. "Minimal Peterson surfaces," *Ukrainsky Geometrichesky Sbornik* **22** (1979), 92–96 (Russian).

229. Lagrange, J. L. "Essai sur une nouvelle methode pour determiner les maxima et minima des formules integreles indefines," *Misc. Soc. Taurin I*, 1759, publ. 1762, pp. 173–195.

230. Ladyzhenskaja, O. A. *Boundary-value Problems of Mathematical Physics*. Nauka, Moscow, 1973 (Russian).

231. Ladyzhenskaja, O. A. and Uralceva, N. N. *Linear and Quasi-linear Equations of Elliptic Type*. Nauka, Moscow, 1964 (Russian).

232. Landau, L. D. and Lifshits, E. M. *The Mechanics of Continuous Media*. Moscow, Gosudarstvennoye Izdatelstvo Tekhnichesko-Teoreticheskoy Literatury, 1953 (Russian).

233. Lawson, H. B. "Local rigidity theorems for minimal hypersurfaces," *Ann. of Math.* **98** (1969), 187–197.

234. Lawson, H. B. "The global behavior of minimal surfaces in $S^n$," *Ann. of Math.* **92** (1970), 224–237.

235. Lawson, H. B. "Complete minimal surfaces in $S^3$," *Ann. of Math.* **92** (1970), 335–374.

236. Lawson, H. B. "The unknottedness of minimal embeddings," *Invent. Math.* **11** (1970), 183–197.

237. Lawson, H. B. "Some intrinsic characterizations of minimal surfaces," *J. Anal. Math.* **24** (1971), 151–161.

238. Lawson, H. B. "The equivariant Plateau problem and interior regularity," *Trans. Amer. Math. Soc.* **173** (1973), 231–249.

239. Lawson, H. B. "The stable homology of a flat torus," *Math. Scand.* **36** (1975), 49–73.

240. Lawson, H. B. and Osserman, R. "Nonexistence, nonuniqueness and irregularity of solutions to the minimal surface system," *Acta Math.* **139** (1977), 1–17.

241. Lawson, H. B. and Simons, J. "On stable currents and their application to global problems in real and complex geometry," *Ann. of Math* **98** (1973), 427–450.

242. Lawson, H. B. and Yau, S. T. "Compact manifolds of nonpositive curvature," *J. Diff. Geom.* **7** (1972), 211–228.

243. Lê Hông Van "Growth of two-dimensional minimal surface," *UMN* **40** (1985), 209–210 (Russian).

244. Lê Hông Van "Calibrations and geometry of minimal currents on symmetric spaces," in: *Trudy seminara po vektornomu i tenzornomu analizu* Vol. 22, 1985, pp. 107–118 (Russian).

245. Lê Hông Van "New examples of GM-surfaces," in: *Geometry, Differential Equations and Mechanics*. Moscow University Press, Moscow, 1966, pp. 102–105 (Russian).

246. Lelong, P. *Fonctions plurisousharmoniques et formes différentielles positives.* Gordon & Breach, Paris-London-New-York, 1968.
247. Lemaire, L. "Applications harmoniques de variétés produits," *Comm. Math. Helv.* **52** (1977), 11–24.
248. Lemaire, L. "Applications harmoniques des surfaces Riemanniennes," *J. Diff. Geom.* **13** (1978), 51–87.
249. Leung, P. F. *On the Stability of Harmonic Maps.* Springer, Berlin-New York, 1982, (Lect. Notes in Math., 949), pp. 122–129.
250. Levy, P. *Leçons d'analyse fonctionelle.* Gauthier-Villars, Paris, 1922.
251. Lévy, P. "Le problème de Plateau," *Matematica. Timisorara* **23** (1948), 1–45.
252. Lewy, H. "A priori limitations for solutions of Monge–Ampere equations. II," *Trans. Amer. Math. Soc.* **41** (1937), 365–374.
253. Lewy, H. and Stampacchia, G. "On the regularity of the solutions of a variational inequality," *Comm. Pure Appl. Math.* **22** (1969), 153–188.
254. Lichnerowich, A. "Applications harmoniques et variétes kahleriennes," *Symposia Mathematica* **III** (1970), 341–402.
255. Lichnerowicz, A. *Global Theory of Connections and Holonomy Groups.* Leyden, Noordhoff, 1976.
256. Lumiste, Ü. "On the theory of two-dimensional minimal surfaces," *Tartu, Riikl Ül. Toimetised* **102** (1961), 3–15, 15–28; **129** (1962), 74–89, 90–102 (Russian).
257. Manturov, O. V. "Homogeneous Riemannian spaces with an irreducible rotation group," in: *Trudy seminara po vektornomy i tenzornomy analizu* **13** (1966), 68–145 (Russian).
258. Manturov, O. V. "On some cohomological properties of Cartan's models of symmetric spaces," in: *Trudy seminara po vektornomy i tenzornomy analizu* Vol. 18. Moscow University Press, Moscow, 1978, pp. 169–175.
259. Masalcev, L. A. "Instability of minimal cones in Lobacevskiĭspace," *Ukrainsky Geometrichesky Sbornik* **21** (1978), 72–81 (Russian).
260. Masalcev, L. A. "Dimensions of stable domains on certain minimal submanifolds of a sphere," *Ukrainsky Geometrichesky Sbornik* **22** (1979), 193–108 (Russian).
261. Masalcev, L. A. "Minimal surfaces in $R^5$, whose Gaussian images are of constant curvature," *Matematicheskiye zametki* **35** (1984), 927–932 (Russian).
262. Matsumotot, M. "Intrinsic character of minimal hypersurfaces in flat spaces," *J. Math. Soc. Jap.* **9** (1957), 146–157.
263. Mazer, E. "La formule de la variation seconde de l'énergie du voisinage d'une application harmonique," *J. Diff. Geom.* **8** (1973), 279–296.
264. Meeks, W. H. "The classification of complete minimal surfaces in $R^3$ with total curvature greater than $-8$," *Duke Math. J.* **48** (1981), 523–535.
265. Meeks, W. H. "The topological uniqueness of minimal surfaces in three-dimensional Euclidean space," *Topology* **20** (1981), 389–410.

266. Meeks, W. H. "Uniqueness theorems for minimal surfaces. III," *J. Math.* **25** (1981), 318–336.

267. Meeks, W. H., Simon, L., and Yau, S. T. "Embedded minimal surfaces, exotic spheres, and manifolds with positive Ricci curvature," *Ann. of Math.* **116** (1982), 621–659.

268. Meeks, W. H. and Yau, S. T. "Topology of three-dimensional manifolds and the embedding problems in minimal surface theory," *Ann. of Math.* **112** (1980), 441–484.

269. Meeks, W. H. and Yau, S. T. "The classical Plateau problem and the topology of three-dimensional manifolds. The embedding of the solution given by Douglas–Morrey and an analytic proof of Dehn's lemma," *Topology* **21** (1982), 409–442.

270. Meeks, W. H. and Yau, S. T. "The existence of embedded minimal surfaces and the problem of uniqueness," *Math. Z.* **179** (1982), 151–168.

271. Mikliukov, V. M. "An estimate of the modulus of a family of curves on a minimal surface and its applications," *Uspekhi Mat. Nauk* **34** (1973), 207–208 (Russian).

272. Mikliukov, V. M. "Some properties of tubular minimal surfaces in $R^n$," *DAN SSSR* **247** (1979), 549–552 (Russian).

273. Mikliukov, V. M. "Capacity and a generalized maximum principle for quasilinear equations of elliptic type," *DAN SSSR* **250** (1980), 1318–1320 (Russian).

274. Mikliukov, V. M. "Some peculiarities of the behavior of the solutions of equations of minimal surface type in unbounded domains," *Matematichesky sbornik* **116** (1981), 72–86 (Russian).

275. Miller, C. E. "The topology of rotation groups," *Ann. of Math.* **57** (1953), 90–114.

276. Milnor, J. *Morse Theory.* Princeton University Press, Princeton, N. J., 1963.

277. Milnor, J. *Lectures on the h-Cobordism Theorem.* Princeton University Press, Princeton, N. J., 1965.

278. Mimura, M. "The homology groups of Lie groups of low rank," *J. Math. Kyoto Univ.* **6** (1967), 131–176.

279. Miranda, M. "Un teorema di esistenza e unicita per il problema dell' area minima in $n$ variabili," *Ann. Scuola Norm. Super. Pisa. Sci. fis. e mat.* **19** (1965), 233–250.

280. Miranda, M. "Sul minimo dell'integrale del gradiente di una funzione," *Ann. Sc. Norm. Sup. Pisa* **XIX** (1965) 627–665.

281. Miranda, M. "Dirichlet problem with $L^1$ data for the nonhomogeneous minimal surface equation," *Indiana Univ. Math. J.* **24** (1974), 227–241.

282. Mishchenko, A. S. and Fomenko, A. T. *Course of Differential Geometry and Topology.* Moscow University Press, Moscow, 1980 (Russian).

283. Morgan, F. "Smooth curve in $R^4$ bounding a continuum of area minimizing surfaces," *Duke Math. J.* **43** (1976), 867–870.

284. Morgan, F. "A smooth curve in $R^3$, bounding a continuum of minimal surfaces." Preprint.

285. Morgan, F. "Almost every curve in $R^3$ bounds a unique area minimizing surface," *Invent. Math.* **45** (1978), 253–297.

286. Mori, S. "Projective manifolds with ample tangent bundles," *Ann. of Math.* **110** (1979), 593–606.

287 Morrey, C. B. "The problem of Plateau on a Riemannian manifold," *Ann. of Math.* **49** (1948), 807–851.

288. Morrey, C. B. "The higher-dimensional Plateau problem on a Riemannian manifold," *Proc. Nat. Acad. Sci. USA* **54** (1965), 1029–1035.

289. Morrey, C. B. *Multiple Integrals in the Calculus of Variations.* Berlin, Springer, Bd. 130, 1966.

290. Morse, M. "The first variation in minimal surface theory," *Duke Math. J.* **6** (1940), 263–289.

291. Morse, M. and Tompkins, C. B. "The existence of minimal surfaces on several critical types," *Ann. of Math.* **40** (1939), 443–472.

292. Morse, M. and Tompkins, C. B. "Unstable minimal surfaces of higher topological types," *Proc. Nat. Acad. Sci. USA* **26** (1940), 713–716.

293. Morse, M. and Tompkins, C. B."Unstable minimal surfaces of higher topological structure," *Duke Math. J.* **8** (1941), 350–375.

294. Morse, M. and Tompkins, C. B. "Minimal surfaces not of minimum type by a new mode of approximation," *Ann. of Math.* **42** (1941), 62–72.

295. Mostow, G. D. "Strong Rigidity of Locally Symmetric Spaces," *Ann. of Math. Studies*, Vol. 78. Princeton University Press, Princeton, 1978.

296. Müntz, C. "Die Lösung des Plateauschen Problems über konvexen Bereichen," *Math. Ann.* **94** (1925), 54–96.

297. Murnaghan, F. *The Theory of Group Representations.* Dover Publications, N. Y., 1963.

298. Narasimhan, R. *Analysis on Real and Complex Manifolds.* Masson & Cie, Paris, North-Holland Publ. Co., Amsterdaam, 1973.

299. Nitsche, J. "On the nonsolvability of Dirichlet's problem for the minimal surface equation," *J. Math. Mech.* **14** (1965), 779–788.

300. Nitsche, J. "On new results in the theory of minimal surfaces," *Bull. Amer. Math. Soc.* **71** (1965), 195–270.

301. Nitsche, J. "Variational problems with inequalities as boundary condition or how to fashion a cheap hat for Giacometti's brother," *Arch. Rational Mech. Anal.* **35** (1969), 83–113.

302. Nitsche, J. "The boundary behavior of minimal surfaces. Kellog's theorem and branch points on the boundary," *Invent. Math.* **8** (1969), 313–333.

303. Nitsche, J. "Minimal surfaces with partially free boundary," *Ann. of Math. Acad. Sci. Fenn., Ser. 2*, **483** (1971), 1–21.

304. Nitsche, J. "A new uniqueness theorem for minimal surfaces," *Arch. R. Mech.* **52** (1973), 319–329.

305. Nitsche, J. *Vorlesungen über Minimalflächen.* Springer-Verlag, Berlin-Heidelberg-New-York, 1975.

306. Nitsche, J. "Nonuniqueness for Plateau's problem. A bifurcation process," *Ann. Acad. Sci. Fenn., Ser. 2A*, **2** (1976), 361–373.

307. *Nonlinear problems in geometry. Proceedings of the Sixth Intern. Symposium, Div. of Math.* The Taniguchi Found. Conf. held at Katata. Tokyo, 1979.

308. Osserman, R. "Minimal surfaces in the large," *Comment. Math. Helv.* **35** (1961), 65–76.

309. Osserman, R. "On complete minimal surfaces," *Arch. Rational Mech. Anal.* **13** (1963), 392–404.

310. Osserman, R. "Global properties of minimal surfaces in $E^3$ and $E^n$," *Ann. of Math.* **80** (1964), 340–364.

311. Osserman, R. "Global properties of classical minimal surfaces," *Duke Math. J.* **32** (1965), 565–573.

312. Osserman, R. "Minimal surfaces," *UMN* **22** (1967), 55–136 (Russian).

313. Osserman, R. "Minimal varieties," *Bull. Amer. Math. Soc.* **75** (1969), 1092–1120.

314. Osserman, R. "A proof of the regularity everywhere of the classical solution to Plateau's problem," *Ann. of Math.* **91** (1970), 550–569.

315. Osserman, R. "The isoperimetric inequality," *Bull. Amer. Math. Soc.* **634** (1978), 1182-1238.

316. Osserman, R. and Schiffer, M. "Doubly-connected minimal surfaces," *Arch. R. Mech.* **58** (1975), 285–307.

317. Otsuki, T. "A construction of closed surfaces of negative curvature in $E^4$," *Math. J. of Okayama Univ.* **3** (1954), 95–108.

318. Otsuki, T. "Minimal hypersurfaces in a Riemannian manifold of constant curvature," *Amer. J. Math.* **92** (1970), 145–173.

319. Otsuki, T. "Minimal submanifolds with $M$-index 2," *J. Diff. Geom.* **6** (1971), 193–211.

320. Otsuki, T. "Minimal submanifolds with $M$-index 2 and generalized Veronese surfaces," *J. Math. Soc. Jap.* **24** (1972), 89–122.

321. Otsuki, T. "Minimal hypersurfaces with three principal curvature fields in $S^{n+1}$," *Kodai Math. J.* **1** (1978), 1–29.

322. Palais, R. "Morse theory on Hilbert manifolds," *Topology* **2** (1963), 299–310.

323. Palais, R. "Homology theory of infinite dimensional manifolds," *Topology* **5** (1966), 1–16.

324. Palais, R. and Smale, S. "A generalized Morse theory," *Bull. Amer. Math. Soc.* **70** (1964), 165–172.

325. Pini, M. "Kugelbilder reeller Minimalflächen im $R^4$," *Math. Z.* **59** (1963), 290–295.

326. Pini, M. "Minimalflächen fester Gaußscher Krümmung," *Math. Z.* **136** (1958), 34–40.

327. Plateau, J. *Statique experimentale et théorique des eiquides soumis aux seules forces moléculaires.* Gauthier-Villars, Paris, 1873.

328. Plateau, J. "Recherches expérimentale & théorique sur les figures d'equilibre d'une masse liquide sans pesanteur," Bruxelles, *Ac. Bull.* **33, 36, 37** (1861–69), 452.

329. Plateau, J. *Mem. de l'Acad. de Belgique*, 1843 bis 1869; *Mém. de l'Acad. de Bruxelles* **16** (1843).

330. Pluzhnikov, A. I. "Harmonic mappings of Riemann surfaces and foliated manifolds," *Matematichesky Sbornik* **113** (1980), 339–347 (Russian).

331. Pluzhnikov, A. I. "Some properties of harmonic mappings in the case of spheres and Lie groups," *DAN SSSR* **268** (1983), 1300–1302.

332. Pluzhnikov, A. I. *Indices of Harmonic Mappings of Spheres in Geometry and Topology in Global Nonlinear Problems.* Veronezh University Press, Voronezh, 1984, pp. 162–166 (Russian).

333. Pluzhnikov, A. I. "Minimization problem for energy functional." Reprint No. 5584–84., VINITI, Moscow, 1984 (Russian).

334. Pluzhnikov, A. I. "Certain geometric properties of harmonic mappings," in: *Trudy seminara po vektornomu i tenzornomu analizu.* Vol. 22, Moscow University Press, Moscow, 1985, pp. 132–147 (Russian).

335. Fomenko, A. T. and Fuchs, D. B. *Course of Homotopic Topology.* Moscow, Nauka, 1989.

336. Pluzhnikov, A. I. *Topological Criterion of Energy Functional Global Minimum Inattainability in New Results in Global Analysis.* Voronezh University Press, Voronezh, 1986, pp. 149–155 (Russian).

337. Pogorelov, A. V. "Minimal hypersurfaces in a spherical space," *DAN SSSR* **206** (1972), 291–292 (Russian).

338. Pogorelov, A. V. "Extrinsic geometry of convex surfaces," *Amer. Math. Soc.* **35** (1973), Providence R. I.

339. Pogorelov, A. V. *Differential Geometry.* Nauka, Moscow, 1974 (Russian).

340. Pogorelov, A. V. "On the stability of minimal surfaces," *DAN SSSR* **260** (1981), 293–295 (Russian).

341. Pontrjagin, L. S. "Homologies in compact Lie groups," *Matematichesky Sbornik* **6** (1939), 389–422.

342. Pontrjagin, L. S. *Topological Groups.* Gordon & Breach, New York-London-Paris, 1966.

343. Poston, T. *The Plateau problem. An invitation to the whole of mathematics. Summer College on global analysis and its applications*, 4 July–25 August 1972, International Centre for Theoretical Physics, Trieste, Italy.

344. Poznyak, E. G. "Isometric imbedding of two-dimensional Riemannian metrics in Euclidean spaces," *Uspekhi Mat. Nauk.* **28** (1973), 47–76 (Russian).

345. Quien, N. "Über die endliche Lösbarkeit des Plateau Problems in Riemannschen Mannigfaltigkeiten," *Manusc. Math.* **39** (1982), 313–338.

346. Radó, T. "Problem of the least area and the problem of Plateau," *Math. Z.* **32** (1930), 763–796.

347. Radó, T. *On the Problem of Plateau.* Springer-Verlag, Berlin, 1933.

348. Raschevsky, P. K. "The theory of spinors," *Uspekhi Mat. Nauk* **10** (1955), 3–110 (Russian).

349. Raschevsky, P. K. *Riemannian Geometry and Tensor Analysis.* Nauka, Moscow, 1967.

350. de Rham, G. *Variétés Différentiables. Formes, Courants, Formes Harmoniques.* Hermann & Cie, Paris, 1955.

351. Reifenberg, E. "Solution of the Plateau problem for $m$-dimensional surfaces of varying topological type," *Acta. Math.* **104** (1960) 1–92.

352. Reifenberg, E. "An epiperimetric inequality related to the analyticity of minimal surfaces," *Ann. of Math.* **80** (1964), 1–14.

353. Reifenberg, E. "On the analyticity of minimal surfaces," *Ann. Math.* **80** (1964), 15–21.

354. Reshetnjak, Ju. G. "A new proof of the theorem of existence of an absolute minimum for two-dimensional problems of the calculus of variations in parametric form," *Sibirsky Matematichesky Zhurnal* **3** (1962), 744–768 (Russian).

355. Ronkin, L. I. "Discrete uniqueness sets for entire functions of exponential type in several variables," *Sibirsky Matematichesky Zhurnal* **19** (1978), 142–152 (Russian).

356. Rohlin, V. A. and Fuchs, D. B. *A Beginning Course in Topology: Geometrical Chapters.* Nauka, Moscow, 1977 (Russian).

357. Rozendorn, E. R. "Weakly irregular surfaces of negative curvature," *Uspekhi Mat. Nauk* **21** (1966), 59–116 (Russian).

358. Rozendorn, E. R. "Investigation of the basic equations of the theory of surfaces in asymptotic coordinates," *Mat. Sb. (N.S.)* **70** (1966), 490–507 (Russian).

359. Rozenknop, I. Z. "Some questions and applications of the homological theory of polynomial ideals," in: *Trudy Moskovskogo Matematicheskogo Obschestva.* Vol. 13, pp. 246–323 (Russian).

360. Sabitov, I. K. "The local structure of Darboux surfaces," *DAN SSSR* **162** (1965), 1001–1004 (Russian).

361. Sabitov, I. K. "Formal solutions of the Hilbert problem for an annulus," *Matematicheskiye Zametki* **12** (1972), 221–232 (Russian).

362. Sabitov, I. K. "A minimal surface as the rotation graph of a sphere," *Matematicheskiye Zametki* **2** (1967), 645–5656 (Russian).

363. Sabitov, I. K. "Minimal surfaces with certain boundary conditions," *Matematichesky sbornik* **76** (1968), 368–389 (Russian).

364. Sacs, J. and Uhlenbeck, K. "The existence of minimal immersions of 2-spheres," *Ann. of Math., Ser. 2*, **113** (1981), 1–24.

365. Sacs, J. and Uhlenbeck, K. "Minimal immersions of closed Riemann surfaces," *Trans. Amer. Math. Soc.* **271** (1982), 639–652.

366. Samelson, H. "Beiträge zur Topologie der Gruppenmanningfaltigkeiten," *Ann. of Math., Ser. 2*, **42** (1941), 1091–1137.

367. Schoen, R. and Uhlenbeck, K. "A regularity theory for harmonic maps," *J. Diff. Geom.* **17** (1982), 307–335.

368. Schoen, R. and Uhlenbeck, K. "Boundary regularity and the Dirichlet problem for harmonic maps," *J. Diff. Geom.* **18** (1983), 253–268.

369. Schoen, R. and Uhlenbeck, K. "Regularity of minimizing harmonic maps into the sphere," *Invent. Math.* **78** (1984), 89–100.

370. Schoen, R. and Yau, S. T. "On univalent harmonic maps between surfaces," *Invent. Math.* **44** (1978), 265–278.

371. Schoen, R. and Yau, S. T. "Existence of incompressible minimal surfaces and the topology of three dimensional manifolds with nonnegative scalar curvature," *Ann. of Math.* **110** (1979), 127–142.

372. Schoen, R. and Yau, S. T. "Compact group actions and the topology of manifolds with nonpositive curvature," *Topology* **18** (1979), 361–380.

373. Schoen, R. and Yau, S. T. "Complete three-dimensional manifolds with positive Ricci curvature and scalar curvature," *Ann. of Math. Stud.* **102** (1982), 209–228.

374. Schwarz, H. A. *Gesammelte Mathematische Abhandlungen*, Bd. 1, Berlin, 1890.

375. Sedov, L. I. *Continuum Mechanics.* Nauka, Moscow, 1976 (Russian).

376. Seifert, H. and Threllfall, W. *Varationsrechnung im Grossen.* Teubner, Leipzig-Berlin, 1938.

377. *Seminar on Differential Geometry.* Edited by Shing-Tung Yau. *Ann. of Math. Studies* **102** Princeton University Press, Princeton, N. J., 1982.

378. Serrin, J. "A priori estimates for solutions of the minimal surface equation," *Arch. R. Mech.* **14** (1963), 376–383.

379. Serrin, J. "On surfaces of constant mean curvature, which span a given space curve," *Math. Z.* **112** (1969), 77–88.

380. Shiffman, M. "The Plateau problem for minimal surfaces of arbitrary topological structure," *Amer. J. Math.* **61** (1939), 853–882.

381. Sigalov, A. G. "Two-dimensional problems of the calculus of variations," *UMN* **6** (1951), 16–101 (Russian).

382. Sigalov, A. G. "Two-dimensional problems of the calculus of variations in nonparametric form," in: *Trudy Moskovskogo Matematicheskogo Obschestva* **2** (1953), pp. 201–233 (Russian).

383. "Two-dimensional problems of the calculus of variations in nonparametric form translated into parametric form," *DAN SSSR* **93** (1953), 405–408 (Russian).

384. Sigalov, A. G. "Variational problems with admissable surfaces of arbitrary topological types," *UMN* **12** (1957), 53–98 (Russian).

385. Simon, L. "Boundary regularity for solutions of the nonparametric least area problem," *Ann. of Math.* **3** (1976), 429–455.

386. Simon, L. "Boundary behavior of solutions of the nonparametric least area problem," *Bull. Aust. Math. Soc.* **26** (1982), 17–27.

387. Simon, L. and Hardt, R. M. "Boundary regularity and embedded solutions for the oriented Plateau problem," *Ann. of Math.* **2** (1979), 439–486.

388. Simon, L. and Hardt, R. M. "Boundary regularity and embedded solutions for the oriented Plateau problem," *Bull. Amer. Math. Soc.* **1** (1979), 263–265.

389. Simons, J. "A note on minimal varieties," *Bull. Amer. Math. Soc.* **73** (1967), 491–495.

390. Simons, J. "Minimal cones, Plateau's problem and the Bernstein conjecture," *Proc. Nat. Acad. Sci. USA* **58** (1967), 410–411.

391. Simons, J. "Minimal varieties in Riemannian manifolds," *Ann. of Math.* **88** (1968), 62–105.

392. Siu, Y. T. "The complex analyticity of harmonic maps and the strong rigidity of compact Kähler manifolds," *Ann. of Math.* **112** (1980), 73–111.

393. Siu, Y. T. and Yau, S. T. "Compact Kähler manifolds of positive bisectional curvature," *Invent. Math.* **59** (1980), 189–204.

394. Smale, S. "Generalized Poincaré's conjecture in dimensions greater than four," *Ann. of Math.* **74** (1961), 391–406.

395. Smale, S. "On the structure of manifolds," *Amer. J. of Math.* **84** (1962), 387–399.

396. Smale, S. "On the Morse index theorem," *J. Math. Mech.* **14** (1965), 1049–1055.

397. Smith, R. "Harmonic mappings of spheres," *Amer. J. of Math.* **97** (1975), 364–385.

398. Smith, R. "The second variation formula for harmonic mappings," *Proc. Amer. Math. Soc.* **47** (1975), 229–236.

399. Steenrod, N. "The topology of fibre bundles," in: *Princeton Mathematical Series*, Vol. 14, Princeton University Press, Princeton, N. J., 1951.

400. Stiefel, E. "Sur les nombres de Betti des groupes de Lie clos," *Topologie algébrique. Colloques Internationaux du Centre National de la Recherche Scientifique* **12** (1949), 97–101. Centre de la Recherche Scientifique, Paris.

401. Struik, D. Outline of a History of Differential Geometry. **55** (1933), pp. 92–120; **58** (1933), pp. 161–191.

402. Sulanke, R. and Wintgen, R. *Differentialgeometrie und Faserbündel.* Birkhäuser Verlag, Basel-Stuttgart, 1972.

403. Suzuki, S. "On homeomorphisms of 3-dimensional handlebody," *Canadian J. Math.* **29** (1977), 111–124.

404. Takeuchi, M. "On Ponrjagin classes of compact symmetric spaces," *J. Fac. Sci. Univ. Tokyo*, Sect. I, V, 9, 4 (1962), pp. 313–328.

405. Thompson, D. A. *On Growth and Form.* London, 1917.

406. Titov, O. V. "Minimal hypersurfaces that are stretched over soft obstructions," *DAN SSSR* **221** (1973), 239–296 (Russian).

407. Titov, O. V. "Minimal hypersurfaces over soft obstacles," *Izvestiya AN SSSR. Mat.* **38** (1974), 374–417 (Russian).

408. Toda, H. *Composition Methods in Homology Groups of Spheres.* Princeton Univ. Press, Princeton, N. J., 1962.

409. Tomi, F. "On the local uniqueness of the problem of least area," *Arch. R. Mech.* **52** (1973), 312–318.

410. Tomi, F. and Tromba, A. J. "Extreme curves bound embedded minimal surfaces of the type of the disc," *Math. Z.* **158** (1978), 137–145.

411. Triscari, D. "Sulle singolarità delle frontiere orientate di misura minima," *Ann. Scuola Norm. Sup.* Pisa. **17** (1963), 349–371.

412. Tromba, A. "On the number of simply-connected minimal surfaces spanning a curve," *Mem Amer. Math. Soc.* **12** (1977).

413. Tromba, A. "A general approach to Morse theory," *J. Differential Geometry* **12** (1977), 47–85.

414. Tuzhilin, A. A. and Fomenko, A. T. "Multidimensional mappings minimal surfaces and soap films," *Vestnik MGU*, Moscow University Press, Moscow, 1986.

415. Tyrin, A. V. "Critical points of multidimensional Dirichlet functional," *Matematichesky Sbornik* **124** (1984), 146–158 (Russian).

416. Tyrin, A. V. "On absence of local minima of multidimensional Dirichlet functional," *Uspekhi Matematicheskikh Nauk* **39** (1984), 133–194 (Russian).

417. Tyrin, A. V. "Dirichlet-type functional minimization problem," in: *Geometry, Differential Equations and Mechanics.* Moscow University Press, Moscow, 1986, pp. 146–150.

418. Uhlenbeck, K. "Minimal 2-spheres and tori in $S^k$." Preprint, 1975.

419. Uhlenbeck, K. "Morse theory by perturbation methods with applications to harmonic maps," *Trans. Amer. Math. Soc.* **267** (1981), 569–583.

420. Uhlenbeck, K. *Minimal Spheres and Other Conformal Variational Problems.* Princeton Univ. Press, Princeton, 1983, *Ann. Math. Studies*, 103.

421. Vikaruk, A. J. "The principle of the argument and the integral curvature of minimal surfaces," *Matematicheskiye Zametki* **15** (1974), 91–700 (Russian).

422. Vikaruk, A. J. "Analogs of Nevanlina theorems for minimal surfaces," *Matematichesky Sbornik* **100** (1976), 555–579 (Russian).

423. Vyshik, M. I. "On strongly elliptic systems of differential equations," *Matematichesky sbornik* **29** (1951), 615–676 (Russian).

424. Whitehead, G. "Generalized homology theories," *Trans. Amer. Math. Soc.* **102** (1962), 227–283.

425. Whitehead, J. "On the groups $\pi_r(Vn, m)$ and spherebundles," *Proc. London Math. Soc.* **48** (1945), 243–291.

426. Whitney, H. *Geometric Integration Theory.* Princeton University Press, Princeton, N. J., 1957.

427. Wolf, J. "Geodesic spheres in Grassmann manifold. III," *J. Math.* **7** (1963), 425–446.

428. Wolf, J. "Elliptic spaces in Grassmann manifold. III," *J. Math.* **7** (1963), 447–462.

429. Xin, Y. "Some results on stable harmonic maps," *Duke Math. J.* **47** (1980), 609–613.

430. Yau, S. T. "On almost minimally elliptic singularities," *Bull. Amer. Math. Soc.* **83** (1977), 362–364.

431. Yau, S. T. "Kohn–Rossi cohomology and its application to the complex Plateau problem," *Ann. of Math.* **113** (1981), 67–110.

432. Yosida, K. "A problem concerning the second fundamental theorem of Lie," *Proc. Imp. Acad. Tokyo* **5** (1927), 152–155.

433. Young, L. *Lectures on the Calculus of Variations and Optimal Control Theory.* Saunders, Philadelphia, 1969.

434. White, B. "Homotopy classes in Sobolev spaces and energy-minimizing maps," *Bull. Amer. Math. Soc.* **13** (1985), 166–168.

435. Pluzhnikov, A. I. "On the Dirichlet functional minima," *DAN SSSR* **290** (1986), 289–293 (Russian).

436. Ivanov, A. O. "Globally minimal symmetric surfaces in Euclidean space," in: *Geometry, Differential Equations and Mechanics.* Moscow University Press, 1986, pp. 69–71 (Russian).

437. Tuzhylin, A. A. "On the bifurcation of some two-dimensional surfaces for a two-parameter contour variation," in: *Geometry, Differential Equations and Mechanics.* Moscow University Press, Moscow, 1986, pp. 140–145 (Russian).

438. Shklyanko, I. V. "Minimal geodesics in symmetric spaces," in: *Geometry, Differential Equations and Mechanics.* Moscow University Press, Moscow, 1986, pp. 159–161 (Russian).

439. Balinskaya, I. S. "Volumes of orbits of smooth Lie group actions," in: *Geometry, Differential Equations and Mechanics.* Moscow University Press, Moscow, 1986, pp. 49–52 (Russian).

440. Balinskaya, I. S. "Volumes of orbits of smooth Lie group actions," in: *Global Analysis and Mathematical Physics.* Voronezh University Press, Voronezh, 1987, pp. 155–159 (Russian).

441. Shklyanko, I. V. "Bifurcations of multidimensional functionals," in: *Global Analysis and Mathematical Physics.* Voronezh University Press, Voronezh, 1987, pp. 178–184 (Russian).

442. Ivanov, A. O. "Minimal cones of large codimensions," in: *Global Analysis and Mathematical Physics.* Voronezh University Press, Voronezh, 1987, pp. 166–171 (Russian).

443. Balinskaya, I. S. "Minimal cones of adjoint action of classical Lie groups," *UMN* **41** (1986) (Russian).

444. Borisovich, A. Yu. "Plateau operator and bifurcation of two-dimensional minimal surfaces," in: *Global Analysis and Mathematical Physics.* Voronezh University Press, Voronezh, 1987, pp. 132–154 (Russian).

445. Borisovich, A. Yu. "Reduction of the minimal surface bifurcation problem to operator equations and search for bifurcations of catenoid, helicoid, Scherk and Ennper surfaces," *UMN* **41** (1986), 165–166 (Russian).

446. Borisovich, A. Yu. "One geometric application of the theorem on a simple bifurcation point," in: *Application of Topology to Modern Analysis.* Voronezh University Press, Voronezh, 1985, pp. 172–174 (Russian).

447. Lê Hồng Van and Fomenko, A. T. "Lagrangian manifolds and Maslov index in minimal-surface theory." *DAN SSSR* **299**, N1 (1988), 42–45 (Russian).

448. Lê Hông Van and Fomenko, A. T. "Volumes of minimal surfaces and curvature tensor of Riemann manifolds." *DAN SSSR* **300**, N6 (1988), 1308–1312 (Russian).

449. Lê Hông Van and Fomenko, A. T. "Criterion for minimality of Lagrangian submanifolds in Kähler manifolds," *Matematicheskiye zametki* **42**, N4 (1987), 559–572 (Russian).

450. Lê Hông Van and Fomenko, A. T. "Nullity of homogeneous spaces and minimal-surface theory," Preprint, 1987.

451. Lê Hông Van . "Minimal surfaces in homogeneous spaces," *Izvestiya AN SSSR* **52**, N2 (1988), 408–423 (Russian).

452. Lê Hông Van . "Absolutely minimal surfaces and calibrations on orbits of adjoint representations of classical Lie groups," *DAN SSSR* **298**, N6 (1987), 1308–1311 (Russian).

453. Tyrin, A. V. "Regularity of Riemann manifolds mappings, minimizing the multidimensional Dirichlet functional," *Matematicheskiy sbornik* **132** 3 (1987), pp. 401–419 (Russian).

454. Tyrin, A. V. "Regularity of harmonic mappings as a consequence of instability property," *DAN SSSR*, 1987 (in print; Russian).

455. Fomenko, T. N. "On efficient construction of retraction of certain spaces onto sphere," in: *Analysis on Manifolds and Differential Equations.* Voronežh University Press, Voronežh, 1986, pp. 165–173 (Russian).

456. Bourbaki, N. *Differential and Analytic Manifolds.*

457. Rassias, T. M. *Foundations of Global Nonlinear Analysis.* Teubner, Leipzig, 1986.

458. Đao Chông Thi and Fomenko, A. T. *Minimal Surfaces and Plateau problem.* Nauka, Moscow, 1987 (Russian).

459. Soboleva, O. Yu. "Minimal surfaces retracting onto their boundaries," in: *Trudy seminara po vektornomu i tenzornomu analizu* **23** (1987). Moscow University Press, Moscow, (Russian).

460. Edmonds, A., Ewing, J. and Kulkarni, R. *Ann. of Math.* **116** (1982), 113–132.

461. Simoes, P. "On a class of minimal cones in $\mathbf{R}^n$, *Bull. Amer. Math. Soc.* **80** (1974), 488-489.

462. Lawlor, G. *The Curvature Creation.* Ph.D. Thesis, Stanford Univ. 1988.

463. Bryant, R. L. " Minimal Lagrangian submanifolds of Kähler–Einstein manifolds," *Lect. Notes in Math.* **N1255** (1987), 1–12.
464. Morgan, F. "The exterior algebra $\Lambda^k \mathbf{R}^n$ and area minimization," *Lin. Alg. App.* **66** (1985), 1–28.
465. Morgan, F. "On the singular structure of three-dimensional area minimizing surfaces," *Trans. Amer. Math. Soc.* **276**, N1 (1983), 137–143.
466. Lê Hông Van . "Minimal Φ-Lagrangian surfaces in almost Hermitian manifolds," *Math. Sbornik* **180**, N7 (1989), 924–936.
467. Fischer-Colbrie, D. "On complete minimal surfaces with finite Morse index in three-manifolds," *Invent. Math.* **82**, N1 (1985), 121–132.

# INDEX